# Mollicutes

Molecular Biology and Pathogenesis

Edited by

Glenn F. Browning

Asia-Pacific Centre for Animal Health
Faculty of Veterinary Science
The University of Melbourne
Victoria
Australia

and

Christine Citti

INRA, École Nationale Vétérinaire de Toulouse and
UMR 1225 Interactions Hôtes-Agents Pathogènes
Toulouse
France

Caister Academic Press

Caister Academic Press
Norfolk, UK

www.caister.com

British Library Cataloguing-in-Publication Data
A catalogue record for this book is available from the British Library

ISBN: 978-1-908230-30-0



Cover design adapted from Figures 12.6 and 13.3

Printed and bound in Great Britain

# Contents

# Contributors

**Nathalie Arricau-Bouvery**
UMR 1332 Biologie du Fruit et Pathologie
INRA–Université de Bordeaux
Villenave d'Ornon
France

nbouvery@bordeaux.inra.fr

**Mitchell F. Balish**
Department of Microbiology
Miami University
Oxford, OH
USA

BalishMF@MiamiOH.edu

**Cécile M. Bébéar**
USC Mycoplasmal and Chlamydial Infections in Humans
INRA–Université de Bordeaux
Bordeaux
France

cecile.bebear@u-bordeaux2.fr

**Laure Béven**
UMR 1332 Biologie du Fruit et Pathologie
INRA–Université de Bordeaux
Villenave d'Ornon
France

laure.beven@bordeaux.inra.fr

**Janet M. Bradbury**
School of Veterinary Science
University of Liverpool
Neston
UK

jmb41@liverpool.ac.uk

**Marc Breton**
UMR 1332 Biologie du Fruit et Pathologie
INRA–Université de Bordeaux
Villenave d'Ornon
France

marc.breton@bordeaux.inra.fr

**Daniel R. Brown**
College of Veterinary Medicine
University of Florida
Gainesville, FL
USA

drbrown@ufl.edu

**Glenn F. Browning**
Asia-Pacific Centre for Animal Health
Faculty of Veterinary Science
The University of Melbourne
Parkville, VIC
Australia

glenfb@unimelb.edu.au

**Christine Citti**
INRA, École Nationale Vétérinaire de Toulouse and UMR 1225 Interactions Hôtes-Agents Pathogènes
Toulouse
France

c.citti@envt.fr

**James M. Daubenspeck**
Department of Genetics
University of Alabama at Birmingham
Birmingham, AL
USA

jametex@uab.edu

**Steven P. Djordjevic**
The ithree Institute
The University of Technology
Sydney, NSW
Australia

steven.djordjevic@uts.edu.au

**Kevin Dybvig**
Department of Genetics
University of Alabama at Birmingham
Birmingham, AL
USA

dybvig@uab.edu

**Birgit Henrich**
Institute of Medical Microbiology and Hospital Hygiene
Heinrich-Heine-University Düsseldorf
Düsseldorf
Germany

birgit.henrich@uni-duesseldorf.de

**Saskia Hogenhout**
Department of Cell and Developmental Biology
The John Innes Centre
Norwich
UK

saskia.hogenhout@bbsrc.ac.uk

**Miriam Hopfe**
Institute of Medical Microbiology and Hospital Hygiene
Heinrich-Heine-University Düsseldorf
Düsseldorf
Germany

miriam.hopfe@gmx.dec

**David S. Jordan**
Department of Genetics
University of Alabama at Birmingham
Birmingham, AL
USA

djordan@uab.edu

**Fabien Labroussaa**
UMR 1332 Biologie du Fruit et Pathologie
INRA–Université de Bordeaux
Villenave d'Ornon
France

flabroussaa@berkeley.edu

**Inna Lysnyansky**
Department of Avian and Fish Diseases
Kimron Veterinary Institute
Bet-Dagan
Israel

innal@moag.gov.il

**Laura McAuliffe**
Veterinary Laboratories Agency
Addlestone
UK

mcauliffelaura@hotmail.com

**Marc S. Marenda**
Asia-Pacific Centre for Animal Health
Faculty of Veterinary Science
The University of Melbourne
Parkville, VIC
Australia

mmarenda@unimelb.edu.au

**Philip F. Markham**
Asia-Pacific Centre for Animal Health
Faculty of Veterinary Science
The University of Melbourne
Parkville, VIC
Australia

pmarkham@unimelb.edu.au

**Meghan May**
Department of Biological Sciences
Molecular Biology, Biochemistry and Bioinformatics Program
Towson University
Towson, MD
USA

mmay@towson.edu

**Makoto Miyata**
Department of Biology
Graduate School of Science
Osaka City University
Osaka
Japan

miyata@sci.osaka-cu.ac.jp

**Daisuke Nakane**
Department of Physics
Faculty of Science
Gakushuin University
Tokyo
Japan

jj20110031@cc.nagasaki-u.ac.jp

**Amir H. Noormohammadi**
Asia-Pacific Centre for Animal Health
Faculty of Veterinary Science
The University of Melbourne
Parkville, VIC
Australia

amirh@unimelb.edu.au

**Joël Renaudin**
UMR 1332 Biologie du Fruit et Pathologie
INRA–Université de Bordeaux
Villenave d'Ornon
France

renaudin@bordeaux.inra.fr

**Colette Saillard**
UMR 1332 Biologie du Fruit et Pathologie
INRA–Université de Bordeaux
Villenave d'Ornon
France

saillard@bordeaux.inra.fr

**Lawrence K. Silbart**
Department of Allied Health Sciences
University of Connecticut
Storrs, CT
USA

lawrence.silbart@uconn.edu

**Steven M. Szczepanek**
Neag Comprehensive Cancer Center
University of Connecticut Health Center
Farmington, CT
USA

szczepanek@uchc.edu

**Jessica L. Tacchi**
The ithree Institute
The University of Technology
Sydney
Australia

jessica.l.tacchi@student.uts.edu.au

**Ken B. Waites**
Department of Pathology
University of Alabama at Birmingham
Birmingham, AL
USA

waiteskb@yahoo.com

**Carl-Ulrich Zimmerman**
Institute of Bacteriology, Mycology and Hygiene
University of Veterinary Medicine Vienna
Vienna
Austria

carl-ulrich.zimmerman@vetmeduni.ac.at

# Current books of interest

# Preface

Because of their extreme simplicity, Mollicutes have attracted considerable attention in recent years as model organisms for studying fundamental aspects of cellular life. The creation of the first synthetic genome was based on a mycoplasma and has raised further interest in this large group of bacteria. However, in conjunction with the progress in deciphering the general biology of *Mollicutes*, the last 10 years have also seen growing understanding of the mechanisms underlying the pathogenesis of disease caused by some Mollicutes in animals and plants. The focus of this book is to review recent progress in these aspects of mycoplasmology.

The increasing availability of complete genomic sequences of Mollicutes has revealed unexpected relationships between many species, prompting us to re-evaluate our understanding of their taxonomy. Perhaps the most remarkable revelation has been the discovery of extensive horizontal genetic transfer between distant species and of considerable genomic plasticity within some species, apparently driven by integrative conjugative elements.

For many years progress in understanding the molecular biology of Mollicutes was hampered by the lack of tools for manipulating their genomes, with most studies relying on only a couple of transposons for mutagenesis. The creation of synthetic plasmids using homologous origins of replication, the improvements of existing transposons, and the development of new selective markers, inducible gene expression systems and methods for creating unmarked mutations have greatly extended our capacity for exploring the genetics of these organisms, offering us valuable tools for functional genomics.

While we are a long way from fully understanding the molecular basis of pathogenesis in infections with Mollicutes, a number of specific virulence factors have now been definitively identified in several of these important pathogens by studying the behaviour of mutants *in vivo*, as well as by examining the biochemical function of purified recombinant proteins *in vitro*. Progress has also been made in our understanding of the molecular basis of transmission of spiroplasmas from their insect vectors to the plants in which they cause economically significant disease.

It has become increasingly clear that many proteins exposed on the surface of mycoplasmas are subjected to considerable post-translational modifications, with complex patterns of cleavage resulting in the derivation of a variety of products from a single coding sequence. This complexity is further enhanced by the capacity of many mycoplasma proteins to serve several, sometimes quite unrelated, roles during infection.

Our comprehension of the mycoplasma glycocalyx has lagged behind that of the proteome, but it is now apparent that both polysaccharides and glycoconjugates are significant components of this extracellular matrix. Progress is now being made in identifying some of the glycosyltransferases involved, but it appears likely that novel enzymes are responsible for carbohydrate biosynthesis in mycoplasmas. There have also been interesting developments in characterization of glycosidases in a number of pathogenic mycoplasmas, and in examining their potential role in infection.

A key focus of studies of the molecular biology of mycoplasmas over the last 20 years has been the exploration of the mechanisms involved in the generation of high frequency phase and antigenic variation in their cell surface proteins. The range of different proteins involved and the variety of mechanisms utilized by different mycoplasmas suggests concurrent independent evolution of these systems in many species. Yet, our increasing understanding of the molecular basis of these systems has not been matched by our elucidation of the role of this variation *in vivo*, in part because of our limited capacity to delete these systems from mycoplasma genomes or to reduce their complexity by genetic manipulation.

Probably the areas where there have been most advances in recent years are the interlinked structures involved in adhesion, gliding motility and the cytoskeleton. Many of the proteins involved, and the interactions between them, have been defined, particularly in the human pathogen *Mycoplasma pneumoniae* and in the highly motile aquatic species *Mycoplasma mobile*, enabling more complex models of the biology and structure of the mycoplasma cell to be developed.

The discovery that many pathogenic mycoplasmas can form biofilms has suggested an additional mechanism that may be involved in persistence and antimicrobial resistance *in vivo*, and this promises to be a field of research that will yield considerable insights into mycoplasmoses in the future.

The persistence of mycoplasmas *in vivo* and the nature of the associated lesions clearly indicate the significance of their interactions with the host immune response in the pathogenesis of the diseases they cause. The mechanisms involved in their capacity to evade and manipulate the attempts made by the host to eliminate them, and the elements of the immune response involved, have now been explored in some depth in several species and some common themes have become apparent.

Over the years, efforts to develop standardized methods for antimicrobial susceptibility testing have facilitated definition of resistance in mycoplasmas. Resistance to multiple classes of antimicrobial drugs has been recognized in several mycoplasma species and in a number of cases the molecular basis of this resistance has been characterized, allowing the development of molecular methods for monitoring the spread of specific resistance genes.

The following chapters provide current reviews of recent research in these aspects of mycoplasmology. We hope it will prove to be a valuable resource not only for researchers in mycoplasmology, but also for those interested in gaining a broader perspective on these surprisingly complex pathogens. Many of us working closely with these organisms believe that, far from offering a simplified model of cellular life, the ongoing detailed study of Mollicutes will reveal a range of novel adaptations to life in a hostile environment, and insights into the evolution of virulence in plant and animal pathogens.

Glenn F. Browning and Christine Citti

# The Contentious Taxonomy of Mollicutes

1

Daniel R. Brown and Janet M. Bradbury

## Abstract

Bacterial systematics is energized by an inherent tension arising from its obligations to respect the dignity of taxa established in the past while maintaining sufficient flexibility to accommodate the advance of knowledge. The taxonomy of Mollicutes has been contentious from its inception, reflecting the challenges in cultivating the organisms axenically, early controversy over what form of life they exemplify, competing systems of nomenclature, chronological misfortune in the discovery of types, evolutionary anomalies, the sometimes enigmatic International Code of Nomenclature of Bacteria, and the usually vague species concept for prokaryotes. Their present taxonomy consequently bears acknowledged imperfections, but recent advances in genome-based systematics offer hope for future resolution of some current controversies.

> The beginning of wisdom is to call things by their right name.
>
> K'ung-fu-tzu (ca. 500 BCE). Zi Lu. *In* Analects, Volume 13, p. 3

## Paths to the present

### Axenic culture

A lethal form of contagious pleuropneumonia of cattle, occurring in epidemics documented in central Europe as early as 1693 (Gordon, 1884), had become a major concern to European agriculture and food hygiene by the early 1800s (Anonymous, 1870; Provost *et al.*, 1987). Although the disease was likely present throughout Western Eurasia and Africa for millennia (Blancou, 1996), Nocard and Leclainche (1898) credited Claude Bourgelat as having made the first detailed clinical differentiation of this disease from rinderpest and tuberculosis in 1765 (see Barberet, 1766). Mid-nineteenth-century epizootics of bovine pleuropneumonia caused massive economic losses in Europe, Great Britain and Australia, and threatened the United States. Based on a variolation-like immunization strategy adapted by Willems (1852), Pasteur (1882) devised a refined pleuropneumonia vaccine consisting of serous exudate from the lungs of calves inoculated with lesion fluid containing the agent of virulence (the 'virus' *sensu lato*; Pasteur, 1882). This 'virus' had never been passaged successfully in broth or solid media, or even been seen with a microscope by staining of infectious material, but there remained great incentive to isolate and culture it axenically due to the high cost of supplying Pasteur's vaccine.

Nocard and Roux (1898a,b) tried the *in situ* culture technique first demonstrated by Metchnikoff *et al.* (1896) to be advantageous for recovering fragile microbes: a pouch made of nitrocellulose, initially containing peptone broth (Martin, 1898) inoculated with a small amount of infectious exudate, was introduced surgically into the peritoneal cavity of a rabbit. When recovered after 2 or 3 weeks the broth had become slightly opalescent, while uninoculated or heat-treated inoculum controls remained transparent. The opalescent fluid contained 'an infinity of small refringent and mobile dots' of indeterminate shape. Whether this represented a pure culture is debatable now, but the significant insight of Nocard and Roux was that the semi-permeable nitrocellulose membrane permitted profound modifications of the original broth by penetration of soluble substances from the rabbit's body that were favourable to the culture. Thus *in situ* culture became unnecessary, and they were able to culture the agent axenically (even if in doubtful purity) first in medium pre-conditioned in nitrocellulose pouches in rabbits or cows, and eventually in unconditioned peptone broth containing 4% v/v rabbit or cow serum (Nocard and Roux, 1898b). This material was able to induce characteristic disease in a few of the cattle they inoculated experimentally, so it seems probable that at least some fraction of the cultures of Nocard and Roux did contain the pleuropneumonia 'virus'.

Bordet (1910) cultured this 'virus' in similar broth and on solid media. He saw with toluidine blue or Giemsa staining of a coarsely filtered (Chamberland F) broth a mixture of filaments, vibrios, spirochaetes, and globules, which he proposed had metamorphosed during prolonged incubation, not an unusual belief among leading microbiologists of the era. Borrel *et al.* (1910) imposed sequential order on this mélange by asserting the globules ('formes d'involution', like the endospores known then from anthrax) must develop morphologically through intermediate oval and vibrio forms to become diplococci, tetracocci or larger 'morulae', and finally streptococci in chains or rings ('formes de multiplication'). These would then become enveloped in mucoid linear, bifurcated, or radiating filaments or pseudo-mycelia, which develop internal nodules before undergoing involution to complete a replicative cycle reminiscent of moulds. Because the apparent radiating filaments ('formes astéroïdes') were so remarkable in appearance, the name *Asterococcus mycoides* was proposed for the agent of pleuropneumonia (Borrel *et al.*, 1910). Martzinovski (1911) attempted to replicate those findings with fresh lung exudate, but after observing only diplococci and streptococci in filtered cultures he concluded that while the microbe might be very polymorphic it ordinarily exists as a coccobacillus, such that the name *Coccobacillus mycoides* would be more appropriate.

The property, 'filterable but not invisible' was the most important reason for Bridré and Donatien (1923, 1925) to associate the 'virus' causing a syndrome of contagious polyarthritis, mastitis and agalactia in sheep and goats (Celli and De Blasi, 1906) with the agent of bovine pleuropneumonia. They bolstered the association by describing the coarsely filtered (Chamberland $L_1$ or $L_2$) agalactia agent grown in broth containing serum as consisting of similar polymorphic granules, spirochaetes, long undulant filaments, and small cocci and streptococci, even though only filaments are actually evident in their published specimens (Bridré and Donatien, 1925). Colonies on solid medium were very small, with opaque centres surrounded by a clear zone, exactly like colonies of the pleuropneumonia agent. They concluded that this was not simply the same germ adapted to different hosts because reciprocal vaccination was not protective (probably the first application of serology in taxonomic characterization of Mollicutes), and predicted that future discoveries would expand this 'group of two'.

In retrospect, it remains doubtful whether the filaments described by Borrel *et al.* (1910) actually represented the agent of pleuropneumonia because of the contamination and other artefacts evident in their specimens (Fig. 1.1). Martzinovski (1911) almost certainly never saw the true 'virus', but with faith that filtration through porcelain produced a pure culture having a multiplicity of forms, an imaginary replication cycle, and by naming the agent of pleuropneumonia both *Asterococcus* and *Coccobacillus* on such thin evidence, a century of taxonomic contention over Mollicutes had begun.

## Nomenclature

Nocard and Roux (1898b) realized that the agent of pleuropneumonia, while just within the limits of optical microscopy, must be much smaller than ordinary microbes. Bordet (1910) did not consider the agent to be anything other than a specimen of vibrio or spirochaete like those already known from cholera and syphilis. Martzinovski (1911), a medical parasitologist trying to reconcile the polymorphism and mould-like features described by Borrell *et al.* (1910) with his own observations, suspected the forms he saw might embody just one stage in the developmental cycle ('l'evolution') of a protozoan. Frosch (1923) did see mixtures of round, oval and polygonal bodies

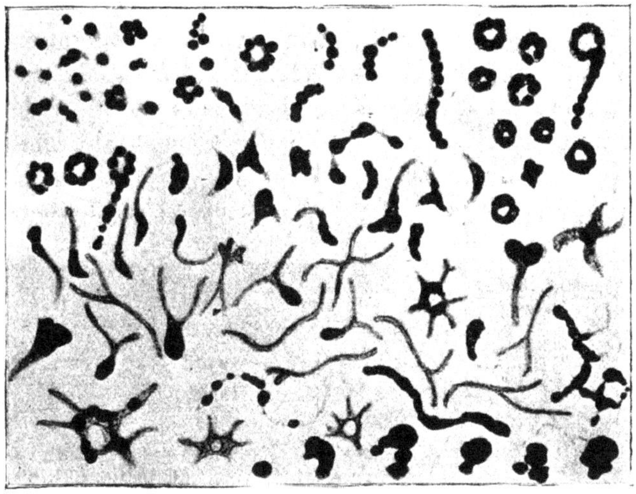

*Fig*. T.— Différentes formes du microbe dessinées d'après des préparations d'âges variés $\frac{5000}{1}$. De haut en bas, les formes sont de plus en plus vieilles.

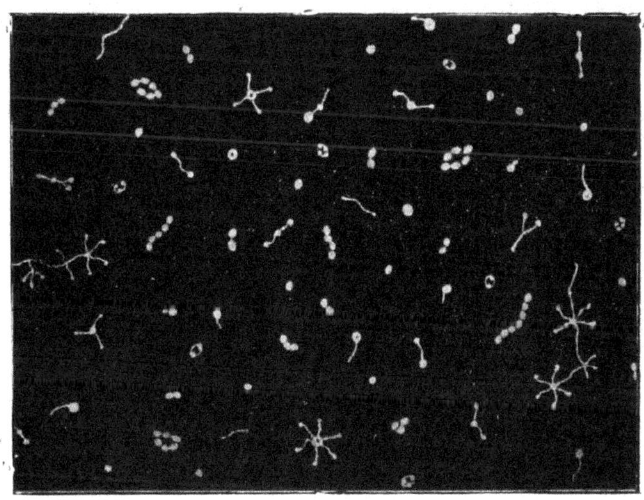

*Fig*. U. M. — Différents aspects du microbe de la péripneumonie à l'ultra-microscope. Même grossissement que ci-dessus. $\frac{5000}{1}$

**Figure 1.1** La morphologie du microbe de la péripneumonie des bovidés (Borrel *et al.*, 1910): a Rorschach test. Upper panel: the developmental cycle of *Asterococcus mycoides* is depicted sequentially from 'formes d'involution' (top left, and bottom right) through 'formes de multiplication' (centre) to 'formes astéroïdes' (bottom left). Lower panel: random depictions of *A. mycoides* (not in copyright; see www.biodiversitylibrary.org).

occurring in rings and filaments, but thought they were either mycelia and conidia or a budding yeast best named *Micromyces peripneumoniae*. The organism now had three names.

To explain the passage of such large particles through filters believed to be impermeable to bacteria, Nowak (1929) re-interpreted the scheme of Borrell *et al.* (1910) as a cycle beginning with elementary bodies consisting of unstainable amorphous agglomerations of 'protoplasm' ('des amas protoplasmiques d'une nature indéterminée') that germinated and budded into filamentous branched mycelia. These later developed strongly staining internal granules ('corpuscles endomycéliens') of an unelucidated nature, then decomposed to form new elementary bodies. The key feature was the plasticity of the acellular protoplasm, a fluid 'gelée vivante' (Dujardin, 1835) with the capacity to coalesce as cells after its passage through porcelain filters, so the name *Mycoplasma peripneumoniae* could reconcile the separate mycelial and plasma-like stages. It's not clear whether Nowak was aware of earlier use of the term 'mycoplasma' to describe fungus-infected protoplasm ('pilzbehaftetes Protoplasma'; Frank, 1890; see Krass and Gardner, 1973), but the organism now had four names. Wróblewski (1931) still insisted that the agent could pass through coarse (Chamberland series $L_1$ through $L_3$) or medium porosity (Berkefeld M) porcelain filters at all stages. He also extended the model of Borrel *et al.* (1910) by accommodating both asexual and sexual reproduction: an initial spore germinates into filaments that develop strongly staining internal endospores and radiating terminal conidia containing weakly staining spermatocysts and strongly staining oogonies ('astres avec exospores'). After producing endospores and exospores, the mycelia and conidia autolyse. The name *Asteromyces peripneumoniae* was proposed in order to emphasize the sexual organs (Wróblewski, 1931). This one organism now had five names and could have exemplified almost any form of microbe. The agent of agalactia was morphologically similar, except that its conidia were imagined to be distinctly ring-shaped, so the name *Anulomyces agalaxiae* seemed appropriate for it (Wróblewski, 1931). Turner (1935) proposed an even more elaborate scheme that encompassed no less than five separate reproductive strategies, and preferred the monumental name *Borrelomyces* 'to perpetuate the name of Borrel, who with his colleagues demonstrated its characteristic morphology'. Excluding obscure Russian references to the *Allococcaceae* (Fischer, 1903) cited by Turner (1935), by 1935 Bridré and Donatien's one 'group' had seven different names.

At this point, Klieneberger (1935) introduced the term 'pleuropneumonia-like organisms' to describe what turned out to be wall-less variants of known *Clostridium, Streptobacillus* and *Proteus* species. These variants resembled the agent of pleuropneumonia mainly in their 'filterable but not invisible' character, but, for reasons difficult to discern from the literature (see Klieneberger-Nobel, 1980) she proposed that these were novel symbionts of the walled bacteria. This caused confusion for more than 20 years while the terms 'pleuropneumonia-like organisms' (PPLO) and 'L (for Lister Institute) forms' were applied to these wall-deficient variants as well as to genuine Mollicutes that were 'pleuropneumonia-like' in their filterability, colonial morphology and nutritional demands but not in their cellular morphology, host range, serological cross reactivity or pathogenicity (Klieneberger, 1935, 1938).

Sabin (1938, 1941a) perpetuated the symbiont hypothesis by proposing that a pleuropneumonia-like microbe, which he thought caused neurological disease in mice, might have developed special pathogenic properties through an association 'far more intimate than that represented by ordinary symbiosis' with toxoplasmas; endosymbiosis within a protozoal vector might explain transmission of this microbe which could otherwise be recovered only from mouse brain. To Sabin, this hinted at the evolutionary origins and correct classification of Mollicutes: the PPLO were like filterable viruses *sensu stricto* (known from foot and mouth disease, rinderpest and yellow fever) in their very small size during some stages of their life cycle, an alleged capacity for intracellular replication, and supposedly irreversible adaptation to serum from a particular animal species during cultivation in cell-free medium. They were not Rickettsiae, which were strict obligate intracellular parasites, but their extreme polymorphism and seemingly variable modes of reproduction ('polygenethodism'; Turner, 1935) which might recapitulate ancestral phylogeny excluded them from the ordinary bacteria then grouped with fungi in the Class *Schizomycetes* according to the Botanical Code of Nomenclature.

Turner (1935) had proposed to place his *Borrelomyces* in the new Order *Borrelomycetales* and Family *Borrelomycetaceae* within the *Schizomycetes*, but with self-admitted trepidation Sabin proposed an entirely new Class *Paramycetes* and Order *Paramycetales*, complete with families *Parasitaceae* for isolates from animals and

*Saprophytaceae* to accommodate Laidlaw and Elford's (1936) environmental isolates. Generic names of the *Parasitaceae* would be according to the species of animal they inhabit, including *Bovimyces, Capromyces, Canomyces, Murimyces* and *Musculomyces* (Sabin, 1941a). Although his proposals were immediately rejected, Sabin's was the first comprehensive attempt to classify the Mollicutes (Fig. 1.2). Sabin's scheme was widely criticized especially for its violation of the rules of bacteriological nomenclature established at the 1936 International Congress of Microbiology under the leadership of Robert Buchanan (see Lapage *et al.*, 1992), and the full-scale revisions proposed within months by Sabin (1941b) were never accepted.

In reaction to the comment that PPLO might have evolved from ordinary bacteria through L or 'soft' forms (Dienes and Weinberger, 1951), Edward (1954) reiterated their similarities to true viruses and countered that ordinary bacteria could just as easily be descendants of PPLO. The Judicial Commission of the International Committee for Bacteriological Nomenclature had had enough: Buchanan *et al.* (1955) rebuked the field, writing that 'the nomenclature of microorganisms placed in the so-called 'pleuropneumonia group' is confused' and offering the choice between *Mycoplasma* or *Borrelomyces* as the only legitimate generic name of the agent of bovine peripneumonia (the Commission itself preferred *Borrelomyces*). Freundt (1955) developed a compromise that grouped previously described organisms as species of *Mycoplasma* comb. nov. in a new order, *Mycoplasmatales,* and family, *Mycoplasmataceae,* along with two new

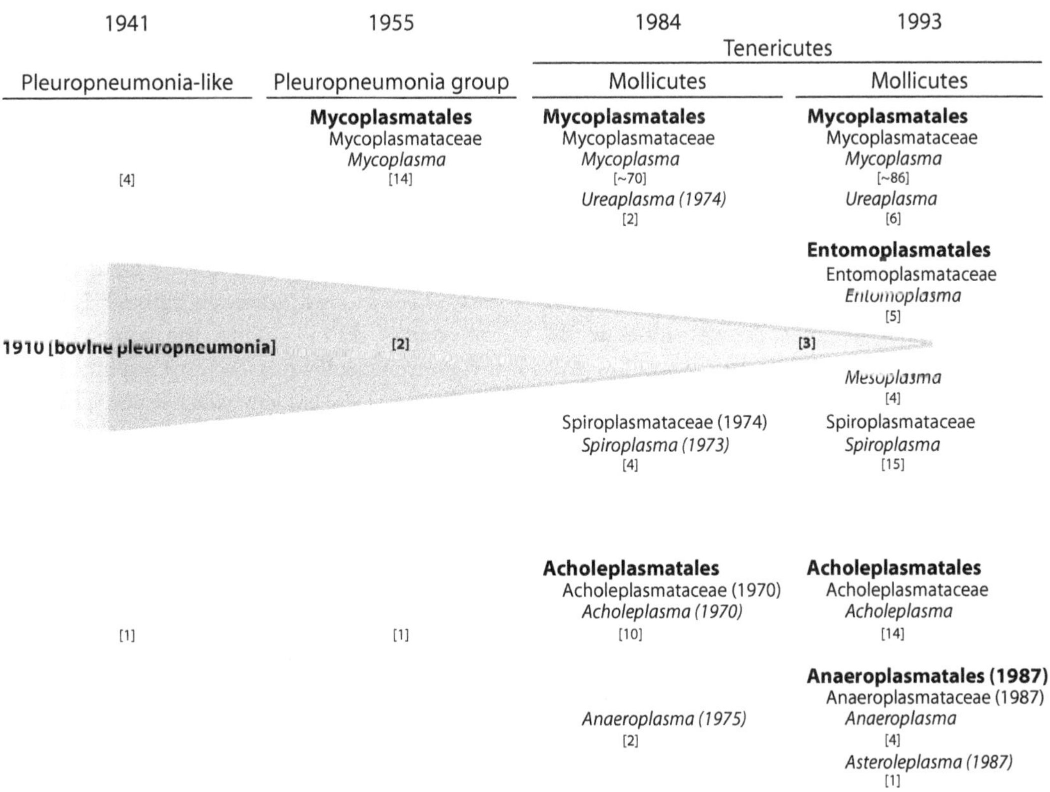


**Figure 1.2** Origins and eclipse of the pleuropneumonia-like organisms. The cumulative number of named organisms (brackets) within taxa recognized at historic milestones are indicated from the perspective of their modern taxonomic relationships. The shaded triangle emphasizes the eccentric taxonomic position of the agent of bovine pleuropneumonia *Mycoplasma mycoides*, which is the type species of Genus *Mycoplasma*, and the diminishing proportion of species closely related to it. This suggests that a more broadly representative neotype of the Order *Mycoplasmatales* can be found.

species of *Borrelomyces*. An alternative ordinal name considered by Freundt was *Molliparietales*, the prefix *molli-* indicating 'soft' cell walls, as suggested by Morton. Edward (1955) preferred Order *Mollicutales* and the generic name *Mycoplasma* for all species. Although assignment to any class remained unspecified, the modern classification of Mollicutes in Order *Mycoplasmatales* Freundt (1955) and Genus *Mycoplasma* Nowak 1929 (Fig. 1.2) was reconciled (Edward and Freundt, 1956), and alternatives waned (e.g. Tulasne and Brisou, 1955; Brisou, 1960). Nearly 50 years after *Asterococcus* and *Coccobacillus*, the agent of bovine pleuropneumonia was simply *Mycoplasma mycoides* (Judicial Commission of the International Committee on Bacteriological Nomenclature, 1958). It would be another 10 years before a Subcommittee on the Taxonomy of Mycoplasmata chose Class *Mollicutes* and 'the absence of a true cell wall and the plasticity of the outer membrane' as the outstanding character of Nowak's 'protoplasmiques' to accommodate the segregation of mycoplasmas from *Schizomycetes* (Edward and Freundt, 1967; Edward *et al.*, 1967); L or 'soft' forms of other bacteria were specifically excluded. Mollicutes became more widely studied especially after 'Eaton's agent' of human primary atypical pneumonia was shown to be a mycoplasma (Eaton *et al.*, 1944; Chanock *et al.*, 1963), and Freundt's *Mycoplasmatales* was subsequently restructured to accommodate new families and genera (Edward and Freundt, 1970; Freundt *et al.*, 1984; Robinson and Freundt, 1987; Fig. 1.2).

### The not so typical Mollicute

The first (1958) International Code of Nomenclature of Bacteria and Viruses emphasized the cultural characteristics of bacteria, so establishment of type cultures, which could be referred to whenever there was doubt about the status of a specimen, became critically important. Type cultures are not necessarily completely typical of a taxon, but they function as points of reference (Sneath, 2003). By 1967, the Genus *Mycoplasma* was legitimately established as the reference for the single order encompassing all Mollicutes, the *Mycoplasmatales*, and the type species of *Mycoplasma* was the agent of bovine pleuropneumonia, *M. mycoides* (Judicial Commission of the International Committee on Bacteriological Nomenclature, 1958). Its historical prominence had been recognized for a long time as being problematic for classification purposes. Sabin (1941a) pointed out that use of the term 'pleuropneumonia group' was frankly 'misleading for it pretends to describe ... a group of microorganisms [which], with one exception, have nothing to do with pleuropneumonia.' Edward admitted that the term 'pleuropneumonia-like organism' is cumbersome and its abbreviation PPLO has meaning only in English, but more importantly it subtly emphasizes a 'distinction between the organism of bovine pleuropneumonia itself and the other species of the group' (Edward, 1954). He was insistent that *M. mycoides* represent the other species, even though the only justification for it being the type was that it was the first to have been isolated and it had been studied the most (Edward and Freundt, 1956). In fact only a subspecies then called *M. mycoides* var. *capri* was much like the agent of bovine pleuropneumonia with regard to its regularly filamentous cellular morphology and serological cross-reactivity, while the 14 other species named at the time, including Bridré and Donatien's equally well-studied agalactia agent, were not. Thus an opportunity to propose a more broadly representative type for the *Mycoplasmatales* was lost, ironically through Edward's criticism of 'PPLO'. When the comprehensive taxonomy of Mollicutes was most recently revised (Tully *et al.*, 1993) *M. mycoides* became subsumed in an anomalous splinter group among the paraphyletic *Entomoplasmataceae*, and only lonesome *Asteroleplasma* had fewer close relatives in the class (Fig. 1.2).

## Current controversies

The modern species concept for Mollicutes, which is based on a combination of 16S rRNA gene sequence analyses, DNA–DNA hybridization tests, directed serology and supplementary phenotypic data, was elaborated in the most recent revision of the minimal standards for description of novel species (Brown *et al.*, 2007). The phylum *Tenericutes* presently accommodates the Class *Mollicutes* with four orders, five named families plus two groups of organisms *incertae sedis*, and eleven genera of Mollicutes including '*Candidatus* Phytoplasma' gen. nov., *Eperythrozoon* and *Haemobartonella* (Krieg *et al.*, 2010; see below). Excluding '*Candidatus*' and illegitimately named organisms there are 117 species in genus *Mycoplasma*, seven in *Ureaplasma*, five in *Eperythrozoon* (confusingly, four of these are legitimate species of *Mycoplasma*), three in *Haemobartonella* (all also legitimately named *Mycoplasma*), six in *Entomoplasma*, 11 in *Mesoplasma*, 37 in *Spiroplasma*, 18 in *Acholeplasma*, four in *Anaeroplasma*, and one in *Asteroleplasma*. In addition, analysis of the 16S rRNA gene sequences of at least ten unique phylotypes detected among the human intestinal microbiota during metagenomic surveys showed they cluster distinctly enough to suggest a previously uncircumscribed order within the class (May *et al.*, 2009). Two important alternatives to the hierarchical Linnaean system are also currently in use. The '*Candidatus* Phytoplasma' organisms are classified in a group system devised by the International Research Programme on Comparative Mycoplasmology (IRPCM) Phytoplasma/Spiroplasma Working Team (2004); 30 candidate species are proposed and many subgroups have been identified. Spiroplasma classification may be truncated, with assignment to one of a very large number of subgroups distinguishable by serological and/or genomic characteristics (Gasparich *et al.*, 2004).

As consequences of its historical development, evolutionary anomalies, the rules of nomenclature of bacteria, and the frankly vague species concept for most prokaryotes, the taxonomy of Mollicutes is acknowledged to bear the following imperfections.

## The *Mycoides* cluster

The Order *Entomoplasmatales* presently accommodates Mollicutes that are regularly associated with arthropod or plant hosts. However, the ecologically, phenotypically and genetically cohesive group called the *mycoides* cluster of mycoplasmas, which includes the type species *M. mycoides* subsp. *mycoides* and four other species of Mollicutes that are regularly associated only with ruminants (*M. mycoides* subsp. *capri*, *Mycoplasma capricolum* subsp. *capricolum*, *M. capricolum* subsp. *capripneumoniae* and *Mycoplasma leachii*), are clearly more closely related by 16S rRNA gene similarity to Mollicutes in the Family *Entomoplasmataceae* in this order than they are to members of the Order *Mycoplasmatales* (see Krieg *et al.*, 2010). Further, the non-helical *Entomoplasmataceae* are situated within the paraphyletic family of helical Mollicutes associated with arthropod or plant hosts, the *Spiroplasmataceae*. Because of these anomalies, the current taxonomic assignment of species in the *mycoides* cluster is controversial (Tully *et al.*, 1993). It seems certain that confusion and peril would result from reassignment to the genus *Entomoplasma* (see Lapage *et al.*, 1992) especially regarding the 'Small Colony' PG1$^{T}$-like strains of *M. mycoides* subsp. *mycoides* and the F38$^{T}$-like strains of *M. capricolum* subsp. *capripneumoniae*, which are highly virulent animal pathogens (Cottew, 1979; McMartin *et al.*, 1980; Cottew *et al.*, 1987). All members of the *mycoides* cluster except *M. leachii* are currently listed in the Terrestrial Animal Health Code of the Office International des Epizooties and subject to strict quarantine regulations in many countries.

## Nomenclature of haemotropic Mollicutes

Certain haemotropic bacteria found on the surface of erythrocytes were once assigned to the genera *Eperythrozoon* ('animals on red blood cells') or *Haemobartonella* (redundantly, 'blood-dwelling *Bartonella*'). Any distinctions between the two genera are tenuous and probably arbitrary (see Kreier and Ristic,1974a,b). These organisms are now understood to be properly affiliated with the *Mycoplasmatales* on the bases of their lack of a cell wall, use of the codon UGA to encode tryptophan, and 16S rRNA gene sequences. Seven species are now legitimately named *Mycoplasma* although their transfer to the *Mycoplasmatales* has not yet been formalized (Neimark *et al.*, 2001, 2002, 2005; Krieg *et al.*, 2010). None of the haemotropic Mollicutes has been cultivated axenically, so no type strains have been established and the designation '*Candidatus*' is still used for new species. Taxonomic reassignments from *Eperythrozoon* to the genus *Mycoplasma* were challenged because the 16S rRNA sequence similarities were thought to be insufficient (Uilenberg *et al.*, 2004, 2006). Establishing a new genus to accommodate the haemoplasmas (Uilenberg *et al.*, 2006) would compound the 16S rRNA-based polyphyly within genus *Mycoplasma* for no reason except to acknowledge their capacity to adhere to erythrocytes, a trait that may occur sporadically throughout the genus. A further complicating factor is that the name *Eperythrozoon* predates *Mycoplasma*, but it would be completely impractical to transfer all mycoplasmas to genus *Eperythrozoon* because only a small minority regularly inhabit the surface of erythrocytes. For the time being the taxonomic status of the haemotropic Mollicutes remains *incertae sedis* (Krieg *et al.*, 2010), but referral to the genus *Mycoplasma* has been embraced by specialists in the molecular biology and ecology of these organisms.

## Possible advantages of a new type species of *Mycoplasma*

The nomenclatural type of a genus is a species, and that of a species is a strain. Appendix 7 of the International Code of Nomenclature of Bacteria explicitly allows for replacement of an unsuitable type strain with a neotype (Lapage *et al.*, 1992), and recommendation 20d regarding the Code's Rule 20 ('Type of a Genus') is to exclude 'species doubtfully referred to the genus'. Although further details regarding neotype species remain enigmatic, the Code seems sufficiently flexible to allow replacement with more appropriate types of any taxon as knowledge advances. The modern taxonomic situation of *M. mycoides* subsp. *mycoides*, which is distantly eccentric to the vast majority of other species currently in the genus, is understandably puzzling to systematists unfamiliar with the complicated history of its nomenclature. Given the taxonomically weak original justification for *M. mycoides* subsp. *mycoides* as a type species, the International Committee on Systematics of Prokaryotes' subcommittee on the taxonomy of *Mollicutes* recently began to consider the possibility of a more broadly representative neotype species for *Mycoplasma* (Firrao and Brown, 2011). In particular, one with a less extreme cellular morphology and degree of virulence, and having fewer restrictions on possession and international transportation might be preferable. One well-known candidate discussed is *Mycoplasma hominis*, the nominal representative of the largest current taxonomic group of mycoplasmas (about 90 species; Johansson and Pettersson, 2002). *M. hominis* is a coccoid to filamentous, arginine-hydrolysing opportunistic pathogen of mucosal, serosal and synovial surfaces of humans. It is also a common tissue culture contaminant. The complete genome of strain PG21$^{T}$ was recently sequenced and annotated (Pereyre *et al.*, 2009). There is as yet no consensus within the subcommittee regarding a neotype species. Some subcommittee members feel strongly that despite the chronological misfortune of its discovery, the precedence of *M. mycoides* should be respected regardless of its imperfections as the representative type.

## The genus *Mesoplasma*

Non-helical Mollicutes that grow in serum-free media supplemented with polyoxyethylene sorbitan are currently assigned to the genus *Mesoplasma* (Tully *et al.*, 1993), which has a sister relationship with the *mycoides* cluster and genus *Entomoplasma* (Krieg *et al.*, 2010). The name mesoplasma connotes a degree of nutritional dependence on exogenous sterols thought to be intermediate between essential and unessential. However, no phylogenetic support for the segregation between *Mesoplasma* and *Entomoplasma* can be found through 16S rDNA sequence similarity analyses. Current species of *Entomoplasma* and *Mesoplasma* are paraphyletic within the *Entomoplasmatales* (Johansson and Pettersson, 2002) but there is no other taxonomic basis to maintain the separation. In retrospect, the degree of dependence on exogenous sterol is not as discriminating a phenotypic character as it was once thought to be for Mollicutes (Gasparich *et al.*, 2004). *Entomoplasma* has priority (Tully *et al.*, 1993), so most extant species of *Mesoplasma* will likely be transferred to that genus, and *Mesoplasma* will become illegitimate. However, *Mesoplasma pleciae* was recently found to be properly affiliated with the genus *Acholeplasma* based on its 16S rRNA sequence similarity and tryptophan codon usage. Therefore, taxonomic transfer of the remaining mesoplasmas cannot occur *en masse* until similar genetic analyses have been completed for each of them (see Brown *et al.*, 2007; Krieg *et al.*, 2010).

## The complexities of *Phytoplasma* systematics

Phytoplasmas constitute a very large monophyletic group of non-helical Mollicutes that colonize plant phloem and insect vectors. They cause many plant disease syndromes including some major diseases subject to international quarantine regulations, but none has been cultivated axenically, so no type strains have been established and a polyphasic taxonomy has been impractical. A taxonomy for the genus was originally based on a 97.5% 16S rRNA sequence similarity cut off between reference strains, as well as considerations of the type of transmission vector, the plant host range or 'behaviour' within a plant, and other molecular evidence of significant genetic or serological diversity (IRPCM Phytoplasma/Spiroplasma Working Team, 2004). The IRPCM team invited a possibly unanticipated degree of taxonomic complexity below the level of genus with this stance: 'Strains in which even minimal differences in the 16S rRNA gene sequence from the reference strain are detected do not 'belong' to the *Candidatus* species, but are "related" to it.' A difference between specimens as slight as a single nucleotide at any position in 16S rDNA might therefore be construed to be taxonomically meaningful. The original 15 groups established in 2004 according to 16S rRNA sequence similarity have consequently multiplied to at least 30 '16Sr' groups, which are further divided into more than 100 subgroups of related strains differentiated on the basis of similarity coefficients calculated through actual or virtual 16S rDNA RFLP typing (Lee *et al.*, 2010b). GenBank presently includes almost 2700 phytoplasma 16S rDNA sequences. The necessity for this degree of taxonomic complexity is not evident to the non-specialist, and seems contrary to the heterogeneity allowed in the current species concept for Mollicutes (Brown *et al.*, 2007), but it is hoped that such exquisitely fine-scale differentiation may eventually be useful in the development of improved diagnosis and control measures for plant diseases caused by closely related but biologically distinct strains of phytoplasmas (Lee *et al.*, 2010a).

## The impending need to accommodate artificial species

The recent synthesis and assembly of entire mycoplasmal chromosomes and their subsequent resurrection as living bacteria (Lartigue *et al.*, 2007, 2009; Gibson *et al.*, 2008, 2010; Benders *et al.*, 2010) demonstrate that *de novo* synthesis of novel Mollicutes arising by artificially directed speciation events is now possible. A new system of nomenclature and classification may be needed for such organisms, because it seems likely that their synthetic genomes, which are expected to derive at least initially from various natural sources in a modular fashion, might exhibit combinations of 16S rRNA sequences, DNA–DNA hybridization values and consequent antigenic and other phenotypic characteristics that would be incompatible with the phylogenetic species concept or taxonomy as currently circumscribed for the *Tenericutes*.

## Genome sequence-based taxonomy: a path to the future?

Contention over the taxonomy of Mollicutes is no greater than current arguments over the prokaryotic species concept itself. Taxonomic groupings of bacteria at the species level are 'right', in the sense meant by K'ung-fu-tzu, when they accurately predict the phenotypic potential and ecology of the strains included in each group (Konstantinos and Tiedje, 2005a; Konstantinos *et al.*, 2006). However, a universal measure of the boundary that circumscribes a species has been hard to establish, in part because the degree of phenotypic and genotypic homogeneity varies among biologically meaningful strain groups. Some prokaryotes may not be referable to any species, or some species may not differ from families or orders (Doolittle and Zhaxybayeva, 2009). The current approach to species delineation based primarily on the extent of genomic DNA–DNA hybridization between strain pairs, supported by diagnostic phenotypic criteria, is technically difficult to implement and tends to result in species that are not sufficiently predictive of phenotype for practical clinical, agricultural and industrial purposes. There is an impractically large amount of phenotypic variation within some species that have high DNA–DNA reassociation values. This is true for example for some assemblages of *Spiroplasma* (see Brown *et al.*, 2007). The DNA–DNA hybridization boundaries are not based on any theoretical rationale, but were chosen *post hoc* to match pre-existing species definitions, even though some strains with high DNA–DNA reassociation values remain assigned to different classical species on the basis of pathogenicity, transmission mechanisms or host range (Konstantinos and Tiedje, 2005a; Achtman and Wagner, 2008). Some Mollicutes have the opposite problem; certain strain pairs within classical species of *Acholeplasma* and *Mycoplasma* have very low DNA–DNA reassociation values, but there is little other justification for subdividing them (see Brown *et al.*, 2007). Strains of the same classical species can vary by as much as 30% in their complement of genes that are not hypothetical or mobile elements, as a result of lateral gene transfer and the consequently reticulate pattern of bacterial evolution (Konstantinos and Tiedje, 2005a; Doolittle and Zhaxybayeva, 2009). Ribosomal RNA sequences can also be too highly conserved to resolve certain strain clusters with patently different phenotypes and ecological niches (Fraser *et al.*, 2009). Mollicute examples of this include the group I spiroplasmas and subspecies within the *mycoides* cluster (Brown *et al.*, 2007; Manso-Silván *et al.*, 2009). Regardless of the criteria used to detect them, many classical species likely contain subpopulations that express different phenotypes, occupy distinct ecological niches, experience different selective pressures and rates of gene exchange, and are therefore evolving independently (Achtman and Wagner, 2008).

Advances in DNA sequencing technology and bioinformatics support a fresh, whole-genome sequence-based approach to bacterial systematics. The holistic concept of average nucleotide identity (ANI) currently seems to be the most promising candidate to replace DNA–DNA hybridization (or subjectively reductionist multilocus sequence typing) as a standard objective metric for bacterial species circumscription (Konstantinos and Tiedje, 2005a; Rosselló-Móra, 2005; Konstantinos *et al.*, 2006; Richter and Rosselló-Móra, 2009). As originally applied in comparisons among 70 fully sequenced bacterial genomes, the Basic Local Alignment Search Tool (BLAST) algorithm was used to calculate the ANI between strain pairs by comparing all genes conserved in genomes of the pair (Konstantinos and Tiedje, 2005a). An alternative approach employs BLAST (Goris *et al.*, 2007) or Maximal Unique Matching (MUMmer) alignment algorithms (Richter and Rosselló-Móra, 2009) to compare whole genomes divided into random 1000-nt-long sequence fragments. Genetic relatedness measured in these ways correlated strongly with 16S rRNA gene sequence identity and DNA–DNA reassociation values; the 70% DNA–DNA hybridization cut off corresponds to an ANI of about 95%. Some pairs with an ANI >94% differed in as many as one-third of their total genes, further evidence that some classical species include doubtfully referred strains, although the majority of intra-species variable genes are hypothetical or mobile and thus may exist only transiently in the species (Konstantinos *et al.*, 2006). As an extension of ANI, the average amino acid identity (AAI) may be useful for comparisons of more distantly related genomes and to objectively circumscribe certain taxa above the species level (Konstantinos and Tiedje, 2005b; Rosselló-Móra, 2005), although uniform criteria for setting boundaries based on AAI may be very difficult to establish.

The number of completely sequenced bacterial genomes is increasing rapidly, but alternative approaches to calculating ANI exist that do not depend on availability of a closed genome sequence for each strain in a pair. Richter and Rosselló-Móra (2009) showed that random sequence coverage of only 50% of each of two strains to be compared, or as little as 20% of a query genome to be compared with a fully sequenced reference genome, was sufficient to obtain ANI values in the range of the species threshold. Partial genome sequencing of strain pairs is also adequate to calculate tetranucleotide frequency correlation coefficients, an alignment-free parameter that correlates well with ANI and is similarly useful for classification at the species level.

The ANI concept is readily applicable to Mollicutes (Rosselló-Móra, 2010) and may indicate a path to future resolution of several current controversies regarding assignments in the class, especially if it (or its reflection in AAI) can eventually be extended at least to the genus level. The proper phylogenetic placement of haemotropic Mollicutes and the status of Genus *Mesoplasma* seem especially amenable to such analyses, and it can be hoped that it will be usefully applicable to phytoplasma groupings. It will certainly be interesting to see whether the ANI and AAI for members of the *mycoides* cluster are higher with the *Mycoplasmatales* or with the *Entomoplasmatales*, and whether an ANI-based taxonomy could resolve the anomalous placement of the *mycoides* cluster within the 16S rDNA-based phylogenetic tree. Genome sequence-based comparisons may objectively narrow the choices of a possible neotype species of *Mycoplasma* based on the most cohesive ANI and AAI within the genus. Finally, an ANI index may be very useful in tracking synthetic organisms that have no phylogenetic relationship with Mollicutes based on evolution by natural descent.

## Acknowledgements

Supported by US Public Health Service grants R01-GM076584-01A1 and -S1 (DRB). This chapter is dedicated to the memory of our friend, Robert F. Whitcomb (1932–2007).

## References

Achtman, M., and Wagner, M. (2008). Microbial diversity and the genetic nature of microbial species. Nat. Rev. Microbiol. 6, 431–440.

Anonymous (1870). Infectious pleuropneumonia. Br. Med. J. *1*, 604–606.

Barberet, D. (1766). Mémoire sur les maladies épidémiques des bestiaux qui a remporté le prix proposé par la Société Royale d'Agriculture de la généralité de Paris pour l'année 1765. (d'Houry, Paris).

Benders, G.A., Noskov, V.N., Denisova, E.A., Lartigue, C., Gibson, D.G., Assad-Garcia, N., Chuang, R.Y., Carrera, W., Moodie, M., Algire, M.A., *et al.* (2010). Cloning whole bacterial genomes in yeast. Nucleic Acids Res. *38*, 2558–2569.

Blancou, J. (1996). Early methods of surveillance and control for contagious bovine pleuropneumonia. Res. Sci. Tech. Off. Int. Epiz. *15*, 1263–1282.

Bordet, J. (1910). La morphologie du microbe de la péripneumonie des bovidés. Ann de l'Institut Pasteur *24*, 161–167.

Borrel, A., Dujardin-Beaumetz, E., Jeantet, P., and Jouan, C. (1910). Le microbe de la péripneumonie. Ann de l'Institut Pasteur *24*, 168–179.

Bridré, J., and Donatien, A. (1923). Le microbe de l'agalaxie contagieuse et sa culture in vitro. Compt. Rend. de l'Acad. des Sci. *177*, 841–843.

Bridré, J., and Donatien, A. (1925). Le microbe de l'agalaxie contagieuse du mouton de la chèvre. Ann de l'Institut Pasteur *39*, 925–179.

Brisou, J. (1960). The L forms and PPLO. In Biology of the Pleuropneumonialike Organisms, Nelson, J.B., ed. (New York Academy of Sciences, New York), pp. 488–492.

Brown, D.R., and Bradbury, J.M. (2007). International committee on systematics of Prokaryotes, subcommittee on the taxonomy of Mollicutes. Minutes of the meetings, 8, 9 and 14 July 2006, Cambridge, UK. Int. J. Syst. Evol. Microbiol. *57*, 192–196.

Brown, D.R., Whitcomb, R.F., and Bradbury, J.M. (2007). Revised minimal standards for description of new species of the class Mollicutes (division Tenericutes). Int. J. Syst. Evol. Microbiol. *57*, 2703–2719.

Buchanan, R.E., Cowan, S.T., and Wikén, T. [Editorial Board] (1955). The status and synonymy of the bacterial generic name *Asterococcus*. Int. Bull. Bact. Nom. Taxon. *5*, 13–20.

Celli, À., and De Blasi, D. (1906). Etiologia dell' agalassia contagiosa delle pecore e capre. Ann. d'Igiene Sper. *16*, 257–293.

Chanock, R.M., Dienes, L., Eaton, M.D., Edward, D.G. ff., Freundt, E.A., Hayflick, L., Hers, J.F. ph., Jensen, K.E., Liu, C., Marmion, B.P., *et al.* (1963). *Mycoplasma pneumoniae*: proposed nomenclature for atypical pneumonia organism (Eaton agent). Science *140*, 662.

Cottew, G.S. (1979). Caprine-ovine mycoplasmas. In The Mycoplasmas, Volume II Human and Animal Mycoplasmas, Tully, J.G., and Whitcomb, R.F., eds. (Academic Press, New York), pp. 103–132.

Cottew, G.S., Breard, A., DaMassa, A.J., Erno, H., Leach, R.H., Lefevre, P.C., Rodwell, A.W., and Smith, G.R. (1987). Taxonomy of the *Mycoplasma mycoides* cluster. Isr. J. Med. Sci. *23*, 632–635.

Dienes, L., and Weinberger, H.J. (1951). The L forms of bacteria. Bacteriol. Rev. *15*, 245–288.

Doolittle, W.F., and Zhaxybayeva, O. (2009). On the origin of prokaryotic species. Genome Res. *19*, 744–756.

Dujardin, F. 1835. Sur les prétendus estomacs des animates infusoires et sur une substance appelée sarcode. Ann. Sci. Nat.: Zool. Biol. Anim. *4*, 343–377.

Eaton, M.D., Meiklejohn, G., and van Herick, W. (1944). Studies on the etiology of primary atypical pneumonia, a filterable agent transmissible to cotton rats, hamsters, and chick embryos. J. Exp. Med. *79*, 649–668.

Edward, D.G. ff. (1954). The pleuropneumonia group of organisms: a review, together with some new observations. J. Gen. Microbiol. *10*, 27–64.

Edward, D.G. ff. (1955). A suggested classification and nomenclature for organisms of the pleuropneumonia group. Int. Bull. Bact. Nom. Taxon. *5*, 85–93.

Edward, D.G. ff., and Freundt, E.A. (1956). A classification and nomenclature of organisms of the pleuropneumonia group. J. Gen. Microbiol. *14*, 197–207.

Edward, D.G. ff., and Freundt, E.A. (1967). Proposal for Mollicutes as name of the class established for the Order *Mycoplasmatales*. Int. J. Syst. Bacteriol. *17*, 267–268.

Edward, D.G. ff., and Freundt, E.A. (1970). Amended nomenclature for strains related to *Mycoplasma laidlawii*. J. Gen. Microbiol. *62*, 1–2.

Edward, D.G. ff., Freundt, E.A., Chanock, R.M., Fabricant, J., Hayflick, L., Lemcke, R.M., Razin, S., Somerson, N.L., and Wittler, R.G. [Subcommittee on the taxonomy of *Mycoplasmatales*] (1967). Recommendations on nomenclature of the Order *Mycoplasmatales*. Science *155*, 1694–1696.

Firrao, G., and Brown, D.R. (2011). International committee on systematics of Prokaryotes. Subcommittee on the taxonomy of Mollicutes. Minutes of the meetings, 11 and 16 July 2010, Chianciano Terme, Italy. Int. J. Syst. Evol. Microbiol. *61*, 695–697.

Fischer, A. (1903). Vorlesungen über Bakterien. (Gustav Fischer, Jena), pp. 58–59.

Frank, B. (1890). Über die pilzsymbiose der leguminosen. Landwirt. Jahrb. *19*, 523–640.

Fraser, C., Alm, E.J., Polz, M.F., Spratt, B.G., and Hanage, W.P. (2009). The bacterial species challenge: making sense of genetic and ecological diversity. Science *323*, 741–746.

Freundt, E.A. (1955). The classification of the pleuropneumonia group of organisms (*Borrelomycetales*). Int. Bull. Bact. Nom. Taxon. *5*, 67–78.

Freundt, E.A., Whitcomb, R.F., Barile, M.F., Razin, S., and Tully, J.G. (1984). Proposal for elevation of the Family *Acholeplasmataceae* to ordinal rank: *Acholeplasmatales*. Int. J. Syst. Bacteriol. *34*, 346–349.

Frosch, P. (1923). Zur morphologie des lungenseucheerregers. Arch. Wiss. Prakt. Tierheilkd. *49*, 273–282.

Gasparich, G.E., Whitcomb, R.F., Dodge, D., French, F.E., Glass, J., and Williamson, D.L. (2004). The genus *Spiroplasma* and its non-helical descendants: phylogenetic classification, correlation with phenotype and roots of the *Mycoplasma mycoides* clade. Int. J. Syst. Evol. Microbiol. *54*, 893–918.

Gibson, D.G., Benders, G.A., Andrews-Pfannkoch, C., Denisova, E.A., Baden-Tillson, H., Zaveri, J., Stockwell, T.B., Brownley, A., Thomas, D.W., Algire, M.A., *et al.* (2008). Complete chemical synthesis, assembly, and cloning of a *Mycoplasma genitalium* genome. Science *319*, 1215–1220.

Gibson, D.G., Glass, J.I., Lartigue, C., Noskov, V.N., Chuang, R.Y., Algire, M.A., Benders, G.A., Montague, M.G., Ma, L., Moodie, M.M., *et al.* (2010). Creation of a bacterial cell controlled by a chemically synthesized genome. Science *329*, 52–56.

Gordon, C.A. (1884). An Epitome of the Reports of the Medical Officers to the Chinese Imperial Maritime Customs Service, from 1871 to 1882: with Chapters on the History of Medicine in China, Materia Medica, Epidemics, Famine, Ethnology, and Chronology in Relation to Medicine and Public Health. (Baillière, Tindall, & Cox, London), p. 361.

Goris, J., Konstantinidis, K.T., Klappenbach, J.A., Coenye, T., Vandamme, P., and Tiedje, J.M. (2007). DNA–DNA hybridization values and their relationship to whole-genome sequence similarities. Int. J. Syst. Evol. Microbiol. *57*, 81–91.

IRPCM Phytoplasma/Spiroplasma Working Team (2004). '*Candidatus* Phytoplasma', a taxon for the wall-less, non-helical prokaryotes that colonize plant phloem and insects. Int. J. Syst. Evol. Microbiol. *54*, 1243–1255.

International Committee on Bacteriological Nomenclature (1958). International Code of Nomenclature of Bacteria and Viruses. (Iowa State College Press, Ames).

Johansson, K.-E., and Pettersson, B. (2002). Taxonomy of Mollicutes. In Molecular Biology and Pathogenicity of Mycoplasmas, Razin, S., and Herrmann, R., eds. (Kluwer Academic/Plenum Publishers, New York), pp. 1–29.

Judicial Commission of the International Committee on Bacteriological Nomenclature (1958). Opinion 22. Status of the generic name *Asterococcus* and conservation of the generic name *Mycoplasma*. Int. Bull. Bact. Nom. Taxon. *8*, 166–168.

Klieneberger, E. (1935). The natural occurrence of pleuropneumonia-like organisms in apparent symbiosis with *Streptobacillus moniliformis* and other bacteria. J. Pathol. Bacteriol. *40*, 93–105.

Klieneberger, E. (1938). Pleuropneumonia-like organisms of diverse provenance: some results of an enquiry into methods of differentiation. J. Hyg. *38*, 458–476.

Klieneberger-Nobel, E. (1980). Memoirs. (Academic Press, London).

Konstantinidis, K.T., and Tiedje, J.M. (2005a). Genomic insights that advance the species definition for prokaryotes. Proc. Nat. Acad. Sci. U.S.A. *102*, 2567–2572.

Konstantinidis, K.T., and Tiedje, J.M. (2005b). Towards a genome-based taxonomy for prokaryotes. J. Bacteriol. *187*, 6258–6264.

Konstantinidis, K.T., Ramette, A., and Tiedje, J.M. (2006). The bacterial species definition in the genomic era. Phil. Trans. R. Soc. B *361*, 1929–1940.

Krass, C.J., and Gardner, M.W. (1973). Etymology of the term mycoplasma. Int. J. Syst. Bacteriol. *23*, 62–64.

Kreier, J.P., and Ristic, M. (1974a). Genus IV. *Haemobartonella* Tyzzer and Weinman 1939, $143^{AL}$. In Bergey's Manual of Determinative Bacteriology, 8th edn, Buchanan, R.E., and Gibbons, N.E., eds. (Williams & Wilkins, Baltimore), pp. 910–912.

Kreier, J.P., and Ristic, M. (1974b). Genus V. *Eperythrozoon* Schilling 1928, $1854^{AL}$. In Bergey's Manual of Determinative Bacteriology, 8th edn, Buchanan, R.E., and Gibbons, N.E., eds. (Williams & Wilkins, Baltimore), pp. 912–914.

Krieg, N.R., Staley, J.T., Brown, D.R., Hedlund, B., Paster, B.J., Ward, N., Ludwig, W., and Whitman, W.B. (2010). Bergey's Manual of Systematic Bacteriology, 2nd edn, Volume 4 (Springer. New York), pp. 567–723.

Laidlaw, P.P., and Elford, W.J. (1936). A new group of filterable organisms. Proc. R. Soc. (London) B *120*, 292–303.

Lapage, S.P., Sneath, P.H.A., Lessel, E.F., Skerman, V.B.D., Seeliger, H.P.R., and Clark, W.A. (1992). International Code of Nomenclature of Bacteria (1990

Revision). Bacteriological Code (American Society for Microbiology, Washington DC).

Lartigue, C., Glass, J.I., Alperovich, N., Pieper, R., Parmar, P.P., Hutchison, C.A. III, Smith, H.O., and Venter, J.C. (2007). Genome transplantation in bacteria: changing one species to another. Science *317*, 632–638.

Lartigue, C., Vashee, S., Algire, M.A., Chuang, R.Y., Benders, G.A., Ma, L., Noskov, V.N., Denisova, E.A., Gibson, D.G., Assad-Garcia, N., *et al.* (2009). Creating bacterial strains from genomes that have been cloned and engineered in yeast. Science *325*, 1693–1696.

Lee, I.-M., Bottner-Parker, K.D., Zhao, Y., Davis, R.E., and Harrison, N.A. (2010a). Phylogenetic analysis and delineation of phytoplasmas based on *secY* gene sequences. Int. J. Syst. Evol. Microbiol. *60*, 2887–2897.

Lee, I.-M., Zhao, Y., Wei, W., and Davis, R.E. (2010b). Combined actual gel and virtual RFLP analyses for identification and classification of phytoplasmas. COST Action FA0807 Workshop WG1-WG4, 16 July 2010, Chianciano Terme, Siena, Italy (abstr.).

Manso-Silván, L., Vilei, E.M., Sachse, K., Djordjevic, S.P., Thiaucourt, F., and Frey, J. (2009). *Mycoplasma leachii* sp. nov. as a new species designation for *Mycoplasma* sp. bovine group 7 of Leach, and reclassification of *Mycoplasma mycoides* subsp. *mycoides* LC as a serovar of *Mycoplasma mycoides* subsp. *capri*. Int. J. Syst. Evol. Microbiol. *59*, 1353–1358.

Martin, L. (1898). Production de la toxine diphtérique. Ann de l'Institut Pasteur *12*, 26–26.

Martzinovski, E.-J. (1911). De l'étiologie de le péripneumonie. Ann de l'Institut Pasteur *25*, 914–917.

May, M., Whitcomb, R.F., and Brown, D.R. (2009). *Mycoplasma* and related organisms. In Practical Handbook of Microbiology, 2nd edn, Goldman, E., and Green, L.H., eds. (CRC Press, Boca Raton, FL), pp. 467–491.

McMartin, D.A., MacOwan, K.J., and Swift, L.L. (1980). A century of classical contagious caprine pleuropneumonia: from original description to aetiology. Br. Vet. J. *136*, 507–515.

Metchnikoff, É., Roux, E., and Taurelli-Salimbeni, A. (1896). Toxine et antitoxine cholérique. Ann. de l'Institut Pasteur *10*, 257–282.

Neimark, H., Johansson, K.-E., Rikihisa, Y., and Tully, J.G. (2001). Proposal to transfer some members of the genera *Haemobartonella* and *Eperythrozoon* to the genus *Mycoplasma* with descriptions of '*Candidatus* Mycoplasma haemofelis', '*Candidatus* Mycoplasma haemomuris', '*Candidatus* Mycoplasma haemosuis' and '*Candidatus* Mycoplasma wenyonii'. Int. J. Syst. Evol. Microbiol. *51*, 891–899.

Neimark, H., Johansson, K.-E., Rikihisa, Y., and Tully, J.G. (2002). Revision of haemotrophic *Mycoplasma* species names. Int. J. Syst. Evol. Microbiol. *52*, 683.

Neimark, H., Peters, W., Robinson, B.L., and Stewart, L.B. (2005). Phylogenetic analysis and description of *Eperythrozoon coccoides*, proposal to transfer to the genus *Mycoplasma* as *Mycoplasma coccoides* comb. nov. and request for an opinion. Int. J. Syst. Evol. Microbiol. *55*, 1385–1391.

Nocard, E., and Leclainche, E. (1898). Les Maladies Microbiennes des Animaux. 2nd edn (Masson and C[ie], Paris), p. 297.

Nocard, E., and Roux, E. (1898a). Le microbe de la péripneumonie. Bull de la Société Centrale de Médecine Vétérinaire *52*, 213.

Nocard, E., and Roux, E. (1898b). Le microbe de la péripneumonie. Ann de l'Institut Pasteur *12*, 240–262.

Nowak, J. (1929). Morphologie, nature et cycle évolutif du microbe de la péripneumonie des bovidés. Ann de l'Institut Pasteur *43*, 1330–1352.

Pasteur, L. (1882). Note sur la péripneumonie contagieuse des bêtes à cornes. Bulletin de la Société d'agriculture de Melun *1882*, 51–58. [Reprinted in Recueil de Médecine Vétérinaire *59*, 1215–1223 (1882), and in L. Pasteur Vallery-Radot, ed. (1939). Oeuvres de Pasteur, Tome 6, Fascicule 1 (Masson and Cie, Paris), pp. 513–522].

Pereyre, S., Sirand-Pugnet, P., Beven, L., Charron, A., Renaudin, H., Barré, A., Avenaud, P., Jacob, D., Couloux, A., Barbe, V., *et al.* (2009). Life on arginine for *Mycoplasma hominis*: clues from its minimal genome and comparison with other human urogenital mycoplasmas. PLoS Genet. *5*, e1000677.

Richter, M., and Rosselló-Móra, R. (2009). Shifting the genomic gold standard for the prokaryotic species definition. Proc. Natl. Acad. Sci. U.S.A. *106*, 19126–19131.

Robinson, I.M., and Freundt, E.A. (1987). Proposal for an amended classification of anaerobic Mollicutes. Int. J. Syst. Bacteriol. *37*, 78–81.

Rosselló-Móra, R. (2005). Updating prokaryotic taxonomy. J. Bacteriol. *187*, 6255–6257.

Rosselló-Móra, R. (2010). Towards a database-based prokaryote taxonomy. 18th Congress of the International Organization for Mycoplasmology, 11–16 July 2010, Chianciano Terme, Siena, Italy (abstr.).

Sabin, A.B. (1938). Identification of the filtrable, transmissible neurolytic agent isolated from toxoplasma-infected tissue as a new pleuropneumonia-like microbe. Science *88*, 575–576.

Sabin, A.B. (1941a). The filterable microörganisms of the pleuropneumonia group. Bact. Rev. *5*, 1–67.

Sabin, A.B. (1941b). The filterable microörganisms of the pleuropneumonia group (appendix to section on classification and nomenclature). Bact. Rev. *5*, 331–335.

Sneath, P.H.A. (2003). A short history of the bacteriological code. Available at: http://www.the-icsp.org/misc/Code_history.htm.

Tulasne, R., and Brisou, J. (1955). Les pleuropneumoniales. Taxonomie des 'pleuropneumonia-like organisms' et des formes L. Ann de l'Institut Pasteur *88*, 237–239.

Tully, J.G., Bové, J.M., Laigret, F., and Whitcomb, R.F. (1993). Revised taxonomy of the class Mollicutes: proposed elevation of a monophyletic cluster of arthropod-associated Mollicutes to ordinal rank (*Entomoplasmatales* ord. nov.), with provision for familial rank to separate species with nonhelical morphology (*Entomoplasmataceae* fam. nov.) from helical species (*Spiroplasmataceae*), and emended descriptions of the Order *Mycoplasmatales,* Family *Mycoplasmataceae.* Int. J. Syst. Bacteriol. *43,* 378–385.

Turner, A.W. (1935). A study of the morphology and life cycles of the organism of *pleuropneumonia contagiosa boum (Borrelomyces peripneumoniæ* nov. gen.) by observation in the living state under dark-ground illumination. J. Pathol. Bacteriol. *41,* 1–32.

Uilenberg, G., Thiaucourt, F., and Jongejan, F. (2004). On molecular taxonomy: what is in a name? Exp. Appl. Acarol. 32, 301–312.

Uilenberg, G., Thiaucourt, F., and Jongejan, F. (2006). *Mycoplasma* and *Eperythrozoon* (*Mycoplasmataceae*). Comments on a recent paper. Int. J. Syst. Evol. Microbiol. *56,* 13–14.

Willems, L. (1852). Mémoire sur la pleuropneumonie épizootique du bétail. (Th. Lesigne, Bruxelles).

Wróblewski, W. (1931). Morphologie et cycle évolutif des microbes de la péripneumonie des bovides et de l'agalaxie contagieuse des chèvres et des moutons. Ann de l'Institut Pasteur 47, 94–114.

# Genomic Mosaics

# 2

Marc S. Marenda


### Abstract

It is generally assumed that the high level of host specialization and the adoption of isolated, parasitic lifestyles that occurred during the reductive evolution of Mollicutes could only result in the evolution of small, stable genomes with a high coding density. This view has been challenged recently by the analyses of an ever-increasing number of whole-genome sequences. The comparative genomics of related mycoplasma species and isolates from the same species have shown that mobile genetic elements and lateral gene transfers have produced mosaic genomes with much more plasticity than previously anticipated. This could have significant consequences on the diagnosis, virulence and taxonomy of these organisms.


## The Mollicutes in the era of genomics

Since the release of the *Haemophilus influenzae* genome in 1995, the number of complete sequences of prokaryotes has increased exponentially, with a doubling time of approximately 20 months for bacteria and 34 months for archea (Koonin and Wolf, 2008). In the class Mollicutes, the rate of genome sequencing has been comparable, albeit slightly slower, to the rates for other microorganisms. As a result, in 2011 approximately 50 full genomes from 30 different species of Mollicutes were available in the public nucleotide databases (Table 2.1).

Mycoplasmas were amongst the first bacteria to be completely sequenced. Early attempts at large-scale genome sequencing of prokaryotes started in the 1990s with *Mycoplasma genitalium* (Peterson *et al.*, 1993) and *M. capricolum* (Bork *et al.*, 1995). Prior to 2000, two mycoplasma species, *M. genitalium* (Fraser *et al.*, 1995) and *M. pneumoniae* (Himmelreich *et al.*, 1996) were amongst the 25 bacterial and archaeal species that had been completely sequenced. This allowed the first genome-scale comparison of two mycoplasma species (Himmelreich *et al.*, 1997). As it became clear that an increasing number of the Mollicutes were going to be completely sequenced, the dedicated online interactive database Molligen (Barre *et al.*, 2004) was developed to facilitate comparative genomic analyses. Remarkably, the early attempts in mycoplasma genomics tackled significant technical challenges, such as sequencing the non-cultivable organisms *Phytoplasma* (Oshima *et al.*, 2004) and *Mesoplasma* (Birren *et al.*, unpublished), and the assembling a genome with a high density of repeated sequences, in *Mycoplasma mycoides* subspecies *mycoides* Small Colony type (Westberg *et al.*, 2004).

Genomics of Mollicutes focused initially on the diversity of species across the different phylogenetic groups of this class. In the period from 2000 to 2005, 14 genomes from 12 different species were added to the sequence databases, representing approximately 5% of the complete prokaryotic genomes available at that time. Since then, the number of completely sequenced Mollicutes has more than doubled, with a further 32 complete genomes added to the nucleotide databases, as well as two nearly complete genomes, *M. fermentans* PG18 (Ishida *et al.*, 2009) and *Spiroplasma citri* (Carle *et al.*, 2010). In addition, two synthetic genomes based on the sequences of *M. genitalium* and *M. mycoides* subspecies *capri* (Gibson *et al.*,

**Table 2.1** Complete Mollicute genomes submitted to Genbank between 1995 and 2011

| RefSeq accession | Species (number of sequenced strains) | Length (bp) | GC% | CDS | Initial Submission in GenBank | Reference |
|---|---|---|---|---|---|---|
| NC_000908 | *Mycoplasma genitalium* G37 (1) | 580,076 | 32% | 475 | 29 October 1995 | Fraser *et al.* (1995) |
| NC_000912 | *Mycoplasma pneumoniae* M129 (2) | 816,394 | 40% | 689 | 15 November 1996 | Himmelreich *et al.* (1996) |
| NC_002162 | *Ureaplasma parvum* serovar 3 strain ATCC 700970 (2) | 751,719 | 26% | 614 | 10 January 2000 | Glass *et al.* (2000) |
| NC_002771 | *Mycoplasma pulmonis* UAB CTIP (1) | 963,879 | 27% | 782 | 16 October 2000 | Chambaud *et al.* (2001) |
| NC_005364 | *Mycoplasma mycoides* subspecies *mycoides* SC strain PG1 (2) | 1,211,703 | 24% | 1017 | 25 June 2001 | Westberg *et al.* (2004) |
| NC_004432 | *Mycoplasma penetrans* HF-2 (1) | 1,358,633 | 26% | 1037 | 25 June 2001 | Sasaki *et al.* (2002) |
| NC_004829 | *Mycoplasma gallisepticum* strain R(low) (3) | 1,012,800 | 32% | 763 | 17 October 2002 | Papazisi *et al.* (2003) |
| NC_005303 | Onion yellows phytoplasma OY-M (2) | 853,092 | 28% | 750 | 27 October 2003 | Oshima *et al.* (2004) |
| NC_006908 | *Mycoplasma mobile* 163K (1) | 777,079 | 25% | 633 | 13 April 2004 | Jaffe *et al.* (2004) |
| NC_006360 | *Mycoplasma hyopneumoniae* 232 (4) | 892,758 | 29% | 691 | 28 April 2004 | Minion *et al.* (2004) |
| NC_006055 | *Mesoplasma florum* L1 (1) | 793,224 | 27% | 682 | 7 June 2004 | - |
| NC_007332 | *Mycoplasma hyopneumoniae* 7448 (4) | 920,079 | 29% | 657 | 13 Sep 2004 | Vasconcelos *et al.* (2005) |
| NC_007295 | *Mycoplasma hyopneumoniae* J (4) | 897,405 | 29% | 657 | 13 Sep 2004 | Vasconcelos *et al.* (2005) |
| NC_007294 | *Mycoplasma synoviae* 53 (1) | 799,476 | 29% | 659 | 13 September 2004 | Vasconcelos *et al.* (2005) |
| NC_007716 | Aster yellows witches'-broom phytoplasma AYWB (2) | 706,569 | 27% | 671 | 6 April 2005 | Bai *et al.* (2006) |
| NC_007633 | *Mycoplasma capricolum* subspecies *capricolum* ATCC 27343 (1) | 1,010,023 | 24% | 812 | 16 September 2005 | - |
| – | *Spiroplasma citri* GII3–3X –77 contigs, 38 annotated- (1) | 1,674,000 | | 1908 | 27 June 2006 | Carle *et al.* (2010) |
| NC_009497 | *Mycoplasma agalactiae* PG2 (2) | 877,438 | 30% | 742 | 11 October 2006 | Sirand-Pugnet *et al.* (2007b) |
| NC_010544 | *Candidatus* Phytoplasma australiense (1) | 879,959 | 27% | 684 | 5 January 2007 | Tran-Nguyen *et al.* (2008) |
| NC_010163 | *Acholeplasma laidlawii* PG-8A (1) | 1,496,992 | 32% | 1380 | 8 June 2007 | - |
| NC_011047 | *Candidatus* Phytoplasma mali (1) | 601,943 | 21% | 479 | 5 November 2007 | Kube *et al.* (2008) |
| AP009608 | *Mycoplasma fermentans* PG18 (3) | 1,004,014 | 27% | 893 | 16 January 2008 | Ishida *et al.* (2009) |
| NC_010503 | *Ureaplasma parvum* serovar 3 strain ATCC 27815 (2) | 751,679 | 26% | 609 | 7 February 2008 | - |
| NC_011025 | *Mycoplasma arthritidis* 158L3–1 (1) | 820,453 | 31% | 631 | 1 April 2008 | Dybvig *et al.* (2008) |
| NC_011374 | *Ureaplasma urealyticum* serovar 10 strain ATCC 33699 (1) | 874,478 | 26% | 646 | 3 October 2008 | - |

| RefSeq accession | Species (number of sequenced strains) | Length (bp) | GC% | CDS | Initial Submission in GenBank | Reference |
|---|---|---|---|---|---|---|
| NC_012806 | *Mycoplasma conjunctivae* HRC/581 (1) | 846,214 | 29% | 692 | 22 October 2008 | Calderon-Copete *et al.* (2009) |
| NC_013511 | *Mycoplasma hominis* ATCC 23114 (1) | 665,445 | 27% | 523 | 12 February 2009 | Pereyre *et al.* (2009) |
| CP001621 | *Mycoplasma mycoides* subspecies *capri* strain GM12 tetM-lacZ (2) | 1,089,202 | 24% | 832 | 14 May 2009 | Lartigue *et al.* (2009) |
| CP001668 | *Mycoplasma mycoides* subspecies *capri* strain GM12 transgenic clone deltatypeIIIres (2) | 1,084,586 | 24% | 830 | 14 May 2009 | Lartigue *et al.* (2009) |
| NC_013948 | *Mycoplasma agalactiae* 5632 (2) | 1,006,702 | 30% | 813 | 13 November 2009 | Nouvel *et al.* (2010) |
| CP001873 | *Mycoplasma gallisepticum* strain F (3) | 977,612 | 31% | 756 | 12 January 2010 | Szczepanek *et al.* (2010) |
| CP001872 | *Mycoplasma gallisepticum* strain R(high) (3) | 1,012,027 | 32% | 766 | 12 January 2010 | Szczepanek *et al.* (2010) |
| NC_014014 | *Mycoplasma crocodyli* MP145 (1) | 934,379 | 27% | 689 | 25 March 2010 | Brown *et al.* (2011) |
| NC_014552 | *Mycoplasma fermentans* JER (3) | 977,524 | 27% | 797 | 26 March 2010 | Rechnitzer *et al.* (2011) |
| CP002077 | *Mycoplasma pneumoniae* FH (2) | 811,088 | 40% | 629 | 10 June 2010 | Krishnakumar *et al.* (2010) |
| NC_014751 | *Mycoplasma leachii* PG50 (1) | 1,008,951 | 24% | 882 | 19 July 2010 | – |
| CP002107 | *Mycoplasma mycoides* subspecies *mycoides* SC strain Gladysdale MU clone SC5 (2) | 1,193,808 | 24% | 1095 | 19 July 2010 | – |
| NC_014448 | *Mycoplasma hyorhinis* HUB-1 (1) | 839,615 | 26% | 654 | 11 August 2010 | Liu *et al.* (2010) |
| NC_014760 | *Mycoplasma bovis* PG45 (2) | 1,003,404 | 29% | 765 | 30 August 2010 | Wise *et al.* (2010) |
| CP002274 | *Mycoplasma hyopneumoniae* 168 (4) | 925,576 | 29% | 695 | 12 October 2010 | Liu *et al.* (2011) |
| NC_014921 | *Mycoplasma fermentans* M64 (3) | 1,118,751 | 27% | 1050 | 27 December 2010 | Shu *et al.* (2011) |
| FR773153 | *Mycoplasma haemofelis* strain Langford 1 (2) | 1,147,259 | 39% | 1545 | 10 January 2011 | Barker *et al.* (2011) |
| NC_015153 | *Mycoplasma suis* KI3806 (2) | 709,270 | 31% | 794 | 17 February 2011 | Oehlerking *et al.* (2011) |
| NC_015155 | *Mycoplasma suis* strain Illinois (2) | 742,431 | 31% | 845 | 11 March 2011 | Guimaraes *et al.* (2011), Messick *et al.* (2011) |
| - | *Mycoplasma haemofelis* strain Ohio2 (2) | 1,152,484 | 39% | 1426 | | Messick *et al.* (2011) |
| NC_015431 | *Mycoplasma mycoides* subspecies *capri* LC strain 95010 (2) | 1,153,998 | 23% | 922 | 12 April 2011 | Thiaucourt *et al.* (2011) |
| NC_015725 | *Mycoplasma bovis* Hubei-1 (2) | 948,121 | 29% | 801 | 05 July 2011 | Li *et al.* (2011) |
| NC_015946 | *Mycoplasma putrefaciens* KS1 (1) | 832,603 | 26% | 614 | 24 August 2011 | – |

2008, 2010) have also been published. A whole range of new species or subspecies of Mollicutes, including *M. alligatoris, M. bovoculi, M. capricolum* subspecies *capripneumoniae, M. flocculare, M. orale, M. ovipneumoniae, M. yeatsii,* and the Beet leafhopper transmitted virescence, Maize bushy stunt and Western X phytoplasmas, are identified in the NCBI genome project database (http://www.ncbi.nlm.nih.gov/genomeprj) under the status 'in progress' or 'draft assembly' in 2011. A key driver of this data expansion is the advent of new sequencing platforms, which have accelerated the pace of genome releases and changed the scale of mycoplasma comparative genomics. Although many species of Mollicutes have been sequenced *de novo* by the conventional Sanger method using shotgun random libraries and manual finishing, next-generation high throughput sequencing, such as 454 (Roche Applied Science), SOLiD (Applied Biosystems) or Genome Analyser (Illumina/Solexa) technologies are now widely used. These methodologies have made the complete or nearly complete sequencing of genomes faster and more affordable. As a result, new mycoplasma genomes have been increasingly determined by hybrid sequencing methods that combine next-generation platforms for bulk data production and Sanger sequencing for finishing and circularisation. This approach has been used successfully to determine the genomes of *M. conjunctivae* HRC/581 (Calderon-Copete *et al.*, 2009), *M. mycoides* subspecies *capri* GM12 (Lartigue *et al.*, 2009), *M. pneumoniae* FH (Krishnakumar *et al.*, 2010), *M. hyorhinis* HUB-1 (Liu *et al.*, 2010), *M. fermentans* JER (Rechnitzer *et al.*, 2011) and *M. hyopneumoniae* 168 (Liu *et al.*, 2011).

The technological leaps offered by high throughput and next generation sequencing platforms have been crucial to delivery of a better insight into the diversity and evolution of the Mollicutes, by allowing interspecies comparative genomics and by showing the remarkable genetic relations between strains within a species, or a cluster of closely related species. Of the 32 Mollicutes that have been completely sequenced so far, eight species (*Mycoplasma agalactiae, Mycoplasma bovis, Mycoplasma haemofelis, M. mycoides* subspecies *capri, Mycoplasma mycoides* subspecies *mycoides* SC, *Mycoplasma pneumoniae, Mycoplasma suis* and *Ureaplasma parvum*) are represented by two strains, two species (*M. fermentans* and *M. gallisepticum*) by three strains and one species (*M. hyopneumoniae*) by four strains. Moreover, many sequencing projects listed at the NCBI genome project database are further revisiting species of mycoplasmas that are already sequenced. The new genome projects often include sequencing of several strains from a species or subspecies at the same time. This trend is likely to increase as the next-generation sequencing platforms continue to improve and the costs of sequencing decrease.

These sequencing efforts have allowed the comparative analysis of genome composition (core and pan-genome, mobilome and gene homology) and architecture (organization, repeated sequences, and gene syntenies). The recent developments of mycoplasma genomics have profoundly and unexpectedly changed our understanding of the evolution, intricate genetic relationships, and propensity for genetic plasticity and exchange within and between many species of Mollicutes. The analysis of the general features of mycoplasma genomes and the discovery and characterization of an unexpectedly rich repertoire of mobile genetic elements in these organisms suggests that, far from being uniform and static, the mycoplasma genomes are prone to a rapid evolution.

## General features of *Mycoplasma* genomes

### Chromosome organization, synteny and genome-based phylogeny

Characteristic features of the mycoplasma genomes have been extensively discussed, including their remarkably small size, low %GC, codon usage (Fadiel *et al.*, 2007), number of rRNA genes (Amikam *et al.*, 1984) and the organization of their OriC regions (Lartigue *et al.*, 2003). More recently, global chromosomal alignments and quantitative estimates of gene order similarity have been used to analyse the evolution and taxonomy of Mollicutes. The alignments of genomes from highly related strains (or species) usually show a high degree of co-linearity between chromosomes, but they also allow the identification of relatively rare, strain-specific sets of genes or regions (see below 'Core genome and Pan genome'), which account for most of the genomic variations.

Although phylogenetic trees of Mollicutes are generally constructed from comparative analyses or multiple sequence alignments of 16S rRNA (Woese *et al.*, 1980; Weisburg *et al.*, 1989) or 5S rRNA genes (Rogers *et al.*, 1985), several alternative approaches based on genomics have been proposed. Evolutionary trees based on the conservation of gene orders have been constructed and compared to conventional trees using nucleotide or amino acid sequence alignments (Markov and Zakharov, 2009). Unsurprisingly, the chromosomes of mycoplasma species that belong to the same 16S RNA clade or sub-group have well conserved gene orders. Alternatively, a subset of 143 highly conserved mycoplasma proteins has been proposed to construct a phylogenetic tree (Oshima and Nishida, 2007) and, despite some differences with the rRNA sequence-based trees, the relative positions of closely related species, such as the human pathogens *M. genitalium* and *M. pneumoniae*, are still conserved. In fact, the proximity of genomic organization in *M. genitalium* and *M. pneumoniae* has been used to test and validate computer programs for whole genome alignments (Delcher *et al.*, 1999; Celamkoti *et al.*, 2004). These studies show that these two species share extensive syntenic regions, which display a conserved order of orthologous genes but are distributed at different positions along their chromosomes. The differences in genomic organization between these two species is attributed to a small number of homologous recombinations between repetitive sequences that have resulted in the translocation of chromosomal regions without inversions (Himmelreich *et al.*, 1997). This is consistent with the hypothesis that large chromosomal inversions can be detrimental to microorganisms because they disturb gene dosage, chromosomal symmetry or gene strand bias, and therefore such inversions are counter-selected during evolution (Rocha, 2004).

However, inversions of large portions of chromosomes have been detected in several instances in mycoplasmas, and such large rearrangements may be more frequent than originally thought. Chromosomal inversions were first discovered by comparative pulsed field gel electrophoresis of strains of *M. hominis* (Ladefoged and Christiansen, 1992) and *Spiroplasma citri* (Ye *et al.*, 1996), and of the closely related species *M. hyopneumoniae* and *M. flocculare* (Blank and Stemke, 2001). With recent advances in mycoplasma genomics, large chromosomal inversions have been characterized more precisely by whole genome alignment analyses. Some of the most prominent inversions have been observed by comparing *M. hyopneumoniae* strain 232 to strains J and 7448 (Vasconcelos *et al.*, 2005), *M. bovis* strain Hubei-1 to *M. agalactiae* strain PG2 (Li *et al.*, 2011), and *M. mycoides* subspecies *capri* strain 95010 to *M. mycoides* subspecies *mycoides* small colony strain PG1 (Thiaucourt *et al.*, 2011). The origin of these large chromosomal inversions can be attributed to various genomic features. In *S. citri*, chromosomal reorganisations appear to be related to the presence of chromosomal copies of the virus SpV1 (Melcher *et al.*, 1999); in *M. bovis* Hubei-1 a 142 kb inversion is associated with the presence of an IS element (Li *et al.*, 2011); and in *M. mycoides* subspecies *capri* strain 95010, one inversion was caused by the presence of two copies of an integrative conjugative element (ICE) in opposite orientations (Thiaucourt *et al.*, 2011). IS elements and phages are well known to induce genomic rearrangements in prokaryotes. The more recently discovered ICEs, also known as conjugative transposons or CONSTINs (see below), are large repeated regions that appear to be present in many mycoplasma species and are also proposed to play a major role in the mosaic structure of these genomes.

Whole genome alignments have been applied to the exploration of the taxonomic relationships between related species of Mollicutes. This has been particularly useful within the 'mycoides' cluster, which occupies a unique position in the phylogenetic tree of Mollicutes. This cluster is composed of highly related mycoplasma species or sub-species that infect ruminants and are quite difficult to classify. A remarkable gene synteny is observed between two species of the cluster that differ in their host range and associated diseases, namely *M. mycoides* subspecies *mycoides* SC, a major respiratory pathogen of cattle, and *M. capricolum* subspecies *capricolum*, which infects sheep and goats at various body sites (Wise *et al.*, 2006). The comparative alignment of two subspecies of *M. mycoides*, *M. mycoides* subspecies *mycoides* SC PG1 and the small ruminant pathogen *M. mycoides* subspecies *capri* 95010, and comparison of them to *M. capricolum* subspecies *capricolum* California Kid, has clearly confirmed the respective taxonomic positions of these species in the mycoides cluster (Thiaucourt *et al.*, 2011). As expected, the highly related *M. mycoides* subspecies *mycoides* SC and *M. mycoides* subspecies *capri*, which were for a long time considered colony size variants of the same subspecies, have extensive syntenic regions with a few large chromosomal inversions. Apart from the mycoides cluster, whole genome alignments of the plant pathogens *Phytoplasma australiense*, *P. asteris* AY-WB and *P. asteris* OY-M show the presence of small regions of gene synteny across the three phytoplasma genomes, and a chromosomal inversion with signs of rearrangements between AY-WB and OY-M (Bai *et al.*, 2006). Overall, the gene order is more conserved between the close relatives AY-WB and OY-M than with the more distant *P. australiense* (Tran-Nguyen *et al.*, 2008). Finally, comparative mapping of bidirectional best hits (BDBH) within pairs of genomes shows a lower conservation of BDBH order for more distantly related species, such as *Mycoplasma pulmonis* from the 'hominis group' and *M. mycoides* subspecies *mycoides* SC from the 'spiroplasma group' (Souza *et al.*, 2007).

## Core genome and pan genome

It has long been considered that mycoplasmas offer some of the most simple genomic models for understanding the fundamental biology of bacteria. They have been conveniently used to analyse the coding potential (Mushegian and Koonin, 1996), metabolic pathways (Pollack *et al.*, 1997; Suthers *et al.*, 2009; Yus *et al.*, 2009) and repertoires of essential or dispensable genes (Hutchison *et al.*, 1999; Glass *et al.*, 2006) in minimal genomes. The comparative analysis of *Mycoplasma* and *Haemophilus* sp. genomes has suggested a minimal set of approximately 250 genes required for life (Koonin, 2000). However, the universal concept of minimal genome remains elusive, even when considering mycoplasmas as a model (Peterson and Fraser, 2001). Nevertheless, the combined analysis of gene homologies, strand biases and other genomic features found across 16 mycoplasma genomes predict that they possess a universal core set of 153 essential genes, and a wider set of approximately 6000 essential genes (Lin and Zhang, 2011).

Establishing a comprehensive catalogue of the genes present in mycoplasma species (the pan-genome) has provided a new perspective of their evolutionary potential. Microbial genomes are continuously sampling exogenous DNA and/or shuffling their own genetic information (Omelchenko *et al.*, 2003). As a consequence, it is likely that the pan-genome of many bacterial species is unlimited, an 'open' pan-genome. In contrast, it is assumed that obligate parasites that live in relatively isolated niches, such as the Mollicutes, tend to have finite or 'closed' pan-genomes (Mira *et al.*, 2010). Pan-genome studies based on genome-centred approaches (Tettelin 2005) or on gene-centred approaches (Lapierre 2009) suggest the existence of a large repertoire of genes that are present in only one member, or in very few members, of a species. These genes represent the 'accessory' gene pool, or ORFans. The ORFan gene pool is a feature of many bacterial lineages, and is usually composed of genes that are mostly associated with plasmids and phages. As these genetic elements appear to be rare in mycoplasmas, the accessory gene pool is difficult to detect. In contrast, the 'character' gene pool is represented by the genes that are shared by some isolates within a species, but the set of these genes present varies from one group of isolates to another. In many bacterial species this pool is the basis of strain-specific attributes, such as virulence, pathotype, or biovar differences. It has been proposed that a typical bacterial genome is composed of 8% core genes (which control the essential aspects of metabolism), 64% character genes and 28% accessory genes (Lapierre 2009). Owing to the reduced size of mycoplasma genomes, the core gene pool is expected to constitute a higher proportion of their genome, and the accessory and character gene pools are more difficult to delineate. Surprisingly, very few studies have addressed the nature of mycoplasma pan-genomes, and whether they are 'closed', as predicted by their lifestyle, or 'open'. It is becoming increasingly apparent that Mollicutes possess both accessory and character gene pools, which confer genomic mosaicism. Thus far, the widest intra-specific mycoplasma pan-genome that has been studied is in *M. hyopneumoniae*, four strains of which have been sequenced. Strain-specific regions were initially found by comparing the genomes of the pathogenic isolates 232 (Minion *et al.*, 2004) and 7448 and the non-pathogenic strain J (Vasconcelos *et al.*, 2005). A total of 59, 53 and 67 CDSs were found to be exclusive to each of strains 232, 7448 and J, respectively. The complete sequencing of *M. hyopneumoniae* strain 168 further expanded the pan-genome of this species by 13 genes (Liu *et al.*, 2011). A significant proportion of these strain-variable genes are carried in an ICE. Similar observations have been made when comparing several *M. fermentans* strains that have been sequenced recently. *M. fermentans* strain JER appears to possess 87 CDSs with no similarity to any counterpart in strain PG18, suggesting quite a large pan-genome (Rechnitzer *et al.*, 2011). However it must be noted that the PG18 genome sequence used in this study was unfinished and therefore this number might be lower if some of the JER sequences are actually present in PG18, which would result in a pan-genome smaller than currently predicted. Regardless, the genome of the fully sequenced strain M64 is 14% larger than that of strain JER, so it is likely that the *M. fermentans* pan-genome is open. Moreover, this difference is mostly due to the presence of phage or ICE-related sequences (Shu *et al.*, 2011). In *M. fermentans*, ICEs have been described in strain PG18 (Calcutt *et al.*, 2002). The impact of ICEs on the mycoplasma pan-genome is potentially very important because these elements share characteristics with conjugative plasmids, as well as with certain phages (Burrus *et al.*, 2002), making them a likely source of the accessory or character gene pools. The sequencing and comparison of two *M. agalactiae* strains, the type strain PG2 (Sirand-Pugnet *et al.*, 2007b) and the field isolate 5632 (Nouvel *et al.*, 2010), suggests the presence of both character and accessory gene pools in this species, with 39 new genes found in strain 5632. Again, most of the accessory gene pool in *M. agalactiae* is composed of ICEs, insertion sequences (IS) and phage-related sequences. The proportion of accessory genes appears negligible in strain PG2, but represents 13% of the genome in strain 5632, which could indicate that *M. agalactiae* possesses an open pan-genome (Nouvel *et al.*, 2010). The same possibility is suggested by the comparison of two *M. bovis* strains, PG45 and Hubei-1, which shows that 51 Hubei-1 genes and 46 PG45 genes are unique to each genome (Li *et al.*, 2011). Finally, an open pan-genome is also suggested by the characterization of numerous phage-related sequence variable motifs (SVMs, see below) in phytoplasma strains (Wei *et al.*, 2008). In contrast, the partial sequencing of nine strains of *Ureaplasma urealyticum* indicates that this species possesses a closed pan-genome, which suggests that the number of new genes found within this species will reach a plateau as more and more genomes are determined (Tettelin *et al.*, 2008). It is too early to ascribe with certainty an open or closed nature to the mycoplasma pan-genomes. However, as sequences of multiple isolates from the same species are added to the databases, this question will be able to be more definitively answered in the near future.

## DNA repeats

Some mycoplasma species, such as *M. mycoides* subsp. *mycoides* SC, *M. mobile* and *M. genitalium*, appear to produce and maintain relatively long sequence duplications on their chromosomes (Teichmann *et al.*, 1998; Jaffe *et al.*, 2004; Bischof *et al.*, 2006). This is also the case for *P. asteris* (Bai *et al.*, 2006) and *P. australiense* (Tran-Nguyen *et al.*, 2008), in which 28% and 24%, respectively, of the ORFs present in the genome occur as multiple copies. Mobile genetic elements such as ICEs, PMUs and ISs constitute abundant classes of repeated coding sequences in mycoplasma genomes (see below). However, much shorter DNA motifs are also detected within coding and intergenic regions and some of these repeats appear to be associated with genomic islands and horizontal gene transfer (HGT) in mycoplasmas.

Repeated genomic sequences contribute to the variability of sequences coding for adhesion or for variable surface proteins in mycoplasmas (Rocha and Blanchard, 2002; Musatovova *et al.*, 2008). Long simple sequence repeats (LSSR) are composed of a recurring single nucleotide or short oligonucleotide, and can lead to slipped-strand mispairing mutations, alterations in DNA topology or genomic rearrangements that ultimately modify the coding potential of micro-organisms. A subset of LSSRs is particularly associated with host adaptation and genome reduction in bacteria (Mrazek *et al.*, 2007) and is overrepresented in certain mycoplasma genomes, including *M. hyopneumoniae, M. gallisepticum* and *M. pulmonis.* (Mrazek, 2006). Surprisingly, these LSSRs are found in many *M. hyopneumoniae* housekeeping genes, including coding sequences for ribosomal proteins, DNA gyrase and tRNA synthetases (Guo and Mrazek, 2008). It is tempting to speculate that the proliferation of LSSRs in ancestral mycoplasma genomes has played a role in their reductive evolution and their obligately parasitic lifestyle, as has been suggested for the evolution of *Mycobacterium leprae* (Guo and Mrazek, 2008). If this hypothesis is correct, the presence of residual LSSRs in some mycoplasma genomes and their absence in others could be explained by species-specific constraints on the long-term maintenance of these repeats.

Clustered regularly interspaced short palindromic repeats (CRISPRs) are another specific group of DNA motifs found in many prokaryotes, including mycoplasmas. CRISPRs are usually present in non-coding regions and are a defence system protecting the integrity of bacterial and archaeal genomes from an overwhelming genetic invasion by foreign DNA such as phages (Garneau *et al.*, 2010; Horvath and Barrangou, 2010). A dedicated CRISPR database of pre-computed analyses of 1324 bacterial genomes using the CRISPRFinder program (Grissa *et al.*, 2007) is available online (http://crispr.u-psud.fr/crispr/) and contains some information on Mollicutes. The repeats are classified into two categories: characteristic CRISPRs and CRISPR-like structures. Characteristic CRISPRs are defined by a repeat length of 23 to 55 bp, a spacer size between the repeats of 25 to 60 bp, ranging from 0.6 to 2.5 times the length of the repeat, less than 60% similarity between spacers and high sequence conservation between the repeats. The CRISPR-like structures possess either only two or three repeats or less conserved repeat sequences, but their presence may still be significant in small genomes such as those of mycoplasmas. The analysis of 27 species of Mollicutes currently included in this database found characteristic CRISPRs in four species, while six other species harbour CRISPR-like structures in intergenic regions (Table 2.2). It is noteworthy that the six mycoplasma species that possess CRISPR or CRISPR-like sequences also harbour phages, ICEs or plasmids. Although the default parameters of the CRISPRFinder tool do not find conventional CRISPRs in three mycoplasma species in which ICEs have been described previously (*M. agalactiae, M. bovis* and *M. conjunctivae*), less stringent search parameters allow the identification of CRISPR-like sequences in these species. Interestingly, these repeats are mainly located in ICEs. The roles and relationships between these different types of sequence repeats is not entirely clear, but CRIPSRs of Mollicutes are likely to attract more attention if the sensitivity of the detection tools can be improved, in particular by taking into account the composition biases and other characteristics of mycoplasmas genomes.

Another category of repeats, PARCELs (palindromic amphipathic repeat coding elements), which show some relatedness to CRISPRs, has been identified recently in several mycoplasma species and other microorganisms (Roske *et al.*, 2010). These palindromic nucleotide repeats are found within the coding regions of proteins with unknown functions that appear to be acquired by HGT between mycoplasmas. Like CRISPRs, some PARCELs are found in putative conjugative genomic islands, Tra islands (see below), which have been identified in the genomes of *M. capricolum* subspecies *capricolum* and *M. mycoides* subspecies *mycoides* SC. The PARCEL motifs include a previously described domain of unknown function (DUF285) within in a number of mycoplasma lipoproteins encoded by genes that have been subject to HGT (Sirand-Pugnet *et al.*, 2007b).

Finally, phytoplasmas contain some special intergenic imperfect palindromic sequences, the phytoplasmal repeated extragenic palindromes or PhREPs (Jomantiene and Davis, 2006), which are associated with sets of genes clustered in sequence-variable mosaics (SVMs, see below) that might also have been acquired by HGT (Jomantiene *et al.*, 2007).

**Table 2.2** CRISPR sequences identified in 27 Mollicutes analysed by CRISPRFinder*

| Species strain | CRISPR_id | Start | End | Spacers | DR consensus |
|---|---|---|---|---|---|
| Characteristic CRISPRs | | | | | |
| *M. arthritidis* 158L3–1 | NC_011025_1 | 539513 | 541733 | 33 | ACTTTTGGACTGTACAATTTTTATATAGAGTAAAGT |
| *M. gallisepticum* R low | NC_004829_1 | 401227 | 401317 | 1 | TTTTTCTTAACAAAATAATTGTACAA |
| | NC_004829_4 | 911020 | 915741 | 71 | GTTTTAGCACTGTACAATACTTGTGTAAGCAATAAC |
| *M. mobile* 163K | NC_006908_1 | 39617 | 43743 | 62 | GTTTAAGAATACATAAGAATGATACTACACCAAAAC |
| *M. synoviae* 53 | NC_007294_1 | 690229 | 690988 | 11 | GTTTTGGGGTTGTACAATTATTTTGTTAAGTAAAAC |
| Questionable CRISPRs present in intergenic regions | | | | | |
| *M. capricolum* subspecies capricolum ATCC 27343 | NC_007633_3 | 239894 | 240002 | 1 | AAGGAATTTCTGTGTATTTACACACTTTTTCCTTA |
| *M. fermentans* JER | NC_014552_1 | 256695 | 256808 | 1 | TGAAAAAAAGTGGTCAAATGACCAGAAATTTTCAATAGAAA |
| *M. fermentans* M64 | NC_014921_1 | 615585 | 615686 | 1 | TTTGACACTTTTGTAAAAGTTTC |
| *M. hyopneumoniae* 232 | NC_006360_2 | 860977 | 861109 | 1 | ATTCTACCAGAAATTAGAAAAATTCAAAGAAGTTTTTA |
| *M hyopneumoniae* J | NC_007295_2 | 867971 | 868103 | 1 | ATTCTACCAGAAATTAGAAAAATTCAAAGAAGTTTTTA |
| *M. leachii* PG50 | NC_014751_3 | 283260 | 283367 | 1 | AAGGAATTTCTGTGTATTTACACACTTTTTCCTT |
| *M. mycoides* subspecies *capri* LC strain 95010 | NC_015431_3 | 251263 | 251441 | 2 | AAGGAATTTCTGTGTATTTACACACTTTTTTC |
| | NC_015431_6 | 777348 | 777499 | 1 | TATTCTACCACTTTTCTACCGGTGGGGGTAGATTAGTGGTAGAAAAC |
| *M. penetrans* HF-2 | NC_004432_1 | 664234 | 664325 | 1 | TTGGTTTTCCAATACTTTTTTCCTAAAAAT |

*http://crispr.u-psud.fr/crispr/.

## Gene decay

Several Mollicutes appear to have been subject to significant gene decay, a surprisingly finding for organisms that have such reduced genomes with high coding capacity, suggesting that at least some species are still undergoing reductive evolution. Gene decay often leads to the formation of pseudogenes (Table 2.3), which is defined by the loss or the alteration of more than 20% of the coding sequence (Ochman and Davalos, 2006). High numbers of pseudogenes are annotated in *P. australiense*, *M. pneumoniae* FH, *M. bovis* PG45 and Hubei-1, *M. mycoides* subspecies *capri* strain GM12, *M. putrefaciens* KS1 and *M. crocodyli* MP145, in which they comprise approximately 5% to 10% of their genomes.

The mechanism underlying gene decay in mycoplasmas could be related to the presence of high numbers of DNA repeats or IS elements (see below). Gene disruptions and decay are found in

**Table 2.3** Pseudogenes annotations in the genomes of Mollicutes

| Species strain | Number of annotated pseudogenes | Total size (nt) | Percentage of genome size |
|---|---|---|---|
| *Candidatus* Phytoplasma australiense | 155 | 90,859 | 10.33 |
| *M. pneumoniae* FH | 128 | 105,331 | 12.99 |
| *M. bovis* PG45 | 61 | 59,293 | 5.91 |
| *M. mycoides* subspecies *capri* strain GM12 tetM-lacZ | 40 | 46,598 | 4.30 |
| *M. putrefaciens* KS1 | 36 | 37,475 | 4.50 |
| *M. crocodyli* MP145 | 31 | 46,922 | 5.02 |
| *M. fermentans* JER | 28 | 18,506 | 1.89 |
| *M. hyorhinis* HUB-1 | 21 | 26,698 | 3.18 |
| *M. fermentans* M64 | 18 | 10,022 | 0.90 |
| *Candidatus* Phytoplasma mali | 18 | 12,639 | 2.10 |
| *M. hyopneumoniae* 7448 | 15 | 15,377 | 1.67 |
| *M. suis* KI3806 | 15 | 12,279 | 1.73 |
| *M. synoviae* 53 | 15 | 12,584 | 1.57 |
| *M. gallisepticum* strain F | 14 | 13,854 | 1.42 |
| *M. gallisepticum* strain R(high) | 14 | 13,335 | 1.32 |
| *M. gallisepticum* strain R(low) | 14 | 13,335 | 1.32 |
| *M. hominis* ATCC 23114 | 14 | 10,687 | 1.61 |
| *M. capricolum* subspecies *capricolum* ATCC 27343 | 13 | 15,752 | 1.56 |
| *M. hyopneumoniae* 168 | 13 | 28,898 | 3.12 |
| *M. agalactiae* 5632 | 12 | 8685 | 0.86 |
| *A. laidlawii* PG-8A | 11 | 15,215 | 1.02 |
| *M. hyopneumoniae* J | 11 | 9354 | 1.04 |
| *U. urealyticum* serovar 10 strain ATCC 33699 | 11 | 18,933 | 2.17 |
| *M. agalactiae* PG2 | 9 | 7887 | 0.90 |
| *M. genitalium* G37 | 6 | 4195 | 0.72 |
| *M. arthritidis* 158L3–1 | 4 | 10,170 | 1.24 |
| *M. suis* strain Illinois | 4 | 5366 | 0.72 |
| *M. leachii* PG50 | 2 | 1594 | 0.16 |
| *M. mycoides* subspecies *capri LC* strain 95010 | 2 | 502 | 0.04 |
| *M. mycoides* subspecies *mycoides SC* strain Gladysdale MU clone SC5 | 1 | 134 | 0.01 |

many species and are particularly documented in *M. mycoides* subspecies *mycoides* SC (Thiaucourt *et al.*, 2011), *S. citri* and *S. kunkelii* (Zhao *et al.*, 2003; Carle *et al.*, 2010), *M. agalactiae* PG2 (Sir-and-Pugnet *et al.*, 2007b) and some phytoplasmas (Davis *et al.*, 2003, 2005b). In these latter organisms, gene decay within plasmids appears to occur at a faster rate than in the chromosome, indicating some potential for selective genomic plasticity (Ishii *et al.*, 2009). Many pseudogenes are found in IS elements or in DNA restriction-modification systems, so they are likely to have a limited biological impact. Furthermore, pseudogenization can affect coding sequences that are redundant in the genome, so any metabolic functions of these disrupted genes are already complemented (Szczepanek *et al.*, 2010; Li *et al.*, 2011). However, in some cases pseudogenization may modify specific cellular functions with possible consequences on pathogen–host relationships, such as the loss of an adhesion factor in *M. bovis* (Thomas *et al.*, 2004), proteolysis in *M. mycoides* subspecies *mycoides* SC (Thiaucourt *et al.*, 2011) and important metabolic pathways in phytoplasmas (Davis *et al.*, 2005b). In other cases pseudogenes appear to be maintained in genomes as a reservoir of coding sequences for proteins subject to high frequency antigenic variation (Noormohammadi *et al.*, 2000; Khiari *et al.*, 2010), or as a reversible feature of coding sequences that control phase variable phenotypes, such as the expression of restriction-modification systems (Pereyre *et al.*, 2009).

## Mobile genetic elements in *Mycoplasma* genomes

As in other bacteria, the mobilome of Mollicutes comprises insertion sequences, phages, plasmids and conjugative transposons. Many of these mobile genetic elements (MGEs) appear to be specific for this class of organisms and share their evolutionary history. Although the existence of MGEs in mycoplasmas has been known for quite some time, their contribution to the genomic mosaicism and fluidity of these organisms has only been appreciated recently. The distribution of the principal MGEs within the different phylogenetic groups of Mollicutes is summarized in the Table 2.4.

### Insertion sequences

Insertion sequences are found in many bacteria, including Mollicutes, and their mobility within and between genomes has been extensively studied (Mahillon and Chandler, 1998). More than 500 annotations of CDSs in the complete genomes of 33 Mollicutes refer to a transposase as the gene product (Table 2.5A), but this is a conservative estimate of the number of ISs in mycoplasmas as additional ISs have been annotated as pseudogenes. Several families of IS elements are found in the genomes of Mollicutes, including the IS*3*, IS*4*, IS*30*, IS*150*, IS*256*, IS*481* and IS*1634* families (Table 2.5B). Members of the mycoides cluster (in particular *M. mycoides* subspecies *mycoides* SC, *M. mycoides* subspecies *capri*) and *M. bovis*, harbour large numbers of IS-related sequences. These sequences represent more than 13% of the complete genome in *M. mycoides* subspecies *mycoides* SC strain PG1 (Westberg *et al.*, 2004). The abundance of IS elements in the genomes of certain mycoplasma species has been applied to genetic typing of strains and epidemiological studies, particularly in mycoplasmas infecting ruminants (March *et al.*, 2000; Pilo *et al.*, 2003; Miles *et al.*, 2005, 2006; Bischof *et al.*, 2006), but also those infecting humans (Hu *et al.*, 1998; Pitcher and Hilbocus, 1998).

The presence of ISs has a significant impact on the genomic organization and coding capacity of mycoplasmas. IS elements are associated with rearrangements or disruptions of genes coding for surface proteins in *M. gallisepticum* (Papazisi *et al.*, 2003; Szczepanek *et al.*, 2010) and DNA restriction-modification in *M. pulmonis* (Dybvig *et al.*, 2007), tandem duplications of chromosomal regions in *M. mycoides* subspecies *mycoides* SC (Bischof *et al.*, 2006), large chromosomal inversions in *M. bovis* (Wise *et al.*, 2010; Li *et al.*, 2011) and formation of paralogous gene clusters in the *M. penetrans* genome (Sasaki *et al.*, 2002). In phytoplasmas, it has been suggested that the transposase Tra5 controls the multiple duplications and rearrangements of gene clusters that are a prominent feature of these genomes (Bai *et al.*, 2006; Arashida *et al.*, 2008). Some transposons also appear to carry virulence genes, as demonstrated by genome analysis of *M. gallisepticum*, in which a region containing two concatenated IS

**Table 2.4** Mobile genetic elements present in Mollicutes

| Species | Host | Main clinical manifestations | Strain | Evidence of mobile element and distribution among strains | | | | | |
|---|---|---|---|---|---|---|---|---|---|
| | | | | | Insertion sequences | | | | |
| | | | | MICE or PMU | Name | Isoform | Family | Plasmid | Virus |
| Spiroplasma group | | | | | | | | | |
| Tomato Big Bud Phytoplasma | Plant | | | | | | | pTBB-peri, pTBBcap (Tran-Nguyen and Gibb, 2006) | |
| *Candidatus* Phytoplasma australiense | Plant | Australian grapevine yellows, papaya dieback<br>Strawberry lethal yellows, strawberry green petal and pumpkin yellow leaf curl | | PMU 1–5 (Tran-Nguyen *et al*., 2008) | IS-like elements | | IS3 | pCPa, (Tran-Nguyen and Gibb, 2006), pPAPh2, pPASb11(Liefting *et al*., 2006) | |
| *Candidatus* Phytoplasma asteris | Plant | | Aster yellows witches' broom (AY-WB) | PMU 1–4 (Bai *et al*., 2006) | Tra5 | | IS3 | pAYWB-I -II, -III, -IV (Bai *et al*., 2006) | |
| | | | Onion yellows mutant (OY-M) | PMU | Tra5 | | IS3 | EcOYW1 (Nishigawa *et al*., 2001)<br>pOYM<br>pOYNIM (Nishigawa *et al*., 2003) | |
| | | | | | | | | pPARG1, pPABN1 (Petrzik *et al*., 2011) | |
| *M. mycoides* subspecies *mycoides SC* | Cattle | Pleuropneumonia | Gladysdale MU clone SC5 | | InsK | | IS150 | | |
| | | | | | Transposase | | IS3 | | |
| | | | | | Putative transposase IS1296 | | IS3 | | |
| | | | | | Transposase | | IS4 | | |

| | | | | | | | | |
|---|---|---|---|---|---|---|---|---|
| | | | PG1 (Westberg *et al.*, 2004) | *traE* homologue (NP_975194) | IS1634, ISMmy1<br>IS1296 | IS1634<br>IS3 | None | putative prophage protein (NP_975197.1) |
| *M. mycoides* subspecies *capri (ex M. mycoides* subspecies *mycoides LC)* | Sheep and goats | Contagious agalactia | GM12 (Lartigue *et al.*, 2009) | ICEM | ISMmy2, ISMmy3 | IS3 | pKMK1 (King and Dybvig, 1992), pADB201 (Bergemann and Finch, 1988) | |
| | | | 95010 (Thiaucourt *et al.*, 2011) | ICEM (two copies) | IS1296, ISMmy2, ISMmy3<br>ISMmy1-like | IS3<br>IS1634 | pMmc-95010 (Thiaucourt *et al.*, 2011) | |
| *M. capricolum* subspecies *capripneumonie* | Sheep and goats | Pleuropneumonia | | | | | | |
| *M. capricolum* subspecies *capricolum* | Sheep and goats | Contagious agalactia | ATCC_27343= California kid. | ICEC | | | | |
| *M. leachii* (ex. sp. bovine group 7) | Cattle | Pneumonia | PG50 | Tra Island encoded by MSB_A0042-MSB_A0073 | IS1296<br>InsK<br>Transposase | IS3<br>IS150<br>IS4 | pBG7AU (Djordjevic *et al.*, 2001) | |
| *M. yeatsii* | Sheep and goats | | GIH | | | | pMyBK1 (Genbank EU429323) | |
| *M. putrefaciens* | Sheep and goats | Contagious agalactia | | | | | | |
| *M. florum* | Insect and plant | | | | | | | |
| *S. citri* | Insect and plant | Citrus stubborn disease | GII3–3X | *traE* homologues (chromosomal and plasmidic) | ISSvi1<br>ISSc1<br>Transposase | IS481<br>IS30,<br>IS3, ISNCY | pBJS-O (Joshi *et al.*, 2005), pSci1–6 (Berho *et al.*, 2006b), pSciA (Saillard *et al.*, 2008) | plectrovirus SPV1-R8A2B (Renaudin *et al.*, 1990); SpV1-C74 (Bébéar *et al.*, 1996); plectrovirus SVGII3 and lambda-like SpV2 (Carle *et al.*, 2010; Cole *et al.*, 1973) |

**Table 2.4** (Continued)

| Species | Host | Main clinical manifestations | Strain | Evidence of mobile element and distribution among strains | | | | | |
|---|---|---|---|---|---|---|---|---|---|
| | | | | | Insertion sequences | | | | |
| | | | | MICE or PMU | Name | Isoform | Family | Plasmid | Virus |
| *S. melliferum* | | | | | | | | | Spiromicrovirus SPV4 (Renaudin *et al.*, 1987); plectrovirus SVTS2 (Sha *et al.*, 2000) |
| *S. kunkelii* | Insect and plant | Maize stunt | | *traE* homologues (chromosomal and plasmidic) | | | | pSKU146 (Davis *et al.*, 2005a) | |
| Pneumoniae group | | | | | | | | | |
| *M. haemofelis* | Cats | Acute haemolytic anaemia | | | | | | | |
| *U. diversum* | Human | | | | | | | | |
| *U. parvum* | Human | Urogenital or respiratory tract infections | ATCC_700970 (Glass *et al.*, 2000) | UU145 (RipX phage integrase) is a BDBH of the ORF1 from ICEC | | | | | |
| *M. penetrans* | Human | Urogenital or Respiratory tract infections | HF-2 (Sasaki *et al.*, 2002) | | IS232-like | | | | |
| | | | | | IS1202 | | | | |
| | | | | | IS1630-like | | IS30 | | |
| | | | | | IS150-like | | | | |
| *M. gallisepticum* | Poultry | Airsacculitis and tracheitis in chickens, sinusitis in turkeys | R low (Papazisi *et al.*, 2003) | | putative Transposases | | | | |
| | | | R high (Szczepanek *et al.*, 2010) | | Putative transposases | | | | |
| | | | F (Szczepanek *et al.*, 2010) | | Putative transposases | | | | |
| *M. genitalium* | Human | Genital infections | | | | | | | |
| *M. pneumoniae* | Human | Respiratory infections | | | | | | | |

| | | | | | | | | |
|---|---|---|---|---|---|---|---|---|
| Hominis group | | | | | | | | |
| *M. mobile* | Fish | unknown | | | | | | |
| *M. arginini* | Cattle | | | | | | | |
| *M. alkalescens* | Cattle | | | | | | | |
| *M. arthritidis* | Mouse/rat | Arthritis | 158L3–1 | | IS1630 fragment | | | Prophage MAV1 (Voelker and Dybvig, 1998) |
| *M. conjunctivae* | Small ruminants | Infectious keratoconjunctivitis | HRC/581 (Calderon-Copete *et al.*, 2009) | ICECJ (two copies, one truncated) | IS1138 | | IS3 | |
| | | | | | IS1630-like | | IS30 | |
| | | | | | IS1634 | | IS1634 | |
| *M. hominis* | Human | | | | IShom | | | MV 20-P (Jansson *et al.*, 1982) |
| *M. hyopneumoniae* | Pig | Enzootic Pneumonia (NB strain J is considered non pathogenic) | 7448 (Vasconcelos *et al.*, 2005) | ICEH | ISMhp1 | | IS1634 | |
| | | | J (Vasconcelos *et al.*, 2005) | ICEH, vestiges | ISMhp1 | | IS1634 | |
| | | | | | IS1221E, IS1221H, IS1221I | IS1221A | IS3 | |
| | | | 232 (Minion *et al.*, 2004) | ICEH, truncated | ISMhp1<br>Two transposase genes annotated (mhp082 and mhp645) | | IS1634 | |
| | | | 168 | | IS-like | | IS1634 | |
| *M. hyorhinis* | Pig | Arthritis | GDL-1 | | IS1221A, IS1221B, IS1221C, IS1221D, IS1221J, IS1221G, IS1221F | IS1221A | IS3 | Phage Hr 1 (Gourlay *et al.*, 1983b) |
| | | | HUB | | IS861<br>InsK | | | |
| *M. bovirhinis* | Cattle | Opportunistic pathogen | | | | | | Phage Br 1 (Gourlay *et al.*, 1983a) |
| *M. pulmonis* | Mouse/rat | Pneumonia | UAB CTIP (Chambaud *et al.*, 2001) | *traE* homologue (NP_326214) | IS1138, IS1138B | IS1138 | IS3 | Phage P1 (Tu *et al.*, 2001) |

**Table 2.4** (Continued)

| Species | Host | Main clinical manifestations | Strain | Evidence of mobile element and distribution among strains | | | | | |
|---|---|---|---|---|---|---|---|---|---|
| | | | | MICE or PMU | Insertion sequences | | | Plasmid | Virus |
| | | | | | Name | Isoform | Family | | |
| *M. synoviae* | Poultry | Airsacculitis and arthritis in chickens | Field strain 53 (Vasconcelos *et al.*, 2005) | | ISMhp1, ISMsy1 | – | IS1634 | | |
| *M. fermentans* | Human* | Infection in immunocompromised patients | PG18 | ICEF (4 copies); (Calcutt *et al.*, 2002) | IS1550 | ISMi1, IS-like Element (ISLE) (Hu *et al.*, 1990; Pitcher and Hilbocus, 1998) | IS3 | | Phage MFV1 (Roske *et al.*, 2004) also present in strain II-29/1 and K7. |
| | | | | | ISMf*1* | – | IS*256* | | |
| | | | | | IS1630 | – | IS30 | | |
| | | | | | Mutator-like transposable elements (MULEs) | – | | | |
| | | | JER | ICEF (two copies) | IS1630-like | – | IS30 | | |
| | | | | | InsK | – | IS150 | | |
| | | | M64 | ICEF (seven copies) | IS*1550* | ISMi1 | IS3 | | Phage MFV1 |
| | | | | | IS*1630* | – | IS30 | | |
| | | | | | ISMf*1* | – | IS*256* | | |
| | | | | | transposase, mutator type | | | | |

| | | | | | | | | |
|---|---|---|---|---|---|---|---|---|
| *M. agalactiae* | Sheep and goats | Contagious agalactia | PG2 (Pilo *et al.*, 2003) | ICEA | ISMag1 | – | IS30 | |
| | | | 5632 | ICEA (three full copies, one truncated copy) | ISMag1, ISMag2 | | IS30 | Homologue of *M. mycoides* SC putative prophage protein (NP_975197.1) |
| *M. bovis* | Cattle | Respiratory infections and mastitis | PG45 | ICEB (two copies) | ISMbov1 | ISMag1 | IS30 | |
| | | | | | ISMbov2 | ISMmy1 | IS1634 | |
| | | | | | ISMbov3 | IS1634 | IS1634 | |
| | | | | | ISMbov4, ISMbov5 | – | IS4 | |
| | | | | | ISMbov6 | – | IS30 | |
| | | | | | ISMbov7 | – | IS3 | |
| | | | Hubei | ICEB (one copy) | | | | |
| *M. orale* | Human | | | | IS-like Elements (ISLEs) with similarity with *M. fermentans* ISLEs (Ditty *et al.*, 1999) | | IS3 | |
| *Acholeplasma laidlawii* | | | | | IS-10 | | | MVL1–3 (Gourlay, 1971; Gourlay *et al.*, 1971; Gourlay and Wyld, 1973) |

**Table 2.5A** Diversity of annotations for IS-related or transposase-related sequences in complete genomes of Mollicutes

| Host species | Annotation type (NB the same sequence may have more than one annotation) | | | | |
|---|---|---|---|---|---|
| | 'CDSs' | 'Gene' | 'Misc. feature' | 'Mobile_ element' | 'Repeat region' |
| *M. mycoides* subspecies *mycoides* SC strain Gladysdale MU clone SC5 | 89 | | | | |
| *M. mycoides* subspecies *mycoides* SC strain PG1 | 73 | | | | 106 |
| *M. bovis* PG45 | 39 | 15 | | 72 | |
| *M. mycoides* subspecies *capri* LC strain 95010 | 35 | | | | |
| *M. bovis* Hubei-1 | 34 | | 13 | | |
| *M. penetrans* HF-2 | 28 | | | | |
| Aster yellows witches'-broom phytoplasma AYWB | 27 | | | | |
| *M. hyorhinis* HUB-1 | 27 | | | | |
| *M. fermentans* M64 | 22 | | 5 | 22 | |
| *M. mycoides* subspecies *capri* strain GM12 | 19 | | | | |
| *M. agalactiae* 5632 | 15 | | | | |
| *M. hyopneumoniae* 168 | 13 | | | | |
| Onion yellows phytoplasma OY-M | 12 | | | | |
| *M. gallisepticum* strain R(high) | 11 | 7 | 4 | | |
| *M. gallisepticum* strain R(low) | 11 | 7 | 4 | | |
| *Candidatus* Phytoplasma australiense | 10 | | 34 | | |
| *M. hyopneumoniae* J | 10 | 2 | | | |
| *M. gallisepticum* strain F | 9 | 7 | 3 | | |
| *M. hyopneumoniae* 7448 | 8 | 3 | | | |
| *M. pulmonis* UAB CTIP | 8 | | | | |
| *M. fermentans* JER | 7 | 20 | | | |
| *M. synoviae* 53 | 6 | 12 | | | |
| *M. leachii* PG50 | 5 | 1 | 1 | | |
| *M. conjunctivae* HRC/581 | 5 | | | | |
| *A. laidlawii* PG-8A | 4 | 2 | | 3 | |
| *U. urealyticum* serovar 10 strain ATCC 33699 | 3 | 1 | | | |
| *M. capricolum* subspecies *capricolum* ATCC 27343 | 3 | | | | |
| *M. hyopneumoniae* 232 | 2 | | | | |
| *Candidatus* Phytoplasma mali | 1 | | | | |
| *M. agalactiae* PG2 | 1 | | | | |
| *M. arthritidis* 158L3–1 | 1 | | | | |
| *M. suis* KI3806 | 1 | | | | |
| *M. putrefaciens* KS1 | | 3 | | | |

elements in the avirulent F strain is replaced by a sequence containing several intact and truncated ISs and the CDS MGA_1107 in the virulent strain $R_{low}$ (Szczepanek *et al.*, 2010). The function of MGA_1107, which may have been acquired by HGT from the avian pathogen *M. synoviae,* is not completely established but an isogenic mutant of this gene induces significantly less tracheal lesions

**Table 2.5B** Insertion sequence or transposase-related sequences in Mollicutes

| IS name | Host species | IS family |
|---|---|---|
| InsK | *M. fermentans* JER | IS150 |
| | *M. hyorhinis* HUB-1 | |
| | *M. leachii* PG50 | |
| | *M. mycoides* subspecies *mycoides SC* strain Gladysdale | |
| IS1138 (Bhugra and Dybvig, 1993) | *M. conjunctivae* HRC/581 | IS3 |
| | *M. pulmonis* UAB CTIP | |
| IS1202 | *M. penetrans* HF-2 | |
| IS1221 (Zheng and McIntosh, 1995) | *M. hyopneumoniae* J | |
| | *M. hyorhinis* HUB-1 | |
| IS1296 (Frey *et al.*, 1995) | *M. leachii* PG50 | IS3 |
| | *M. mycoides* subspecies *capri* LC strain 95010 | |
| | *M. mycoides* subspecies *mycoides* SC strain Gladysdale | |
| | *M. mycoides* subspecies *mycoides* SC strain PG1 | |
| IS150 | *M. penetrans* HF-2 | |
| IS1550 (Hu and Yang, 2001) | *M. fermentans* M64 | IS3 |
| | *M. fermentans* PG18 | |
| IS1630 (Calcutt *et al.*, 1999) | *M. fermentans* JER | IS30 |
| | *M. fermentans* M64 | |
| | *M. fermentans* PG18 | |
| | *M. penetrans* HF-2 | |
| IS1634 (Vilei *et al.*, 1999)<br>ISMmy1 (Westberg *et al.*, 2002) | *M. bovis* PG45 | IS1634 |
| | *M. mycoides* subspecies *capri* LC strain 95010 | |
| | *M. mycoides* subspecies *mycoides* SC strain PG1 | |
| | *M. conjunctivae* HRC/581 | |
| | *M. alligatoris* A21JP (Brown *et al.*, 2011) | |
| IS232 | *M. penetrans* HF-2 | |
| IS861 | *M. hyorhinis* HUB-1 | |
| ISHom | *M. hominis* ATCC 23114 | |
| ISMag1 (Pilo *et al.*, 2003) | *M. agalactiae* PG2 | IS30 |
| ISMag2 (Nouvel *et al.*, 2010) | *M. agalactiae* 5632 | IS30 |
| ISMbov1 (Thomas *et al.*, 2005) | *M. bovis* PG45 | IS30 |
| ISMbov2 (Thomas *et al.*, 2005) | *M. bovis* PG45 | IS1634 |
| ISMbov3 (Thomas *et al.*, 2005) | *M. bovis* PG45 | IS1634 |
| ISMbov4 (Lysnyansky *et al.*, 2009) | *M. bovis* PG45 | IS4?? |
| ISMbov5 (Lysnyansky *et al.*, 2009) | *M. bovis* PG45 | IS4 |
| ISMbov6 (Lysnyansky *et al.*, 2009) | *M. bovis* PG45 | IS30 |
| ISMbov7 (Lysnyansky *et al.*, 2009) | *M. bovis* PG45 | IS3 |
| ISMf1 (Calcutt *et al.*, 2002) | *M. fermentans* M64 | IS256 |
| | *M. fermentans* PG18 | |
| ISMhp1 (Vasconcelos *et al.*, 2005) | *M. hyopneumoniae* 168 | IS1634 |
| | *M. hyopneumoniae* 232 | |
| | *M. hyopneumoniae* 7448 | |

**Table 2.5B** (Continued)

| IS name | Host species | IS family |
|---|---|---|
| ISMhp1 (Vasconcelos *et al*., 2005) (continued) | *M. hyopneumoniae* J | IS1634 |
| | *M. synoviae* 53 | |
| | *M. conjunctivae* HRC/581 | |
| ISMmy2 | *M. mycoides* subspecies *capri* LC strain 95010 | IS3 |
| | *M. mycoides* subspecies *capri* strain GM12 | |
| ISMmy3 | *M. mycoides* subspecies *capri LC* strain 95010 | IS3 |
| | *M. mycoides* subspecies *capri* strain GM12 | |
| ISMsy1 | *M. synoviae* 53 | |
| putative transposase | *M. mycoides* subspecies *mycoides* SC strain Gladysdale | IS3 |
| putative transposase | *M. leachii* PG50 | IS4 |
| | *M. mycoides* subspecies *mycoides* SC strain Gladysdale | |
| Tra 5 | Aster yellows witches'-broom phytoplasma AY | IS3 |
| | Onion yellows phytoplasma OY | |

in chickens than the wild type, demonstrating a role of sequences carried by or associated with IS elements in the pathogenicity of *M. gallisepticum*.

Comparison of *M. mycoides* subspecies *mycoides* PG1 to its close relative *M. mycoides* subspecies *capri* 95010 shows that the main difference between the two genomes is a higher number of ISs in PG1 that have resulted in gene decay and loss of protein coding capacity (Thiaucourt *et al.*, 2011). *S. citri*, which belongs to the same phylogenetic group as the mycoides cluster, also harbours large numbers of IS elements, with 69 full or truncated copies of transposase genes from 4 IS families, including the families IS*30* and IS*431*, which are associated with phage sequences (Carle *et al.*, 2010). At the other end of the spectrum, some species, such as *M. arthritidis* 158L3–1, which harbours only a short fragment of IS1630, contain very few complete or truncated ISs. Several species do not have any IS elements in their genomes, including *M. genitalium*, *M. pneumoniae* and *M. mobile*. The repertoires and numbers of IS elements can also vary widely within a species, as exemplified by *M. agalactiae*, with strain PG2 only having one complete transposase gene, MAG_3410, while strain 5632 contains 12 copies of IS*Mag1* (Pilo *et al.*, 2003) and 3 copies IS*Mag2*, an IS*30*-like element (Nouvel *et al.*, 2010).

The distribution and sequence homologies of IS elements across different species suggests a complex evolutionary history and, in some cases, the occurrence of HGT. Some IS elements are common to phylogenetically close mycoplasmas that infect the same hosts. Indeed, the presence of IS*1221* in the pig pathogens *M. hyorhinis* and *M. hyopneumoniae* (Zheng and McIntosh, 1995), IS*1296* in the ruminant pathogens *M. mycoides* subspecies *mycoides* SC, *M. mycoides* subspecies *capri* (Thiaucourt *et al.*, 2011) and *M. leachii* (Frey *et al.*, 1995), and the isoforms IS*Mbov1*/IS*Mag1* in *M. bovis* and *M. agalactiae* (Thomas *et al.*, 2005) is not surprising given the similar habitats of these species. Identical or highly similar IS elements can also be found in species that are usually isolated from different hosts but belong to the same phylogenetic group, such as IS*1138* in the ruminant pathogen *M. conjunctivae* and the rodent pathogen *M pulmonis*, IS*Mhp1* in the porcine pathogen *M. hyopneumoniae* and the avian pathogen *M. synoviae*, and IS*1630* sequences (Calcutt *et al.*, 1999) in the human pathogen *M. fermentans* and the rat pathogen *M. arthritidis*, which all belong to the hominis phylogenetic group. However, phylogenetically distant mycoplasmas can also possess highly related IS elements. This is mainly observed for mycoplasmas that share the same hosts, such as the mycoplasmas infecting ruminants, with IS*Mmy1*/IS*Mbov2* and IS*1634*/IS*Mbov3* found in *M. mycoides* and *M. bovis* (Westberg *et al.*, 2002; Thomas *et al.*, 2005; Li *et al.*, 2011), IS*1634* in *M.*

*conjunctivae* and *M. mycoides* (Vilei *et al.*, 1999; Calderon-Copete *et al.*, 2009), and IS*Mbov1*/IS*Mag1* in *M. conjunctivae* and *M. bovis* (Calderon-Copete *et al.*, 2009), and the mycoplasmas infecting humans, with insertion sequence-like elements (ISLE) found in *M. fermentans* and *M. orale* (Ditty *et al.*, 2003). It is also noteworthy that some IS elements related to the IS3 family are found in phytoplasmas (Lee *et al.*, 2005) as well as in spiroplasmas and other members of the mycoides cluster. This could be explained by the occurrence of HGT between phylogenetically distant mycoplasmas during the co-infection of their plant and ruminant hosts.

## Plasmids

Endogenous plasmids are rarely found in Mollicutes. They have only been detected in the ruminant pathogens *M. mycoides*, (Bergemann and Finch, 1988; Dybvig and Khaled, 1990; Thiaucourt *et al.*, 2011), *M. leachii* (Djordjevic *et al.*, 2001) and *M. yeatsii*, and in the plant pathogens *S. citri* (Joshi *et al.*, 2005; Berho *et al.*, 2006b; Saillard *et al.*, 2008), *S. kunkelii* (Davis *et al.*, 2005a) and several phytoplasmas (Oshima *et al.*, 2001; Nishigawa *et al.*, 2001, 2002b; Liefting *et al.*, 2006; Tran-Nguyen and Gibb, 2006; Petrzik *et al.*, 2011). Plasmid sequences from spiroplasmas and phytoplasmas are unrelated even though both groups of organisms are able to infect plants.

Compared to those in other eubacteria, the plasmids of Mollicutes that have been sequenced thus far are relatively modest in size, ranging from 1 to 35 kb and harbouring 1 to 25 CDSs (Table 2.6). Many of these plasmids are cryptic, although some of the *S. citri* plasmids encode adhesins and appear to have a role in insect transmission (Berho *et al.*, 2006a,b). The plasmids of *Spiroplasma* species also harbour genes potentially encoding a partition system ParA/Soj (Breton *et al.*, 2008) and components of conjugative machinery, TraE/TrsE and TraG/TraD (Bai *et al.*, 2004a; Davis *et al.*, 2005a; Saillard *et al.*, 2008), which have homologues in mycoplasma ICEs (see below). Phylogenetic analysis of the spiroplasma ParA/Soj chromosomal and plasmid sequences shows that they have different origins. It is not known if the plasmids of *Spiroplasma* species are all transmissible, but it is tempting to speculate that some ICEs of mycoplasmas have recombined with conjugative plasmids, possibly from insect-transmitted organisms.

In phytoplasmas, the replication of most known plasmids appears to follow a rolling circle model (Oshima *et al.*, 2001; Nishigawa *et al.*, 2002b), although alternative mechanisms that involve a putative DNA primase may also exist (Tran-Nguyen and Gibb, 2006). The rolling circle replication mechanism is also found in ICEs and certain phages. Interestingly, a group of insect transmitted, ssDNA plant viruses, the geminiviruses, also use the rolling circle mechanism for their replication and may have evolved from recombination between a phytoplasmal plasmid and a plant RNA virus (Krupovic *et al.*, 2009), demonstrating the potential for genomic permeation between Mollicutes and other life forms (Gibbs *et al.*, 2006). The rolling circle mode of replication is also suspected to occur in the plasmids pMKM1, pADB201, pMmc-95010 and pBG7AU, based on the presence of the conserved domain pfam01719 and sequence homology between the putative *rep* genes of these plasmids and their counterparts in other bacterial plasmids, including pE194 from *Staphylococcus* species and pLS1 from *Lactococcus* species (Xu *et al.*, 1991). Rolling circle replication requires single strand DNA binding (ssb) proteins to stabilize the replication intermediates. The *ssb* genes can be localized on the plasmid itself, as in some phytoplasma plasmids, or on the bacterial chromosome. It is notable that many chromosomally integrated MICEs also encode ssb proteins. However, the plasmids of spiroplasmas appear to have a different replication system, most probably based on a theta mechanism using a cis-acting protein, PE, that does not share sequence homology with other known replication initiator proteins. The presence of sequences similar to DnaA boxes, but that seem to be dispensable for replication, indicate that these plasmids might represent a new family of extrachromosomal replicons of unknown origin in Mollicutes (Breton *et al.*, 2008).

The plasmids from the mycoides cluster have significant similarities. Sequence alignment analyses of pBG7AU from *M. leachii* (Djordjevic *et al.*, 2001), pADB201 from *M. mycoides* subspecies *mycoides* (Bergemann *et al.*, 1989), pKMK1 (King

**Table 2.6** Features of phages and plasmids isolated from Mollicutes

| Sequence type | Host organism | Genbank Description | Size (bp) | CDS | Gene | GC% | Accession |
|---|---|---|---|---|---|---|---|
| Plasmid | *M. leachii* | *Mycoplasma* sp. 'bovine group 7' plasmid pBG7AU, complete sequence. | 1022 | 2 | 2 | 30.7 | NC_002569 |
| Plasmid | *M. mycoides* subspecies *mycoides* strain GC1176–2 | *Mycoplasma mycoides mycoides* plasmid pADB201, complete sequence | 1717 | 2 | 2 | 29.5 | NC_001382 |
| Plasmid | *M. mycoides* subspecies *capri* strain 95010 | *Mycoplasma mycoides* subspecies *capri* LC strain 95010 plasmid pMmc-95010,complete sequence | 1840 | 0 | 0 | 29.2 | NC_015407 |
| Plasmid | *M. mycoides* subspecies *capri* strain GM12 | *Mycoplasma mycoides* strain GM12 plasmid pKMK1 replication protein gene, complete cds | 1875 | 1 | 0 | 29.0 | M81470 |
| Plasmid | *Candidatus* Phytoplasma asteris strain OY | Onion yellows phytoplasma plasmid pOYNIM, complete sequence | 3062 | 3 | 3 | 23.8 | NC_012090 |
| Plasmid | *Candidatus* Phytoplasma asteris | Tomato big bud phytoplasma plasmid pTBBperi, complete sequence | 3319 | 4 | 4 | 26.8 | NC_010920 |
| Plasmid | *M. yeatsii* | *Mycoplasma yeatsii* plasmid pMyBK1, complete sequence | 3432 | 2 | 2 | 28.0 | NC_011102 |
| Plasmid | *Candidatus* Phytoplasma australiense | *Candidatus* Phytoplasma australiense plasmid pCPa, complete sequence | 3773 | 4 | 4 | 28.4 | NC_010918 |
| Plasmid | *Candidatus* Phytoplasma asteris strain Nanyang | Paulownia witches'-broom phytoplasma plasmid pPaWBNy-2, complete sequence | 3837 | 5 | 5 | 25.9 | NC_010406 |
| Plasmid | *Candidatus* Phytoplasma asteris strain OY | Onion yellows phytoplasma plasmid pOYM, complete sequence | 3932 | 5 | 5 | 24.3 | NC_012089 |
| Plasmid | *Candidatus* Phytoplasma asteris | Aster yellows witches'-broom phytoplasma AYWB plasmid pAYWB-I, complete sequence | 3972 | 5 | 5 | 25.6 | NC_007717 |
| Plasmid | *Candidatus* Phytoplasma asteris | Aster yellows witches'-broom phytoplasma AYWB plasmid pAYWB-II, complete sequence | 4009 | 4 | 4 | 23.9 | NC_007718 |
| Plasmid | *Candidatus* Phytoplasma asteris | Tomato big bud phytoplasma plasmid pTBBcap, partial sequence | 4092 | 3 | 0 | 30.8 | DQ119296 |
| Plasmid | *Candidatus* Phytoplasma asteris | Aster yellows phytoplasma plasmid pJHW, complete sequence | 4278 | 2 | 2 | 23.9 | NC_003353 |
| Plasmid | *Candidatus* Phytoplasma asteris | Aster yellows witches'-broom phytoplasma AYWB plasmid pAYWB-IV, complete sequence | 4316 | 6 | 6 | 24.4 | NC_007720 |
| Plasmid | *Candidatus* Phytoplasma asteris | '*Rehmannia glutinosa*' phytoplasma plasmid pPARG1, complete sequence | 4371 | 7 | 7 | 24.5 | NC_014123 |
| Plasmid | *Candidatus* Phytoplasma asteris | Chinaberry witches' broom phytoplasma plasmid pCWBFq, complete sequence | 4446 | 6 | 6 | 26.5 | NC_015473 |
| Plasmid | *Candidatus* Phytoplasma asteris strain Nanyang | Paulownia witches' broom phytoplasma plasmid pPaWBNy-1, complete sequence | 4485 | 6 | 6 | 24.8 | NC_010405 |

| Sequence type | Host organism | Genbank Description | Size (bp) | CDS | Gene | GC% | Accession |
|---|---|---|---|---|---|---|---|
| Plasmid | *Candidatus* Phytoplasma asteris strain OY | Onion yellows phytoplasma plasmid unknown, complete sequence | 5045 | 6 | 6 | 24.3 | NC_006903 |
| Plasmid | *Candidatus* Phytoplasma asteris | Aster yellows witches'-broom phytoplasma AYWB plasmid pAYWB-III, complete sequence. | 5104 | 7 | 7 | 21.8 | NC_007719 |
| Plasmid | *Candidatus* Phytoplasma asteris strain OY | Onion yellows phytoplasma plasmid EcOYW1, complete sequence. | 7005 | 7 | 7 | 21.9 | NC_012088 |
| Plasmid | *S. citri* | *Spiroplasma citri* plasmid pSciA, complete sequence | 7790 | 8 | 8 | 21.3 | NC_007386 |
| Plasmid | *S. citri* | *Spiroplasma citri* plasmid pSci1, complete sequence | 12989 | 12 | 12 | 28.7 | NC_007387 |
| Plasmid | *S. citri* | *Spiroplasma citri* plasmid pBJS-O, complete sequence | 13374 | 10 | 10 | 28.9 | NC 007101 |
| Plasmid | *S. citri* | *Spiroplasma citri* plasmid pSci2, complete sequence | 14407 | 10 | 12 | 29.0 | NC_007388 |
| Plasmid | *S. kunkelii* CR2–3x | *Spiroplasma kunkelii* CR2–3x plasmid pSKU146, complete sequence | 14615 | 18 | 18 | 28.1 | NC_006400 |
| Plasmid | *S. citri* | *Spiroplasma citri* plasmid pSci3, complete sequence | 19325 | 14 | 16 | 27.8 | NC_007389 |
| Plasmid | *S. citri* | *Spiroplasma citri* plasmid pSci4, complete sequence | 20224 | 17 | 19 | 28.6 | NC_007390 |
| Plasmid | *S. citri* | *Spiroplasma citri* plasmid pSci5, complete sequence | 27778 | 24 | 26 | 27.0 | NC_007391 |
| Plasmid | *S. citri* | *Spiroplasma citri* plasmid pSci6, complete sequence | 35318 | 25 | 37 | 25.6 | NC_007392 |
| Phage | *S. melliferum* | *Spiroplasma* phage 4, complete genome | 4421 | 9 | 9 | 32.1 | NC_003438 |
| Phage | *S. melliferum* | *Spiroplasma* phage SVTS2, complete genome | 6825 | 13 | 13 | 22.7 | NC_001270 |
| Phage | *S. citri* | *Spiroplasma* phage 1-C74, complete genome | 7768 | 13 | 13 | 23.2 | NC_003793 |
| Phage | *S. citri* strain GII3 | *Spiroplasma* phage SVGII3, complete genome | 7878 | 12 | 2 | 23.0 | AJ969242 |
| Phage | *S. citri* | *Spiroplasma* phage 1-R8A2B, complete genome | 8273 | 12 | 12 | 22.9 | NC_001365 |
| Phage | *M. pulmonis* strain UAB 6510 | *Mycoplasma* phage P1, complete genome | 11660 | 11 | 11 | 26.8 | NC_002515 |
| Phage | *Acholeplasma laidlawii* | *Acholeplasma* phage L2, complete genome | 11965 | 14 | 14 | 32.0 | NC_001447 |
| Phage | *M. arthritidis* | *Mycoplasma* phage MAV1, complete genome | 15644 | 15 | 15 | 29.0 | NC_001942 |
| Phage | *M. fermentans* strainII-29/1 | *Mycoplasma* phage phiMFV1, complete genome | 15757 | 18 | 9 | 25.0 | AY583234 |
| Phage | *M. fermentans* strain K7 | *Mycoplasma* phage phiMFV1, complete genome. | 18855 | 19 | 10 | 24.8 | AY583236 |
| Phage | *M. fermentans* strain PG18 | *Mycoplasma* phage phiMFV1, complete genome. | 23300 | 26 | 18 | 25.9 | AY583235 |

and Dybvig, 1992) and pMmc-95010 (Thiaucourt *et al.*, 2011) from *M. mycoides* subspecies *capri*, and to a lesser extend pMyBK1 from *M. yeatsii* (Genbank EU429323), suggest a common origin with subsequent rearrangements for all these plasmids. Intermolecular recombinations have been documented in some phytoplasma plasmids, resulting in chimaeric extrachromosomal elements (Nishigawa *et al.*, 2002a), and *S. citri* plasmids (Joshi *et al.*, 2005), resulting in mosaic structures, with a number of the same CDSs present in several plasmids (Breton *et al.*, 2008).

## Phages

As with endogenous plasmids, only a few phages have been described in Mollicutes (Table 2.6). The earliest descriptions of viruses refer to MLV1, -2 and -3 in *Acholeplasma laidlawii* (Gourlay, 1971; Gourlay *et al.*, 1971; Gourlay and Wyld, 1973) and lambda-like phage in *Spiroplasma* species (Cole *et al.*, 1973). In *Mycoplasma* species, only a few phages have been described, including Br1 in *M. bovirhinis* (Gourlay *et al.*, 1983a), Hr1 in *M. hyorhinis* (Gourlay *et al.*, 1983b), MV20-P in *M. hominis* (Jansson *et al.*, 1982), P1 in *M. pulmonis* (Dybvig *et al.*, 1987), and the closely related MAV1 in *M. arthritidis* (Voelker and Dybvig, 1998) and MFV1 in *M. fermentans* (Roske *et al.*, 2004).

The plasmavirus MLV2 from *A. laidlawii* is a double stranded DNA virus (Nowak and Maniloff, 1979) that is associated with lysogeny (Dybvig and Maniloff, 1983), while another phage of *A. laidlawii*, L172, has a single stranded DNA genome (Dybvig *et al.*, 1985). The plant pathogen *S. citri* also contains several single stranded phages that belong to the plectroviruses, namely SpV1-R8A2B (Renaudin *et al.*, 1990), SpV1-C74 (Bébéar *et al.*, 1996) and SVGII3 (Carle *et al.*, 2010), while *S. melliferum* harbours a similar but divergent plectrovirus, SVTS2 (Sha *et al.*, 2000), and a spirovirus, SPV4 (Renaudin *et al.*, 1987). A linear double stranded DNA structure is proposed for the picovirinae phage P1 (Zou *et al.*, 1995) and the unclassified phage MAV1 (Voelker and Dybvig, 1998).

Little is known about the mechanisms used by these phages to infect their host and replicate. The variable surface protein VsaA has been proposed to act as a receptor on *M. pulmonis* for the phage P1 (Dybvig *et al.*, 1988). The phage L2 of *A. laidlawii* contains DnaA boxes in its putative replication sites (Maniloff *et al.*, 1994) and the phage P1 of *M. pulmonis* is believed to encode a DNA polymerase (Tu *et al.*, 2001). Most of the other putative P1 proteins have unknown functions, apart from a structural protein with collagen-like repeated motifs that could act as a tail fibre component (Tu *et al.*, 2001). Putative excisionases and integrases are encoded by the phages L2 (Maniloff *et al.*, 1994), MFV1 of *M. fermentans* (Voelker and Dybvig, 1999) and MAV1 of *M. arthritidis* (Roske *et al.*, 2004). MAV1 and MFV1 also encode a putative DNA replication protein, as well as a DNA methylase and restriction endonuclease (Voelker and Dybvig, 1999; Roske *et al.*, 2004). The integration sites of MAV1 (TATTTTT) and MFV1 (TTTTTA) are very similar (Roske *et al.*, 2004) and reminiscent of the conjugative transposon Tn916 transposition site (Voelker and Dybvig, 1998).

In *S. citri* the viral sequences are integrated at multiple sites in the chromosome and often contain truncated CDSs (Carle *et al.*, 2010). Spiroplasma viruses may be related to certain IS elements and their insertion has resulted in gene disruptions and genome reorganisations (Melcher *et al.*, 1999). Genomic mosaic structures can be caused by phages. In *M. fermentans*, copies of prophages have integrated non-coding intergenic regions or pseudogenes, which in some cases are associated with concatenated IS elements (Roske *et al.*, 2004). In several strains of *M. arthritidis*, rearrangements and deletions have affected the organization of prophage copies themselves (Washburn *et al.*, 2004).

The phenotypes or functions conferred by mycoplasma phages appear to be diverse, but their significance has been somewhat difficult to appreciate. In *Acholeplasma* species, phages have been related to the formation of giant cells and/or lysis and alteration of colony morphology (Brunner *et al.*, 1980). The virulence of *M. arthritidis* in rats is enhanced by the presence of MAV1 (Voelker *et al.*, 1995), although the main virulence gene candidate, *vir*, appears to control an exclusion mechanism that prevents further phage infection

(Clapper *et al.*, 2004). Genes encoding surface exposed membrane proteins or lipoproteins are carried by several phages (Maniloff *et al.*, 1994; Clapper *et al.*, 2004; Roske *et al.*, 2004), but their exact functions remain to be established.

Phages or phage remnants are also present in phytoplasmas, where they have clustered into sequence variable mosaics (SVMs), which constitute a unique and characteristic feature of these genomes. Some of the gene products of SVMs show homology with phage proteins and with transposases, suggesting that SVMs are acquired by HGT (Jomantiene *et al.*, 2007). These SVMs are responsible for most of the non-syntenic regions and sequence variation between strains, but are not present in the phylogenetically related *Acholeplasma* species, suggesting that phage invasions have played a critical role in the evolution and differentiation of phytoplasmas (Wei *et al.*, 2008).

**Table 2.7** Features of mycoplasma integrative conjugative elements (MICEs)

| Host organism | Tag_locus | Size (bp) | CDS | Gap | Gene | ICE GC% | Host GC% |
|---|---|---|---|---|---|---|---|
| *M. agalactiae* PG2 | MAG_4060–3860 | 23,583 | 23 | 0 | 23 | 27.4 | 29.7 |
| *M. agalactiae* 5632 | MAGa2980–3220 | 28,787 | 26 | 0 | 26 | 27.5 | 29.6 |
| *M. agalactiae* 5632 | MAGa4850–5060 | 28,809 | 23 | 0 | 24 | 27.3 | 29.6 |
| *M. agalactiae* 5632 | MAGa7100–6880 | 28,627 | 24 | 0 | 25 | 27.4 | 29.6 |
| *M. fermentans* PG18 | MBIO_0266–0285 | 21,082 | 22 | 1 | 22 | 22.6 | 26.8 |
| *M. fermentans* PG18 | MBIO_0551–0567 | 23,223 | 19 | 0 | 19 | 28.8 | 26.8 |
| *M. bovis* PG45 | MBOVPG45_0213–0183 | 38,154 | 28 | 0 | 34 | 27.0 | 29.3 |
| *M. bovis* PG45 | MBOVPG45_0495–0479 | 23,657 | 19 | 0 | 20 | 27.4 | 29.3 |
| *M. capricolum* subspecies *capricolum* ATCC 27343 | MCAP_0554–0571 | 25,067 | 18 | 0 | 20 | 23.7 | 23.8 |
| *M. conjunctivae* HRC/581 | MCJ_003520–003720 | 28,487 | 23 | 0 | 24 | 25.2 | 28.5 |
| *M. fermentans* JER | MFE_02750–02960 | 24,367 | 24 | 0 | 25 | 24.7 | 26.9 |
| *M. fermentans* JER | MFE 04970–04820 | 23,917 | 20 | 0 | 21 | 24.8 | 26.9 |
| *M. fermentans* M64 | MFeM64YM_0096–0116 | 22,727 | 23 | 0 | 25 | 22.7 | 26.9 |
| *M. fermentans* M64 | MFeM64YM_0227–0247 | 22,578 | 25 | 0 | 26 | 22.7 | 26.9 |
| *M. fermentans* M64 | MFeM64YM_0393–0370 | 26,079 | 26 | 0 | 27 | 28.3 | 26.9 |
| *M. fermentans* M64 | MFeM64YM_0462–0443 | 24,934 | 22 | 0 | 24 | 28.5 | 26.9 |
| *M. fermentans* M64 | MFeM64YM_0485–0464 | 23,327 | 25 | 0 | 26 | 22.8 | 26.9 |
| *M. fermentans* M64 | MFeM64YM_0791–0810 | 23,422 | 22 | 0 | 22 | 28.7 | 26.9 |
| *M. fermentans* M64 | MFeM64YM_0857–0838 | 22,746 | 22 | 0 | 22 | 28.9 | 26.9 |
| *M. hyopneumoniae* 168 | MHP168_0223–0237 | 23,442 | 19 | 0 | 20 | 25.5 | 28.5 |
| *M. hyopneumoniae* 168 | MHP168_689–677 | 17,111 | 13 | 0 | 14 | 27.7 | 28.5 |
| *M. hyopneumoniae* 232 | mhp521–538 | 22,483 | 19 | 0 | 19 | 27.5 | 28.6 |
| *M. hyopneumoniae* 7448 | MHP7448_0424–0412 | 24,653 | 17 | 0 | 18 | 25.8 | 28.5 |
| *M. mycoides* subspecies capri LC strain 95010 | MLC_2080–2280 | 31,497 | 23 | 0 | 23 | 23.6 | 23.8 |
| *M. mycoides* subspecies capri LC strain 95010 | MLC_3070–2890 | 31,078 | 22 | 0 | 22 | 23.6 | 23.8 |
| *M. bovis* Hubei-1 | MMB_0360–0378 | 23,274 | 21 | 0 | 21 | 27.3 | 29.3 |
| *M. mycoides* subspecies *capri* strain GM12 | MMCAP1_0552–0573 | 30,993 | 23 | 0 | 24 | 23.7 | 23.9 |

## Conjugative elements

### Integrative conjugative elements

Mycoplasma integrative conjugative elements (MICEs) are clusters of genes present in at least seven different species, with 1–7 copies present in individual strains (Table 2.7 and Fig. 2.1). MICEs belong to a large group of mobile genetic elements, the integrative conjugative elements (ICEs), also known as conjugative transposons in some bacteria (Salyers *et al.*, 1995; Burrus *et al.*, 2002), or conjugative self-transmissible integrating elements (CONSTINs) in *Vibrio cholerae* (Hochhut and Waldor, 1999). One of the most studied ICEs is the promiscuous transposon Tn916 from *Enterococcus faecalis,* which confers resistance to antimicrobials (Roberts and Mullany, 2011) and has been shown be transferable *in vitro* to a number of Mollicutes (Flannagan *et al.*, 1994; Clewell *et al.*, 1995), although it has not been naturally found in any mycoplasma sequenced genome so far.. The importance of ICEs in the evolution of microorganisms has been appreciated only recently, as it is becoming more apparent that these elements are widely distributed in prokaryotes, in which they represent a major source of HGT by conjugation (Wozniak and Waldor, 2010; Guglielmini *et al.*, 2011). In mycoplasmas, MICEs are excised from their chromosomal site at low frequency, to create a circular extrachromosomal intermediate. The excision involves short inverted repeats located at the

A

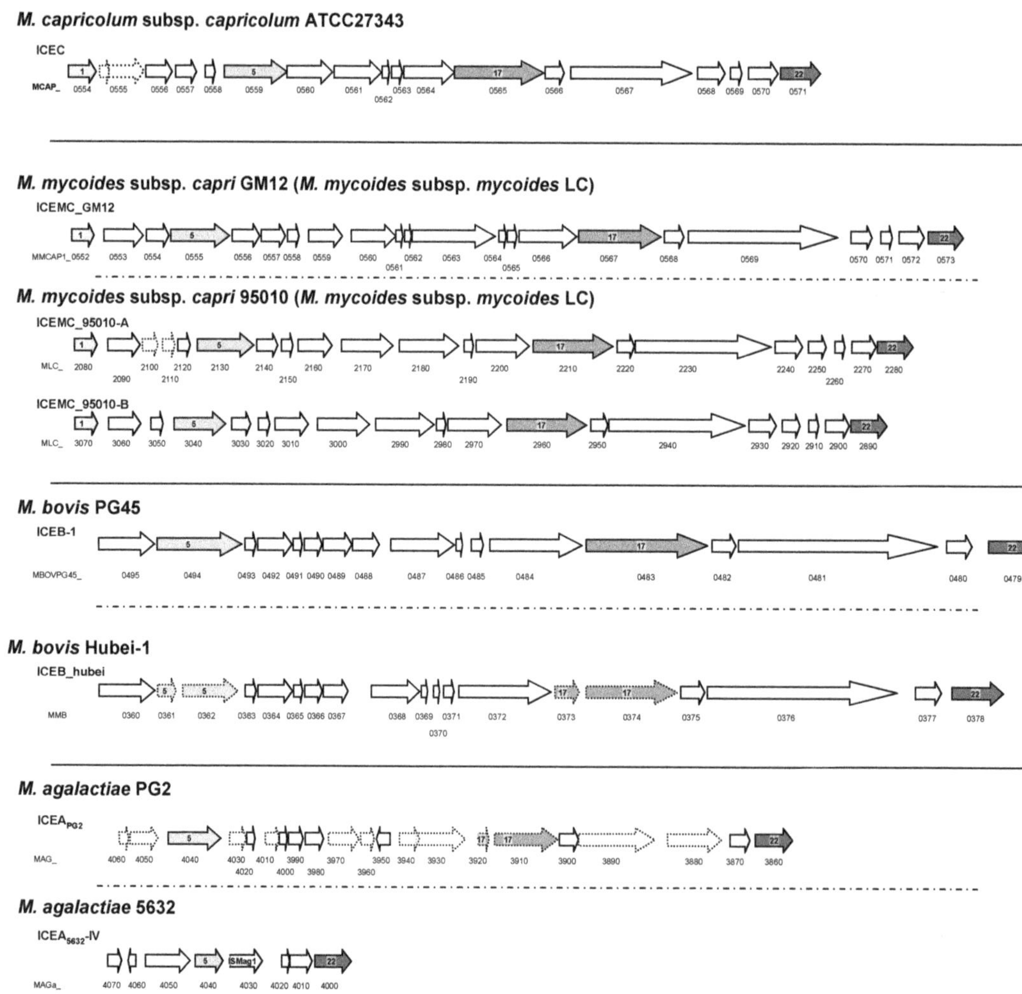

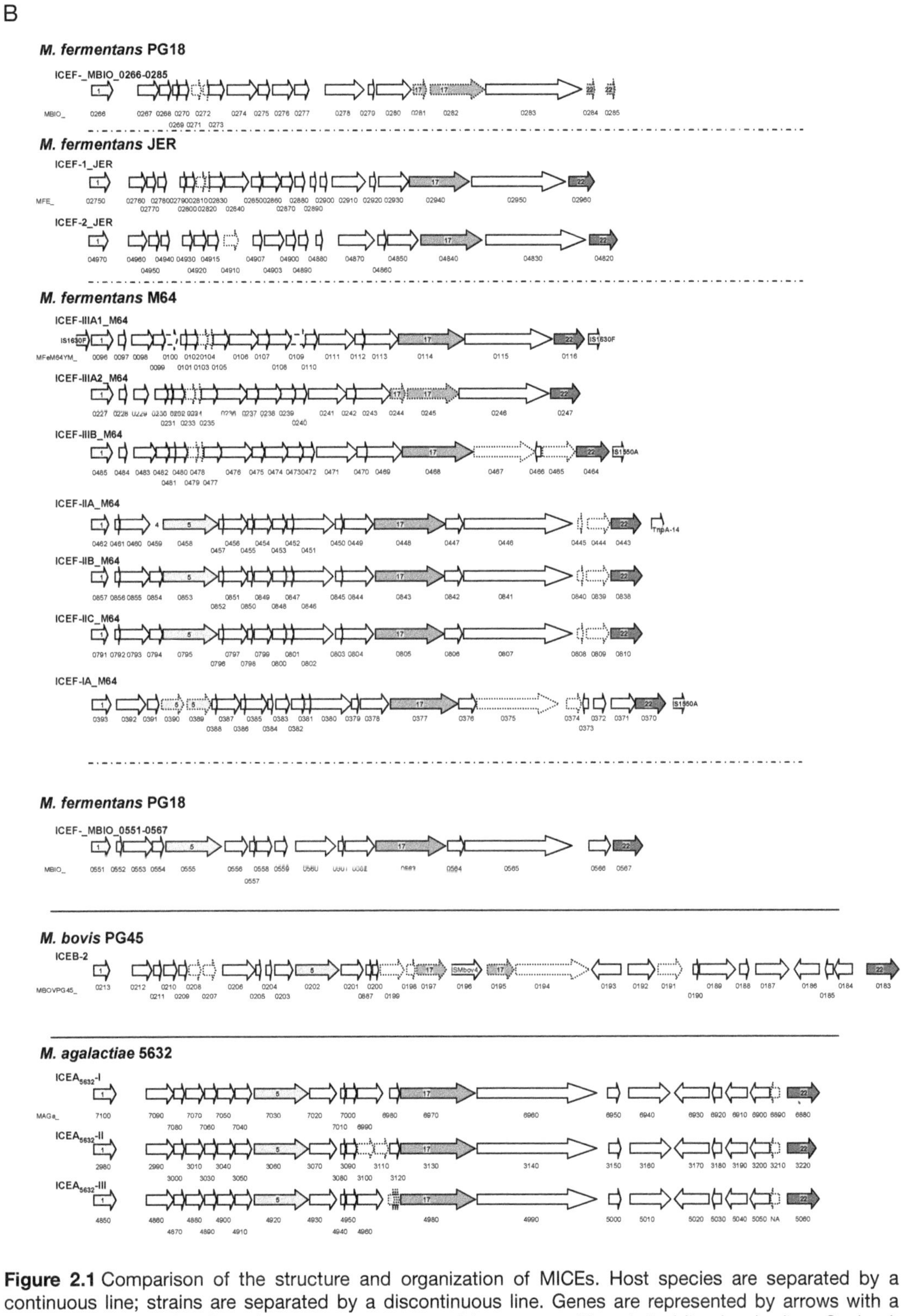


**Figure 2.1** Comparison of the structure and organization of MICEs. Host species are separated by a continuous line; strains are separated by a discontinuous line. Genes are represented by arrows with a continuous line border. Pseudogenes are represented by arrows with a dotted line border line. Genbank unique locus tag numbers are indicated under each gene. The highly conserved ORFs 1, 5, 17 and 22 are shaded in grey and referenced by the same number. (A) *M. mycoides* type MICEs; (B) *M. fermentans* type MICEs; (C) *M. hyopmeumoniae* type MICEs. (Continued overleaf.)

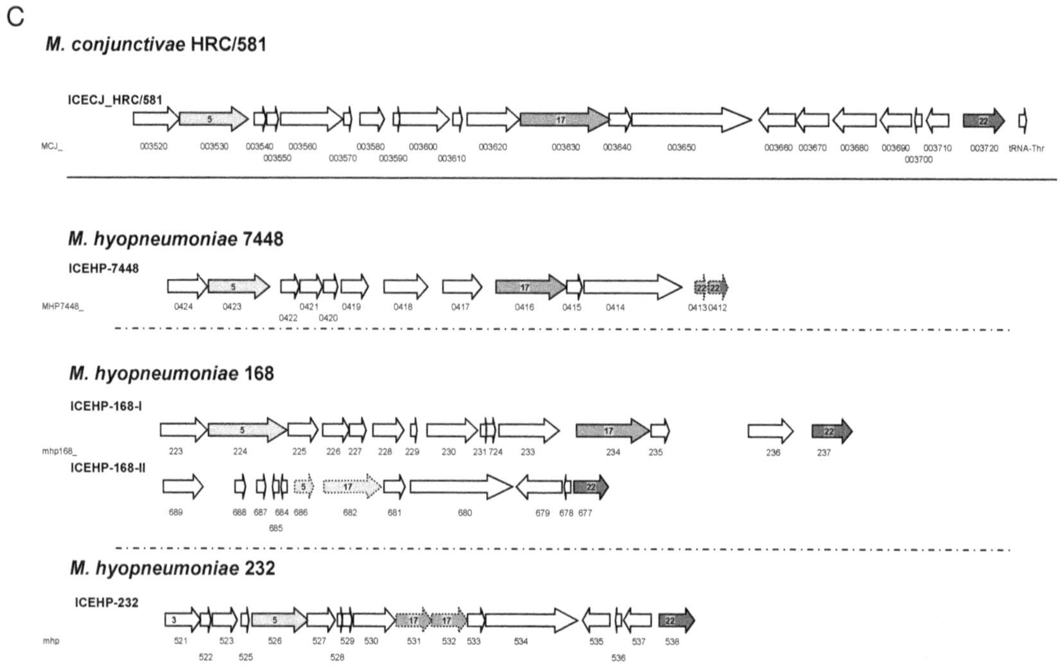


**Figure 2.1** (Continued)

ends of the element and has been demonstrated by PCR amplification of the juxtaposed termini of the circular form and the empty chromosomal locus after excision in *M. fermentans* (Calcutt *et al.*, 2002), *M. agalactiae* (Marenda *et al.*, 2006) and *M. hyopneumoniae* (Vasconcelos *et al.*, 2005; Pinto *et al.*, 2007). Based on observations in other bacteria, the extrachromosomal MICE intermediate is thought to be non-replicative and is most probably transferred to a recipient cell by conjugative machinery as a single stranded DNA molecule with the help of ssb proteins. Meanwhile, the remaining strand template in the donor cell is replicated by a rolling circle mechanism. The conjugative system comprises two conserved ATPases, TraE/VirB4/TrsE and TraG/VirD4/TraD (Alvarez-Martinez and Christie, 2009; Smillie *et al.*, 2010). Homologues of these two proteins are encoded by most MICEs (in which they correspond to the ORFs-17 and -5, respectively, in Fig. 2.1), but also by *S. citri* plasmids. Once in the recipient cell, it is thought that the transferred MICE eventually integrates into the host chromosome after synthesis of its complementary strand. The integration of a MICE circular form results in the duplication of short DNA motifs that are present on each side of the element as direct repeats (Calcutt *et al.*, 2002).

The functions encoded by MICEs are not clearly established, apart from a few genes that are likely to be involved in the conjugative transfer of DNA (ssb, TraG, TraE). In *M. fermentans* ICEF encodes a lipoprotein, P57 (Calcutt *et al.*, 2002), and it is therefore tempting to speculate that MICEs may play a role in host adaptation by conferring new surface properties. Unlike in the mycoplasma phages MAV1 and MFV1, none of the proteins encoded on MICEs shows similarity with excisionase or integrase gene products. However, in addition to the *traE* and *traG* homologues, a number of genes of unknown function are conserved in MICEs (see Fig. 2.1) and could encode a novel MICE-specific excisionase or integrase. Interestingly, some MICEs encode DNA methylases and plasmid partition systems that have homologues in mycoplasma phages and plasmids. Clearly, the ICEs, plasmids and phages of Mollicutes have intricate relationships that may have reciprocally influenced their gene structures.

Typical MICEs contain approximately 20 genes and generally each copy spans 20–30 kb (Table 2.7), although shorter versions, and even apparently isolated MICE coding sequences, can also be found in some genomes. Southern blot, PCR and genomic analyses indicate that the distribution and organization of MICEs varies extensively between isolates of the same species. This strain-to-strain variability of distribution is particularly apparent in *M. fermentans* (Calcutt *et al.*, 2002), *M. agalactiae*, *M. bovis* (Marenda *et al.*, 2005) and *M. hyopneumoniae* (Pinto *et al.*, 2007). Within a species, some isolates may completely lack MICEs, others may harbour a single MICE, for example *M. bovis* strain Hubei (Li *et al.*, 2011), or multiple MICE copies, for example *M. bovis* strain PG45 (Wise *et al.*, 2010). Based on their general organization, MICEs can be temptatively classified into three groups: *M. mycoides* type MICEs, *M. fermentans* type MICEs and *M. hyopmeumoniae* type MICEs (Fig. 2.1A–C).

In *M. pulmonis* strain UAB CTIP, a strain apparently devoid of MICEs, the coding sequence MYPU_3830 is homologous to the conserved ORF17 (*traE*) of ICEF from *M. fermentans* PG18 (Calcutt *et al.*, 2002). This is puzzling, because no other MICE-related sequences are present in this genome. It is possible that this sequence is a remnant of a conjugative plasmid, as *S. citri* plasmids harbour sequences homologous to *traE*, or a larger MICE that has almost completely disappeared through gene decay, leaving only an isolated *traE* homologue in the *M. pulmonis* chromosome. Gene decay clearly occurs in MICEs and can induce significant rearrangements and sequence loss in these elements (Fig. 2.1). This was first described after the sequencing of a MICE locus containing a large number of pseudogenes in *M. agalactiae* strain PG2 (Sirand-Pugnet *et al.*, 2007b). In addition, one of the four MICE copies identified in *M. agalactiae* 5632, $ICEA_{5632}$-IV, is unusually short and appears to be a vestigial form of a larger MICE, similar to the *M. agalactiae* $ICE_{PG2}$ but displaying a more advanced stage of decay (Nouvel *et al.*, 2010). In the other MICEs present in strain 5632, pseudogenes have been created by stop codons that split CDS16 (MAGa_3100 and MAGa_3110) in the $ICEA_{5632}$-II copy and CDS27 in the $ICEA_{5632}$-III copy (Nouvel *et al.*, 2010). Ongoing MICE gene decay is also apparent in *M. bovis*, in which an IS element (MBOVPG45_0196) has interrupted the *traE* gene in ICEB-2. Finally, in *M. capricolum* subspecies *capricolum*, *M. conjunctivae* and *M. hyopneumoniae*, some massive MICE decay is suggested by the presence of small clusters of conserved MICE CDS homologues, in addition to a larger MICE copy within the chromosome (Fig. 2.1).

With the notable exception of MICEs found in the members of the mycoides cluster, the %GC of most MICEs clearly differs from that of the host chromosome (Table 2.7), suggesting that these elements have been acquired by HGT. In some cases, a single strain harbours differently organized copies of MICEs, as shown for example in Fig. 2.1 by the MICE copies of *M fermentans* M64 (Shu *et al.*, 2011) or *M. agalactiae* 5632 (Nouvel *et al.*, 2010). In *M. fermentans* M64, two distinct groups of MICEs can be defined based on their different GC% and gene organization. Similarly, in *M. agalactiae* and *M. bovis* the MICE copies $ICEA_{PG2}$ and ICEB-1 show a distinct organization from ICEB-2 and $ICEA_{5632}$I-III (Fig. 2.1A and B). This strongly suggests that multiple events of HGT from different donors have occurred in these species.

An intriguing CDS of unknown function encoding a phage-associated protein of the GepA family (COG3600), Ps3, is present in each of the 3 large MICE copies in *M. agalactiae* 5632 (MAGa6950, MAGa3150, MAGa5000) and in one of the two MICE copies in *M. bovis* PG45 (MBOVPG45_0188). Homologues of *ps3* are also found in non-MICE loci in *M. agalactiae* strains PG2 (MAG_6440) and 5632 (MAGa7400) and in *M. bovis* strains PG45 (MBOVPG45_0501) and Hubei (MMB_0353). Apart from *M. agalactiae* and *M. bovis*, single or paralogous copies of *ps3* are overwhelmingly found in the mycoides cluster, namely in *M. mycoides* subspecies *capri* strains GM12 (MMCAP1_0417) and 95010 (MLC_4580), *M. mycoides* subspecies *mycoides* SC strains Gladysdale (MMS_A0221 and MMS_A0526) and PG1 (MSC_0194 and MSC_0478), *M. leachii* strains PG50 (MSB_A0506) and 99/014/6 (MLEA_003070), and *M capricolum* subspecies *capricolum* ATCC 27343 (MCAP_0493). The *ps3* gene is also present in *Candidatus* Phytoplasma australiense (PAa_0794 and PAa_0404) and *Ureaplasma* species. The role of *ps3* is not known, but its association with prophages of various bacteria, its wide occurrence in the mycoides cluster, which is rich in other mobile elements, and its localization both inside and outside MICEs is intriguing and illustrates the intricate evolutionary relationships between various categories of mobile elements (see 'Tra islands' below).

### Tra islands in the mycoides cluster

The chromosomes of *M. leachii* PG50 and *M. capricolum* subspecies *capricolum* ATCC 27343 contain, respectively, one and two clusters of genes, known as Tra islands. In *M leachii*, the Tra island spans from MSB_A0073 to MSB_A0042, while in *M capricolum* subspecies *capricolum* Tra island I spans from MCAP_0274 to MCAP_0315 and Tra island II spans from MCAP_0167 to MCAP_0187.

Most of the genes in Tra islands have unknown functions, but several phage- and plasmid-related proteins are annotated in these regions based on sequence homologies. These include DNA helicases, rep_2 plasmid replication-initiation proteins, SoJ/ParA plasmid partition proteins and ssb proteins. Moreover, the *M. leachii* Tra island encodes a hypothetical protein of the *M. yeatsii* plasmid pMyBK1, and the *M. capricolum* subspecies *capricolum* Tra island II encodes a putative DNA methylase. Several of the other putative proteins encoded by the Tra islands contain transmembrane domains, specific variable repeats, originally found in the lipoprotein LppQ of *M. mycoides* subspecies *mycoides* SC, that are rich in hydrophobic and aromatic amino acids and are thought to participate to pore formation (Abdo *et al.*, 2000), and the related PARCEL motifs also found in other non-MICE sequences (Roske *et al.*, 2010).

All Tra islands contain a copy of IS3, which is located at one end of the element, as well as several homologues of conserved MICE genes, in particular *traE* (MSB_A0052 in *M. leachii* and MCAP_0301 and MCAP_0179 in *M. capricolum* subspecies *capricolum*). Although they encode TraE homologues, the Tra islands are clearly different from the MICEs. Nonetheless, they are proposed to represent integrative units acquired by HGT, like the MICEs. In *M. mycoides* subspecies *mycoides* SC, a syntenic region of strains PG1 (MSC_0185–0211) and Gladysdale (MMS_A0212–0219) is reminiscent of a Tra island, although it is not annotated as such in these genomes and shows some divergence from the *M. leachii* and *M. capricolum* elements. An IS element from the IS3 family, IS*1296* in PG1 (MSC_0211) and IS*150* in Gladysdale (MMS_A0239) is located at one end of the region. The region also harbours a copy of *traE* (MSC_0191 in PG1 and MMS_A0218 in Gladysdale), and the phage-associated gene *ps3* (MSC_0194 and MMS_A0221) also found in certain MICEs. The comparison of Tra islands and MICEs suggests a complex history of HGT within the mycoides cluster, most probably involving phages, plasmids and conjugative transposons. Regardless of their origin, the MICEs and Tra islands clearly contribute to the mosaicism of mycoplasma genomes by integrating at different sites of the host chromosome, inducing numerous genome rearrangements, inversions and recombination, and carrying gene paralogues (Calcutt *et al.*, 2002; Nouvel *et al.*, 2010).

## Phytoplasma potential mobile units

Phytoplasma genomes contain 'potential mobile units' (PMUs) similar to composite replicative transposons that could be a driver of genomic instability and rearrangements (Bai *et al.*, 2006). Many PMUs are considered to be genomic islands and they encode a putative transposase, Tra5, which appears to participate in gene duplications, recombination and transposition events (Arashida *et al.*, 2008). In *Phytoplasma australiense*, 143 ORFs (12.1% of the genome) covering 106,682 bp of the chromosome are thought to form five PMUs based on genetic organization and sequence homologies (Tran-Nguyen *et al.*, 2008). Two of these PMUs are present as single copies, while there are two, three or five copies of each of the others. A PMU characterized in the genome of Aster Yellows phytoplasma strain Witches' Broom (AY-WB) is 20 kb in size and contains 21 genes encoding some predicted DNA replication and membrane-targeted proteins (Toruno *et al.*, 2010). It is present as a linear copy integrated into the chromosome, and as an extrachromosomal circular copy. This behaviour is remarkably similar to the excision/integration of MICEs. The role of PMUs in virulence is not established, but a few observations suggest that these elements are involved in some aspects of phytopathogenicity. During the transmission cycle of the phytoplasma, the circular copy appears more abundant and there is a higher level of gene expression from these regions in the insect vector than in the plant host (Toruno *et al.*, 2010). The PMUs encode a number of putative membrane proteins or secreted proteins that are proposed virulence factors. One of them, SAP11, contains a nuclear localization signal and is transferred into the nucleus of the infected plant cells (Bai *et al.*, 2009). SAP11 appears to facilitate insect transmission of the phytoplasma by modifying defence pathways in the infected plant (Sugio *et al.*, 2011).

## The impact of horizontal gene transfer on Mycoplasma genomics

### Experimental evidence

Horizontal gene transfer has rarely been observed under laboratory conditions with mycoplasmas. Natural competence has not been demonstrated, although artificial transformation has been achieved with a range of suicide plasmids or replicating plasmids of natural or artificial origin for many species of Mollicutes, including *A. laidlawii* (Dybvig and Cassell, 1987), *M. pulmonis* (Dybvig and Alderete, 1988; Cordova *et al.*, 2002), *M. mycoides* subspecies *mycoides* (King and Dybvig, 1991, 1994), *S. citri* (Renaudin *et al.*, 1995; Breton *et al.*, 2008), *M. capricolum* subspecies *capricolum* (Janis *et al.*, 2005), *M. agalactiae* (Chopra-Dewasthaly *et al.*, 2005a,b), *M. gallisepticum* (Cao *et al.*, 1994) and *M. imitans* (Lee *et al.*, 2008). However, transformation of mycoplasmas with foreign DNA is often difficult to achieve. This has been attributed to the presence of restriction endonucleases in the recipient cells (Voelker and Dybvig, 1996; Lartigue *et al.*, 2009), which appear to be widespread in mycoplasmas and may interfere with HGT. It is worth noting that many mycoplasma MGEs encode restriction-modification systems.

Indirect evidence suggests that HGT, more particularly by conjugation mechanisms, is more frequent and important than previously thought. Few studies have explored conjugation in Mollicutes. The exchange and recombination of chromosomal genetic markers between mycoplasma cells was observed in *M. pulmonis* (Teachman *et al.*, 2002), but whether these transfers occur by conjugation or cell fusion remains unclear. Conjugative transfer of the ICE Tn916 from *E. coli* to *M. pulmonis* (Dybvig and Cassell, 1987) and *E. faecalis* to *M. hominis* (Roberts and Kenny, 1987), *M. arthritidis* (Voelker and Dybvig, 1996) and *M. gallisepticum* (Ruffin *et al.*, 2000), has been described. The significance of these findings for the evolution of Mollicutes is difficult to appreciate, because the donors and recipients occupy different niches and might never, or very rarely, come in contact within the host. Nevertheless, these studies demonstrate the ability of mycoplasmas to acquire genes by conjugative transfer.

### Genomics of mycoplasmas suggests massive horizontal gene transfer

The statistical analysis of %GC and codon usage in bacterial genomes suggests that mycoplasmas have widely variable proportions of horizontally acquired genes, ranging from 14.47% in *M. genitalium* to 5.93% in *M. pneumoniae* (Garcia-Vallve *et al.*, 2000). Moreover, genomic analyses based on hidden Markov models reveal the presence of PARCEL motifs in mycoplasma proteins that most likely originate from HGT (Roske *et al.*, 2010). The precise identification of the nature and position of sequences acquired by HGT in mycoplasma genomes is sometimes difficult to establish, and must combine results from homology searches, BDBH analysis, gene order conservation analysis and comparisons of the topology of phylogenetic trees. However, several cases of HGT have been tentatively identified, on the basis of comparative genomics, between species of Mollicutes that belong to different phylogenetic groups but share the same hosts (Sirand-Pugnet *et al.*, 2007a).

Early analyses of spiroplasma and phytoplasma draft genomes allowed the discovery of eight homologous coding sequences that were present in both of these organisms, but were either absent from any of the six other fully sequenced *Mycoplasma* and *Ureaplasma* genomes, or more similar to each other than to any mycoplasma or ureaplasma coding sequence. This suggests the limited occurrence of HGT between spiroplasmas and phytoplasmas, which belong to two distinct phylogenetic groups but both infect plants and are insect-transmitted (Bai *et al.*, 2004b). However, as only *M. genitalium*, *M. pneumoniae*, *U. urealyticum*, *M. pulmonis*, *M. penetrans* and *M. gallisepticum* genomes were available at that time, the database of proteins used in this study did not contain any representative of the mycoides group, which is closely related to spiroplasmas, nor any *Acholeplasma* species, which are related to the phytoplasmas. Therefore, it would be interesting to revisit these results by including these genomes as controls. Nevertheless, the approach used in this study showed that large scale genomics can be used to detect HGT between Mollicutes using a relatively simple method.

The comparative genomics of *M. gallisepticum* (*pneumoniae* group) and *M. synoviae* (*hominis* group) suggests that 14 regions were horizontally transferred between these organisms, representing at least 2.58% and 3.21% of their respective genomes (Vasconcelos *et al.*, 2005). The acquired sequences encode mostly hypothetical proteins of unknown function, as well as a few sequences encoding potential virulence factors, such as haemagglutinins and a sialidase. Apart from a few IS elements that have also been exchanged, no conjugative systems, plasmids or phages appear to be present in these genomes. Although the extent of HGT between these two organisms appears relatively modest, it is interesting to note that they belong to two different phylogenetic groups but infect the same hosts.

More than 10% of the *M. hominis* coding sequences have their most similar counterpart in a member of another phylogenetic group, namely the pneumoniae, spiroplasma or phytoplasma groups, rather than in the hominis group, making these genes likely HGT candidates (Pereyre *et al.*, 2009). It is difficult to determine the origins for all of them, but at least 24 genes (from the 537 predicted CDSs) clearly show characteristics compatible with HGT, mostly with the human pathogen *Ureaplasma* species. These include numerous genes of unknown function, as well as DNA restriction and modification systems and IS elements, but also putative surface proteins that may play a role in the host–pathogen relationship. Moreover, some key metabolic enzymes of *M. hominis* appear to have been acquired from other bacteria (Pereyre *et al.*, 2009). Although no plasmids or MICEs have been described in *M. hominis*, a phage has been reported (Jansson *et al.*, 1982) and could explain the acquisition of foreign genetic material in this species.

An even more dramatic HGT has occurred between the mycoides cluster (spiroplasma group) and *M. agalactiae* and *M. bovis* (hominis group). More than 130 genes of *M. agalactiae* PG2, representing 18% of the genome, are thought to have been acquired from a *M. mycoides*-related organism. These genes display their best (and often only) similarity with orthologues from the mycoides cluster, while the rest of the *M. agalactiae* genes show higher sequence similarities with orthologues from the same phylogenetic group, such as *M. pulmonis*. Moreover, many regions acquired in *M. agalactiae* by HGT are organized in clusters that have the same gene order as in the mycoides cluster (Sirand-Pugnet *et al.*, 2007b), and some of these genes contain repeats of a domain of unknown function (DUF285) that is included in a PARCEL (Roske *et al.*, 2010). A high level of HGT from the mycoides cluster to the *M. agalactiae*/*M. bovis* taxon has been further confirmed by the analysis of another *M. agalactiae* strain (Nouvel *et al.*, 2010) and the *M. bovis* strains Hubei-1 (Li *et al.*, 2011) and PG45 (Wise *et al.*, 2010), as well as *M. mycoides* subspecies *capri* (Thiaucourt *et al.*, 2011). Again, it is notable that, despite their distant phylogenetic relationship, *M. agalactiae*, *M. bovis* and the members of the mycoides cluster are all major pathogens of ruminants.

Although most HGT appears to occur between species of Mollicutes, rare acquisition of genes from non-mycoplasma species is also suspected in several species, including *M. mobile* (Jaffe *et al.*, 2004), *M. arthritidis* (Dybvig *et al.*, 2008) and *M. hominis* (Pereyre *et al.*, 2009), because of the presence of genes with atypical phylogenies and with a high sequence similarity to Gram-positive or proteobacterial genes that cannot be explained by sequence divergence during evolution. Comparative genomics also suggest that, while the *ruvA* gene of *Mycoplasma* species and *Ureaplasma* species is, unsurprisingly, related to a Gram-positive ancestor, the *ruvB* gene, which belongs to the same operon, it was acquired from a Gram-negative organism after the divergence of the mycoplasma branch from other Gram-positive bacteria (Omelchenko *et al.*, 2003).

## Conclusions

Genomic studies of Mollicutes indicate that, despite their highly specialized host range and parasitic lifestyle, these organisms are prone to significant lateral gene transfer and have a large potential for rearrangement and mosaicism in their small genomes. This is in contrast to the generally accepted hypothesis that highly evolved parasites tend to have stable genomes with little scope for acquisition of new genetic material. Although some the mobilome of mycoplasmas has been known for a long time, its importance as a driver of the evolution of Mollicutes has long been underestimated. This is probably because the conventional MGEs, such as phages and conjugative plasmids, appeared to be relatively rare and confined to specific phylogenetic groups. Moreover, the acquisition of genetic material by mycoplasmas is rarely observed in the laboratory, possibly because of the activity of restriction-modification systems. Within the class Mollicutes, the spiroplasma/mycoides cluster stands out as having the most pervasive and diverse mobilome. The discovery of original, Mollicute-specific MGEs, such as MICEs, Tra islands and PMUs, and the presence of more widely distributed molecular markers of HGT such as PARCELs and DUF285 motifs, has shed a new light on the evolution of this class of pathogens. As the boundaries and relationships between phages, conjugative plasmids and ICEs of Mollicutes is better understood, it will be interesting to dissect the structural, molecular and genetic mechanisms that allow these organisms to exchange genetic material and to shuffle their genomes into complex mosaics. However, the most important question to address will probably be the elucidation of the novel functions encoded by the mycoplasma mobilome that control the physiology and virulence of these organisms and may explain their successful adaptation to a remarkably wide variety of hosts.

## References

Abdo, E.M., Nicolet, J., and Frey, J. (2000). Antigenic and genetic characterization of lipoprotein LppQ from *Mycoplasma mycoides* subsp. mycoides SC. Clin. Diagn. Lab. Immunol. *7*, 588–595.

Alvarez-Martinez, C.E., and Christie, P.J. (2009). Biological diversity of prokaryotic type IV secretion systems. Microbiol. Mol. Biol. Rev. *73*, 775–808.

Amikam, D., Glaser, G., and Razin, S. (1984). Mycoplasmas (Mollicutes) have a low number of rRNA genes. J. Bacteriol. *158*, 376–378.

Arashida, R., Kakizawa, S., Hoshi, A., Ishii, Y., Jung, H.Y., Kagiwada, S., Yamaji, Y., Oshima, K., and Namba, S. (2008). Heterogeneic dynamics of the structures of multiple gene clusters in two pathogenetically different lines originating from the same phytoplasma. DNA Cell Biol. *27*, 209–217.

Bai, X., Fazzolari, T., and Hogenhout, S.A. (2004a). Identification and characterization of traE genes of *Spiroplasma kunkelii*. Gene *336*, 81–91.

Bai, X., Zhang, J., Holford, I.R., and Hogenhout, S.A. (2004b). Comparative genomics identifies genes shared by distantly related insect-transmitted plant pathogenic Mollicutes. FEMS Microbiol. Lett. *235*, 249–258.

Bai, X., Zhang, J., Ewing, A., Miller, S.A., Jancso Radek, A., Shevchenko, D.V., Tsukerman, K., Walunas, T., Lapidus, A., Campbell, J.W., *et al.* (2006). Living with genome instability: the adaptation of phytoplasmas to diverse environments of their insect and plant hosts. J. Bacteriol. *188*, 3682–3696.

Bai, X., Correa, V.R., Toruno, T.Y., Ammar el, D., Kamoun, S., and Hogenhout, S.A. (2009). AY-WB phytoplasma secretes a protein that targets plant cell nuclei. Mol. Plant Microbe Interact. *22*, 18–30.

Barker, E.N., Helps, C.R., Peters, I.R., Darby, A.C., Radford, A.D., and Tasker, S. (2011). Complete genome sequence of Mycoplasma haemofelis, a hemotropic mycoplasma. J. Bacteriol. *193*, 2060–2061.

Barre, A., de Daruvar, A., and Blanchard, A. (2004). MolliGen, a database dedicated to the comparative genomics of Mollicutes. Nucleic Acids Res. *32*, D307–310.

Bébéar, C.-M., Aullo, P., Bové, J.-M., and Renaudin, J. (1996). *Spiroplasma citri* Virus SpV1: characterization of viral sequences present in the spiroplasmal host chromosome. Curr. Microbiol. *32*, 134–140.

Bergemann, A.D., and Finch, L.R. (1988). Isolation and restriction endonuclease analysis of a mycoplasma plasmid. Plasmid *19*, 68–70.

Bergemann, A.D., Whitley, J.C., and Finch, L.R. (1989). Homology of mycoplasma plasmid pADB201 and staphylococcal plasmid pE194. J. Bacteriol. *171*, 593–595.

Berho, N., Duret, S., Danet, J.L., and Renaudin, J. (2006a). Plasmid pSci6 from *Spiroplasma citri* GII-3 confers insect transmissibility to the non-transmissible strain S. citri 44. Microbiology *152*, 2703–2716.

Berho, N., Duret, S., and Renaudin, J. (2006b). Absence of plasmids encoding adhesion-related proteins in non-insect-transmissible strains of *Spiroplasma citri*. Microbiology *152*, 873–886.

Bhugra, B., and Dybvig, K. (1993). Identification and characterization of IS1138, a transposable element from *Mycoplasma pulmonis* that belongs to the IS3 family. Mol. Microbiol. *7*, 577–584.

Bischof, D.F., Vilei, E.M., and Frey, J. (2006). Genomic differences between type strain PG1 and field strains of *Mycoplasma mycoides* subsp. mycoides small-colony type. Genomics *88*, 633–641.

Blank, W.A., and Stemke, G.W. (2001). A physical and genetic map of the Mycoplasma flocculare ATCC 27716 chromosome reveals large genomic inversions when compared with that of *Mycoplasma hyopneumoniae* strain J(T). Int. J. Syst. Evol. Microbiol. *51*, 1395–1399.

Bork, P., Ouzounis, C., Casari, G., Schneider, R., Sander, C., Dolan, M., Gilbert, W., and Gillevet, P.M. (1995). Exploring the Mycoplasma capricolum genome: a minimal cell reveals its physiology. Mol. Microbiol. *16*, 955–967.

Breton, M., Duret, S., Arricau-Bouvery, N., Beven, L., and Renaudin, J. (2008). Characterizing the replication and stability regions of *Spiroplasma citri* plasmids identifies a novel replication protein and expands the genetic toolbox for plant-pathogenic spiroplasmas. Microbiology *154*, 3232–3244.

Brown, D.R., Farmerie, W.G., May, M., Benders, G.A., Durkin, A.S., Hlavinka, K., Hostetler, J., Jackson, J., Johnson, J., Miller, R.H., *et al.* (2011). Genome sequences of Mycoplasma alligatoris A21JP2T and *Mycoplasma crocodyli* MP145T. J. Bacteriol. *193*, 2892–2893.

Brunner, H., Schaar, H., and Krauss, H. (1980). Giant cell formation in *Acholeplasma laidlawii*, strain JA1. Zentralbl. Bakteriol. A *248*, 120–128.

Burrus, V., Pavlovic, G., Decaris, B., and Guedon, G. (2002). Conjugative transposons: the tip of the iceberg. Mol. Microbiol. *46*, 601–610.

Calcutt, M.J., Lavrrar, J.L., and Wise, K.S. (1999). IS1630 of *Mycoplasma fermentans*, a novel IS30-type insertion element that targets and duplicates inverted repeats of variable length and sequence during insertion. J. Bacteriol. *181*, 7597–7607.

Calcutt, M.J., Lewis, M.S., and Wise, K.S. (2002). Molecular genetic analysis of ICEF, an integrative conjugal element that is present as a repetitive sequence in the chromosome of *Mycoplasma fermentans* PG18. J. Bacteriol. *184*, 6929–6941.

Calderon-Copete, S.P., Wigger, G., Wunderlin, C., Schmidheini, T., Frey, J., Quail, M.A., and Falquet, L. (2009). The Mycoplasma conjunctivae genome sequencing, annotation and analysis. BMC Bioinformatics *10* (Suppl. 6), S7.

Cao, J., Kapke, P.A., and Minion, F.C. (1994). Transformation of *Mycoplasma gallisepticum* with Tn916, Tn4001, and integrative plasmid vectors. J. Bacteriol. *176*, 4459–4462.

Carle, P., Saillard, C., Carrere, N., Carrere, S., Duret, S., Eveillard, S., Gaurivaud, P., Gourgues, G., Gouzy, J., Salar, P., *et al.* (2010). Partial chromosome sequence of *Spiroplasma citri* reveals extensive viral invasion and important gene decay. Appl. Environ. Microbiol. *76*, 3420–3426.

Celamkoti, S., Kundeti, S., Purkayastha, A., Mazumder, R., Buck, C., and Seto, D. (2004). GeneOrder3.0: software for comparing the order of genes in pairs of small bacterial genomes. BMC Bioinformatics *5*, 52.

Chambaud, I., Heilig, R., Ferris, S., Barbe, V., Samson, D., Galisson, F., Moszer, I., Dybvig, K., Wróblewski, H., Viari, A., *et al.* (2001). The complete genome sequence of the murine respiratory pathogen *Mycoplasma pulmonis*. Nucleic Acids Res. *29*, 2145–2153.

Chopra-Dewasthaly, R., Marenda, M., Rosengarten, R., Jechlinger, W., and Citti, C. (2005a). Construction of the first shuttle vectors for gene cloning and homologous recombination in *Mycoplasma agalactiae*. FEMS Microbiol. Lett. *253*, 89–94.

Chopra-Dewasthaly, R., Zimmermann, M., Rosengarten, R., and Citti, C. (2005b). First steps towards the genetic manipulation of *Mycoplasma agalactiae* and *Mycoplasma bovis* using the transposon Tn4001mod. Int. J. Med. Microbiol. *294*, 447–453.

Clapper, B., Tu, A.H., Elgavish, A., and Dybvig, K. (2004). The vir gene of bacteriophage MAV1 confers resistance to phage infection on *Mycoplasma arthritidis*. J. Bacteriol. *186*, 5715–5720.

Clewell, D.B., Flannagan, S.E., and Jaworski, D.D. (1995). Unconstrained bacterial promiscuity: the Tn916-Tn1545 family of conjugative transposons. Trends Microbiol. *3*, 229–236.

Cole, R.M., Tully, J.G., Popkin, T.J., and Bove, J.M. (1973). Morphology, ultrastructure, and bacteriophage infection of the helical mycoplasma-like organism (*Spiroplasma citri* gen. nov., sp. nov.) cultured from 'stubborn' disease of citrus. J. Bacteriol. *115*, 367–384.

Cordova, C.M., Lartigue, C., Sirand-Pugnet, P., Renaudin, J., Cunha, R.A., and Blanchard, A. (2002). Identification of the origin of replication of the *Mycoplasma pulmonis* chromosome and its use in oriC replicative plasmids. J. Bacteriol. *184*, 5426–5435.

Davis, R.E., Jomantiene, R., Zhao, Y., and Dally, E.L. (2003). Folate biosynthesis pseudogenes, PsifolP and PsifolK, and an O-sialoglycoprotein endopeptidase gene homolog in the phytoplasma genome. DNA Cell Biol. *22*, 697–706.

Davis, R.E., Dally, E.L., Jomantiene, R., Zhao, Y., Roe, B., Lin, S., and Shao, J. (2005a). Cryptic plasmid pSKU146 from the wall-less plant pathogen Spiroplasma kunkelii encodes an adhesin and components of a type IV translocation-related conjugation system. Plasmid *53*, 179–190.

Davis, R.E., Jomantiene, R., and Zhao, Y. (2005b). Lineage-specific decay of folate biosynthesis genes suggests ongoing host adaptation in phytoplasmas. DNA Cell Biol. *24*, 832–840.

Delcher, A.L., Kasif, S., Fleischmann, R.D., Peterson, J., White, O., and Salzberg, S.L. (1999). Alignment of whole genomes. Nucleic Acids Res. *27*, 2369–2376.

Ditty, S.E., Connolly, M.A., Li, B.J., and Lo, S.C. (1999). *Mycoplasma orale* has a sequence similar to the insertion-like sequence of M. fermentans. Mol. Cell. Probes *13*, 183–189.

Ditty, S.E., Li, B., Zhang, S., Zou, N., and Lo, S.C. (2003). Characterization of an IS element in *Mycoplasma orale* that is highly homologous to M. fermentans ISLE. Curr. Microbiol. *46*, 302–306.

Djordjevic, S.R., Forbes, W.A., Forbes-Faulkner, J., Kuhnert, P., Hum, S., Hornitzky, M.A., Vilei, E.M., and Frey, J. (2001). Genetic diversity among Mycoplasma species bovine group 7: clonal isolates from an outbreak of polyarthritis, mastitis, and abortion in dairy cattle. Electrophoresis *22*, 3551–3561.

Dybvig, K., and Alderete, J. (1988). Transformation of *Mycoplasma pulmonis* and *Mycoplasma hyorhinis*: transposition of Tn916 and formation of cointegrate structures. Plasmid *20*, 33–41.

Dybvig, K., and Cassell, G.H. (1987). Transposition of Gram-positive transposon Tn916 in *Acholeplasma laidlawii* and *Mycoplasma pulmonis*. Science *235*, 1392–1394.

Dybvig, K., and Khaled, M. (1990). Isolation of a second cryptic plasmid from *Mycoplasma mycoides* subsp. *mycoides*. Plasmid *24*, 153–155.

Dybvig, K., and Maniloff, J. (1983). Integration and lysogeny by an enveloped mycoplasma virus. J. Gen. Virol. *64 (Pt 8)*, 1781–1785.

Dybvig, K., Nowak, J.A., Sladek, T.L., and Maniloff, J. (1985). Identification of an enveloped phage, mycoplasma virus L172, that contains a 14-kilobase single-stranded DNA genome. J. Virol. *53*, 384–390.

Dybvig, K., Liss, A., Alderete, J., Cole, R.M., and Cassell, G.H. (1987). Isolation of a virus from *Mycoplasma pulmonis*. Isr. J. Med. Sci. *23*, 418–422.

Dybvig, K., Alderete, J., Watson, H.L., and Cassell, G.H. (1988). Adsorption of mycoplasma virus P1 to host cells. J. Bacteriol. *170*, 4373–4375.

Dybvig, K., Cao, Z., French, C.T., and Yu, H. (2007). Evidence for type III restriction and modification systems in *Mycoplasma pulmonis*. J. Bacteriol. *189*, 2197–2202.

Dybvig, K., Zuhua, C., Lao, P., Jordan, D.S., French, C.T., Tu, A.H., and Loraine, A.E. (2008). Genome of *Mycoplasma arthritidis*. Infect. Immun. *76*, 4000–4008.

Fadiel, A., Eichenbaum, K.D., El Semary, N., and Epperson, B. (2007). Mycoplasma genomics: tailoring the genome for minimal life requirements through reductive evolution. Front. Biosci. *12*, 2020–2028.

Flannagan, S.E., Zitzow, L.A., Su, Y.A., and Clewell, D.B. (1994). Nucleotide sequence of the 18-kb conjugative transposon Tn916 from Enterococcus faecalis. Plasmid *32*, 350–354.

Fraser, C.M., Gocayne, J.D., White, O., Adams, M.D., Clayton, R.A., Fleischmann, R.D., Bult, C.J., Kerlavage, A.R., Sutton, G., Kelley, J.M., *et al.* (1995). The minimal gene complement of *Mycoplasma genitalium*. Science *270*, 397–403.

Frey, J., Cheng, X., Kuhnert, P., and Nicolet, J. (1995). Identification and characterization of IS1296 in *Mycoplasma mycoides* subsp. mycoides SC and presence in related mycoplasmas. Gene *160*, 95–100.

Garcia-Vallve, S., Romeu, A., and Palau, J. (2000). Horizontal gene transfer in bacterial and archaeal complete genomes. Genome Res. *10*, 1719–1725.

Garneau, J.E., Dupuis, M.E., Villion, M., Romero, D.A., Barrangou, R., Boyaval, P., Fremaux, C., Horvath, P., Magadan, A.H., and Moineau, S. (2010). The CRISPR/Cas bacterial immune system cleaves bacteriophage and plasmid DNA. Nature *468*, 67–71.

Gibbs, M.J., Smeianov, V.V., Steele, J.L., Upcroft, P., and Efimov, B.A. (2006). Two families of rep-like genes that probably originated by interspecies recombination are represented in viral, plasmid, bacterial, and parasitic protozoan genomes. Mol. Biol. Evol. *23*, 1097–1100.

Gibson, D.G., Benders, G.A., Andrews-Pfannkoch, C., Denisova, E.A., Baden-Tillson, H., Zaveri, J., Stockwell, T.B., Brownley, A., Thomas, D.W., Algire, M.A., *et al.* (2008). Complete chemical synthesis, assembly, and cloning of a *Mycoplasma genitalium* genome. Science *319*, 1215–1220.

Gibson, D.G., Glass, J.I., Lartigue, C., Noskov, V.N., Chuang, R.Y., Algire, M.A., Benders, G.A., Montague, M.G., Ma, L., Moodie, M.M., *et al.* (2010). Creation of a bacterial cell controlled by a chemically synthesized genome. Science *329*, 52–56.

Glass, J.I., Assad-Garcia, N., Alperovich, N., Yooseph, S., Lewis, M.R., Maruf, M., Hutchison, C.A., 3rd, Smith, H.O., and Venter, J.C. (2006). Essential genes of a minimal bacterium. Proc. Natl. Acad. Sci. U.S.A. *103*, 425–430.

Glass, J.I., Lefkowitz, E.J., Glass, J.S., Heiner, C.R., Chen, E.Y., and Cassell, G.H. (2000). The complete sequence of the mucosal pathogen *Ureaplasma urealyticum*. Nature *407*, 757–762.

Gourlay, R.N. (1971). Mycoplasmatales virus-laidlawii 2, a new virus isolated from *Acholeplasma laidlawii*. J. Gen. Virol. *12*, 65–67.

Gourlay, R.N., and Wyld, S.G. (1973). Isolation of mycoplasmatales virus-laidlawii 3, a new virus infecting *Acholeplasma laidlawii*. J. Gen. Virol. *19*, 279–283.

Gourlay, R.N., Bruce, J., and Garwes, D.J. (1971). Characterization of Mycoplasmatales virus laidlawii 1. Nat. New Biol. *229*, 118–119.

Gourlay, R.N., Wyld, S.G., and Garwes, D.J. (1983a). Some properties of mycoplasma virus Br 1. Arch. Virol. *75*, 1–15.

Gourlay, R.N., Wyld, S.G., and Poulton, M.E. (1983b). Some characteristics of mycoplasma virus Hr 1, isolated from and infecting *Mycoplasma hyorhinis*. Brief report. Arch. Virol. *77*, 81–85.

Grissa, I., Vergnaud, G., and Pourcel, C. (2007). CRISPRFinder: a web tool to identify clustered regularly interspaced short palindromic repeats. Nucleic Acids Res. *35*, W52–57.

Guglielmini, J., Quintais, L., Garcillan-Barcia, M.P., de la Cruz, F., and Rocha, E.P. (2011). The repertoire of ICE in prokaryotes underscores the unity, diversity, and ubiquity of conjugation. PLoS Genet. *7*, e1002222.

Guimaraes, A.M., Santos, A.P., SanMiguel, P., Walter, T., Timenetsky, J., and Messick, J.B. (2011). Complete genome sequence of *Mycoplasma suis* and insights into its biology and adaption to an erythrocyte niche. PLoS One *6*, e19574.

Guo, X., and Mrazek, J. (2008). Long simple sequence repeats in host-adapted pathogens localize near genes encoding antigens, housekeeping genes, and pseudogenes. J. Mol. Evol. *67*, 497–509.

Himmelreich, R., Hilbert, H., Plagens, H., Pirkl, E., Li, B.C., and Herrmann, R. (1996). Complete sequence analysis of the genome of the bacterium *Mycoplasma pneumoniae*. Nucleic Acids Res. *24*, 4420–4449.

Himmelreich, R., Plagens, H., Hilbert, H., Reiner, B., and Herrmann, R. (1997). Comparative analysis of the genomes of the bacteria *Mycoplasma pneumoniae* and *Mycoplasma genitalium*. Nucleic Acids Res. *25*, 701–712.

Hochhut, B., and Waldor, M.K. (1999). Site-specific integration of the conjugal Vibrio cholerae SXT element into prfC. Mol. Microbiol. *32*, 99–110.

Horvath, P., and Barrangou, R. (2010). CRISPR/Cas, the immune system of bacteria and archaea. Science *327*, 167–170.

Hu, W.S., and Yang, C.C. (2001). IS1550 from *Mycoplasma fermentans* is transposable in Escherichia coli. FEMS Microbiol. Lett. *198*, 159–164.

Hu, W.S., Hayes, M.M., Wang, R.Y., Shih, J.W., and Lo, S.C. (1998). High-frequency DNA rearrangements in the chromosomes of clinically isolated *Mycoplasma fermentans*. Curr. Microbiol. *37*, 1–5.

Hu, W.S., Wang, R.Y., Liou, R.S., Shih, J.W., and Lo, S.C. (1990). Identification of an insertion-sequence-like genetic element in the newly recognized human pathogen Mycoplasma incognitus. Gene *93*, 67–72.

Hutchison, C.A., Peterson, S.N., Gill, S.R., Cline, R.T., White, O., Fraser, C.M., Smith, H.O., and Venter, J.C. (1999). Global transposon mutagenesis and a minimal Mycoplasma genome. Science *286*, 2165–2169.

Ishida, N., Irikura, D., Matsuda, K., Sato, S., Sone, T., Tanaka, M., and Asano, K. (2009). Molecular cloning and expression of a novel cholinephosphotransferase involved in glycoglycerophospholipid biosynthesis of *Mycoplasma fermentans*. Curr. Microbiol. *58*, 535–540.

Ishii, Y., Oshima, K., Kakizawa, S., Hoshi, A., Maejima, K., Kagiwada, S., Yamaji, Y., and Namba, S. (2009). Process of reductive evolution during 10 years in plasmids of a non-insect-transmissible phytoplasma. Gene *446*, 51–57.

Jaffe, J.D., Stange-Thomann, N., Smith, C., DeCaprio, D., Fisher, S., Butler, J., Calvo, S., Elkins, T., FitzGerald, M.G., Hafez, N., *et al.* (2004). The complete genome and proteome of *Mycoplasma mobile*. Genome Res. *14*, 1447–1461.

Janis, C., Lartigue, C., Frey, J., Wróblewski, H., Thiaucourt, F., Blanchard, A., and Sirand-Pugnet, P. (2005). Versatile use of oriC plasmids for functional genomics of Mycoplasma capricolum subsp. capricolum. Appl. Environ. Microbiol. *71*, 2888–2893.

Jansson, E., Backman, A., Hakkarainen, K., von Bonsdorff, C.H., Seniusova, B., and Miettinen, A. (1982). Isolation and preliminary characterisation of mycoplasmavirus 20-P. Med. Biol. *60*, 116–120.

Jomantiene, R., and Davis, R.E. (2006). Clusters of diverse genes existing as multiple, sequence-variable mosaics in a phytoplasma genome. FEMS Microbiol. Lett. *255*, 59–65.

Jomantiene, R., Zhao, Y., and Davis, R.E. (2007). Sequence-variable mosaics: composites of recurrent transposition characterizing the genomes of phylogenetically diverse phytoplasmas. DNA Cell Biol. *26*, 557–564.

Joshi, B.D., Berg, M., Rogers, J., Fletcher, J., and Melcher, U. (2005). Sequence comparisons of plasmids pBJS-O of *Spiroplasma citri* and pSKU146 of S. kunkelii: implications for plasmid evolution. BMC Genomics *6*, 175.

Khiari, A.B., Gueriri, I., Mohammed, R.B., and Mardassi, B.B. (2010). Characterization of a variant vlhA gene of Mycoplasma synoviae, strain WVU 1853, with a highly divergent haemagglutinin region. BMC Microbiol. *10*, 6.

King, K.W., and Dybvig, K. (1991). Plasmid transformation of *Mycoplasma mycoides* subspecies mycoides is promoted by high concentrations of polyethylene glycol. Plasmid *26*, 108–115.

King, K.W., and Dybvig, K. (1992). Nucleotide sequence of *Mycoplasma mycoides* subspecies *mycoides* plasmid pKMK1. Plasmid *28*, 86–91.

King, K.W., and Dybvig, K. (1994). Mycoplasmal cloning vectors derived from plasmid pKMK1. Plasmid *31*, 49–59.

Koonin, E.V. (2000). How many genes can make a cell: the minimal-gene-set concept. Ann. Rev. Genomics Hum. Genet. *1*, 99–116.

Koonin, E.V., and Wolf, Y.I. (2008). Genomics of bacteria and archaea: the emerging dynamic view of the prokaryotic world. Nucleic Acids Res. *36*, 6688–6719.

Krishnakumar, R., Assad-Garcia, N., Benders, G.A., Phan, Q., Montague, M.G., and Glass, J.I. (2010). Targeted chromosomal knockouts in *Mycoplasma pneumoniae*. Appl. Environ. Microbiol. *76*, 5297–5299.

Krupovic, M., Ravantti, J.J., and Bamford, D.H. (2009). Geminiviruses: a tale of a plasmid becoming a virus. BMC Evol. Biol. *9*, 112.

Kube, M., Schneider, B., Kuhl, H., Dandekar, T., Heitmann, K., Migdoll, A.M., Reinhardt, R., and Seemuller, E. (2008). The linear chromosome of the plant-pathogenic mycoplasma 'Candidatus Phytoplasma mali'. BMC Genomics *9*, 306.

Ladefoged, S.A., and Christiansen, G. (1992). Physical and genetic mapping of the genomes of five *Mycoplasma hominis* strains by pulsed-field gel electrophoresis. J. Bacteriol. *174*, 2199–2207.

Lartigue, C., Blanchard, A., Renaudin, J., Thiaucourt, F., and Sirand-Pugnet, P. (2003). Host specificity of Mollicutes oriC plasmids: functional analysis of replication origin. Nucleic Acids Res. *31*, 6610–6618.

Lartigue, C., Vashee, S., Algire, M.A., Chuang, R.Y., Benders, G.A., Ma, L., Noskov, V.N., Denisova, E.A., Gibson, D.G., Assad-Garcia, N., *et al.* (2009). Creating bacterial strains from genomes that have been cloned and engineered in yeast. Science *325*, 1693–1696.

Lee, I.M., Zhao, Y., and Bottner, K.D. (2005). Novel insertion sequence-like elements in phytoplasma strains of the aster yellows group are putative new members of the IS3 family. FEMS Microbiol. Lett. *242*, 353–360.

Lee, S.W., Browning, G.F., and Markham, P.F. (2008). Development of a replicable oriC plasmid for *Mycoplasma gallisepticum* and Mycoplasma imitans, and gene disruption through homologous recombination in *M. gallisepticum*. Microbiology *154*, 2571–2580.

Li, Y., Zheng, H., Liu, Y., Jiang, Y., Xin, J., Chen, W., and Song, Z. (2011). The complete genome sequence of *Mycoplasma bovis* strain Hubei-1. PLoS One *6*, e20999.

Liefting, L.W., Andersen, M.T., Lough, T.J., and Beever, R.E. (2006). Comparative analysis of the plasmids from two isolates of '*Candidatus Phytoplasma australiense*'. Plasmid *56*, 138–144.

Lin, Y., and Zhang, R.R. (2011). Putative essential and core-essential genes in *Mycoplasma* genomes. Sci. Rep. *1*, 53.

Liu, W., Fang, L., Li, S., Li, Q., Zhou, Z., Feng, Z., Luo, R., Shao, G., Wang, L., Chen, H., *et al.* (2010). Complete genome sequence of *Mycoplasma hyorhinis* strain HUB-1. J. Bacteriol. *192*, 5844–5845.

Liu, W., Feng, Z., Fang, L., Zhou, Z., Li, Q., Li, S., Luo, R., Wang, L., Chen, H., Shao, G., *et al.* (2011). Complete genome sequence of *Mycoplasma hyopneumoniae* strain 168. J. Bacteriol. *193*, 1016–1017.

Lysnyansky, I., Calcutt, M.J., Ben-Barak, I., Ron, Y., Levisohn, S., Methe, B.A., and Yogev, D. (2009). Molecular characterization of newly identified IS3, IS4 and IS30 insertion sequence-like elements in *Mycoplasma bovis* and their possible roles in genome plasticity. FEMS Microbiol. Lett. *294*, 172–182.

Mahillon, J., and Chandler, M. (1998). Insertion sequences. Microbiol. Mol. Biol. Rev. *62*, 725–774.

Maniloff, J., Kampo, G.J., and Dascher, C.C. (1994). Sequence analysis of a unique temperature phage: mycoplasma virus L2. Gene *141*, 1–8.

March, J.B., Clark, J., and Brodlie, M. (2000). Characterization of strains of *Mycoplasma mycoides* subsp. *mycoides* small colony type isolated from recent outbreaks of contagious bovine pleuropneumonia in Botswana and Tanzania: evidence for a new biotype. J. Clin. Microbiol. *38*, 1419–1425.

Marenda, M.S., Sagne, E., Poumarat, F., and Citti, C. (2005). Suppression subtractive hybridization as a basis to assess *Mycoplasma agalactiae* and *Mycoplasma bovis* genomic diversity and species-specific sequences. Microbiology *151*, 475–489.

Marenda, M., Barbe, V., Gourgues, G., Mangenot, S., Sagne, E., and Citti, C. (2006). A new integrative conjugative element occurs in *Mycoplasma agalactiae* as chromosomal and free circular forms. J. Bacteriol. *188*, 4137–4141.

Markov, A., and Zakharov, I. (2009). Evolution of gene orders in mycoplasmas (Bacteria, Firmicutes, Mollicutes). Russian J. Genet. *45*, 781–787.

Melcher, U., Sha, Y., Ye, F., and Fletcher, J. (1999). Mechanisms of spiroplasma genome variation associated with SpV1-like viral DNA inferred from sequence comparisons. Microb. Comp. Genomics *4*, 29–46.

Messick, J.B., Santos, A.P., and Guimaraes, A.M. (2011). Complete genome sequences of two hemotropic mycoplasmas, Mycoplasma haemofelis strain Ohio2 and *Mycoplasma suis* strain Illinois. J. Bacteriol. *193*, 2068–2069.

Miles, K., McAuliffe, L., Persson, A., Ayling, R.D., and Nicholas, R.A. (2005). Insertion sequence profiling of UK *Mycoplasma bovis* field isolates. Vet. Microbiol. *107*, 301–306.

Miles, K., Churchward, C.P., McAuliffe, L., Ayling, R.D., and Nicholas, R.A. (2006). Identification and differentiation of European and African/Australian strains of *Mycoplasma mycoides* subspecies mycoides small-colony type using polymerase chain reaction analysis. J. Vet. Diagn. Invest. *18*, 168–171.

Minion, F.C., Lefkowitz, E.J., Madsen, M.L., Cleary, B.J., Swartzell, S.M., and Mahairas, G.G. (2004). The genome sequence of *Mycoplasma hyopneumoniae* strain 232, the agent of swine mycoplasmosis. J. Bacteriol. *186*, 7123–7133.

Mira, A., Martin-Cuadrado, A.B., D'Auria, G., and Rodriguez-Valera, F. (2010). The bacterial pan-genome: a new paradigm in microbiology. Int. Microbiol. *13*, 45–57.

Mrazek, J. (2006). Analysis of distribution indicates diverse functions of simple sequence repeats in Mycoplasma genomes. Mol. Biol. Evol. *23*, 1370–1385.

Mrazek, J., Guo, X., and Shah, A. (2007). Simple sequence repeats in prokaryotic genomes. Proc. Natl. Acad. Sci. U.S.A. *104*, 8472–8477.

Musatovova, O., Kannan, T.R., and Baseman, J.B. (2008). Genomic analysis reveals *Mycoplasma pneumoniae* repetitive element 1-mediated recombination in a clinical isolate. Infect. Immun. *76*, 1639–1648.

Mushegian, A.R., and Koonin, E.V. (1996). A minimal gene set for cellular life derived by comparison of complete bacterial genomes. Proc. Natl. Acad. Sci. U.S.A. *93*, 10268–10273.

Nishigawa, H., Miyata, S., Oshima, K., Sawayanagi, T., Komoto, A., Kuboyama, T., Matsuda, I., Tsuchizaki, T., and Namba, S. (2001). *In planta* expression of a protein encoded by the extrachromosomal DNA of a phytoplasma and related to geminivirus replication proteins. Microbiology *147*, 507–513.

Nishigawa, H., Oshima, K., Kakizawa, S., Jung, H.Y., Kuboyama, T., Miyata, S., Ugaki, M., and Namba, S. (2002a). Evidence of intermolecular recombination between extrachromosomal DNAs in phytoplasma: a trigger for the biological diversity of phytoplasma? Microbiology *148*, 1389–1396.

Nishigawa, H., Oshima, K., Kakizawa, S., Jung, H.Y., Kuboyama, T., Miyata, S., Ugaki, M., and Namba, S. (2002b). A plasmid from a non-insect-transmissible line of a phytoplasma lacks two open reading frames that exist in the plasmid from the wild-type line. Gene *298*, 195–201.

Nishigawa, H., Oshima, K., Miyata, S.-i., Ugaki, M., and Namba, S. (2003). Complete set of extrachromosomal

DNAs from three pathogenic lines of onion yellows phytoplasma and use of PCR to differentiate each line. J. Gen. Plant Pathol. *69*, 194–198.

Noormohammadi, A.H., Markham, P.F., Kanci, A., Whithear, K.G., and Browning, G.F. (2000). A novel mechanism for control of antigenic variation in the haemagglutinin gene family of mycoplasma synoviae. Mol. Microbiol. *35*, 911–923.

Nouvel, L.X., Sirand-Pugnet, P., Marenda, M.S., Sagne, E., Barbe, V., Mangenot, S., Schenowitz, C., Jacob, D., Barre, A., Claverol, S., *et al.* (2010). Comparative genomic and proteomic analyses of two *Mycoplasma agalactiae* strains: clues to the macro- and micro-events that are shaping mycoplasma diversity. BMC Genomics *11*, 86.

Nowak, J.A., and Maniloff, J. (1979). Physical characterization of the superhelical DNA genome of an enveloped mycoplasmavirus. J. Virol. *29*, 374–380.

Ochman, H., and Davalos, L.M. (2006). The nature and dynamics of bacterial genomes. Science *311*, 1730–1733.

Oehlerking, J., Kube, M., Felder, K.M., Matter, D., Wittenbrink, M.M., Schwarzenbach, S., Kramer, M.M., Hoelzle, K., and Hoelzle, L.E. (2011). Complete genome sequence of the hemotrophic *Mycoplasma suis* strain KI3806. J. Bacteriol. *193*, 2369–2370.

Omelchenko, M.V., Makarova, K.S., Wolf, Y.I., Rogozin, I.B., and Koonin, E.V. (2003). Evolution of mosaic operons by horizontal gene transfer and gene displacement *in situ*. Genome Biol. *4*, R55.

Oshima, K., and Nishida, H. (2007). Phylogenetic relationships among mycoplasmas based on the whole genomic information. J. Mol. Evol. *65*, 249–258.

Oshima, K., Kakizawa, S., Nishigawa, H., Kuboyama, T., Miyata, S., Ugaki, M., and Namba, S. (2001). A plasmid of phytoplasma encodes a unique replication protein having both plasmid- and virus-like domains: clue to viral ancestry or result of virus/plasmid recombination? Virology *285*, 270–277.

Oshima, K., Kakizawa, S., Nishigawa, H., Jung, H.Y., Wei, W., Suzuki, S., Arashida, R., Nakata, D., Miyata, S., Ugaki, M., *et al.* (2004). Reductive evolution suggested from the complete genome sequence of a plant-pathogenic phytoplasma. Nat. Genet. *36*, 27–29.

Papazisi, L., Gorton, T.S., Kutish, G., Markham, P.F., Browning, G.F., Nguyen, D.K., Swartzell, S., Madan, A., Mahairas, G., and Geary, S.J. (2003). The complete genome sequence of the avian pathogen *Mycoplasma gallisepticum* strain R(low). Microbiology *149*, 2307–2316.

Pereyre, S., Sirand-Pugnet, P., Beven, L., Charron, A., Renaudin, H., Barre, A., Avenaud, P., Jacob, D., Couloux, A., Barbe, V., *et al.* (2009). Life on arginine for *Mycoplasma hominis*: clues from its minimal genome and comparison with other human urogenital mycoplasmas. PLoS Genet. *5*, e1000677.

Peterson, S.N., and Fraser, C.M. (2001). The complexity of simplicity. Genome Biol. *2*, COMMENT2002.

Peterson, S.N., Hu, P.C., Bott, K.F., and Hutchison, C.A., 3rd (1993). A survey of the *Mycoplasma genitalium* genome by using random sequencing. J. Bacteriol. *175*, 7918–7930.

Petrzik, K., Krawczyk, K., and Zwolinska, A. (2011). Two high-copy plasmids found in plants associated with strains of 'Candidatus Phytoplasma asteris'. Plasmid *66*, 122–127.

Pilo, P., Fleury, B., Marenda, M., Frey, J., and Vilei, E.M. (2003). Prevalence and distribution of the insertion element ISMag1 in *Mycoplasma agalactiae*. Vet. Microbiol. *92*, 37–48.

Pinto, P.M., de Carvalho, M.O., Alves-Junior, L., Brocchi, M., and Schrank, I.S. (2007). Molecular analysis of an Integrative Conjugative Element, ICEH, present in the chromosome of different strains of *Mycoplasma hyopneumoniae*. Genet. Mol. Biol. *30*, 256–263.

Pitcher, D., and Hilbocus, J. (1998). Variability in the distribution and composition of insertion sequence-like elements in strains of *Mycoplasma fermentans*. FEMS Microbiol. Lett. *160*, 101–109.

Pollack, J.D., Williams, M.V., and McElhaney, R.N. (1997). The comparative metabolism of the Mollicutes (Mycoplasmas): the utility for taxonomic classification and the relationship of putative gene annotation and phylogeny to enzymatic function in the smallest free-living cells. Crit. Rev. Microbiol. *23*, 269–354.

Rechnitzer, H., Brzuszkiewicz, E., Strittmatter, A., Liesegang, H., Lysnyansky, I., Daniel, R., Gottschalk, G., and Rottem, S. (2011). Genomic features and insights into the biology of *Mycoplasma fermentans*. Microbiology *157*, 760–773.

Renaudin, J., Pascarel, M.C., and Bove, J.M. (1987). Spiroplasma virus 4: nucleotide sequence of the viral DNA, regulatory signals, and proposed genome organization. J. Bacteriol. *169*, 4950–4961.

Renaudin, J., Aullo, P., Vignault, J.C., and Bove, J.M. (1990). Complete nucleotide sequence of the genome of *Spiroplasma citri* virus SpV1-R8A2 B. Nucleic Acids Res. *18*, 1293.

Renaudin, J., Marais, A., Verdin, E., Duret, S., Foissac, X., Laigret, F., and Bove, J.M. (1995). Integrative and free *Spiroplasma citri* oriC plasmids: expression of the Spiroplasma phoeniceum spiralin in *Spiroplasma citri*. J. Bacteriol. *177*, 2870–2877.

Roberts, A.P., and Mullany, P. (2011). Tn916-like genetic elements: a diverse group of modular mobile elements conferring antibiotic resistance. FEMS Microbiol. Rev. *35*, 856–871.

Roberts, M.C., and Kenny, G.E. (1987). Conjugal transfer of transposon Tn916 from *Streptococcus faecalis* to *Mycoplasma hominis*. J. Bacteriol. *169*, 3836–3839.

Rocha, E.P. (2004). Order and disorder in bacterial genomes. Curr. Opin. Microbiol. *7*, 519–527.

Rocha, E.P., and Blanchard, A. (2002). Genomic repeats, genome plasticity and the dynamics of Mycoplasma evolution. Nucleic Acids Res. *30*, 2031–2042.

Rogers, M.J., Simmons, J., Walker, R.T., Weisburg, W.G., Woese, C.R., Tanner, R.S., Robinson, I.M., Stahl, D.A., Olsen, G., Leach, R.H., *et al.* (1985). Construction of the mycoplasma evolutionary tree from 5S rRNA sequence data. Proc. Natl. Acad. Sci. U.S.A. *82*, 1160–1164.

Roske, K., Calcutt, M.J., and Wise, K.S. (2004). The *Mycoplasma fermentans* prophage phiMFV1: genome organization, mobility and variable expression of an encoded surface protein. Mol. Microbiol. *52*, 1703–1720.

Roske, K., Foecking, M.F., Yooseph, S., Glass, J.I., Calcutt, M.J., and Wise, K.S. (2010). A versatile palindromic amphipathic repeat coding sequence horizontally distributed among diverse bacterial and eucaryotic microbes. BMC Genomics *11*, 430.

Ruffin, D.C., van Santen, V.L., Zhang, Y., Voelker, L.L., Panangala, V.S., and Dybvig, K. (2000). Transposon mutagenesis of *Mycoplasma gallisepticum* by conjugation with enterococcus faecalis and determination of insertion site by direct genomic sequencing. Plasmid *44*, 191–195.

Saillard, C., Carle, P., Duret-Nurbel, S., Henri, R., Killiny, N., Carrere, S., Gouzy, J., Bove, J.M., Renaudin, J., and Foissac, X. (2008). The abundant extrachromosomal DNA content of the *Spiroplasma citri* GII3-3X genome. BMC Genomics *9*, 195.

Salyers, A.A., Shoemaker, N.B., Stevens, A.M., and Li, L.Y. (1995). Conjugative transposons: an unusual and diverse set of integrated gene transfer elements. Microbiol. Rev. *59*, 579–590.

Sasaki, Y., Ishikawa, J., Yamashita, A., Oshima, K., Kenri, T., Furuya, K., Yoshino, C., Horino, A., Shiba, T., Sasaki, T., *et al.* (2002). The complete genomic sequence of Mycoplasma penetrans, an intracellular bacterial pathogen in humans. Nucleic Acids Res. *30*, 5293–5300.

Sha, Y., Melcher, U., Davis, R.E., and Fletcher, J. (2000). Common elements of spiroplasma plectroviruses revealed by nucleotide sequence of SVTS2. Virus Genes *20*, 47–56.

Shu, H.W., Liu, T.T., Chan, H.I., Liu, Y.M., Wu, K.M., Shu, H.Y., Tsai, S.F., Hsiao, K.J., Hu, W.S., and Ng, W.V. (2011). Genome sequence of the repetitive-sequence-rich *Mycoplasma fermentans* strain M64. J. Bacteriol. *193*, 4302–4303.

Sirand-Pugnet, P., Citti, C., Barre, A., and Blanchard, A. (2007a). Evolution of Mollicutes: down a bumpy road with twists and turns. Res. Microbiol. *158*, 754–766.

Sirand-Pugnet, P., Lartigue, C., Marenda, M., Jacob, D., Barre, A., Barbe, V., Schenowitz, C., Mangenot, S., Couloux, A., Segurens, B., *et al.* (2007b). Being pathogenic, plastic, and sexual while living with a nearly minimal bacterial genome. PLoS Genet. *3*, e75.

Smillie, C., Garcillan-Barcia, M.P., Francia, M.V., Rocha, E.P., and de la Cruz, F. (2010). Mobility of plasmids. Microbiol. Mol. Biol. Rev. *74*, 434–452.

Souza, R.C., de Almeida, D.F., Zaha, A., Morais, D.A.D., and de Vasconcelos, A. (2007). In search of essentiality: Mollicute-specific genes shared by twelve genomes. Genet. Mol. Biol. *30*, 169–173.

Sugio, A., Kingdom, H.N., MacLean, A.M., Grieve, V.M., and Hogenhout, S.A. (2011). Phytoplasma protein effector SAP11 enhances insect vector reproduction by manipulating plant development and defense hormone biosynthesis. Proc. Natl. Acad. Sci. U.S.A. *108*, E1254–1263.

Suthers, P.F., Dasika, M.S., Kumar, V.S., Denisov, G., Glass, J.I., and Maranas, C.D. (2009). A genome-scale metabolic reconstruction of *Mycoplasma genitalium*, iPS189. PLoS Comput. Biol. *5*, e1000285.

Szczepanek, S.M., Tulman, E.R., Gorton, T.S., Liao, X., Lu, Z., Zinski, J., Aziz, F., Frasca, S., Jr., Kutish, G.F., and Geary, S.J. (2010). Comparative genomic analyses of attenuated strains of *Mycoplasma gallisepticum*. Infect. Immun. *78*, 1760–1771.

Teachman, A.M., French, C.T., Yu, H., Simmons, W.L., and Dybvig, K. (2002). Gene transfer in *Mycoplasma pulmonis*. J. Bacteriol. *184*, 947–951.

Teichmann, S.A., Park, J., and Chothia, C. (1998). Structural assignments to the *Mycoplasma genitalium* proteins show extensive gene duplications and domain rearrangements. Proc. Natl. Acad. Sci. U.S.A. *95*, 14658–14663.

Tettelin, H., Riley, D., Cattuto, C., and Medini, D. (2008). Comparative genomics: the bacterial pan-genome. Curr. Opin. Microbiol. *11*, 472–477.

Thiaucourt, F., Manso-Silvan, L., Salah, W., Barbe, V., Vacherie, B., Jacob, D., Breton, M., Dupuy, V., Lomenech, A.M., Blanchard, A., *et al.* (2011). *Mycoplasma mycoides*, from 'mycoides Small Colony' to 'capri'. A microevolutionary perspective. BMC Genomics *12*, 114.

Thomas, A., Linden, A., Mainil, J., Dizier, I., Baseman, J.B., Kannan, T.R., Fleury, B., Frey, J., and Vilei, E.M. (2004). The p40* adhesin pseudogene of *Mycoplasma bovis*. Vet. Microbiol. *104*, 213–217.

Thomas, A., Linden, A., Mainil, J., Bischof, D.F., Frey, J., and Vilei, E.M. (2005). *Mycoplasma bovis* shares insertion sequences with *Mycoplasma agalactiae* and *Mycoplasma mycoides* subsp. mycoides SC: evolutionary and developmental aspects. FEMS Microbiol. Lett. *245*, 249–255.

Toruno, T.Y., Music, M.S., Simi, S., Nicolaisen, M., and Hogenhout, S.A. (2010). Phytoplasma PMU1 exists as linear chromosomal and circular extrachromosomal elements and has enhanced expression in insect vectors compared with plant hosts. Mol. Microbiol. *77*, 1406–1415.

Tran-Nguyen, L.T., and Gibb, K.S. (2006). Extrachromosomal DNA isolated from tomato big bud and Candidatus Phytoplasma australiense phytoplasma strains. Plasmid *56*, 153–166.

Tran-Nguyen, L.T., Kube, M., Schneider, B., Reinhardt, R., and Gibb, K.S. (2008). Comparative genome analysis of 'Candidatus Phytoplasma australiense' (subgroup tuf-Australia I; rp-A) and 'Ca. Phytoplasma asteris' Strains OY-M and AY-WB. J. Bacteriol. *190*, 3979–3991.

Tu, A.H., Voelker, L.L., Shen, X., and Dybvig, K. (2001). Complete nucleotide sequence of the mycoplasma virus P1 genome. Plasmid *45*, 122–126.

Vasconcelos, A.T., Ferreira, H.B., Bizarro, C.V., Bonatto, S.L., Carvalho, M.O., Pinto, P.M., Almeida, D.F., Almeida, L.G., Almeida, R., Alves-Filho, L., *et al.* (2005). Swine and poultry pathogens: the complete genome sequences of two strains of *Mycoplasma hyopneumoniae* and a strain of *Mycoplasma synoviae*. J. Bacteriol. *187*, 5568–5577.

Vilei, E.M., Nicolet, J., and Frey, J. (1999). IS1634, a novel insertion element creating long, variable-length direct repeats which is specific for *Mycoplasma mycoides* subsp. mycoides small-colony type. J. Bacteriol. *181*, 1319–1323.

Voelker, L.L., and Dybvig, K. (1996). Gene transfer in *Mycoplasma arthritidis*: transformation, conjugal transfer of Tn916, and evidence for a restriction system recognizing AGCT. J. Bacteriol. *178*, 6078–6081.

Voelker, L.L., and Dybvig, K. (1998). Characterization of the lysogenic bacteriophage MAV1 from *Mycoplasma arthritidis*. J. Bacteriol. *180*, 5928–5931.

Voelker, L.L., and Dybvig, K. (1999). Sequence analysis of the *Mycoplasma arthritidis* bacteriophage MAV1 genome identifies the putative virulence factor. Gene 233, 101–107.

Voelker, L.L., Weaver, K.E., Ehle, L.J., and Washburn, L.R. (1995). Association of lysogenic bacteriophage MAV1 with virulence of *Mycoplasma arthritidis*. Infect. Immun. 63, 4016–4023.

Washburn, L.R., Miller, E.J., Mukherjee, S., and Dannenbring, D. (2004). *Mycoplasma arthritidis* bacteriophage MAV1 prophage integration, deletions, and strain-related polymorphisms. Plasmid 52, 31–47.

Wei, W., Davis, R.E., Jomantiene, R., and Zhao, Y. (2008). Ancient, recurrent phage attacks and recombination shaped dynamic sequence-variable mosaics at the root of phytoplasma genome evolution. Proc. Natl. Acad. Sci. U.S.A. *105*, 11827–11832.

Weisburg, W.G., Tully, J.G., Rose, D.L., Petzel, J.P., Oyaizu, H., Yang, D., Mandelco, L., Sechrest, J., Lawrence, T.G., Van Etten, J., *et al.* (1989). A phylogenetic analysis of the mycoplasmas: basis for their classification. J. Bacteriol. *171*, 6455–6467.

Westberg, J., Persson, A., Pettersson, B., Uhlen, M., and Johansson, K.E. (2002). ISMmy1, a novel insertion sequence of *Mycoplasma mycoides* subsp. mycoides small colony type. FEMS Microbiol. Lett. *208*, 207–213.

Westberg, J., Persson, A., Holmberg, A., Goesmann, A., Lundeberg, J., Johansson, K.E., Pettersson, B., and Uhlen, M. (2004). The genome sequence of *Mycoplasma mycoides* subsp. mycoides SC type strain PG1T, the causative agent of contagious bovine pleuropneumonia (CBPP). Genome Res. *14*, 221–227.

Wise, K.S., Foecking, M.F., Roske, K., Lee, Y.J., Lee, Y.M., Madan, A., and Calcutt, M.J. (2006). Distinctive repertoire of contingency genes conferring mutation-based phase variation and combinatorial expression of surface lipoproteins in Mycoplasma capricolum subsp. capricolum of the *Mycoplasma mycoides* phylogenetic cluster. J. Bacteriol. *188*, 4926–4941.

Wise, K.S., Calcutt, M.J., Foecking, M.F., Roske, K., Madupu, R., and Methe, B.A. (2010). Complete genome sequence of *Mycoplasma bovis* type strain PG45 (ATCC 25523). Infect. Immun. *79*, 982–983.

Woese, C.R., Maniloff, J., and Zablen, L.B. (1980). Phylogenetic analysis of the mycoplasmas. Proc. Natl. Acad. Sci. U.S.A. *77*, 494–498.

Wozniak, R.A., and Waldor, M.K. (2010). Integrative and conjugative elements: mosaic mobile genetic elements enabling dynamic lateral gene flow. Nat. Rev. Microbiol. *8*, 552–563.

Xu, F.F., Pearce, L.E., and Yu, P.L. (1991). Genetic analysis of a lactococcal plasmid replicon. Mol. Gen. Genet. *227*, 33–39.

Ye, F., Melcher, U., Rascoe, J.E., and Fletcher, J. (1996). Extensive chromosome aberrations in *Spiroplasma citri* Strain BR3. Biochem. Genet. *34*, 269–286.

Yus, E., Maier, T., Michalodimitrakis, K., van Noort, V., Yamada, T., Chen, W.H., Wodke, J.A., Guell, M., Martinez, S., Bourgeois, R., *et al.* (2009). Impact of genome reduction on bacterial metabolism and its regulation. Science *326*, 1263–1268.

Zhao, Y., Hammond, R.W., Jomantiene, R., Dally, E.L., Lee, I.M., Jia, H., Wu, H., Lin, S., Zhang, P., Kenton, S., *et al.* (2003). Gene content and organization of an 85-kb DNA segment from the genome of the phytopathogenic Mollicute Spiroplasma kunkelii. Mol. Genet. Genomics *269*, 592–602.

Zheng, J., and McIntosh, M.A. (1995). Characterization of IS1221 from *Mycoplasma hyorhinis*: expression of its putative transposase in Escherichia coli incorporates a ribosomal frameshift mechanism. Mol. Microbiol. *16*, 669–685.

Zou, N., Park, K., and Dybvig, K. (1995). Mycoplasma virus P1 has a linear, double-stranded DNA genome with inverted terminal repeats. Plasmid 33, 41–49.

# Molecular Genetic Tools for Mollicutes

3

Joël Renaudin, Marc Breton and Christine Citti


## Abstract

The class Mollicutes represents a group of wall-less bacteria with small genome size and thus possesses limited metabolic capabilities. As a result of series of genome reduction and gene acquisition through horizontal gene transfer, Mollicutes have evolved a parasitic life style with the ability to colonize a wide diversity of hosts, suggesting that these organisms may use a variety of virulence mechanisms. Along with the rapid accumulation of genome sequence data, significant advances have been made in the last decade in designing new genetic tools for Mollicutes. These include new selective markers, new inducible gene expression systems, and new replicative vectors for plant pathogenic spiroplasmas as well as improved Tn*4001* derivatives for random mutagenesis or gene delivery, extended *oriC* plasmid-based strategies for new mycoplasma species for gene expression or targeted gene disruption, the production of unmarked mutations and the use of fluorescent protein genes as reporters. Beyond the description of these new genetic tools, this review also highlights their input to the concomitant advances that have been made in understanding the intricate and complex Mollicute–host interactions.


## Introduction

The class Mollicutes is composed of more than 100 species that form a coherent phylogenetic group having arisen from low G + C Gram-positive bacteria through regressive evolution (Weisburg *et al.*, 1989). Many Mollicute species are known as significant pathogens of a wide range of hosts including man, animals and plants. With their small-size genome and their lack of a cell-wall, these bacteria represent an interesting and unique model for studying both reductive genome evolution (Sirand-Pugnet *et al.*, 2007) and the minimal gene set required for sustaining free-living cells (Glass *et al.*, 2006).

While advances in genome sequencing is generating a growing body of data for an increasing number of Mollicute species, the functional, detailed genetic analyses of these particular organisms are still lagging far behind those of most of the other bacteria studied. Gene vectors engineered in conventional bacteria failed to function in mycoplasmas and, beside the difficulties inherent to cultivating these fastidious organisms, the development of specific genetic tools has been complicated by a number of factors. These include the paucity of natural plasmids for most mycoplasma species, the lack of chemically defined growth media that prevented the use of auxotrophic markers, the scarcity of selective, antibiotic-resistance markers, and the use of UGA as a tryptophan codon instead of the universal, opal stop codon.

In spite of these limitations, significant progresses have been made following the first use of the staphylococcal transposon Tn*4001* that was shown to randomly transpose in mycoplasma genomes. Since then, a number of sophisticated Tn*4001* derivatives have been customized and are currently broadly used in several mycoplasma species for random mutagenesis, and to a lesser extend for promoter trapping and gene delivery.

Another significant step forward was accomplished when parts of the *Spiroplasma citri* chromosomal origin of replication, *oriC*, were combined to *colE1*-derived *E. coli* replicons and to selective markers, resulting in the first pioneer vectors of a long series. Expression of cloned genes had since been achieved in a growing number of Mollicute species using artificial plasmids with homologous or heterologous *oriC* regions. While *oriC*-based plasmids also offer very useful tools for improving mycoplasma transformation protocols, they have been valuable in targeted-gene disruption by homologous recombination. More recent developments have taken advantages of the sequencing of S. *citri* natural plasmids which were customized into efficient replicative vectors for plant pathogenic spiroplasmas.

Several reviews on molecular genetics of Mollicutes have been published (Dybvig and Voelker, 1996; Razin *et al.*, 1998; Renaudin, 2002; Renaudin and Lartigue, 2005; Halbedel and Stulke, 2007). This chapter intends to review the current content of the 'genetic tool box' for Mollicutes (Table 3.1) with the exception of the latest development in genome transplantation and manipulation in yeast (Lartigue *et al.*, 2007; Lartigue *et al.*, 2009; Benders *et al.*, 2010; Montague *et al.*, 2012).

## Transformation of Mollicutes

### Transformation methods

Successful transfer of exogenous DNA into Mollicutes was first reported when *A. laidlawii* was transfected with mycoplasma virus (L2) DNA in the presence of polyethylene glycol (PEG). PEG-mediated transfection with viral DNAs has also been reported in spiroplasmas however electroporation proved to be more efficient in promoting transfer of spiroplasma virus (SpV1) RF DNA into S. *citri* cells (Stamburski *et al.*, 1991; Gasparich *et al.*, 1993).

Currently, there is not one general method to introduce DNA into Mollicutes and both the PEG-method and electroporation are routinely used to transform Mollicutes. The method of choice depends on both the particular species (and strains) being studied and the DNA molecule to be introduced as previously reviewed (Dybvig and Voelker, 1996; Renaudin, 2002). In addition, the restriction and modification systems which are prevalent in Mollicutes may also represent barriers in gene transfer (King and Dybvig, 1994b; Voelker and Dybvig, 1996). In the recent report of *M. mycoides* subsp. *capri* genome transplantation from yeast to *M. capricolum* the restriction barrier has been circumvented through inactivation of the single restriction enzyme of the recipient cells and protection of the donor genome by *in vitro* methylation, using *M. capricolum* extracts (Lartigue *et al.*, 2009). Despite successful transformation of increasing number of Mollicutes species including *A. laidlawii*, the human pathogens *M. pneumoniae* and *M. genitalium*, the animal pathogens *M. agalactiae, M. arthritidis, M. bovis, M. capricolum, M. gallisepticum, M. mycoides* subsp. *mycoides, M. mycoides* subsp. *capri, M. pulmonis,* and the plant pathogenic spiroplasmas *Spiroplasma citri, Spiroplasma kunkelii,* and *Spiroplasma phoeniceum,* many other Mollicute species such as the important pathogens *M. hyopneumoniae* and *Ureaplasma* sp. are still not amenable to genetic manipulation. Besides transformation with purified DNA, transformation of *M. arthritidis* has also been achieved through conjugal transfer of Tn*916* from *Enterococcus faecalis* (Voelker and Dybvig, 1996). Though the conjugal transfer of Tn*916* from *Streptococcus faecalis* to *M. hominis* has been reported (Roberts and Kenny, 1987), this mycoplasma has not been genetically manipulated further.

### Selectable markers

Owing to the lack of a cell wall, Mollicutes are intrinsically resistant to β-lactams and antimicrobials that target the cell wall. In addition, a conserved mutation in the β-subunit of RNA polymerase is responsible for their insensitivity to rifampicin. Consequently, molecular genetics of these organisms suffer the paucity of antibiotic resistance genes that can be used as selection markers. Even though the erythromycin resistance gene from pAMβ1 has been used for selecting *A. laidlawii* and *M. mycoides* transformants (Dybvig, 1989; King and Dybvig, 1994a), the tetracycline-resistance gene (*tetM*) of Tn*916* (Clewell and Gawron-Burke, 1986) and the gentamycin resistance gene (*aacA-aphD*) of Tn*4001* (Rouch *et al.*, 1987) are by far the most commonly used for genetic manipulation of Mollicutes.

**Table 3.1** Mollicutes species for which genetic tools are available

| Mollicutes species[1] | Phylogenetic group | Host tropism | Disease/ pathology | Transformation methods | Replicon[2] | Transposon mutagenesis[3] | Mutagenesis by homologous recombination[4] |
|---|---|---|---|---|---|---|---|
| *M. pulmonis* | Hominis | Rodent | Respiratory disease | Electroporation, PEG | OriC | Y | Y |
| *M. agalactiae* | Hominis | Small ruminants | Contagious agalactia | Electroporation | OriC | Y | Y |
| *M. bovis* | Hominis | Large ruminants | Respiratory disease, arthritis, mastitis | Electroporation | OriC | Y | N |
| *M. arthritidis* | Hominis | Rodent | Arthritis | PEG | N | Y | N |
| *M. pneumoniae* | Pneumoniae | Human | Respiratory disease | Electroporation | N | Y | Y |
| *M. genitalium* | Pneumoniae | Human | Genital infections | Electroporation PEG | N | Y | Y |
| *M. gallisepticum* | Pneumoniae | Avian | Respiratory disease | Electroporation, PEG | OriC | Y | Y |
| *M. imitans* | Pneumoniae | Avian | Respiratory disease in mixed infections | Electroporation | OriC | N | N |
| *M. mycoides* subsp. *mycoides* | Spiroplasma | Small ruminants | Respiratory disease | PEG | OriC | Y | N |
| *M. mycoides* subsp. *capri* | Spiroplasma | Small ruminants | Contagious agalactia | PEG | pKMK1, OriC | Y | Y |
| *M. capricolum* subsp. *capricolum* | Spiroplasma | Small ruminants | Contagious agalactia | PEG | pKMK1, OriC | Y | N |
| *S. kunkelii* | Spiroplasma | Insect and Plants | Maize stunt | Electroporation | pSci | N | N |
| *S. phoeniceum* | Spiroplasma | Insect and Plants | Periwinkle yellowing | Electroporation | pSci | N | N |
| *S. citri* | Spiroplasma | Insect and Plants | Citrus stubborn, Horseradish brittle root | Electroporation, PEG | OriC, pSci, SpV1-RF | Y | Y |

[1]Acholeplasma were omitted from this table because of the scarcity of the data. [2]OriC plasmids and derivatives (OriC), *M. mycoides* plasmid pKMK1and derivatives (pKMK1), *S. citri* plasmids pSci and derivatives (pSci), Spiroplasma virus1 replicative form and derivatives (SpV1-RF). [3]Insertion of Tn4001 or derivatives into the genome (Y) or not done yet as of 2011 (N). [4]Gene disruption through homologous recombination (Y) or not done yet as of 2011 (N).

With the exception of clinical isolates of human urogenital species that were found to carry the *tetM* gene (Bébéar and Kempf, 2005) and a few fast-growing spiroplasma species such as *S. melliferum*, *S. apis*, and *S. floricola* (S. Duret and J. Renaudin, unpublished data), Mollicutes are generally sensitive to tetracycline. The *tetM* gene has conferred high level resistance in all Mollicute species into which it has been introduced including *A. laidlawii*, *M. agalactiae*, *M. arthritidis*, *M. capricolum*, *M. gallisepticum*, *M. genitalium*, *M. mycoides* subsp. *mycoides*, *M. mycoides* subsp. *capri*, *M. pulmonis*, *S. citri*, *S. kunkelii* and *S. phoeniceum*.

Similarly, when introduced through transposition of *Tn4001* or transformation with a replicative plasmid, the *aacA-aphD* gene of

Tn*4001* has conferred resistance to gentamycin in various mycoplasmas as well as in spiroplasmas with relatively low rates of spontaneous resistance. Because several mycoplasma species such as *M. fermentans* and *M. penetrans* are naturally resistant to gentamycin, Tn*4001* derivatives have been constructed that contain the *Staphylococcus aureus* chloramphenicol acetyl transferase (*cat*) gene instead of the *aacA-aphD* gene and were used to transform *M. pneumoniae* (Hahn *et al.*, 1999). As well, attempts to introduce transposon Tn*4001* into *M. arthritidis* and *M. pulmonis* have been unsuccessful, possibly due to functional failure of the gentamicin resistance determinant. In contrast, the *cat* and *tetM* genes have been shown to function in these mycoplasma species (Dybvig *et al.*, 2000). In *M. genitalium*, expression of the *aacA-aphD* gene was found to correlate with a growth impairment of the resistant strains whereas expression of *tetM* did not (Pich *et al.*, 2006). The *cat* gene from *E. coli* Tn*9* was also found to confer chloramphenicol resistance when introduced into *S. citri* by transformation with an *oriC*-plasmid (Duret *et al.*, 2005). In this case, transformation frequency was significantly lower than those observed with the *aacA-aphD* and *tetM* markers (S. Duret and J. Renaudin, unpublished data).

Fluoroquinolones are broad-spectrum antibiotics that are active against mycoplasmas. The main targets of quinolones are DNA gyrase and topoisomerase IV, both of which are essential for cell viability. Accordingly, mycoplasma strains have acquired quinolone resistance due to mutations in the quinolone resistance determining region (QRDR) of the target *gyrA*, *gyrB*, *parC* and *parE* genes (Bébéar and Kempf, 2005). However, in contrast to the situation in *Enterobacteriacae*, in which plasmid-encoded genes confer quinolone resistance through protection of the targeted proteins (topoisomerases), or modification or active efflux of the antibiotic (Rodriguez-Martinez *et al.*, 2011), no such quinolone resistance genes have been reported so far in Mollicutes.

Recently, resistance to puromycin has been described as a new selectable marker for manipulation of mycoplasma genomes (Algire *et al.*, 2009). Puromycin is a potent inhibitor of translation by disrupting the peptide transfer on ribosomes and mycoplasmas are unable to grow in its presence even at low concentrations (less than 3 μg/ml). Resistance to puromycin is conferred by the puromycin N-acetyltransferase (PAC) gene from *Streptomyces alboniger* that was shown to function in all mycoplasma species in which it had been introduced. These species include *M. gallisepticum*, *M. genitalium*, and *M. pneumoniae* that have been made puromycin resistant through transposition of a modified Tn*4001*, and *M. capricolum* and *M. mycoides* subsp. *capri*, which were transformed by a replicative *oriC* plasmid carrying the *pac* gene downstream of the *S. citri* spiralin gene promoter (Algire *et al.*, 2009).

Although levels of spontaneous resistance are highly dependent on Mollicute species, it is generally thought that spontaneous resistance arises more frequently with erythromycin and gentamycin than with chloramphenicol, tetracycline, and puromycin. Basically, four antibiotic/resistance gene combinations including tetracycline/*tetM*, chloramphenicol/*cat*, gentamycin/*aacA-aphD*, and puromycin/*pac* are currently usable as selectable markers in Mollicutes. With a few exceptions such as the *aacA-aphD* gene in *M. arthritidis* and *M. pulmonis* (Dybvig *et al.*, 2000), these markers are of general use in Mollicutes provided that the resistance genes are efficiently expressed. The promoter for spiralin, an abundant protein from *S. citri*, has been used for expressing foreign genes in *S. citri* (Duret *et al.*, 2005; Renaudin and Lartigue, 2005; Breton *et al.*, 2010b). Interestingly, the spiralin gene promoter also proved to be efficient in driving gene expression in various mycoplasma species including *M. agalactiae*, *M. capricolum*, *M. gallisepticum*, *M. genitalium*, *M. mycoides subsp. capri*, *M. pneumoniae*, and *M. yeatsii* (Janis *et al.*, 2005; Algire *et al.*, 2009; Breton *et al.*, 2010b, 2012; Kent *et al.*, 2012).

Therefore expression of the *cat*, *aacA-aphD*, *tetM*, and *pac* genes under the control of the *S. citri* spiralin gene promoter provides a panel of selectable markers for Mollicutes. Together with the option of producing unmarked mutations (see 'Production of unmarked mutations' below), the increasing number of selectable markers offers the possibility of generating double knock-out mutants and performing complementation studies in a wide range of Mollicute species.

## Random mutagenesis through transposition

Random mutagenesis through transposon insertion for production of mutants' libraries is a powerful tool for identifying genes related to a particular phenotype.

In Mollicutes, transposition was first described in *A. laidlawii* and *M. pulmonis*, in which insertion of Tn*916* into the host chromosome conferred resistance to tetracycline (Dybvig and Cassell, 1987). Since then, Tn*916* has been introduced into a variety of Mollicutes species by conjugal transfer from *Enterococcus faecalis* donor cells, as well as by PEG-mediated transformation and electroporation with plasmid pAM120 as previously reviewed (Dybvig and Voelker, 1996; Renaudin, 2002; Halbedel and Stulke, 2007). However integration of Tn*916* is known to occur at preferred hot spots rather than randomly (Whitley and Finch, 1989; Scott *et al.*, 1994; Nelson *et al.*, 1997) and, despite its conjugative properties, its large size has strongly limited its usefulness as a genetic tool.

In contrast, transposon Tn*4001* from *Staphylococcus aureus* and its various derivatives have been successfully used in various Mollicute species, not only for insertional mutagenesis but also as gene vectors. Tn*4001* is a composite transposon made of two copies of IS256 flanking the *aacA-aphD* gene which confers resistance to gentamycin, kanamycin and tobramycin, and its integration results in a 8-bp direct repeat (Lyon *et al.*, 1984; Rouch *et al.*, 1987; Byrne *et al.*, 1989). Tn*4001* mutagenesis has first been applied in the human mycoplasmas *M. genitalium* and *M. pneumoniae* to generate cytadherence-deficient mutants (Hedreyda *et al.*, 1993; Hedreyda and Krause, 1995; Reddy *et al.*, 1996) and also in the plant pathogenic Mollicute *S. citri* leading to the identification of genes involved in motility and pathogenicity (Foissac *et al.*, 1997a,b; Jacob *et al.*, 1997). Mutagenesis using Tn*4001* and derivatives was later extended to other Mollicute species including *M. agalactiae*, *M. arthritidis*, *M. bovis*, *M. mycoides*, and *M. pulmonis* (Dybvig *et al.*, 2000; Teachman *et al.*, 2002; Chopra-Dewasthaly *et al.*, 2005b; French *et al.*, 2008; Janis *et al.*, 2008) to generate insertional mutant libraries. In particular, Tn*4001* and its derivatives Tn*4001T*, Tn*4001tet*, and Tn*4001TF1* have been used in global transposon mutagenesis experiments with the aim to define the set of essential genes (i.e. those for which no knock-out mutant could be generated) in *M. genitalium* (Hutchison *et al.*, 1999; Glass *et al.*, 2006; Pich *et al.*, 2006) and *M. pulmonis* (French *et al.*, 2008; Dybvig *et al.*, 2010).

In early experiments using the full-length Tn*4001*, transposition into secondary sites within the chromosome were observed as well as independent insertion of IS256 possibly due to inadequate control of the transposase expression in Mollicutes. To circumvent these difficulties, a major improvement has been the use of mini-transposons, in which the transposase gene was placed outside the transposable elements (Fig. 3.1A), preventing further excision of the transposon from the primary insertion site (Pour-El *et al.*, 2002; Zimmerman and Herrmann, 2005; Pich *et al.*, 2006). A further advance was then made when combining the pMT85 mini-transposon with the $\gamma\delta$ TnpR/res recombination system to produce unmarked mutations in *M. mycoides* subsp. *mycoides* (Janis *et al.*, 2008). This system has been successfully used to generate a marker-less mutant in which the gene encoding the main lipoprotein LppQ was disrupted. Recently, the combination of large-scale mini-transposon mutagenesis with an assay based on co-culture with HeLa cells has been used for high-throughput screening of a *M. agalactiae* mutant library (Baranowski *et al.*, 2010). In this study, characterization of selected growth-deficient mutants led to the finding that the NIF locus, encoding proteins presumably involved in [Fe–S] cluster biosynthesis, is critical for proliferation in the presence of HeLa cells while dispensable for axenic growth.

Transposon insertion sites mapping. Several strategies have been used to map the transposon insertion sites. One of these strategies is based on the use of a Tn*4001* derivative comprising sequences capable of replication and selection in *E. coli* (Fig. 3.1A). It consists of cloning the junctions through circularization of relevant restriction fragments as replicative plasmids (Knudtson and Minion, 1993; Hasselbring *et al.*, 2006). The other two strategies are reverse PCR amplification (Teachman *et al.*, 2002; French *et al.*, 2008) and direct sequencing from genomic DNA templates

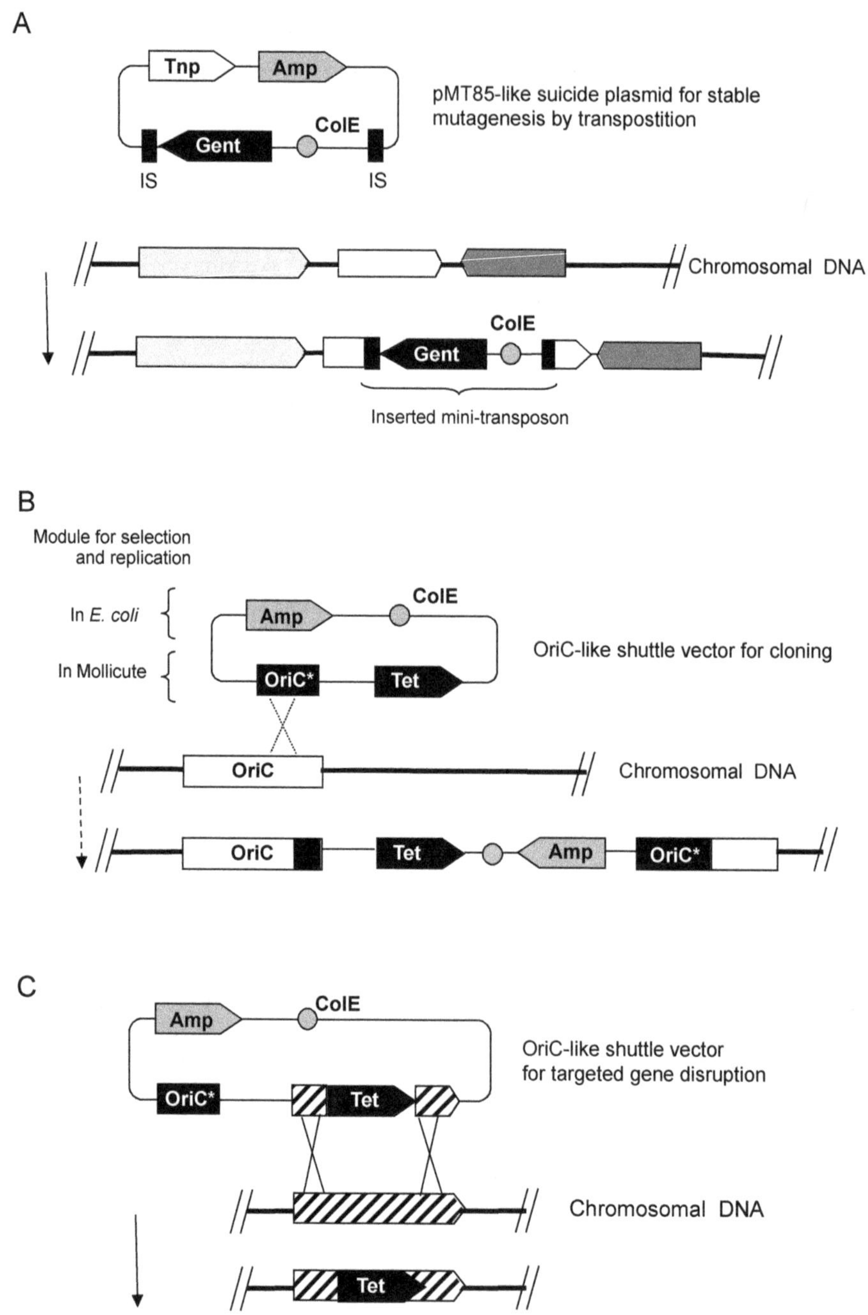


**Figure 3.1** Schematic representing the molecular tools used in Mollicutes for gene disruption and gene cloning. (A) Suicide pMT85-like plasmids carry a transposon composed of an antibiotic selectable marker flanked by two IS elements, which transposase (Tnp) has been placed outside of the IS. pMT85-like plasmids were used for random, stable mutagenesis by transposition to generate mutant libraries in a number of Mollicutes. ColE: *E. coli* origin of replication used for the cloning, in the Gram-negative bacteria, of the chromosomal regions that flank the transposon. Panel B. OriC-like shuttle vectors were used for gene-cloning or gene-complementation, with the chromosomal integration of the vector occurring from time to time by homologous recombination at the chromosomal OriC region. The frequency of this event has been reduced by using only parts of the OriC (OriC*). (B) OriC-like vectors were also used for targeted gene-disruption based on homologous recombination.

(Glass *et al.*, 2006; Baranowski *et al.*, 2010), both of which can be used regardless of the transposon used. In addition, a 'haystack mutagenesis' strategy has been described that allow for isolating any viable gene specific mutant from a transposon mutant library, provided that each dispensable gene is disrupted at a desired confidence level (Halbedel *et al.*, 2006). As an example, a *M. pneumoniae hprK* mutant was isolated from an ordered collection of 2976 individual transposon mutants. A similar strategy has been used successfully to isolate a *M. pneumoniae* MPN474 gene mutant and a *M. mycoides* subsp. *mycoides lppQ* mutant (Hegermann *et al.*, 2008; Janis *et al.*, 2008).

## Targeted mutagenesis through homologous recombination

Targeted gene disruption through homologous recombination has been an alternate means for generating mutants to assess the role of individual genes in various processes such as pathogenesis. In Mollicutes, the first report of gene inactivation through homologous recombination was in *A. laidlawii* (Dybvig and Woodard, 1992). In this study, disruption of the *recA* gene was obtained from one crossover recombination between the chromosome and a suicide plasmid carrying an internal fragment of the *recA* gene. As a result the *recA*-disrupted mutant was proved deficient in DNA repair.

Gene disruption through homologous recombination has also been achieved in various mycoplasma species including *M. gallisepticum* (Cao *et al.*, 1994; Lee *et al.*, 2008), *M. genitalium* (Dhandayuthapani *et al.*, 1999, 2001; Burgos *et al.*, 2008), *M. pulmonis* (Cordova *et al.*, 2002) and *M. agalactiae* (Chopra-Dewasthaly *et al.*, 2008), as well as in the plant pathogen *S. citri* (Duret *et al.*, 1999, 2003; Gaurivaud *et al.*, 2000b; Lartigue *et al.*, 2002) (Fig. 3.1C). However, in agreement with the fact that Mollicutes lack a substantial part of the standard recombination and repair mechanisms that are present in *E. coli* and *B. subtilis* (Rocha *et al.*, 2005), the frequency of recombination is generally very low in these organisms. Allelic exchange through double-crossover recombination with a suicide plasmid is the method of choice to construct null mutants; in Mollicutes however its use has long time been limited to *M. genitalium*.

In *S. citri* (Duret *et al.*, 1999), as well as in *M. pulmonis* (Cordova *et al.*, 2002), *M. gallisepticum* (Lee *et al.*, 2008), and *M. agalactiae* (Chopra-Dewasthaly *et al.*, 2008), gene disruption was achieved using a single crossover recombination strategy and a replicative *oriC* plasmid as the disruption vector (see '*Spiroplasma citri oriC* plasmids as gene expression vectors', '*Spiroplasma citri oriC* plasmids as gene disruption vectors' and 'Mycoplasma *oriC* plasmids' below).

Very recently, targeted chromosomal knockouts created using homologous recombination have been reported in *M. pneumoniae* and in *M. mycoides* subsp. *capri* (Allam *et al.*, 2010; Krishnakumar *et al.*, 2010). In *M. pneumoniae*, complete deletion of MPN345 and MPN607 has been achieved by a double-crossover recombination using a suicide plasmid carrying the antibiotic resistance marker gene flanked by 1-kbp regions surrounding the target gene. Despite the fact that a majority of the antibiotic-resistant colonies resulted from a single crossover recombination, selection of the double crossover recombination event only required the screening of a limited number of colonies, in contrast to transposon mutagenesis.

In *M. mycoides* subsp. *capri*, targeting disruption of the *clpA* gene has been improved by increasing the recombination frequency through expression of the *recA* gene from *E. coli* (Allam *et al.*, 2010). In this study, inclusion of the *E. coli recA* in the disruption vector resulted in a 140-fold increase in the percentage of the desired mutants, i.e. those resulting from a double-crossover recombination. These data offer the possibility of double-crossover recombination in other Mollicutes species and are valuable especially for those lacking a functional *recA* gene such as *S. citri*.

## Production of unmarked mutations

Gene disruption and complementation experiments play an important role in functional genetic studies and frequently rely on the use of the integration of antibiotic resistance cassettes into the chromosome. As there are very few antibiotic resistance markers available for direct selection of Mollicute tranformants (see 'Selectable markers' above), efficient removal of the marker offers the option of using the same antibiotic resistance marker to generate multiple mutants or complement mutations. Based on earlier studies in mycobacteria (Malaga *et al.*, 2003) the feasibility of using the $\gamma\delta$ TnpR/*res* recombination system to produce unmarked mutations in the Mollicutes, was first demonstrated in *S. citri*. In an arginine deiminase (*arcA*-disrupted) mutant, the *tetM* gene flanked by the *res* sequences was efficiently excised from the chromosome following expression of the TnpR resolvase carried by an *oriC* plasmid (Duret *et al.*, 2005). Interestingly the transposon $\gamma\delta$ TnpR/*res* recombination system proved to be of wider use in Mollicutes, provided that the TnpR resolvase can be expressed into the host strain. By combining the $\gamma\delta$ TnpR/*res* recombination system to the mini transposon pMT85, a *lppQ*-disrupted mutant of *M. mycoides* subsp. *mycoides* was obtained free of any antibiotic-resistance marker (Janis *et al.*, 2008).

## Vectors for gene expression/disruption in Mollicutes

Bacterial gene vectors are usually based on self-replicating, extrachromosomal elements. However, whereas many plasmids have been described in spiroplasmas and in phytoplasmas, very few have been described in mycoplasmas (Bergemann *et al.*, 1989; King and Dybvig, 1992). In the past, gene vectors derived from the *M. mycoides* plasmid pKMK1 have been constructed and introduced into both *M. mycoides* subsp. *capri* and in *M. capricolum* (King and Dybvig, 1994a,b). As well, attempts to construct replicative, gene expression vectors from the replicative form of the *S. citri* virus SpV1 have encountered limited success due to insert instability (Stamburski *et al.*, 1991; Marais *et al.*, 1993)

The most successful approach has been the construction of artificial plasmids based on the chromosomal replication origin *oriC* as reviewed in (Renaudin, 2002; Renaudin and Lartigue, 2005; Halbedel and Stulke, 2007) (Fig. 3.1B and C). Structural features of the Mollicute *oriC* regions, including gene organization in the vicinity of the *dnaA* gene as well as the number and positions of DnaA boxes have been previously described previously (Renaudin and Lartigue, 2005). More recently, detailed descriptions of the *oriC* regions of *M. agalactiae, M. arthritidis, M. hominis,* and *M. mobile,* were reported along with genome sequencing (Jaffe *et al.*, 2004; Sirand-Pugnet *et al.*, 2007; Dybvig *et al.*, 2008; Pereyre *et al.*, 2009).

### *Spiroplasma citri oriC* plasmids as gene expression vectors

The so-called *oriC* plasmids have been first described in *S. citri* by combining the chromosomal *oriC* region, comprising the *dnaA* gene and flanking DnaA boxes, with a DNA fragment from Tn*916* carrying the tetracycline resistance gene *tetM* (Ye *et al.*, 1994). Further combination with a *colE1*-derived *E. coli* replicon has resulted in the *S. citri*/*E. coli* shuttle plasmid pBOT1 (Renaudin *et al.*, 1995). Construction of deletion derivatives revealed that the minimal sequences exhibiting *oriC* activity correspond to the DnaA box region downstream of *dnaA* (Lartigue *et al.*, 2002). In the pBOT1-derivative pSD4, the Tn*916* fragment has been replaced by the *tetM* gene driven by the *S. citri* spiralin gene promoter (Lartigue *et al.*, 2002). These shuttle plasmids have been successfully used as gene expression vectors for functional complementation in *S. citri* mutants, leading to the characterization of genes involved in motility and pathogenicity (Jacob *et al.*, 1997; Gaurivaud *et al.*, 2000a; Boutareaud *et al.*, 2004). Similarly to the situation in the Gram-positive bacterium *Bacillus subtilis,* in which the *oriC* plasmids are known to be incompatible, the *S. citri oriC* plasmids may not be stably maintained as free extrachromosomal elements but, instead, integrate into the chromosome by one crossover recombination at the *oriC* region. Once integrated, the plasmid is stably maintained regardless of the presence or absence of selection. As a result, expression of cloned genes is stably maintained not only *in vitro* but also *in vivo*, in the spiroplasma hosts (the leafhopper vector and the host plant), in which selection cannot be applied.

### *Spiroplasma citri oriC* plasmids as gene disruption vectors

In *S. citri*, the frequency of recombination is very low, probably due to the absence of a functional *recA* gene (Marais *et al.*, 1996). As a result, allelic exchange through double crossovers between the chromosome and a suicide plasmid could not be achieved. Taking advantage of their integrative properties, *oriC* plasmids have been used as disruption vectors for producing *S. citri* mutants by gene targeting through homologous recombination. In comparison to suicide plasmids, the use of replicative, *oriC* plasmids both extends the time during which recombination can occur and makes to profit *oriC* plasmid incompatibility for selecting recombination events. The motility gene *scm1* was inactivated by transformation with a pBOT1 derivative carrying an internal *scm1* gene fragment (Duret *et al.*, 1999). A similar strategy based on pBOT1-derived disruption vector has been used to produce fructose operon mutants by targeting genes *fruA* (fructose-PTS permease enzyme II) and/or *fruk* (1-phosphofructokinase) (Gaurivaud *et al.*, 2000b). In these studies however, plasmid recombination occurred more frequently at the *oriC* than at the target gene. Interestingly, using *oriC* disruption vectors in which the *oriC* region was reduced to the minimal sequences required for replication significantly decreased the frequency of plasmid recombination at the *oriC* and concomitantly increased the frequency of recombination at the target gene (Lartigue *et al.*, 2002). These plasmids (pC1/2) were then preferentially used for further gene targeting experiments including disruption of *ptsG* (glucose-permease enzyme IICB) to produce the *S. citri* mutant GII3-Glc unable to use glucose (André *et al.*, 2005). *OriC* plasmid recombination has also been used to construct a translational fusion between GFP and spiralin, the major membrane protein of *S. citri* (Duret *et al.*, 2003).

In most cases plasmid recombination at the target gene required extensive passaging of the transformants. *S. citri oriC* plasmids were further engineered to improve selection of rare recombination events, the rationale being to make expression of the tetracycline resistance marker dependent on plasmid recombination at the target gene (Lartigue *et al.*, 2002). Plasmid pGOT1 contains the reduced *S. citri oriC* region and two distinct selection markers: one, the gentamycin resistance gene, is constitutively expressed whereas the other, the tetracycline resistance gene, cannot be expressed from the free plasmid because it lacks a promoter and is located immediately downstream of a transcription terminator (Renaudin and Lartigue, 2005). In this system, the tetracycline resistance gene can only be expressed once inserted into the chromosome downstream of a promoter. Owing to low recombination frequency, direct selection of the recombinants by plating the transformed spiroplasmas onto tetracycline medium failed according to the fact that, in the presence of tetracycline, pGOT1 behaves as a suicide plasmid. In contrast, selection of recombinants could be achieved in two steps. Firstly, spiroplasmal transformants carrying the free plasmid are selected for their resistance to gentamycin. Secondly, the recombinants in which recombination has occurred at the target gene are selected by plating gentamycin-resistant transformants in the presence of tetracycline. Such an easy screening of very large number of transformants allows selection of rare recombination events and proved especially helpful when short sequences are targeted. Indeed, whereas disruption of *ptsG* could be achieved by using pC1/2 as the vector, disruption of the *crr* gene (glucose-PTS permease enzyme IIA) could not, probably because of the small size of the homologous region (220 bp compared with 1400 bp for *ptsG*). In contrast, inactivation of *crr* was achieved when the 220-bp fragment was carried by pGOT1, leading to the finding that in *S. citri*, the glucose- and trehalose-PTS enzymes II function with a single IIA component (André *et al.*, 2003; Duret *et al.*, 2005). A further advance has been the combination of *oriC* plasmids and the $\gamma\delta$ TnpR/res recombination system to produce unmarked mutations in *S. citri* (Duret *et al.*, 2005) (see also below).

### Mycoplasma *oriC* plasmids

In mycoplasmas, *oriC* plasmids were first constructed in *M. pulmonis* (Cordova *et al.*, 2002), following a strategy similar to that used for *S. citri*. The putative replication region, i.e. the *dnaA* gene and its flanking DnaA box regions, was combined to a *colE1*-derived *E. coli* replicon carrying the *tetM* gene driven by the spiralin gene promoter to yield pMPO1. Successful transformation of *M. pulmonis* by pMPO1 proved the putative replication region to contain sequences with autonomous replication activity. In contrast to the situation in *S. citri*, the two DnaA box regions are required for plasmid replication in *M. pulmonis*. In addition the DnaA box regions must be separated as no replication occurred when they were located side by side (Cordova *et al.*, 2002). Like the *S. citri oriC* plasmids pBOT1 and pSD4, pMPO1 may integrate into the host chromosome through single crossover recombination at *oriC*.

Development of *oriC* plasmids has also been achieved for mycoplasmas species of the *M. mycoides* cluster: *M. capricolum*, *M. mycoides* subsp. *capri*, and *M. mycoides* subsp. *mycoides* (Lartigue *et al.*, 2003). The host-specificity of the *oriC* plasmids was addressed by transforming *S. citri*, three species of the *M. mycoides* cluster and *M. pulmonis* with homologous and heterologous *oriC* plasmids. All five Mollicutes were successfully transformed with the homologous *oriC* plasmids, indicating that the *dnaA* gene region does represent the functional replication origin of the chromosome. However, despite the strong similarities of the replication regions, the ability to replicate heterologous plasmids varies significantly with the species. Plasmids pMCO3, pMYCO1, pMYSO1, and pSD4 having respectively the *oriC* from *M. capricolum*, *M. mycoides* subsp. *capri*, *M. mycoides* subsp. *mycoides*, and *S. citri*, all replicate in *M. capricolum*. In contrast, the *M. capricolum oriC* plasmid pMCO3 does not replicate in the closely related species *M. mycoides* subsp. *capri* and *M. mycoides* subsp. *mycoides*, suggesting that cis-acting *oriC* elements carried by the plasmids are not the only determinants of host specificity (Lartigue *et al.*, 2003). Whereas the host range of *oriC* plasmids is generally restricted to closely related species, the replication of the *S. citri* plasmid pSD4 in *M. capricolum* is one of the very few examples of *oriC* plasmids that replicate in bacteria of distinct genera.

Development of shuttle *oriC* plasmids (*E. coli*/*Mycoplasma* sp.) has been further extended to other species including *M. agalactiae* (Chopra-Dewasthaly *et al.*, 2005a), and *M. gallisepticum* and *M. imitans* (Lee *et al.*, 2008). As for *S. citri*, *oriC* plasmids have been used in mycoplasmas for expression of heterologous proteins and, thanks to their integrative property, for gene inactivation through homologous recombination. In *M. capricolum* subsp. *capricolum*, successful expression of the *E. coli* β-galactosidase and of the spiroplasma membrane protein spiralin has been achieved using pMCO3-derived *oriC* plasmids carrying the corresponding genes (Janis *et al.*, 2005). Interestingly, disruption of the *lppA* lipoprotein gene has been obtained with a heterologous *S. citri oriC* plasmid. In contrast, no integration into the target gene was observed when a plasmid harbouring the homologous, *M. capricolum oriC* was used. One hypothesis that might explain these results is that the low replication rate of the heterologous *oriC* plasmid might favour the selection of the rare recombinant cells (Janis *et al.*, 2005). In *M. agalactiae*, *oriC* plasmids have been constructed that may replicate as free extrachromosomal elements or integrate into the host chromosome by one crossover recombination (Chopra-Dewasthaly *et al.*, 2005a). These data emphasize the usefulness of *oriC*-based vectors for genetic manipulation of *M. agalactiae*. As an example, homologous recombination with an *oriC*-based vector has been used to inactivate the *xer1* gene, encoding a putative site-specific recombinase adjacent to the variable protein locus *vpma*. The results showed that further Vpma switching was abolished and that the *xer1*-disrupted, phase-locked mutant expressed a single Vpma product (Chopra-Dewasthaly *et al.*, 2008). Interestingly, whereas the transposon-based delivery of the wild-type *xer1* failed to restore the phenotypic switching (possibly due to side effects of transposon insertion), successful complementation of the mutant was achieved by using an *oriC* plasmid carrying the wild type gene (Chopra-Dewasthaly *et al.*, 2008).

## Transposon-based gene delivery

Despite the development of artificial *oriC* plasmids in several species, there is not such a tool as one single vector that would function in every mycoplasma species. To overcome the lack of cloning vectors in some species such as in *M. genitalium* and *M. pneumoniae*, the Tn*4001* has been further engineered for gene expression and used to complement chromosomal spontaneous mutations. To that purpose *Sma*I and *Bam*HI cloning sites have been introduced into one IS256 arm of Tn*4001*, creating Tn*4001mod* (Knudtson and Minion, 1993), which has been then used as the delivery vector to express cytadherence-associated proteins in *M. pneumoniae* mutants (Hahn *et al.*, 1996; Fisseha *et al.*, 1999; Romero-Arroyo *et al.*, 1999). Tn*4001* derivatives have served as an alternative to cloning vectors in many mycoplasma species including *M. agalactiae, M. arthritidis, M. bovis, M. gallisepticum, M. pneumoniae,* and *M. pulmonis* (Knudtson and Minion, 1993; Romero-Arroyo *et al.*, 1999; Dybvig *et al.*, 2000; Chopra-Dewasthaly *et al.*, 2005b; Horino *et al.*, 2009; Burgos *et al.*, 2012) but hold several disadvantages. These are the low transformation efficiencies (less than $10^{-6}$) and possible unwanted effects due to the localization of the transposon insertion that may lead to misinterpretation for complementation studies. For instance, complementation of *M. pneumoniae* HMW2-deficient mutant I-2 with wild-type *hmw2* via transposon delivery restored cytadherence in most of the transformants but not all most likely because transposon insertion disrupts other genes also involved in adherence or is located in a position that alters the expression of the wild-type allele (Fisseha *et al.*, 1999). Besides complementation, transposon-based delivery has also been used to introduce *lacZ*-fusions into *M. pneumoniae* for *in vivo* promoter analyses (Halbedel and Stulke, 2006). More recently, a novel Tn*4001* derivative construct has enabled the expression and translocation of alkaline phosphatase as a recombinant lipoprotein in *M. gallisepticum* (Panicker *et al.*, 2012).

## *Spiroplasma citri* pSci plasmids

Many spiroplasma species and most *S. citri* strains possess plasmids, a few of which have been sequenced (Davis *et al.*, 2005; Joshi *et al.*, 2005; Saillard *et al.*, 2008). *S. citri* strain GII3 contains seven plasmids namely pSciA and pSci1–6, that share extensive regions of sequence homology among themselves, as well as with pBJS-O from *S. citri* BR3 and pSKU146 from *S. kunkelii* CR2 (Saillard *et al.*, 2008). These plasmids all possess two conserved genes (*pE* and *soj*), that were shown to be essential for replication and stability, respectively (Breton *et al.*, 2008). Based on the lack of similarity of protein PE with any known replication protein, the *S. citri* plasmids certainly represent a new replicon family. Using the *tetM* gene as the selection marker, the *S. citri* plasmids and their shuttle (*S. citri*/*E. coli*) derivatives have been efficiently introduced into various *S. citri* strains by electrotransformation (Berho *et al.*, 2006; Breton *et al.*, 2008). Similar PE replicating plasmids have been detected in several spiroplasma species of group I including all three plant pathogenic spiroplasmas, *S. citri, S. kunkelii,* and *S. phoeniceum*, indicating that these plasmids might have a broad host-range. From these studies, it is clear that *S. citri* plasmids and their shuttle derivatives hold considerable advantages as gene vectors: (i) they transform *S. citri* at relatively high frequencies ($10^{-4}$–$10^{-3}$ transformant/CFU/μg), including strains such as R8A2 which could not (or very poorly) be transformed by *oriC* plasmids; (ii) the doubling times of the transformants are not significantly affected as compared to the wild-type; (iii) depending on the needs, they are stably maintained or not based on the presence of *soj* and, in contrast to *oriC* plasmids, never integrate into the chromosome and (iv) their host range is not restricted to *S. citri* but extends to the three plant pathogenic spiroplasmas in which expression of cloned genes has been demonstrated (Breton *et al.*, 2008).

Plasmid curing/replacement through incompatibility in *S. citri*. Regardless of their high homology, the seven *S. citri* GII3 plasmids coexist within one cell suggesting that suppression of their incompatibility may reside in the variability of the replication region. Indeed, in presence of selection pressure, transformation of *S. citri* GII3 by a given selectable pSci results in the nearly specific loss of its native counterpart. Depending on the presence or absence of *soj* which is involved in segregational stability, further propagation of the transformant in absence of selection pressure results in stable maintenance or in rapid loss of the plasmid, respectively. A strategy of plasmids curing/replacement based on incompatibility of their replication regions has been used to produce a collection of plasmid mutants, leading to the identification of plasmid sequences that are essential for *S. citri* transmission by its leafhopper vector (Breton *et al.*, 2010a).

## Mycoplasma natural plasmids for heterologous gene expression

Plasmids have been identified in a limited number of mycoplasma species, namely *M. mycoides* subsp. capri, M. leachii, and M. yeatsii, all of which belongs or are closely related to the *M. mycoides* cluster (Breton *et al.*, 2012). Attempts to use natural plasmids as cloning vectors have been first successful when the erythromycin resistant determinant ermr was expressed in *M. mycoides* subsp. capri and *M. capricolum* by using *M. mycoides* subsp. capri pKMK1 derivatives as gene vectors (King and Dybvig, 1994a). However, heterologous gene expression in mycoplasmas has mostly been achieved using oriC plasmid- or transposon-based vectors (see 'Mycoplasma *oriC* plasmids' and 'Transposon-based gene delivery' above). Recently natural plasmid pMyBK1 from *M. yeatsii* was described as the first member of a new replicon family as its replication protein did not share any similarity with known replication proteins (Breton *et al.*, 2012; Kent *et al.*, 2012). pMyBK1-based shuttle (*E. coli/M. yeatsii*) vectors were found to transform *M. yeatsii* at high frequency and stably maintained for multiple generations. One such vector has been successfully used for expressing the *M. fermentans malp* gene in *M. yeatsii* (Kent *et al.*, 2012). Interestingly the host-range of pMyBK1-derivatives is not restricted to *M. yeatsii*. The pMyBK1 derivatives pCM-K3/4 were successfully introduced into M. *putrefaciens* KS1 TS, *M. leachii* PG50 TS, and *M. capricolum* subsp. capricolum California kid TS, and *M. mycoides* subsp. capri. Heterologous expression of the spiralin gene, encoding the major surface protein of *S. citri*, was demonstrated in *M. yeatsii* and *M. capricolum* subsp. capricolum indicating that pMyBK1 derivatives can be used as expression vectors in mycoplasma species of veterinary importance (Breton *et al.*, 2012).

## Heterologous expression of Mollicute genes in *Escherichia coli*

Biochemical characterization, activity assays, and preparation of antibodies usually require relatively large amounts of highly purified proteins. In most cases, it has been achieved by overexpressing recombinant proteins in *E. coli*. Although heterologous expression of UGA-containing genes has been achieved in several Mollicutes species, the use of a Mollicute host for protein production is not yet possible. In contrast to *E. coli*, none of these Mollicutes species has been specially engineered for high yield production of intact proteins and there is no vector with an inducible system that would over-express a given protein at a given time. In addition, Mollicutes grow at lower rates and titres than *E. coli*, and the serum-containing medium in which they grow would make protein purification uneasy and expensive.

Yet, heterologous expression of Mollicute proteins in *E. coli* has considerably been limited by the fact that most Mollicutes, with the exceptions of Acholeplasmas and Phytoplasmas, use UGA as a tryptophan codon instead of a translational stop signal (Yamao *et al.*, 1985; Renaudin *et al.*, 1986; Blanchard, 1990; Muto and Ushida, 2002). Consequently, only those rare Mollicute proteins having no UGA-encoded tryptophan such as the *S. citri* spiralin or the *M. pneumoniae* HPr kinase/phosphatase could be fully translated in *E. coli* (Mouchès *et al.*, 1985; Chevalier *et al.*, 1990; Steinhauer *et al.*, 2002). The attempts to overcome this difficulty by using a UGA, opal suppressor strain of *E. coli* has encountered limited success as it proved poorly efficient (Smiley and Minion, 1993).

Alternatively, TGA codons of Mollicute genes can be mutated into the TGG tryptophan codon to allow full-length expression in *E. coli* (Knudtson *et al.*, 1997). This can be time-consuming when using sequential replacement of TGA codons by standard site-directed mutagenesis but PCR has turned out to offer a rapid method for introducing a series of mutations into cloned DNA (Ito *et al.*, 1991). In Mollicutes, various PCR-based site-directed mutagenesis methods have been used for replacing TGA codons. In *M. hyopneumoniae*, site-directed mutagenesis of 14 genes, with up to three TGA to TGG substitutions per gene, was achieved using the overlap extension-PCR method (Simionatto *et al.*, 2009). Also, based on the combined chain reaction method (Bi and Stambrook, 1998), a multiple-mutation reaction strategy has been successfully used for overproducing the glycerol kinase GlpK of *M. pneumoniae* in *E. coli* (Hames *et al.*, 2005). In this study up to nine opal codons were replaced simultaneously, and it was indicated that the method could be even more effective provided that high quality primers are used.

## Gene expression in Mollicutes

### Identification of Mollicutes' promoters

Along with DNA sequencing and annotation, many Mollicute promoters have been predicted based on their similarities with the −10 and −35 consensus sequences of eubacterial promoters that are recognized by the RNA polymerase associated to the general sigma factor. Some of them have been experimentally characterized through determination of the mRNA starts (Taschke and Herrmann, 1986; Stamburski *et al.*, 1990; Williamson *et al.*, 1991; Weiner *et al.*, 2000; Horino *et al.*, 2003; Musatovova *et al.*, 2003; Halbedel *et al.*, 2007).

DNA sequences with transcription promoter activity have also been identified by using promoter-probe vectors, in which they are inserted upstream of a promoterless reporter gene. In Mollicutes, the use of the *lacZ*-encoded β-galactosidase as a reporter system was first demonstrated in *A. oculi* and *M. gallisepticum* (Knudtson and Minion, 1993). A promoter-less *lacZ* gene was inserted into one IS*256* of Tn*4001* so that, after transposition of Tn*4001lac* into the host chromosome, the *lacZ* gene could be transcribed from adjacent promoters and translated to a functional β-galactosidase. Using such an approach, screening of *M. gallisepticum* transposon mutants carrying *lacZ* fusions in the chicken tracheal ring organ culture system has revealed the induction of a pMGA gene encoding a cell surface adhesin involved in haemagglutination (Bearson *et al.*, 2003). Transposon delivery of *lacZ* fusions was also used to investigate the regulation of the pMGA gene expression (Liu *et al.*, 2000, 2002). Based on β-galactosidase activity, pMGA gene expression is regulated not only by the number of trinucleotide GAA repeats located in the promoter region but also by the region surrounding the repeats (Liu *et al.*, 2002).

A similar strategy based on an integrative plasmid rather than a transposon has been used for constructing a promoter probe vector in *A. oculi* (Knudtson and Minion, 1994). Successful expression of functional, *lacZ*-encoded β-galactosidase has been achieved in many other Mollicute species including *M. agalactiae* (Czurda *et al.*, 2010), *M. arthritidis* (Dybvig *et al.*, 2000), *M. capricolum* (Janis *et al.*, 2005), *M. genitalium* (Lluch-Senar *et al.*, 2007), *M. pneumoniae* (Halbedel and Stulke, 2006), *M. pulmonis* (Dybvig *et al.*, 2000), and *S. citri* (W. Maccheroni and J. Renaudin, unpublished data). In *M. pneumoniae*, the *lacZ*-based, reporter plasmid pGP353 has been constructed and used to determine transcription promoter activity (Halbedel and Stulke, 2006). Fusions of promoter sequences to a promoter-less *lacZ* were inserted into the pMT85 minitransposon (Zimmerman and Herrmann, 2005) which were then inserted into the mycoplasma chromosome through transposition. Promoter efficiency was then determined *in vivo* by measuring β-galactosidase activity. A similar strategy was used to study the requirements of *M. pneumoniae* RNA polymerase for promoter recognition and results revealed that while mutations affecting the −10 region strongly inferred with gene expression, the −35 regions seems of minor importance (Halbedel *et al.*, 2007).

A promoter-probe vector containing the promoterless *lacZ* gene combined to a minitransposon conferring tetracycline resistance has also been constructed to analyse *M. genitalium* transcription *in vivo* (Lluch-Senar *et al.*, 2007). In this study the β-galactosidase activity of transposon mutants was measured to quantify the transcription levels of the disrupted genes. Unexpectedly, the results showed that a significant proportion of the detected activity corresponded to the *lacZ* gene being opposite to the coding frame, suggesting the occurrence of antisense transcripts that could be involved in controlling gene expression (Lluch-Senar *et al.*, 2007).

Recently, a smart excision assay based on the versatile use of the *lacZ* reporter system has been developed to demonstrate that recombinase Xer1 mediates site specific DNA inversions and excisions in *M. agalactiae* (Czurda *et al.*, 2010).

## Promoters for expression of cloned genes in Mollicutes

Despite the functional characterization of a number of Mollicutes promoters, only few have been commonly used for constructing gene expression vectors. Among these, the *S. citri* spiralin gene promoter (Chevalier *et al.*, 1990) has been extensively used in *S. citri* plasmid constructs for driving expression of antibiotic resistance genes [such as *tetM*, *cat*, *aacA-aphD* (Renaudin, 2002; Duret *et al.*, 2005)], reporter genes [such as *lacZ* and GFP, W. Maccheroni and J. Renaudin, unpublished data (Duret *et al.*, 2003)], the resolvase gene *tnpR* (Duret *et al.*, 2005), as well as the genes encoding the TetR repressor and the I-*SceI* meganuclease, respectively (Breton *et al.*, 2010b, 2011). The spiralin gene promoter controls the expression of the most abundant protein of *S. citri*, the spiralin, suggesting that it is a 'strong' promoter and has turned out to be highly valuable for other Mollicutes species. For instance, introduction of the *tetM* gene under the control of the spiralin gene promoter has been shown to confer tetracycline resistance in various mycoplasma species, including *M. pulmonis* (Cordova *et al.*, 2002), *M. capricolum*, *M. mycoides* subsp. *capri*, and *M. mycoides* subsp. *mycoides* (Lartigue *et al.*, 2003). In addition, the spiralin gene promoter was shown to drive the expression of (i) the TetR repressor in *M. agalactiae* (Breton *et al.*, 2010b), (ii) the β-galactosidase and spiralin in *M. capricolum* subsp. *capricolum* (Janis *et al.*, 2005), and (iii) the puromycin resistance gene in *M. gallisepticum*, *M. genitalium*, and *M. pneumoniae* (Algire *et al.*, 2009). Similarly, the transcription promoter of the P40 immunodominant protein of *M. agalactiae* has been used for complementation studies in this mycoplasma (Baranowski *et al.*, 2010). In this study, transposon knockout mutants displaying a growth-deficient phenotype in cell culture were successfully complemented through transformation with an *oriC*-based plasmid carrying the NIF locus under the control of the *p40* gene promoter. Other mycoplasma promoters such as those of *M. pneumoniae p65* adhesin and *tuf* genes have been used for controlling expression of EGFP/EYFP fusion proteins in this organism (Balish *et al.*, 2003; Kenri *et al.*, 2004). Taken together, these experiments suggest that several Mollicute promoters may be of general use for constructing expression vectors. However, their relative strength in driving expression of cloned genes in a given Mollicute has not been yet determined.

## Inducible gene expression systems for Mollicutes

Inducible gene expression systems are critical for studying gene function. In fact, the availability of an inducible promoter system provides a wide range of applications, such as production of conditional knockout mutants (Kamionka *et al.*, 2005), conditional expression of toxins (Carroll *et al.*, 2007), plasmid removal through inducible counter-selection (Bae and Schneewind, 2006), titration of gene expression in the host cell (Carroll *et al.*, 2005; Corrigan *et al.*, 2007), and conditional gene silencing (Blokpoel *et al.*, 2005), that all rely on the tight control of the promoter in the absence of inducer (Corrigan and Foster, 2009; Williams *et al.*, 2010). In Mollicutes, very few transcription regulators have been identified, suggesting that most of the regulatory systems have been lost during their reductive evolution. Recently however, a comprehensive study of *M. pneumoniae* transcripts under various conditions has revealed an unanticipated complexity of the transcriptome that cannot be explained by the presence of the eight transcription regulators identified in this organism (Guell *et al.*, 2009). Basically studies on transcription regulation in these organisms are still limited (Weiner *et al.*, 2003; Musatovova *et al.*, 2006; Halbedel *et al.*, 2007; Lluch-Senar *et al.*, 2007; Schafer *et al.*, 2007; Madsen *et al.*, 2008). The presence of inducible promoters has been suggested in various Mollicutes. In *S. citri*, transcription levels of the glucose and fructose operons were found to increase in the presence of the relevant sugars (Gaurivaud *et al.*, 2001; André *et al.*, 2005). Also, it has been shown by using *lacZ* transcriptional fusions that transcription of the *ackA* and *ldh* genes from *M. pneumoniae* is induced by glucose and glycerol, respectively (Halbedel *et al.*, 2007). In these cases, however, the genes are still expressed at a basal level in the absence of inducer, and no repressor has been identified. In *M. genitalium*, expression of heat-shock proteins *dnaK clpB* and *lon* (that possess a CIRCE element in their promoter regions) is up-regulated in response to elevated temperatures, suggesting that the temperature-dependent HrcA-CIRCE regulatory mechanism is functional. However, direct evidence that the HrcA repressor binds to the CIRCE sequences of heatshock protein genes has not been provided (Musatovova *et al.*, 2006). Alternatively, various regulated promoter systems have been considered for a broad use in Gram-positive bacteria, of which the tetracycline inducible systems are among the most versatile (Stieger *et al.*, 1999; Blokpoel *et al.*, 2005; Ehrt *et al.*, 2005; Kamionka *et al.*, 2005; Corrigan and Foster, 2009). The main advantage of using tetracycline inducible systems is that tetracycline penetrates most eukaryotic cells allowing for sensitive regulation also in bacteria that reside inside the host-cell. The first report of tetracycline-inducible gene expression in Mollicutes was in the plant pathogen *S. citri* (Breton *et al.*, 2010b). An *S. citri* plasmid carrying the spiralin gene under the control of the tetracycline-inducible promoter *xyl/tetO*$_2$ (originated from *B. subtilis*) and the TetR repressor gene was introduced into the spiralin-less mutant GII3–9a3. In the absence of tetracycline, constitutive expression of the TetR repressor almost completely abolished expression of spiralin from the *xyl/tetO*$_2$ promoter, whereas addition of tetracycline into the medium induced high-level of spiralin expression. As expected, inducible spiralin expression was also detected *in vivo*, in the *S. citri*-infected leafhoppers fed on tetracycline-containing medium as well as in the infected plants watered with tetracycline. A similar construct inserted into the *M. agalactiae* chromosome through transposition demonstrated that the TetR-P*xyl/tetO*$_2$ also functions in this ruminant pathogen. From these studies, it is expected that the usefulness of tetracycline-inducible promoter systems will no longer be restricted to a few of Mollicutes species.

## Expression of fluorescent proteins in Mollicutes

Intrinsically fluorescent proteins such as the green fluorescent protein (GFP), the yellow fluorescent protein (YFP), and derivatives are widely used to study protein localization, protein–protein interactions, cell division and gene expression in a variety of organisms. In Mollicutes, expression of recombinant fluorescent proteins has been reported first in *M. pneumoniae* (Balish *et al.*, 2003). In this study, the subcellular location of the cytadherence associated protein HMW2 was determined through fusion of its coding gene with that of EGFP. The constructs were introduced into mycoplasma cells through transposition with Tn*4001*mod. Localization of the fluorescent fusion protein confirmed HMW2 as a component of the attachment organelle. A similar fluorescent-protein tagging technique was used to further determine the subcellular localization of *M. pneumoniae* proteins P65, HMW2, P41, and P24 encoded by the cytadherence regulatory locus, showing that, in contrast to P65 and HMW2 which localized at the distal end, P41 and P24 preferentially localized at the proximal end of the attachment organelle suggesting that they are cytoskeletal proteins (Kenri *et al.*, 2004). Also in *M. pneumoniae*, a gene fusion with the monomeric fluorescent protein (mRFP1) has served as a tool to demonstrate that a small, cysteine-rich peptide was expressed from its own promoter (Zimmerman and Herrmann, 2005). In *S. citri*, a spiralin–GFP fusion protein has been constructed through homologous recombination with an *oriC*-plasmid. However, despite clear-cut detection of the fusion protein, the GFP sequences proved to be unstable during passaging of the spiroplasmal transformant (Duret *et al.*, 2003).

## I-*Sce*I-mediated counter-selection and intramolecular recombination in *Spiroplasma citri*

Although allelic exchange is straightforward in many bacteria, it remains very inefficient in Mollicutes with the outcome of this event, if occurring, difficult to isolate because of its low frequency. Counter-selectable markers are often helpful for construction of such mutants especially in microorganisms for which genetic tools are poorly developed (Reyrat *et al.*, 1998). So far, none of the most popular counter-selectable markers has been shown to function in the Mollicutes. Recently in *S. citri*, plasmid deletion has been achieved by using the *Saccharomyces cerevisiae* meganuclease I-*Sce*I (Breton *et al.*, 2011). The *S. citri* plasmid pSci1NT-I, a *S. citri* GII3 pSci1 derivative carrying the *tetM* selection marker and a unique I-*Sce*I recognition site, was introduced into *S. citri* 44. Further expression in the spiroplasmal transformant of the I-*Sce*I endonuclease gene under the control of the spiralin gene promoter resulted in the rapid loss of the plasmid. These data indicates that plasmid vectors having a unique I-*Sce*I site can easily be counter-selected in any Mollicute, provided that expression of I-*Sce*I is achieved. In *S. citri* and *M. agalactiae*, for which tetracycline-inducible gene expression has been reported (Breton *et al.*, 2010b), engineering a strain for conditional expression of I-*Sce*I would provide a smart counter-selection system for vectors having an I-*Sce*I site. In addition, a pMT85 derivative has been engineered that allows random delivery of the I-*Sce*I recognition within the spiroplasma genome (M. Breton and J. Renaudin, unpublished data).

By creating DNA double-strand breaks at specific sites, meganucleases are also known as potent stimulators of homologous recombination and have been used for the construction of targeted gene deletions in a wide range of organisms (Doyon *et al.*, 2008; Flannagan *et al.*, 2008; Yu *et al.*, 2008; Shukla *et al.*, 2009). Therefore, beyond counter-selection of plasmid vectors, meganuclease-based genetic tools are anticipated to be relevant to other Mollicute species.

## Discussion

Since the first successful introduction of exogenous DNA into Mollicutes in the 1990s, tremendous progresses have been made towards expanding the Mollicute genetic tool box. Systems for random and targeted mutagenesis along with gene expression vectors for complementation studies, as well as inducible gene expression systems, promoter trapping vectors, plasmid counter-selection, and the possibility of unmarked mutations, are becoming available for an increasing number of Mollicute species. The finest tool for manipulating mycoplasma genomes has turned out to be the yeast (Lartigue *et al.*, 2009; Benders *et al.*, 2010) though this new 'technology' might not serve all purposes and still requires further tuning to be routinely implemented in other Mollicute species.

For pathogenic spiroplasmas, recent findings have enriched the genetic tool box with *S. citri* natural plasmids that better perform than their artificial counterparts, once customized. With the accumulation of genomic data, the genetic diversity within a single mycoplasma species appears to be greater than first expected and the general thought that mycoplasmas are deprived of extrachromosomal elements starts fading away. Besides the recent report of the high prevalence of plasmids in mycoplasma species belonging or closely related to the *M. mycoides* cluster (Breton *et al.*, 2012; Kent *et al.*, 2012), a number of recent studies are describing the presence of putative, complex, conjugative elements or phages occurring as integrated in the host-genome and/or as free circular forms (Calcutt *et al.*, 2002; Marenda *et al.*, 2006; Roske *et al.*, 2010). The distribution of these elements among strains of a single species is one source for diversification and suggests a dynamic circulation. While deciphering the structure and function of these extrachromosomal elements might provide new clues in understanding evolutionary or phenotypic traits, it might also lead the way to designing novel genetic tools.

The current genetic tool box for Mollicutes together with the advances in genome sequencing now offer the possibility of addressing major questions related to functional genomics for a number of Mollicute species. For those relevant to animal, plant or human diseases, it is a crucial step towards the better understanding of factors involved in pathogenesis provided that efforts are made in the future to improve or adapt experimental, *in vivo* models for high-throughput screening.

## Acknowledgements

Researches, performed in the Mollicutes' teams of UMR1332 and UMR1225, have been supported by INRA and by the Université de Bordeaux or by the Ecole Nationale Véterinaire de Toulouse (ENVT), respectively. Funding supports also include grants from the Ministère de l'Enseignement Supérieur et de la Recherche and the Agence Nationale de la Recherche (ANR). We thank past and present coworkers for their contributions.

## References


Algire, M.A., Lartigue, C., Thomas, D.W., Assad-Garcia, N., Glass, J.I., and Merryman, C. (2009). New selectable marker for manipulating the simple genomes of Mycoplasma species. Antimicrob. Agents Chemother. *53*, 4429–4432.

Allam, A.B., Reyes, L., Assad-Garcia, N., Glass, J.I., and Brown, M.B. (2010). Enhancement of targeted homologous recombination in *Mycoplasma mycoides* subsp. capri by inclusion of heterologous recA. Appl. Environ. Microbiol. *76*, 6951–6954.

André, A., Maccheroni, W., Doignon, F., Garnier, M., and Renaudin, J. (2003). Glucose and trehalose PTS permeases of *Spiroplasma citri* probably share a single IIA domain, enabling the spiroplasma to adapt quickly to carbohydrate changes in its environment. Microbiology *149*, 2687–2696.

André, A., Maucourt, M., Moing, A., Rolin, D., and Renaudin, J. (2005). Sugar import and phytopathogenicity of *Spiroplasma citri*: glucose and fructose play distinct roles. Mol. Plant Microbe Interact. *18*, 33–42.

Bae, T., and Schneewind, O. (2006). Allelic replacement in Staphylococcus aureus with inducible counter-selection. Plasmid *55*, 58–63.

Balish, M.F., Santurri, R.T., Ricci, A.M., Lee, K.K., and Krause, D.C. (2003). Localization of *Mycoplasma pneumoniae* cytadherence-associated protein HMW2 by fusion with green fluorescent protein: implications for attachment organelle structure. Mol. Microbiol. *47*, 49–60.

Baranowski, E., Guiral, S., Sagne, E., Skapski, A., and Citti, C. (2010). Critical role of dispensable genes in *Mycoplasma agalactiae* interaction with mammalian cells. Infect. Immun. *78*, 1542–1551.

Bearson, S.M., Collier, S.D., Bearson, B.L., and Branton, S.L. (2003). Induction of a mycoplasma gallisepticum pMGA gene in the chicken tracheal ring organ culture model. Avian Dis. *47*, 745–749.

Bébéar, C., and Kempf, I. (2005). Antimicrobial therapy and antimicrobial resistance. In Mycoplasmas: Pathogenesis, Molecular Biology, and Emerging Strategies for Control, Blanchard, A., and Browning, G.F., eds. (Horizon Scientific Press, Norwich, UK), pp. 535–568.

Benders, G.A., Noskov, V.N., Denisova, E.A., Lartigue, C., Gibson, D.G., Assad-Garcia, N., Chuang, R.Y., Carrera, W., Moodie, M., Algire, M.A., *et al.* (2010). Cloning whole bacterial genomes in yeast. Nucleic Acids Res. *38*, 2558–2569.

Bergemann, A.D., Whitley, J.C., and Finch, L.R. (1989). Homology of mycoplasma plasmid pADB201 and staphylococcal plasmid pE194. J. Bacteriol. *171*, 593–595.

Berho, N., Duret, S., Danet, J.L., and Renaudin, J. (2006). Plasmid pSci6 from *Spiroplasma citri* GII-3 confers insect transmissility to the non-transmissible *S. citri* strain 44. Microbiology *152*, 2703–2716.

Bi, W., and Stambrook, P.J. (1998). Site-directed mutagenesis by combined chain reaction. Anal. Biochem. *256*, 137–140.

Blanchard, A. (1990). *Ureaplasma urealyticum* urease genes; use of a UGA tryptophan codon. Mol. Microbiol. *4*, 669–676.

Blokpoel, M.C., Murphy, H.N., O'Toole, R., Wiles, S., Runn, E.S., Stewart, G.R., Young, D.B., and Robertson, B.D. (2005). Tetracycline-inducible gene regulation in mycobacteria. Nucleic Acids Res. *33*, e22.

Boutareaud, A., Danet, J.L., Garnier, M., and Saillard, C. (2004). Disruption of a gene predicted to encode a solute binding protein of an ABC transporter reduces transmission of *Spiroplasma citri* by the leafhopper *Circulifer haematoceps*. Appl. Environ. Microbiol. *70*, 3960–3967.

Breton, M., Duret, S., Arricau-Bouvery, N., Béven, L., and Renaudin, J. (2008). Characterizing the replication and stability regions of *Spiroplasma citri* plasmids identifies a novel replication protein and expands the genetic toolbox for plant pathogenic spiroplasmas. Microbiology *154*, 3232–3244.

Breton, M., Duret, S., Danet, J.L., Dubrana, M.P., and Renaudin, J. (2010a). Sequences essential for transmission of *Spiroplasma citri* by its leafhopper vector, *Circulifer haematoceps*, revealed by plasmid curing and replacement based on incompatibility. Appl. Environ. Microbiol. *76*, 3198–3205.

Breton, M., Sagne, E., Duret, S., Beven, L., Citti, C., and Renaudin, J. (2010b). First report of a tetracycline-inducible gene expression system for Mollicutes. Microbiology *156*, 198–205.

Breton, M., Duret, S., Béven, L., Dubrana, M.-P., and Renaudin, J. (2011). I-*Sce*I-mediated plasmid deletion and intra-molecular recombination in *Spiroplasma citri*. J. Microbiol. Meth. *84*, 216–222.

Breton, M., Tardy, F., Dordet-Frisoni, E., Sagné, E., Mick, V., Renaudin, J., Sirand-Pugnet, P., Citti, C., and Blanchard, A. (2012). Distribution and diversity of mycoplasma plasmids, lessons from cryptic genetic elements. BMC Microbiol. *12*, 257–271.

Burgos, R., Pich, O.Q., Querol, E., and Pinol, J. (2008). Deletion of the *Mycoplasma genitalium* MG_217 gene modifies cell gliding behaviour by altering terminal organelle curvature. Mol. Microbiol. *69*, 1029–1040.

Burgos, R., Wood, G.E., Young, L., Glass, J.I., and Totten, P.A. (2012). RecA mediates MgpB and MgpC phase and antigenic variation in *Mycoplasma genitalium*, but plays a minor role in DNA repair. Mol. Microbiol. *85*, 669–683.

Byrne, M.E., Rouch, D.A., and Skurray, R.A. (1989). Nucleotide sequence analysis of IS256 from the Staphylococcus aureus gentamicin-tobramycin-kanamycin-resistance transposon Tn4001. Gene *81*, 361–367.

Calcutt, M.J., Lewis, M.S., and Wise, K.S. (2002). Molecular genetic analysis of ICEF, an integrative conjugal element that is present as a repetitive sequence in the chromosome of *Mycoplasma fermentans* PG18. J. Bacteriol. *184*, 6929–6941.

Cao, J., Kapke, P.A., and Minion, F.C. (1994). Transformation of *Mycoplasma gallisepticum* with Tn*916*, Tn*4001*, and integrative plasmid vectors. J. Bacteriol. *176*, 4459–4462.

Carroll, P., Muttucumaru, D.G., and Parish, T. (2005). Use of a tetracycline-inducible system for conditional expression in *Mycobacterium tuberculosis* and *Mycobacterium smegmatis*. Appl. Environ. Microbiol. *71*, 3077–3084.

Carroll, P., Brown, A.C., Hartridge, A.R., and Parish, T. (2007). Expression of *Mycobacterium tuberculosis* Rv1991c using an arabinose-inducible promoter demonstrates its role as a toxin. FEMS Microbiol. Lett. *274*, 73–82.

Chevalier, C., Saillard, C., and Bové, J.M. (1990). Organization and nucleotide sequences of the *Spiroplasma citri* genes for ribosomal protein S2, elongation factor Ts, spiralin, phosphofructokinase, pyruvate kinase, and an unidentified protein. J. Bacteriol. *172*, 2693–2703.

Chopra-Dewasthaly, R., Marenda, M., Rosengarten, R.a., Jechlinger, W., and Citti, C. (2005a). Construction of the first shuttle vectors for gene cloning and homologous recombination in *Mycoplasma agalactiae*. FEMS Microbiol. Lett. *253*, 89–94.

Chopra-Dewasthaly, R., Zimmermann, M., Rosengarten, R., and Citti, C. (2005b). First steps towards the genetic manipulation of *Mycoplasma agalactiae* and *Mycoplasma bovis* using the transposon Tn4001mod. Int. J. Med. Microbiol. *294*, 447–453.

Chopra-Dewasthaly, R., Citti, C., Glew, M.D., Zimmermann, M., Rosengarten, R., and Jechlinger, W. (2008). Phase-locked mutants of *Mycoplasma agalactiae*: defining the molecular switch of high-frequency Vpma antigenic variation. Mol. Microbiol. *67*, 1196–1210.

Clewell, D.B., and Gawron-Burke, C. (1986). Conjugative transposons and the dissemination of antibiotic resistance in streptococci. Annu. Rev. Microbiol. *40*, 635–659.

Cordova, C.M., Lartigue, C., Sirand-Pugnet, P., Renaudin, J., Cunha, R.A., and Blanchard, A. (2002). Identification of the origin of replication of the *Mycoplasma pulmonis* chromosome and its use in *oriC* replicative plasmids. J. Bacteriol. *184*, 5426–5435.

Corrigan, R.M., and Foster, T.J. (2009). An improved tetracycline-inducible expression vector for *Staphylococcus aureus*. Plasmid *61*, 126–129.

Corrigan, R.M., Rigby, D., Handley, P., and Foster, T.J. (2007). The role of Staphylococcus aureus surface protein SasG in adherence and biofilm formation. Microbiology *153*, 2435–2446.

Czurda, S., Jechlinger, W., Rosengarten, R., and Chopra-Dewasthaly, R. (2010). Xer1-mediated site-specific DNA inversions and excisions in *Mycoplasma agalactiae*. J. Bacteriol. *192*, 4462–4473.

Davis, R.E., Dally, E.L., Jomantiene, R., Zhao, Y., Roe, B., Lin, S., and Shao, J. (2005). Cryptic plasmid pSKU146 from the wall-less plant pathogen *Spiroplasma kunkelii* encodes an adhesin and components of a type IV translocation-related conjugation system. Plasmid *53*, 179–190.

Dhandayuthapani, S., Rasmussen, W.G., and Baseman, J.B. (1999). Disruption of gene mg218 of *Mycoplasma genitalium* through homologous recombination leads to an adherence-deficient phenotype. Proc. Natl. Acad. Sci. U.S.A. *96*, 5227–5232.

Dhandayuthapani, S., Blaylock, M.W., Bebear, C.M., Rasmussen, W.G., and Baseman, J.B. (2001). Peptide methionine sulfoxide reductase (MsrA) is a virulence determinant in *Mycoplasma genitalium*. J. Bacteriol. *183*, 5645–5650.

Doyon, Y., McCammon, J.M., Miller, J.C., Faraji, F., Ngo, C., Katibah, G.E., Amora, R., Hocking, T.D., Zhang, L., Rebar, E.J., *et al.* (2008). Heritable targeted gene disruption in zebrafish using designed zinc-finger nucleases. Nat. Biotechnol. *26*, 702–708.

Duret, S., Danet, J.L., Garnier, M., and Renaudin, J. (1999). Gene disruption through homologous recombination in *Spiroplasma citri*: an *scm*1-disrupted motility mutant is pathogenic. J. Bacteriol. *181*, 7449–7456.

Duret, S., Berho, N., Danet, J.L., Garnier, M., and Renaudin, J. (2003). Spiralin is not essential for helicity, motility, or pathogenicity but is required for efficient transmission of *Spiroplasma citri* by its leafhopper vector *Circulifer haematoceps*. Appl. Environ. Microbiol. *69*, 6225–6234.

Duret, S., André, A., and Renaudin, J. (2005). Specific gene targeting in *Spiroplasma citri*: improved vectors and production of unmarked mutations using site-specific recombination. Microbiology *151*, 2793–2803.

Dybvig, K. (1989). Transformation of *Acholeplasma laidlawii* with streptococcal plasmids pVA868 and pVA920. Plasmid *21*, 155–160.

Dybvig, K., and Cassell, G.H. (1987). Transposition of Gram-positive transposon Tn916 in *Acholeplasma laidlawii* and *Mycoplasma pulmonis*. Science *235*, 1392–1394.

Dybvig, K., and Voelker, L.L. (1996). Molecular biology of mycoplasmas. Annu. Rev. Microbiol. *50*, 25–57.

Dybvig, K., and Woodard, A. (1992). Construction of recA mutants of *Acholeplasma laidlawii* by insertional inactivation with a homologous DNA fragment. Plasmid *28*, 262–266.

Dybvig, K., French, C.T., and Voelker, L.L. (2000). Construction and use of derivatives of transposon Tn*4001* that function in *Mycoplasma pulmonis* and *Mycoplasma arthritidis*. J. Bacteriol. *182*, 4343–4347.

Dybvig, K., Zuhua, C., Lao, P., Jordan, D.S., French, C.T., Tu, A.H., and Loraine, A.E. (2008). Genome of *Mycoplasma arthritidis*. Infect. Immun. *76*, 4000–4008.

Dybvig, K., Lao, P., Jordan, D.S., and Simmons, W.L. (2010). Fewer essential genes in mycoplasmas than previous studies suggest. FEMS Microbiol. Lett. *311*, 51–55.

Ehrt, S., Guo, X.V., Hickey, C.M., Ryou, M., Monteleone, M., Riley, L.W., and Schnappinger, D. (2005). Controlling gene expression in mycobacteria with anhydrotetracycline and Tet repressor. Nucleic Acids Res. *33*, e21.

Fisseha, M., Gohlmann, H.W., Herrmann, R., and Krause, D.C. (1999). Identification and complementation of frameshift mutations associated with loss of cytadherence in *Mycoplasma pneumoniae*. J. Bacteriol. *181*, 4404–4410.

Flannagan, R.S., Linn, T., and Valvano, M.A. (2008). A system for the construction of targeted unmarked gene deletions in the genus Burkholderia. Environ. Microbiol. *10*, 1652–1660.

Foissac, X., Danet, J.L., Saillard, C., Gaurivaud, P., Laigret, F., Pare, C., and Bové, J.M. (1997a). Mutagenesis by insertion of Tn*4001* into the genome of *Spiroplasma citri*: Characterization of mutants affected in plant pathogenicity and transmission to the plant by the leafhopper vector *Circulifer haematoceps*. Mol. Plant Microbe Interact. *10*, 454–461.

Foissac, X., Saillard, C., and Bové, J.M. (1997b). Random insertion of transposon Tn*4001* in the genome of *Spiroplasma citri* strain GII3. Plasmid *37*, 80–86.

French, C.T., Lao, P., Loraine, A.E., Matthews, B.T., Yu, H., and Dybvig, K. (2008). Large-scale transposon mutagenesis of *Mycoplasma pulmonis*. Mol. Microbiol. *69*, 67–76.

Gasparich, G.E., Hackett, K.J., Stamburski, C., Renaudin, J., and Bove, J.M. (1993). Optimization of methods for transfecting *Spiroplasma citri* strain R8A2 hp with the spiroplasma virus Spv1 replicative form. Plasmid *29*, 193–205.

Gaurivaud, P., Danet, J.L., Laigret, F., Garnier, M., and Bové, J.M. (2000a). Fructose utilization and phytopathogenicity of *Spiroplasma citri*. Mol. Plant Microbe Interact. *13*, 1145–1155.

Gaurivaud, P., Laigret, F., Verdin, E., Garnier, M., and Bové, J.M. (2000b). Fructose operon mutants of *Spiroplasma citri*. Microbiology *146*, 2229–2236.

Gaurivaud, P., Laigret, F., Garnier, M., and Bové, J.M. (2001). Characterization of FruR as a putative activator of the fructose operon of *Spiroplasma citri*. FEMS Microbiol. Lett. *198*, 73–78.

Glass, J.I., Assad-Garcia, N., Alperovich, N., Yooseph, S., Lewis, M.R., Maruf, M., Hutchison, C.A., 3rd, Smith, H.O., and Venter, J.C. (2006). Essential genes of a minimal bacterium. Proc. Natl. Acad. Sci. U.S.A. *103*, 425–430.

Guell, M., van Noort, V., Yus, E., Chen, W.H., Leigh-Bell, J., Michalodimitrakis, K., Yamada, T., Arumugam, M., Doerks, T., Kuhner, S., *et al.* (2009). Transcriptome complexity in a genome-reduced bacterium. Science *326*, 1268–1271.

Hahn, T.W., Krebes, K.A., and Krause, D.C. (1996). Expression in *Mycoplasma pneumoniae* of the recombinant gene encoding the cytadherence-associated protein HMW1 and identification of HMW4 as a product. Mol. Microbiol. *19*, 1085–1093.

Hahn, T.W., Mothershed, E.A., Waldo, R.H., 3rd, and Krause, D.C. (1999). Construction and analysis of a modified Tn4001 conferring chloramphenicol resistance in *Mycoplasma pneumoniae*. Plasmid *41*, 120–124.

Halbedel, S., and Stulke, J. (2006). Probing in vivo promoter activities in *Mycoplasma pneumoniae*: a system for generation of single-copy reporter constructs. Appl. Environ. Microbiol. *72*, 1696–1699.

Halbedel, S., and Stulke, J. (2007). Tools for the genetic analysis of Mycoplasma. Int. J. Med. Microbiol. *297*, 37–44.

Halbedel, S., Busse, J., Schmidl, S.R., and Stulke, J. (2006). Regulatory protein phosphorylation in *Mycoplasma pneumoniae*. A PP2C-type phosphatase serves to dephosphorylate HPr(Ser-P). J. Biol. Chem. *281*, 26253–26259.

Halbedel, S., Eilers, H., Jonas, B., Busse, J., Hecker, M., Engelmann, S., and Stulke, J. (2007). Transcription in *Mycoplasma pneumoniae*: analysis of the promoters of the ackA and ldh genes. J. Mol. Biol. *371*, 596–607.

Hames, C., Halbedel, S., Schilling, O., and Stulke, J. (2005). Multiple-mutation reaction: a method for simultaneous introduction of multiple mutations into the glpK gene of *Mycoplasma pneumoniae*. Appl. Environ. Microbiol. *71*, 4097–4100.

Hasselbring, B.M., Page, C.A., Sheppard, E.S., and Krause, D.C. (2006). Transposon mutagenesis identifies genes associated with *Mycoplasma pneumoniae* gliding motility. J. Bacteriol. *188*, 6335–6345.

Hedreyda, C.T., and Krause, D.C. (1995). Identification of a possible cytadherence regulatory locus in *Mycoplasma pneumoniae*. Infect. Immun. *63*, 3479–3483.

Hedreyda, C.T., Lee, K.K., and Krause, D.C. (1993). Transformation of *Mycoplasma pneumoniae* with Tn*4001* by electroporation. Plasmid *30*, 170–175.

Hegermann, J., Halbedel, S., Dumke, R., Regula, J., Gabdoulline, R.R., Mayer, F., Stulke, J., and Herrmann, R. (2008). The acidic, glutamine-rich Mpn474 protein of *Mycoplasma pneumoniae* is surface exposed and covers the complete cell. Microbiology *154*, 1185–1192.

Horino, A., Sasaki, Y., Sasaki, T., and Kenri, T. (2003). Multiple promoter inversions generate surface antigenic variation in Mycoplasma penetrans. J. Bacteriol. *185*, 231–242.

Horino, A., Kenri, T., Sasaki, Y., Okamura, N., and Sasaki, T. (2009). Identification of a site-specific tyrosine recombinase that mediates promoter inversions of phase-variable mpl lipoprotein genes in Mycoplasma penetrans. Microbiology *155*, 1241–1249.

Hutchison, C.A., Peterson, S.N., Gill, S.R., Cline, R.T., White, O., Fraser, C.M., Smith, H.O., and Venter, J.C. (1999). Global transposon mutagenesis and a minimal Mycoplasma genome. Science *286*, 2165–2169.

Ito, W., Ishiguro, H., and Kurosawa, Y. (1991). A general method for introducing a series of mutations into cloned DNA using the polymerase chain reaction. Gene *102*, 67–70.

Jacob, C., Nouzieres, F., Duret, S., Bové, J.M., and Renaudin, J. (1997). Isolation, characterization, and complementation of a motility mutant of *Spiroplasma citri*. J. Bacteriol. *179*, 4802–4810.

Jaffe, J.D., Stange-Thomann, N., Smith, C., DeCaprio, D., Fisher, S., Butler, J., Calvo, S., Elkins, T., FitzGerald, M.G., Hafez, N., *et al.* (2004). The complete genome and proteome of *Mycoplasma mobile*. Genome Res. *14*, 1447–1461.

Janis, C., Lartigue, C., Frey, J., Wróblewski, H., Thiaucourt, F., Blanchard, A., and Sirand-Pugnet, P. (2005). Versatile use of *oriC* plasmids for functional genomics of *Mycoplasma capricolum* subsp. capricolum. Appl. Environ. Microbiol. *71*, 2888–2893.

Janis, C., Bischof, D., Gourgues, G., Frey, J., Blanchard, A., and Sirand-Pugnet, P. (2008). Unmarked insertional mutagenesis in the bovine pathogen *Mycoplasma mycoides* subsp. mycoides SC: characterization of a lppQ mutant. Microbiology *154*, 2427–2436.

Joshi, B.D., Berg, M., Rogers, J., Fletcher, J., and Melcher, U. (2005). Sequence comparisons of plasmids pBJS-O of *Spiroplasma citri* and pSKU146 of *S. kunkelii:* implications for plasmid evolution BMC Genomics *6*, 175–185.

Kamionka, A., Bertram, R., and Hillen, W. (2005). Tetracycline-dependent conditional gene knockout in Bacillus subtilis. Appl. Environ. Microbiol. *71*, 728–733.

Kenri, T., Seto, S., Horino, A., Sasaki, Y., Sasaki, T., and Miyata, M. (2004). Use of fluorescent-protein tagging to determine the subcellular localization of mycoplasma pneumoniae proteins encoded by the cytadherence regulatory locus. J. Bacteriol. *186*, 6944–6955.

Kent, B.N., Foecking, M.F., and Calcutt, M.J. (2012). Development of a novel plasmid as a shuttle vector for heterologous gene expression in *Mycoplasma yeatsii*. J. Microbiol. Methods *91*, 121–127.

King, K.W., and Dybvig, K. (1992). Nucleotide sequence of *Mycoplasma mycoides* subspecies Mycoides plasmid pKMK1. Plasmid *28*, 86–91.

King, K.W., and Dybvig, K. (1994a). Mycoplasmal cloning vectors derived from plasmid pKMK1. Plasmid *31*, 49–59.

King, K.W., and Dybvig, K. (1994b). Transformation of *Mycoplasma capricolum* and examination of DNA restriction-modification in *Mycoplasma capricolum* and *Mycoplasma mycoides* subsp. *mycoides*. Plasmid *31*, 308–311.

Knudtson, K.L., and Minion, F.C. (1993). Construction of Tn4001lac derivatives to be used as promoter probe vectors in mycoplasmas. Gene *137*, 217–222.

Knudtson, K.L., and Minion, F.C. (1994). Use of lac gene fusions in the analysis of Acholeplasma upstream gene regulatory sequences. J. Bacteriol. *176*, 2763–2766.

Knudtson, K.L., Manohar, M., Joyner, D.E., Ahmed, E.A., and Cole, B.C. (1997). Expression of the superantigen *Mycoplasma arthritidis* mitogen in Escherichia coli and characterization of the recombinant protein. Infect. Immun. *65*, 4965–4971.

Krishnakumar, R., Assad-Garcia, N., Benders, G.A., Phan, Q., Montague, M.G., and Glass, J.I. (2010). Targeted chromosomal knockouts in *Mycoplasma pneumoniae*. Appl. Environ. Microbiol. *76*, 5297–5299.

Lartigue, C., Duret, S., Garnier, M., and Renaudin, J. (2002). New plasmid vectors for specific gene targeting in *Spiroplasma citri*. Plasmid *48*, 149–159.

Lartigue, C., Blanchard, A., Renaudin, J., Thiaucourt, F., and Sirand-Pugnet, P. (2003). Host specificity of Mollicutes oriC plasmids: functional analysis of replication origin. Nucleic Acids Res. *31*, 6610–6618.

Lartigue, C., Glass, J.I., Alperovich, N., Pieper, R., Parmar, P.P., Hutchison, C.A., 3rd, Smith, H.O., and Venter, J.C. (2007). Genome transplantation in bacteria: changing one species to another. Science *317*, 632–638.

Lartigue, C., Vashee, S., Algire, M.A., Chuang, R.Y., Benders, G.A., Ma, L., Noskov, V.N., Denisova, E.A., Gibson, D.G., Assad-Garcia, N., *et al.* (2009). Creating bacterial strains from genomes that have been cloned and engineered in yeast. Science *325*, 1693–1696.

Lee, S.W., Browning, G.F., and Markham, P.F. (2008). Development of a replicable *oriC* plasmid for *Mycoplasma gallisepticum* and *Mycoplasma imitans*, and gene disruption through homologous recombination in *M. gallisepticum*. Microbiology *154*, 2571–2580.

Liu, L., Dybvig, K., Panangala, V.S., van Santen, V.L., and French, C.T. (2000). GAA trinucleotide repeat region regulates M9/pMGA gene expression in *Mycoplasma gallisepticum*. Infect. Immun. *68*, 871–876.

Liu, L., Panangala, V.S., and Dybvig, K. (2002). Trinucleotide GAA repeats dictate pMGA gene expression in *Mycoplasma gallisepticum* by affecting spacing between flanking regions. J. Bacteriol. *184*, 1335–1339.

Lluch-Senar, M., Vallmitjana, M., Querol, E., and Pinol, J. (2007). A new promoterless reporter vector reveals antisense transcription in *Mycoplasma genitalium*. Microbiology *153*, 2743–2752.

Lyon, B.R., May, J.W., and Skurray, R.A. (1984). Tn*4001*: A gentamicin and kanamycin resistance transposon in *Staphylococcus aureus*. Mol. Gen. Genet. *193*, 554–556.

Madsen, M.L., Puttamreddy, S., Thacker, E.L., Carruthers, M.D., and Minion, F.C. (2008). Transcriptome changes in *Mycoplasma hyopneumoniae* during infection. Infect. Immun. *76*, 658–663.

Malaga, W., Perez, E., and Guilhot, C. (2003). Production of unmarked mutations in mycobacteria using site-specific recombination. FEMS Microbiol. Lett. *219*, 261–268.

Marais, A., Bove, J.M., Dallo, S.F., Baseman, J.B., and Renaudin, J. (1993). Expression in *Spiroplasma citri* of an epitope carried on the G- fragment of the cytadhesin-P1-gene from *Mycoplasma pneumoniae*. J. Bacteriol. *175*, 2783–2787.

Marais, A., Bové, J.M., and Renaudin, J. (1996). *Spiroplasma citri* virus SpV1-derived cloning vector: Deletion formation by illegitimate and homologous recombination in a spiroplasmal host strain which probably lacks a functional *rec*A gene. J. Bacteriol. *178*, 862–870.

Marenda, M., Barbe, V., Gourgues, G., Mangenot, S., Sagne, E., and Citti, C. (2006). A new integrative conjugative element occurs in *Mycoplasma agalactiae* as chromosomal and free circular forms. J. Bacteriol. *188*, 4137–4141.

Montague, M.G., Lartigue, C., and Vashee, S. (2012). Synthetic genomics: potential and limitations. Curr. Opin. Biotechnol. *23*, 659–665.

Mouchès, C., Candresse, T., Barroso, G., Saillard, C., Wroblewski, H., and Bové, J.M. (1985). Gene for spiralin, the major membrane-protein of the helical Mollicute *Spiroplasma citri* – Cloning and expression in *Escherichia coli*. J. Bacteriol. *164*, 1094–1099.

Musatovova, O., Dhandayuthapani, S., and Baseman, J.B. (2003). Transcriptional starts for cytadherence-related operons of *Mycoplasma genitalium*. FEMS Microbiol. Lett. *229*, 73–81.

Musatovova, O., Dhandayuthapani, S., and Baseman, J.B. (2006). Transcriptional heat shock response in the smallest known self-replicating cell, *Mycoplasma genitalium*. J. Bacteriol. *188*, 2845–2855.

Muto, A., and Ushida, C. (2002). Transcription and translation. In Molecular Biology and Pathogenicity of Mycoplasmas, Razin, S., and Hermann, R., eds. (Kluwer Academic, New York), pp. 323–345.

Nelson, K.E., Richardson, D.L., and Dougherty, B.A. (1997). Tn916 transposition in Haemophilus influenzae Rd: preferential insertion into noncoding DNA. Microb. Comp. Genomics 2, 313–321.

Panicker, I.S., Kanci, A., Chiu, C.J., Veith, P.D., Glew, M.D., Browning, G.F., and Markham, P.F. (2012). A novel transposon construct expressing PhoA with potential for studying protein expression and translocation in *Mycoplasma gallisepticum*. BMC Microbiol. *12*, 138.

Pereyre, S., Sirand-Pugnet, P., Beven, L., Charron, A., Renaudin, H., Barre, A., Avenaud, P., Jacob, D., Couloux, A., Barbe, V., *et al.* (2009). Life on arginine for *Mycoplasma hominis*: clues from its minimal genome and comparison with other human urogenital mycoplasmas. PLoS Genet. 5, e1000677.

Pich, O.Q., Burgos, R., Planell, R., Querol, E., and Pinol, J. (2006). Comparative analysis of antibiotic resistance gene markers in *Mycoplasma genitalium*: application to studies of the minimal gene complement. Microbiology *152*, 519–527.

Pour-El, I., Adams, C., and Minion, F.C. (2002). Construction of mini-Tn4001tet and its use in *Mycoplasma gallisepticum*. Plasmid 47, 129–137.

Razin, A., Yogev, D., and Naot, Y. (1998). Molecular biology and pathogenicity of mycoplasma. Microbiol. Mol. Biol. Rev. *62*, 1094–1156.

Reddy, S.P., Rasmussen, W.G., and Baseman, J.B. (1996). Isolation and characterization of transposon Tn4001-generated, cytadherence-deficient transformants of *Mycoplasma pneumoniae* and *Mycoplasma genitalium*. FEMS Immunol. Med. Microbiol. *15*, 199–211.

Renaudin, J. (2002). Extrachromosomal elements and gene transfer. In Molecular Biology and Pathogenicity of Mycoplasmas, Razin, S., and Hermann, R., eds. (Kluwer Academic, New York), pp. 347–370.

Renaudin, J., and Lartigue, C. (2005). *OriC* plasmids as gene vectors for Mollicutes. In Mycoplasmas: pathogenesis, Molecular Biology, and Emerging Strategies for Control, Blanchard, A., and Browning, G., eds. (Horizon Scientific Press, Norwich, UK), pp. 3–30.

Renaudin, J., Pascarel-Devilder, M.C., Saillard, C., Chevalier, C., and Bové, J.M. (1986). Chez les spiroplasmes le codon UGA n'est pas non-sens et semble coder pour le tryptophane. C. R. Acad. Sci. Paris *303*, 539–540.

Renaudin, J., Marais, A., Verdin, E., Duret, S., Foissac, X., Laigret, F., and Bové, J.M. (1995). Integrative and free *Spiroplasma citri oriC* plasmids: expression of the *Spiroplasma phoeniceum* spiralin in *Spiroplasma citri*. J. Bacteriol. *177*, 2870–2877.

Reyrat, J.M., Pelicic, V., Gicquel, B., and Rappuoli, R. (1998). Counterselectable markers: untapped tools for bacterial genetics and pathogenesis. Infect. Immun. *66*, 4011–4017.

Roberts, M.C., and Kenny, G.E. (1987). Conjugal transfer of transposon Tn916 from *Streptococcus faecalis* to *Mycoplasma hominis*. J. Bacteriol. *169*, 3836–3839.

Rocha, E., Sirand-Pugnet, P., and Blanchard, A. (2005). Genome analysis: recombination, repair and recombinational hotspots. In Mycoplasmas: Molecular Biology, Pathogenicity and Strategies for Control, Blanchard, A., and Browning, G.F., eds. (Horizon Scientific Press, Norwich, UK), pp. 31–73.

Rodriguez-Martinez, J.M., Cano, M.E., Velasco, C., Martinez-Martinez, L., and Pascual, A. (2011). Plasmid-mediated quinolone resistance: an update. J. Infect. Chemother. *17*, 149–182.

Romero-Arroyo, C.E., Jordan, J., Peacock, S.J., Willby, M.J., Farmer, M.A., and Krause, D.C. (1999). *Mycoplasma pneumoniae* protein P30 is required for cytadherence and associated with proper cell development. J. Bacteriol. *181*, 1079–1087.

Roske, K., Foecking, M.F., Yooseph, S., Glass, J.I., Calcutt, M.J., and Wise, K.S. (2010). A versatile palindromic amphipathic repeat coding sequence horizontally distributed among diverse bacterial and eucaryotic microbes. BMC Genomics *11*, 430.

Rouch, D.A., Byrne, M.E., Kong, Y.C., and Skurray, R.A. (1987). The *aacA-aphD* gentamycin and kanamycin resistance determinant of Tn*4001* from *Staphylococcus aureus*: expression and nucleotide sequence analysis. J. Gen. Microbiol. *133*, 3039–3052.

Saillard, C., Carle, P., Duret-Nurbel, S., Henri, R., Killiny, N., Carrère, S., Gouzy, J., Bové, J.M., Renaudin, J., and Foissac, X. (2008). The abundant extrachromosomal content of *Spiroplasma citri* strain GII3-3X. BMC Genomics 9, 195–207.

Schafer, E.R., Oneal, M.J., Madsen, M.L., and Minion, F.C. (2007). Global transcriptional analysis of *Mycoplasma hyopneumoniae* following exposure to hydrogen peroxide. Microbiology *153*, 3785–3790.

Scott, J.R., Bringel, F., Marra, D., Van Alstine, G., and Rudy, C.K. (1994). Conjugative transposition of Tn916: preferred targets and evidence for conjugative transfer of a single strand and for a double-stranded circular intermediate. Mol. Microbiol. *11*, 1099–1108.

Shukla, V.K., Doyon, Y., Miller, J.C., DeKelver, R.C., Moehle, E.A., Worden, S.E., Mitchell, J.C., Arnold, N.L., Gopalan, S., Meng, X., *et al.* (2009). Precise genome modification in the crop species Zea mays using zinc-finger nucleases. Nature 459, 437–441.

Simionatto, S., Marchioro, S.B., Galli, V., Luerce, T.D., Hartwig, D.D., Moreira, A.N., and Dellagostin, O.A. (2009). Efficient site-directed mutagenesis using an overlap extension-PCR method for expressing *Mycoplasma hyopneumoniae* genes in Escherichia coli. J. Microbiol. Methods 79, 101–105.

Sirand-Pugnet, P., Lartigue, C., Marenda, M., Jacob, D., Barre, A., Barbe, V., Schenowitz, C., Mangenot, S., Couloux, A., Segurens, B., *et al.* (2007). Being pathogenic, plastic, and sexual while living with a nearly minimal bacterial genome. PLoS Genet. 3, 744–758.

Smiley, B.K., and Minion, F.C. (1993). Enhanced read-through of opal (UGA) stop codons and production of *Mycoplasma pneumoniae* P1 epitopes in *E. coli*. Gene *134*, 33–40.

Stamburski, C., Renaudin, J., and Bove, J.M. (1990). Characterization of a promoter and a transcription terminator of *Spiroplasma melliferum* virus Spv4. J. Bacteriol. *172*, 5586–5592.

Stamburski, C., Renaudin, J., and Bové, J.M. (1991). First step toward a virus-derived vector for gene cloning and expression in spiroplasmas, organisms which read UGA as a tryptophan codon – synthesis of chloramphenicol acetyltransferase in *Spiroplasma citri*. J. Bacteriol. *173*, 2225–2230.

Steinhauer, K., Jepp, T., Hillen, W., and Stulke, J. (2002). A novel mode of control of *Mycoplasma pneumoniae* HPr kinase/phosphatase activity reflects its parasitic lifestyle. Microbiology *148*, 3277–3284.

Stieger, M., Wohlgensinger, B., Kamber, M., Lutz, R., and Keck, W. (1999). Integrational plasmids for the tetracycline-regulated expression of genes in Streptococcus pneumoniae. Gene *226*, 243–251.

Taschke, C., and Herrmann, R. (1986). Analysis of transcription and processing signals of the 16S-23S rRNA operon of *Mycoplasma hyopneumoniae*. Mol. Gen. Genet. *205*, 434–441.

Teachman, A.M., French, C.T., Yu, H., Simmons, W.L., and Dybvig, K. (2002). Gene transfer in *Mycoplasma pulmonis*. J. Bacteriol. *184*, 947–951.

Voelker, L.L., and Dybvig, K. (1996). Gene transfer in *Mycoplasma arthritidis*: transformation, conjugal transfer of Tn916, and evidence for a restriction system recognizing AGCT. J. Bacteriol. *178*, 6078–6081.

Weiner, J., 3rd, Herrmann, R., and Browning, G.F. (2000). Transcription in *Mycoplasma pneumoniae*. Nucleic Acids Res. *28*, 4488–4496.

Weiner, J., 3rd, Zimmerman, C.U., Gohlmann, H.W., and Herrmann, R. (2003). Transcription profiles of the bacterium *Mycoplasma pneumoniae* grown at different temperatures. Nucleic Acids Res. *31*, 6306–6320.

Weisburg, W.G., Tully, J.G., Rose, D.L., Petzel, J.P., Oyaizu, H., Yang, D., Mandelco, M., Schrest, J., Lawrence, T.G., van Etten, J., *et al.* (1989). A phylogenetic analysis of the mycoplasmas: basis for their classification. J. Bacteriol. *171*, 6455–6467.

Whitley, J.C., and Finch, L.R. (1989). Location of sites of transposon Tn916 insertion in the *Mycoplasma mycoides* genome. J. Bacteriol. *171*, 6870–6872.

Williams, K.J., Joyce, G., and Robertson, B.D. (2010). Improved mycobacterial tetracycline inducible vectors. Plasmid *64*, 69–73.

Williamson, D.L., Renaudin, J., and Bové, J.M. (1991). Nucleotide sequence of the *Spiroplasma citri* fibril protein gene. J. Bacteriol. *173*, 4353–4362.

Yamao, F., Muto, A., Kawauchi, Y., Iwami, M., Iwagami, S., Azumi, Y., and Osawa, S. (1985). UGA is read as tryptophan in *Mycoplasma capricolum*. Proc. Natl. Acad. Sci. U.S.A. *82*, 2306–2309.

Ye, F., Renaudin, J., Bove, J.M., and Laigret, F. (1994). Cloning and sequencing of the replication origin (*oriC*) of the *Spiroplasma citri* chromosome and construction of autonomously replicating artificial plasmids. Curr. Microbiol. *29*, 23–29.

Yu, B.J., Kang, K.H., Lee, J.H., Sung, B.H., Kim, M.S., and Kim, S.C. (2008). Rapid and efficient construction of markerless deletions in the Escherichia coli genome. Nucleic Acids Res. *36*, e84.

Zimmerman, C.U., and Herrmann, R. (2005). Synthesis of a small, cysteine-rich, 29 amino acids long peptide in *Mycoplasma pneumoniae*. FEMS Microbiol. Lett. 253, 315–321.

# Identification and Characterization of Virulence Genes in Mycoplasmas

4

Glenn F. Browning, Amir H. Noormohammadi and Philip F. Markham

## Abstract

The pathogenesis of most mycoplasmoses is predominantly attributable to the immunopathological response of the host to the persistent presence of these pathogens. Therefore, virulence genes in mycoplasmas are probably best defined as those that are not necessary for growth *in vitro*, but that are required for optimal colonization of, persistence in or pathological effects on the host, including those genes whose primary function appears to be ensuring optimal nutrient acquisition *in vivo*. While the full array of virulence genes is far from being defined in any one mycoplasmal pathogen, there has been significant progress in recent years in defining the roles of adhesins, invasion, toxin production, immune evasion, immunostimulation and immunosuppression in some mycoplasmas, and in characterizing some of the genes involved in these processes. In addition, it has become clear that several systems for acquisition and utilization of specific nutrients are required for optimal pathogenicity, in some cases because these nutrients contribute to the generation of toxic metabolites, while in others presumably because they facilitate persistence in a nutrient limited environment. Most recently some of the mechanisms used to regulate transcription of virulence genes have been identified, and the complexity of post-translational control of virulence factors is beginning to be revealed. These recent findings have demonstrated that mycoplasmas are far more complex pathogens than their superficially simple genomes would suggest.

## Introduction

The definition of virulence genes is an area of contention in microbiology. At one extreme virulence genes may be regarded as only those that encode products inducing a direct pathological effect in host tissue, while at the other they can be regarded as any gene that is required for optimal survival *in vivo* but that is not essential *in vitro*. These two distinct definitions are particularly problematic in pathogens like the mycoplasmas, which generally do not appear to exert a direct pathological effect on the tissues of their host and are also highly adapted host specific parasites. Pathological effects of mycoplasmas, for the most part, appear to result from the host response to infection, rather than from the effects of any one gene product on host tissues. In addition, the high level of adaptation to a parasitic lifestyle results in a requirement for media containing high concentrations of complex components, which in some cases can be relatively host specific. For these reasons, in most cases investigators have regarded mycoplasma virulence genes as any gene that is not required for growth in conventional culture media used for that species, but that is required for optimal colonization of, persistence in or pathological effects on the host. This definition is most likely to enable progress towards understanding the pathogenesis of the diseases these organisms induce in their hosts.

In a recent review, Gyles (2011) suggested that there were a number of recurring themes in the pathogenesis of bacterial infections in animals. Of those he identified, the themes most relevant to understanding mycoplasmoses are adherence, invasion, toxin production, mimicry, evading the immune system, multifunctional proteins and protein secretion. Although there are a couple of species in which toxins have been identified, there is little evidence that toxins are a typical contributor to mycoplasmosis across multiple species, of attack on the immune system or of a role for iron acquisition in virulence (even though there is evidence that iron deprivation has a significant effect on the mycoplasma transcriptome; Madsen *et al.*, 2006), the other common themes identified by Gyles, in the pathogenesis of most mycoplasmoses. However, given the highly specialized niches occupied by the mycoplasmas, it seems reasonable to include an additional theme for these pathogens, efficient scavenging of complex nutrients from the host. It also seems likely that the unique mechanisms involved in motility in mycoplasmas may also play a role in virulence, although the close relationship between adhesion and motility is likely to make distinction of the roles of these two functions difficult.

An important adjunct to studies of the role of virulence genes in pathogenesis is an understanding of their regulation during the course of infection, as efficient regulation ensures that these genes are only expressed at times during the infection cycle that their products will assist the pathogen.

Global mutagenesis studies have now been performed on several mycoplasma species (Glass *et al.*, 2006; French *et al.*, 2008), with the aim of defining the minimal essential set of genes for independent life *in vitro*, but a secondary outcome of these studies is the definition of a set of genes in each of these pathogens that are likely to be utilized predominantly during infection of the host, and thus a subset of genes that can be expected to include many of the virulence genes. In addition to these studies, there have been several published reports using *in vivo* selection to identify mutants carrying transposon insertions in genes likely to be required specifically for efficient colonization and persistence *in vivo* (Hudson *et al.*, 2006; Szczepanek *et al.*, 2010). A notable feature of mycoplasmoses is that, in general, the lesions induced by different species in different hosts are quite similar, with the most distinguishing feature being the host range and, to some extent, the tissue tropism of the different species (some of the species in the mycoides cluster, which commonly cause significant tissue necrosis, being a significant exception). This suggests that many of the genes responsible for pathology *in vivo* are likely to be similar in different species, and that the most dramatic differences between different pathogenic species are most likely to be found in the gene products used for colonization. This is borne out by our current understanding of virulence genes in mycoplasmas, with the greatest interspecies variation being seen in the genes playing a role in adhesion (Table 4.1).

## Adhesins

Adherence to mucosal epithelial cells is clearly a prerequisite for successful colonization of a host and subsequent disease. The proteins involved in adhesion have been studied in considerable detail in some mycoplasmas, and are reviewed elsewhere in this volume. However, while much is understood of the proteins involved in adhesion in some members of the pneumoniae phylogenic group, and increasingly more of the complex interactions between these proteins (Balish, 2006), very little is known about adherence in most other mycoplasmas, and outside the pneumoniae group there are no recognizable homologues of the proteins involved in adherence in *Mycoplasma pneumoniae*. This, and the clearly distinct morphological appearance of many species in the pneumoniae group, many of which have a characteristic terminal bleb, or attachment organelle, suggests that the proteins and mechanisms involved in adherence may differ significantly in different mycoplasma species. Indeed, homologues of a putative adhesin found in *Mycoplasma synoviae*, the variable lipoprotein VlhA, has only been identified in this species and two other avian mycoplasmas, in which it appears to have been acquired by horizontal gene transfer (Noormohammadi *et al.*, 1998; Markham *et al.*, 1999). That such an apparently fundamental process for a mucosal inhabitant has not been highly conserved throughout evolution is surprising, but possibly reflects the intense selective pressure likely to be exerted on structures of crucial importance by the host mucosal immune response. The possible involvement of a number of proteins with other

**Table 4.1** Virulence genes in mycoplasmas

| Gene name | Species | Homologues in other mycoplasma species | Function of product | Role demonstrated | | Reference |
|---|---|---|---|---|---|---|
| | | | | *In vitro* | *In vivo* | |
| *gapA* | *M. gallisepticum* | P1 (*M. pneumoniae*), MgPa (*M. genitalium*) | Adhesin | Yes | Yes | Papazisi *et al.* (2002) |
| *crmA* | *M. gallisepticum* | ORF6 (*M. pneumoniae*) | Cytadhesin accessory protein | Yes | Yes | Papazisi *et al.* (2002) |
| *lpd* | *M. gallisepticum* | *lpd* (many species) | Subunit of pyruvate dehydrogenase | | Yes | Hudson *et al.* (2006) |
| *mslA* | *M. gallisepticum* | Multiple paralogues in *M. pneumoniae*, homologues in most species | Unknown | | Yes | Szczepanek *et al.* (2010) |
| *gtsABC* | *M. mycoides* ss *mycoides SC* | *gtsABC* (many species) | Glycerol transport | Yes | Yes | Vilei and Frey (2001) |
| *mam* | *M. arthritidis* | | Superantigen and DNase | Yes | Yes | Luo *et al.* (2008) |
| *glpO* | *M. mycoides* ss *mycoides SC* | *glpO* (many species) | Glycerol-3-phosphate oxidase – hydrogen peroxide production | Yes | Yes | Pilo *et al.* (2005) |
| *P1* | *M. pneumoniae* | MgPa (*M. genitalium*), *gapA* (*M. pneumoniae*) | Cytadhesin | Yes | Yes | Baseman *et al.* (1982) |
| *hmw1* | *M. pneumoniae* | *hmw1* (*M. genitalium, M. gallisepticum*) | Cytadhesin accessory protein | Yes | Yes | Krause *et al.* (1982) |
| *hmw2* | *M. pneumoniae* | *hmw2* (*M. genitalium, M. gallisepticum*) | Cytadhesin accessory protein | Yes | Yes | Krause *et al.* (1997) |
| *hmw3* | *M. pneumoniae* | *hmw3* (*M. genitalium*) | Cytadhesin accessory protein | Yes | Yes | Krause *et al.* (1982) |
| *P30* | *M. pneumoniae* | *P32* (*M. genitalium*), *mgc2* (*M. gallisepticum*) | Cytadhesin | Yes | | Baseman *et al.* (1987) |
| *mpn142* (*orf6*) | *M. pneumoniae* | *crmA* (*M. gallisepticum*) *mgpC* (*M. genitalium*) | Cytadhesin accessory proteins P40 and P90 | Yes | Yes | Waldo *et al.* (2005) |
| *glpD* | *M. pneumoniae* | *glpD* (many species) | Glycerol-3-phosphate oxidase – hydrogen peroxide production | Yes | | Hames *et al.* (2009) |
| *mpn372* | *M. pneumoniae* | | Vacuolating toxin | Yes | | Kannan *et al.* (2005) |
| *nifS* | *M. agalactiae* | *nifS* (several species) | Putative cysteine desulfurase | Yes | | Baranowski *et al.* (2010) |

functions moonlighting as adhesins is discussed in detail elsewhere in this volume.

## Invasion

A number of mycoplasma cells have now been shown to be capable of invading cells (Baseman *et al.*, 1995; Winner *et al.*, 2000; Much *et al.*, 2002; Yavlovich *et al.*, 2004a,b; van der Merwe *et al.*, 2010; Buim *et al.*, 2011), at least *in vitro*, but as yet there has been no characterization of the mechanisms involved. In one *in vitro* model invasion has been shown to be enhanced by protease treatment of the mycoplasma cells, suggesting that protease cleavage of a cell surface protein may activate an adhesin or enhance exposure of an adhesin, facilitating more effective binding (Kornspan *et al.*, 2010).

It is clear that a number of mycoplasmas are able to invade at the site of infection and spread systemically, as they establish foci of infection at sites such as the joints and meninges.

## Toxin production

While there is limited evidence for the role of typical bacterial exotoxins in the pathogenesis of most mycoplasmoses, there is clear evidence that some mycoplasmas can exert a cytotoxic effect on host cells. The clearest evidence for the release of a cytotoxic factor playing a role in the virulence of several mycoplasmas is the production of hydrogen peroxide during the catabolism of glycerol. However, recent studies on the human pathogen *M. pneumoniae* have identified a unique cytotoxin that may play a significant role in the disease it causes.

### Production of hydrogen peroxide

Glycerol metabolism has been shown to play a significant role in the cytotoxic effects of two mycoplasma species, *M. mycoides* subspecies *mycoides* small colony type and *M. pneumoniae* (Pilo *et al.*, 2005; Hames *et al.*, 2009). While the *glyF* and *glyK* genes, encoding the glycerol uptake facilitator and the glycerol kinase, respectively, appear to be essential, at least in *M. pneumoniae*, *glyD*, which encodes glycerol-3-phosphate oxidase, is dispensable and also probably plays a significant role in virulence, as its activity results in generation of hydrogen peroxide, which has a significant cytotoxic effect, at least *in vitro* (Hames *et al.*, 2009). In contrast to other bacteria, in which these glycerol utilization genes are only expressed in the absence of preferred carbon sources and the concurrent presence of glycerol, they are constitutively expressed in *M. pneumoniae*. Glycerol-3-phosphate is generated from glycerophospholipids, which are particularly abundant in the respiratory tract, by deacylation by lipases to release glycerophosphodiesters, from which glycerol-3-phosphate is released by the glycerophosphodiesterase GlpQ (Schmidl *et al.*, 2011). Thus GlpQ is required for production of hydrogen peroxide from glycerophospholipids and can also be classified as a virulence determinant. A paralogue of GlpQ in *M. pneumoniae* does not appear to have either glycerophosphodiesterase activity or to play a role in gene regulation in the presence of glycerol (see below).

While glycerol-3-phosphate oxidase is a cytoplasmic protein in *M. pneumoniae*, GlpO, the glycerol-3-phosphate oxidase of *M. mycoides* subspecies *mycoides* small colony type, which is a much more prolific producer of hydrogen peroxide, is located in the membrane and at least partially exposed on the cell surface (Pilo *et al.*, 2005). The greater glycerol-3-phosphate oxidase activity of *M. mycoides* subspecies *mycoides* small colony type is linked to the presence of a highly efficient ABC glycerol transporter, GtsABC, that is not present in *M. pneumoniae*. Naturally occurring mutants of *M. mycoides* subspecies *mycoides* small colony type, in which the 3′ half of *gtsB* and all of *gtsC* are deleted (apparently as a result of transposon-mediated mutagenesis), are much less virulent than strains with a functional transporter, and produce much less hydrogen peroxide in the presence of glycerol (Djordjevic *et al.*, 2001; Bischof *et al.*, 2008). Furthermore, antibodies against the glycerol-3-phosphate oxidase and GtsB are able to neutralize hydrogen peroxide production and cytotoxicity *in vitro* (Vilei and Frey, 2001; Bischof *et al.*, 2008, 2009). *In vitro* studies suggest that the cytotoxic effect is mediated by translocation of hydrogen peroxide into eukaryotic cells, and that this translocation requires efficient attachment to the target cells (Bischof *et al.*, 2008).

### CARDS toxin

Recent studies have identified an ADP-ribosylating and vacuolating toxin in *M. pneumoniae* that has been named the community-acquired respiratory distress syndrome (CARDS) toxin (Kannan and Baseman, 2006). Homologues of the gene for the toxin, *mpn372*, can be identified in the human mycoplasma *M. penetrans* (MYPE9110) and the avian mycoplasma *M. iowae*, although not in other members of the pneumoniae phylogenic group. All three homologues contain a pertussis toxin subunit 1 domain at their amino terminal end. Recombinant CARDS toxin induces formation of coalescing vacuoles in CHO and HeLa cells, and mutations that abolish ADP-ribosylation activity result in a dramatic decrease in this effect. The toxin can slow ciliary activity and induce vacuolation and nuclear fragmentation in the epithelium of baboon tracheal organ cultures. It appears to be a surface exposed membrane protein and can bind lung surfactant A (Kannan *et al.*, 2005). Studies on the *M. penetrans* homologue have shown that it can also induce vacuolation in HeLa cells, and that the carboxyl end of this protein is required for binding to cells in culture and subsequent internalization of the protein (Johnson *et al.*, 2009). Exposure of the respiratory tract of BalbC mice to recombinant CARDS results in a transient proinflammatory cytokine response, with increased expression of IL-1, IL-6, IL-12p40, IL-17 and TNF-alpha, as well as increased expression of G-CSF, KC, MCP-1, MIP-1alpha and MIP-1beta. These changes are accompanied by vacuolation of the bronchial and bronchiolar epithelium, cilial damage and, at higher doses, infiltration of neutrophils and lymphocytes around pulmonary and bronchial blood vessels, and a neutrophilic exudate into the alveoli (Hardy *et al.*, 2009). The respiratory tract response to CARDS toxin is characterized by an eosinophilia between day 7 and 14 after exposure, although the inflammatory response persists for up to 56 days after a single exposure. The eosinophilia coincides with a dramatic increase in expression of the T helper 2 chemokines CCL17 and CCL22 and the effector cytokines IL-4 and IL-13 (Medina *et al.*, 2012). A baboon exposed to recombinant CARDS toxin also developed lymphocytic bronchiolitis (Hardy *et al.*, 2009). Recombinant CARDS toxin also induces mucus metaplasia in naive BalbC mice (Hardy *et al.*, 2009). However, mucus metaplasia (Wawegama *et al.*, 2012) and eosinophilia (Bakshi *et al.*, 2006) are also seen in the lungs of animals infected with mycoplasmas that do not have homologues of the CARDS toxin gene, indicating that this is not the only factor produced by mycoplasmas capable of inducing increased mucus production. Mice exposed to CARDS toxin develop prolonged airway obstruction and hyperreactivity, and exhibit increased airway resistance and decreased lung compliance after methacholine challenge for up to 14 days. These changes are suggestive of allergic airway hyperresponsiveness and have been shown to be dependent on the presence of CD4-positive T-cells (Medina *et al.*, 2012). CARDS toxin expression appears to be up-regulated *in vivo*, and is induced *in vitro* by infection of cultured mammalian cells (Kannan *et al.*, 2010).

## Mimicry

There have been a number of reports implicating antigenic mimicry in the pathogenesis of mycoplasmoses. Up to 25% of people with pulmonary disease due to *M. pneumoniae* also suffer from extrapulmonary disease, most frequently of neurological origin, and autoimmune responses have been suggested as a cause, although the mycoplasma antigens that might be responsible have not been identified. However, it has been cautioned that, as many of the original studies preceded the availability of highly sensitive methods for detection, at least some of these extrapulmonary manifestations could result from unrecognised localization after systemic invasion from the respiratory tract (Waites and Talkington, 2004). That this is likely to be the case is reinforced by the clear evidence that extrapulmonary sequelae in a number of mycoplasmal pathogens of domestic animals are clearly a result of systemic invasion and localization. There are a number of reports implicating *M. pneumoniae* as a sensitizing agent in the aetiology of some cases of Guillain-Barré-Strohl syndrome, an acute peripheral neuropathy (Kuwabara, 2004). The autoreactive antibodies that appear to be induced bind to galactocerebroside, although the mycoplasma antigen responsible for inducing these antibodies has not been identified (Ang *et al.*, 2002). More recently, autoreactive antibodies induced by the *M. pneumoniae* pyruvate dehydrogenase E2 subunit have been implicated in the aetiology of primary biliary cirrhosis (Berg *et al.*, 2009).

## Immune evasion

The high prevalence of multigene families in mycoplasmas, particularly encoding products exposed on the cell surface, has been interpreted as an immune evasion strategy. This area of mycoplasma molecular biology has been reviewed in detail elsewhere in this volume. The capacity of some mycoplasmas to invade cells has also been interpreted as facilitating evasion from immune attack (McGowin *et al.*, 2009; van der Merwe *et al.*, 2010; Buim *et al.*, 2011), although intracellular invasion has thus far only been demonstrated *in vivo* for the avian pathogen *M. gallisepticum* (Vogl *et al.*, 2008) and there is no definitive evidence of its role in immune evasion. The capacity to invade cells appears to be correlated with adhesion in *M. gallisepticum* (Vogl *et al.*, 2008), but the genes required have not yet been elucidated.

## Immunostimulation

A prominent feature of mycoplasmoses is the proliferation of lymphoid cells in infected tissues. While in one specific case the stimulus for immunopathology is clearly a specific superantigen (see below), in most species the response appears to be linked (Metzger *et al.*, 1995), at least partially, to the unusual structure of mycoplasma lipoproteins, which are only diacylated, on the sulphydryl group of the amino terminal cysteine. Mycoplasma lipoproteins have been shown to bind to Toll-like receptors 2 and 6, resulting in high levels of activation of monocytes and macrophages, as well as cultured fibroblasts, with subsequent release of pro-inflammatory cytokines. This response also induces expression of ICAM-1, an important intercellular adhesion molecule for leucocytes, and thus is likely to promote transmigration of inflammatory cells into infected tissue (Okusawa *et al.*, 2004). While the diacylated glycerol is clearly required for activity, it has also been shown that the peptide sequence also exerts an influence on the level of activation, suggesting that specific lipoproteins may be more potent in the induction of immunopathology in mycoplasmoses. It has been suggested that mycoplasma lipoproteins should be regarded as endotoxins, as their actions and effects are very similar to those of lipopolysaccharide.

## Immunosuppression

A number of studies have suggested that infection with some mycoplasmas can be immunosuppressive. Evidence for this immunosuppressive effect includes the exacerbation of disease caused by other bacterial and viral pathogens during coinfection, as well as studies showing that the immunosuppressive effects of infection can be reversed by enhancing cytokine expression. There have been a number of studies showing that infection of pigs with *M. hyopneumoniae* can have a profound effect on susceptibility to several viral and bacterial pathogens. Experimental infection of pigs with *M. hyopneumoniae* 21 days before, at the same time as, or 10 days after infection with the arterivirus porcine reproductive and respiratory syndrome virus results in more severe lesions and a longer duration of pneumonia typical of that caused by porcine reproductive and respiratory syndrome virus, even when the mycoplasmal pneumonia is minimal (Thacker *et al.*, 1999). Pigs infected with *M. hyopneumoniae* had slower viral clearance than pigs infected with either pathogen alone, and increased levels of IL-8 and IL-1β in bronchoalveolar lavage fluid (Thanawongnuwech *et al.*, 2004). Similarly, pleuropneumonia caused by *Actinobacillus pleuropneumoniae* was more severe in pigs that had been experimentally infected 4 weeks earlier with *M. hyopneumoniae* (Marois *et al.*, 2009). While these enhanced pathological effects in the respiratory tract could result from exacerbation of inflammation, infection with *M. hyopneumoniae* 2 weeks before infection with porcine circovirus type 2 (PCV2) has also been shown to increase the severity of both pulmonary and lymphoid lesions caused by this virus (Opriessnig *et al.*, 2004). Furthermore, concentrations of PCV2 in serum are higher in pigs previously infected with *M. hyopneumoniae,* and lymphoid depletion is greater in the spleen and tonsils of coinfected pigs, suggesting that the effects are systemic.

Systemic immunosuppression in response to respiratory mycoplasmosis has also been seen in chickens experimentally infected with *M. gallisepticum* (Muneta *et al.*, 2008). Splenic lymphocytes from infected birds have significantly reduced responses, as measured by interferon gamma production, to concanavalin A and lipopolysaccharide, although not to *M. gallisepticum* whole cell proteins. *M. mycoides* subspecies *mycoides* small colony type, *Mycoplasma ovipneumoniae* and *M. bovis* inhibit mitogen-induced lymphocyte blastogenesis *in vitro* (Vanden Bush and Rosenbusch, 2004; Dedieu and Balcer-Rodrigues, 2006; Shahzad *et al.*, 2010). In *M. mycoides* subspecies *mycoides* small colony type and *M. ovipneumoniae* this immunosuppressive effect has been shown to be exerted only by viable mycoplasma cells, and to extend across all lymphocyte subsets. In contrast, the effect of *M. bovis* has been shown to be associated with a 26 amino acid peptide derived from the carboxyl end of one of the variable surface lipoproteins, VspL (Vanden Bush and Rosenbusch, 2004). In addition, macrophage responsiveness to lipopolysaccharide has been shown to result from previous exposure to the mycoplasma lipopeptide TLR-2 ligand (Sato *et al.*, 2002). The immunosuppressive effect of infection with *M. gallisepticum* is reversed in a recombinant strain expressing interferon gamma (Muneta *et al.*, 2008).

## Multifunctional proteins

In recent years a number of mycoplasma proteins, particularly those exposed on the cell surface, have been shown to have multiple functions. This is perhaps not surprising in organisms with a minimal genome, as the use of a single protein to fulfil multiple roles is likely to contribute significant genomic economy. The roles and significance of multifunctional proteins are reviewed in detail elsewhere in this volume.

## Protein secretion

In many bacterial pathogens, specialized mechanisms for protein export are associated with delivery of virulence proteins into host cells. Mycoplasmas are notably deficient in specialized protein secretion pathways, with annotated mechanisms limited to the Sec pathway. However, it is evident that there is significant post-translational cleavage of many mycoplasma surface proteins, which may enable the delivery of a range of modified proteins into the immediate environment of the mycoplasma cell. The post-translational modification of proteins in mycoplasmas has been dealt with in detail elsewhere in this volume.

## Scavenging complex nutrients

The complex nutritional requirements of mycoplasmas when cultured *in vitro*, and their relative dearth of biosynthetic capacity, suggest that they require efficient mechanisms for acquiring complex nutrients from their surroundings. These mechanisms are likely to include an ability to degrade macromolecules into importable subunits, as well as efficient binding proteins and transport mechanisms. While there are a relatively large number of recognizable ABC transporter operons in mycoplasma genomes, the specific substrates for most of these transporters are unknown. Many of these operons include a predicted lipoprotein, and it is presumed that in many cases these lipoproteins are likely to fulfil the role of the periplasmic binding proteins of Gram-negative bacteria – increasing the efficiency of import by binding substrates with high affinity and delivering them to the transport system for import. The *oppA-D* operon has been shown to encode an oligopeptide transporter, and the lipoprotein encoded by *oppA* to bind oligopeptides, although its significance in survival *in vivo* has not yet been demonstrated (Henrich *et al.*, 1999). However, it is notable that this operon has been subject to horizontal gene transfer between *M. synoviae* and *M. gallisepticum*, with the result that *M. gallisepticum* possesses two copies of the operon. The presence of frameshift mutations in one operon and the absence of the proximal end of the operon in the other suggests that neither of these operons is fully functional, and that they therefore complement each other (Papazisi *et al.*, 2003; Vasconcelos *et al.*, 2005; Sirand-Pugnet *et al.*, 2007). Another conserved ABC transporter of uncertain significance includes two lipoprotein genes. One of these genes has been shown to function as a exonuclease, suggesting that the operon may play a role in importation of nucleotides into the cell (Schmidt *et al.*, 2007; Browning *et al.*, 2011). The second lipoprotein has been named *mslA* and has been shown to be required for efficient colonization of the respiratory tract by *M. gallisepticum* (Szczepanek *et al.*, 2010). While the function of this virulence gene is yet to be determined, it seems likely that it may also play a role in efficient importation of nucleotides into the mycoplasma cell. The presence of multiple paralogues of this gene in several species, in particular in *M. pneumoniae*, in which there are 10 paralogues (Himmelreich *et al.*, 1996; Dandekar *et al.*, 2000), suggests that it may be under significant selective pressure by the immune response, although there is little evidence for variation in expression of these multiple paralogues *in vitro* (Hallamaa *et al.*, 2006, 2008).

## Other virulence factors

### MAM superantigen

One specialized virulence factor that, thus far, appears unique to a specific species, is the *Mycoplasma arthritidis* superantigen, also known as the *M. arthritidis* mitogen, or MAM. Comprehensive studies by Cole and his colleagues over many years have shown that MAM is a typical superantigen, binding directly to MHC class II molecules outside the antigen-binding groove, where it is recognized by the variable regions of the T-cell receptor, without MHC restriction. However, MAM is unusual among the bacterial superantigens in having a defined enzymatic function, as a DNase (Diedershagen *et al.*, 2007). In addition, it has a strong preference for class II H2-E or HLA-DR, can dimerize MHC molecules (Zhao *et al.*, 2004), binds the CDR3 region of the TCR as well as the V beta chain, and interacts directly with Toll-like receptor (TLR)-2 and TLR-4 on macrophages, exerting a significant influence on cytokine responses (Mu *et al.*, 2005). When both TLR-2 and TLR-4 are present, MAM induces down regulation of TLR-2 and induction of IL-6 and IL-17 production (Mu *et al.*, 2011). When only TLR-2 is present the cytokine response is dominated by IL-1β and an inflammatory type 1 T-cell response. Mice expressing TLR-2 and not TLR-4 are highly susceptible to lethal toxic shock when infected with *M. arthritidis*. In contrast, mice that express both TLR-2 and TLR-4 and develop a type 2 cytokine response predominantly develop arthritis (Mu *et al.*, 2011). Mice that are highly susceptible to toxic shock can be protected by blockade of the co-stimulatory B7-1 molecule on antigen presenting cells, which switches the effect of MAM from a type 1 to a type 2 cytokine response (Mu *et al.*, 2006). This also results in increased susceptibility to development of arthritis. Mutagenesis studies that generated strains of *M. arthritidis* that either overproduced MAM or did not produce it have shown that production of MAM is correlated with the induction of fatal toxic shock in DBA/2J mice, but not with induction of arthritis in CBA/J or DBA/2J mice (Luo *et al.*, 2008).

### Pyruvate dehydrogenase complex

Although metabolic enzymes are not typically regarded as virulence factors, the dihydrolipoamide dehydrogenase gene, which is not essential *in vitro*, has been shown to play a significant role in the capacity of *M. gallisepticum* to colonize the respiratory tract of chickens and to induce lesions in the upper and lower regions of the tract (Hudson *et al.*, 2006). While disruption of this gene did not eliminate pyruvate dehydrogenase activity *in vitro*, it significantly reduced it, and presumably the capacity of this enzyme complex to contribute to production of ATP by the cell. This might be expected to reduce the capacity of a pathogen to grow as efficiently in a nutrient limited environment.

## Regulation of virulence genes

There have been few studies of the regulation of virulence genes in mycoplasmas. Genomic studies have failed to identify many of the regulatory systems common to pathogenic bacteria, such as the two component regulatory systems, and very few transcriptional regulators (for example, only three have been identified in *M. pneumoniae*) (Himmelreich *et al.*, 1996; Dandekar *et al.*, 2000). The relative paucity of identifiable regulatory genes has led to the suggestion that either mycoplasmas reside in such a stable environment that they require few changes in gene expression and thus have much less need for complex regulatory networks, or that they must possess novel regulatory mechanisms that are not apparent in other bacteria. It has also been suggested that they may achieve more regulation post-translationally, possibly by differential protein cleavage, or by other modifications, such as phosphorylation and dephosphorylation. The direct effect exerted by these regulatory networks on virulence is yet to be examined, but their effect on known virulence genes suggests that they probably contribute to virulence.

### Protein phosphorylation and acetylation

A study of the phosphoproteome of *M. pneumoniae* identified 63 proteins, or around 10% of the total proteome, that were phosphorylated. This is a much higher proportion than is seen in other bacteria, suggesting that phosphorylation and dephosphorylation may be relatively more important in the *Mollicutes* (Schmidl *et al.*, 2010). While around 33% of the phosphorylated proteins were part of the central metabolic pathways and thus could not be regarded as playing a role in virulence, one of the only two annotated protein kinases in the *M. pneumoniae* genome, PrkC, has been shown by mutagenesis studies to phosphorylate the adhesion related proteins HMW3 and P41, as well as the cell surface protein MPN474, and the uncharacterised protein MPN256. The same proteins are dephosphorylated by the only annotated protein phosphatase in *M. pneumoniae,* PrpC. All but one of the remaining phosphoproteins, including three of unknown function, MPN349, MPN555 and MPN677, are either modified by unidentified kinases or are autophosphorylated, suggesting there is much to be understood about the mechanisms and effects of protein phosphorylation in mycoplasmas.

Recent investigations have also revealed extensive lysine acetylation of mycoplasma proteins and suggest that these modifications may play major regulatory roles. Proteomic studies of *M. pneumoniae* mutants, in which the two known protein kinases and the known phosphatase have been disrupted, have revealed effects not only on protein phosphorylation, but also on lysine acetylation (van Noort *et al.*, 2012). Mutation of the *pknB* kinase gene was found to reduce lysine acetylation, while mutation of the *hypK* kinase gene increases acetylation. Similarly, mutation of either of the two putative N-acetyltransferase genes in the genome, *mpn027* or *mpn114,* has effects on both protein acetylation and phosphorylation. Among the proteins that appear to be reciprocally regulated by these two post-translational modification systems are a number of the cytadherence proteins, as well as other proteins of unknown function. Notably, very few lipoproteins are modified by phosphorylation or acetylation, suggesting that these systems regulate intracellular proteins. A number of the sites of modification are predicted to influence protein–protein interfaces, and particularly oligomerization.

### Regulatory RNA

Recent studies of the *M. pneumoniae* transcriptome have suggested that much of the regulation of gene expression in mycoplasmas may be effected by regulatory RNA molecules. Detailed *in vitro* transcriptomic studies of *M. pneumoniae* have revealed that a high proportion of coding sequences are covered by anti-sense transcripts, suggesting that anti-sense RNA may be a significant regulatory mechanism in mycoplasmas (Guell *et al.*, 2009). In addition, these studies found evidence for production of alternative transcripts within operons under different environmental conditions, implying that both alternative promoters and terminators may be regulating transcription. Although the specific role of these recently discovered regulatory mechanisms in virulence is yet to be demonstrated, it is clear from earlier studies on transcription in the *hmw* cytadherence gene operon (Waldo *et al.*, 1999) that internal activation and termination in operons is likely to play a part in control of expression of virulence genes during infection and the recent transcriptomic studies suggest that it is much more widespread than previously recognized.

### Transcriptional regulation by GlpQ

In *M. pneumoniae* the glycerophosphodiesterase GlpQ appears to be involved in transcriptional regulation of expression of the glycerol uptake facilitator, as well as three uncharacterised lipoproteins and a putative metal ion ABC transporter (Schmidl *et al.*, 2011). This regulation may be associated with a palindromic element adjacent to or overlying the −10 box of the promoter (for repression) or upstream of the promoter (for activation), although direct involvement of GlpQ and these elements in regulation is yet to be established definitively.

## Challenges for the future

The development of a greater array of tools for manipulation of mycoplasmas over the last 15 years has enhanced our capacity to dissect the role of specific genes in virulence, although there has still only been limited application of these tools to large scale dissections of virulence. In many species complementation remains problematic, so definitive identification of virulence genes often remains difficult. Furthermore a large proportion of the genes in mycoplasma genomes, and particularly those likely to be located on the cell surface, are hypothetical or have a putative function. Thus far, use of large scale mutagenesis studies to explore virulence *in vivo* have been limited to *M. gallisepticum*. It is notable that the two genes identified in the studies using *in vivo* screening were involved in catabolism or, putatively, in nutrient transport. Neither of these genes was essential for survival *in vivo*, but they were required for optimal persistence and to cause severe disease. Thus screening for virulence genes cannot focus only on genes essential for survival in the infected host, but rather needs to also examine genes that may be required for optimal pathogen function *in vivo*. This is particularly important in studies using *in vivo* survival as a selection tool to identify mutants within a pool that may have transposon insertions in virulence genes. It is probable that quantitative screening of pools of mutants recovered from infected hosts using high throughput sequencing may yield more information about mycoplasma virulence genes. Transposon-directed insertion site sequencing of mutants of *E. coli* O157:H7 recovered from experimentally infected calves has been used to determine the relative fitness of mutants in the pool and, as a result, identify many more genes influencing fitness *in vivo* than had been identified previously using signature-tagged mutagenesis (Eckert *et al.*, 2011).

A further major challenge in the years ahead will be dissecting not only the contribution of specific genes to virulence, but also defining the biochemical function(s) of these genes so that the mechanisms involved in their contribution can be appreciated. While systems biology approaches may provide some insights into function, there is limited capacity to apply high throughput techniques to determining the biochemical function of proteins, so this is likely to be the point of slowest progress in developing greater understanding of virulence in mycoplasmas, although one of the more significant hurdles in expressing recombinant mycoplasma proteins for functional analyses, the use of TGA as a tryptophan codon, can now be readily overcome using gene synthesis. Initial identification of the role of the P1 adhesin in *M. pneumoniae* resulted from studies of attenuated strains generated using nitrosoguanidine as vaccine candidates (Lipman *et al.*, 1969). Several attenuated vaccine strains that have been generated using the same mutagen are now widely available for control of mycoplasmoses in domestic animals (Whithear *et al.*, 1990; Whithear, 1996; Markham *et al.*, 1998; Browning *et al.*, 2012). It is clear that the crucial attenuating mutation in one of these attenuated strains is not within the P1 homologue in *M. gallisepticum* (Shil *et al.*, 2011), and *M. synoviae* does not possess a P1 homologue. Thus close examination of these attenuated mutants is likely to reveal additional genes that contribute to the virulence of mycoplasmas. A final source of candidate virulence genes that require further investigation are those genes identified by transcriptional profiling to be significantly up or down regulated either *in vivo* (Madsen *et al.*, 2008), during culture in association with various host cell lines (Skapski *et al.*, 2011), or under nutritionally restrictive conditions mimicking those seen *in vivo* (Madsen *et al.*, 2006).

## References

Ang, C.W., Tio-Gillen, A.P., Groen, J., Herbrink, P., Jacobs, B.C., Van Koningsveld, R., Osterhaus, A.D., Van der Meche, F.G., and van Doorn, P.A. (2002). Cross-reactive anti-galactocerebroside antibodies and *Mycoplasma pneumoniae* infections in Guillain-Barre syndrome. J. Neuroimmunol. *130*, 179–183.

Bakshi, C.S., Malik, M., Carrico, P.M., and Sellati, T.J. (2006). T-bet deficiency facilitates airway colonization by *Mycoplasma pulmonis* in a murine model of asthma. J. Immunol. *177*, 1786–1795.

Balish, M.F. (2006). Subcellular structures of mycoplasmas. Front. Biosci. *11*, 2017–2027.

Baranowski, E., Guiral, S., Sagne, E., Skapski, A., and Citti, C. (2010). Critical role of dispensable genes in *Mycoplasma agalactiae* interaction with mammalian cells. Infect. Immun. *78*, 1542–1551.

Baseman, J.B., Cole, R.M., Krause, D.C., and Leith, D.K. (1982). Molecular basis for cytadsorption of *Mycoplasma pneumoniae*. J. Bacteriol. *151*, 1514–1522.

Baseman, J.B., Morrison-Plummer, J., Drouillard, D., Puleo-Scheppke, B., Tryon, V.V., and Holt, S.C. (1987). Identification of a 32-kilodalton protein of *Mycoplasma pneumoniae* associated with hemadsorption. Isr. J. Med. Sci. 23, 474–479.

Baseman, J.B., Lange, M., Criscimagna, N.L., Giron, J.A., and Thomas, C.A. (1995). Interplay between mycoplasmas and host target cells. Microb. Pathog. *19*, 105–116.

Berg, C.P., Kannan, T.R., Klein, R., Gregor, M., Baseman, J.B., Wesselborg, S., Lauber, K., and Stein, G.M. (2009). Mycoplasma antigens as a possible trigger for the induction of antimitochondrial antibodies in primary biliary cirrhosis. Liver Int. *29*, 797–809.

Bischof, D.F., Janis, C., Vilei, E.M., Bertoni, G., and Frey, J. (2008). Cytotoxicity of *Mycoplasma mycoides* subsp. *mycoides* small colony type to bovine epithelial cells. Infect. Immun. *76*, 263–269.

Bischof, D.F., Vilei, E.M., and Frey, J. (2009). Functional and antigenic properties of GlpO from *Mycoplasma mycoides* subsp. *mycoides* SC: characterization of a flavin adenine dinucleotide-binding site deletion mutant. Vet. Res. *40*, 35.

Browning, G.F., Marenda, M.S., Noormohammadi, A.H., and Markham, P.F. (2011). The central role of lipoproteins in the pathogenesis of mycoplasmoses. Vet. Microbiol. *153*, 44–50.

Browning, G.F., Kanci, A., and Markham, P.F. (2012). Developing attenuated vaccines to control mycoplasmoses. Microbiol. Aust. *32*, 121–122.

Buim, M.R., Buzinhani, M., Yamaguti, M., Oliveira, R.C., Mettifogo, E., Ueno, P.M., Timenetsky, J., Santelli, G.M., and Ferreira, A.J. (2011). *Mycoplasma synoviae* cell invasion: elucidation of the Mycoplasma pathogenesis in chicken. Comp. Immunol. Microbiol. Infect. Dis. *34*, 41–47.

Dandekar, T., Huynen, M., Regula, J.T., Ueberle, B., Zimmermann, C.U., Andrade, M.A., Doerks, T., Sanchez-Pulido, L., Snel, B., Suyama, M., *et al.* (2000). Re-annotating the *Mycoplasma pneumoniae* genome sequence: adding value, function and reading frames. Nucleic Acids Res. *28*, 3278–3288.

Dedieu, L., and Balcer-Rodrigues, V. (2006). Viable *Mycoplasma mycoides* ssp. *mycoides* small colony-mediated depression of the bovine cell responsiveness to the mitogen concanavalin A. Scand. J. Immunol. *64*, 376–381.

Diedershagen, M., Overbeck, S., Arlt, S., Plumakers, B., Lintges, M., and Rink, L. (2007). *Mycoplasma arthritidis*-derived superantigen (MAM) displays DNase activity. FEMS Immunol. Med. Microbiol. *49*, 266–271.

Djordjevic, S.R., Forbes, W.A., Forbes-Faulkner, J., Kuhnert, P., Hum, S., Hornitzky, M.A., Vilei, E.M., and Frey, J. (2001). Genetic diversity among *Mycoplasma* species bovine group 7: clonal isolates from an outbreak of polyarthritis, mastitis, and abortion in dairy cattle. Electrophoresis *22*, 3551–3561.

Eckert, S.E., Dziva, F., Chaudhuri, R.R., Langridge, G.C., Turner, D.J., Pickard, D.J., Maskell, D.J., Thomson, N.R., and Stevens, M.P. (2011). Retrospective application of transposon-directed insertion site sequencing to a library of signature-tagged mini-Tn5Km2 mutants of *Escherichia coli* O157:H7 screened in cattle. J. Bacteriol. *193*, 1771–1776.

French, C.T., Lao, P., Loraine, A.E., Matthews, B.T., Yu, H., and Dybvig, K. (2008). Large-scale transposon mutagenesis of *Mycoplasma pulmonis*. Mol. Microbiol. *69*, 67–76.

Glass, J.I., Assad-Garcia, N., Alperovich, N., Yooseph, S., Lewis, M.R., Maruf, M., Hutchison, C.A., 3rd, Smith, H.O., and Venter, J.C. (2006). Essential genes of a minimal bacterium. Proc. Natl. Acad. Sci. U.S.A. *103*, 425–430.

Guell, M., van Noort, V., Yus, E., Chen, W.H., Leigh-Bell, J., Michalodimitrakis, K., Yamada, T., Arumugam, M., Doerks, T., Kuhner, S., *et al.* (2009). Transcriptome complexity in a genome-reduced bacterium. Science *326*, 1268–1271.

Gyles, C.L. (2011). Relevance in pathogenesis research. Vet. Microbiol. *153*, 2–12.

Hallamaa, K.M., Browning, G.F., and Tang, S.L. (2006). Lipoprotein multigene families in *Mycoplasma pneumoniae*. J. Bacteriol. *188*, 5393–5399.

Hallamaa, K.M., Tang, S.L., Ficorilli, N., and Browning, G.F. (2008). Differential expression of lipoprotein genes in *Mycoplasma pneumoniae* after contact with human lung epithelial cells, and under oxidative and acidic stress. BMC Microbiol. *8*, 124.

Hames, C., Halbedel, S., Hoppert, M., Frey, J., and Stulke, J. (2009). Glycerol metabolism is important for cytotoxicity of *Mycoplasma pneumoniae*. J. Bacteriol. *191*, 747–753.

Hardy, R.D., Coalson, J.J., Peters, J., Chaparro, A., Techasaensiri, C., Cantwell, A.M., Kannan, T.R., Baseman, J.B., and Dube, P.H. (2009). Analysis of pulmonary inflammation and function in the mouse and baboon after exposure to *Mycoplasma pneumoniae* CARDS toxin. PLoS One *4*, e7562.

Henrich, B., Hopfe, M., Kitzerow, A., and Hadding, U. (1999). The adherence-associated lipoprotein P100, encoded by an *opp* operon structure, functions as the oligopeptide-binding domain OppA of a putative oligopeptide transport system in *Mycoplasma hominis*. J. Bacteriol. *181*, 4873–4878.

Himmelreich, R., Hilbert, H., Plagens, H., Pirkl, E., Li, B.C., and Herrmann, R. (1996). Complete sequence analysis of the genome of the bacterium *Mycoplasma pneumoniae*. Nucleic Acids Res. *24*, 4420–4449.

Hudson, P., Gorton, T.S., Papazisi, L., Cecchini, K., Frasca, S., Jr., and Geary, S.J. (2006). Identification of a virulence-associated determinant, dihydrolipoamide dehydrogenase (*lpd*), in *Mycoplasma gallisepticum* through in vivo screening of transposon mutants. Infect. Immun. *74*, 931–939.

Johnson, C., Kannan, T.R., and Baseman, J.B. (2009). Characterization of a unique ADP-ribosyltransferase of *Mycoplasma penetrans*. Infect. Immun. *77*, 4362–4370.

Kannan, T.R., and Baseman, J.B. (2006). ADP-ribosylating and vacuolating cytotoxin of *Mycoplasma pneumoniae* represents unique virulence determinant among bacterial pathogens. Proc. Natl. Acad. Sci. U.S.A. *103*, 6724–6729.

Kannan, T.R., Provenzano, D., Wright, J.R., and Baseman, J.B. (2005). Identification and characterization of human surfactant protein A binding protein of *Mycoplasma pneumoniae*. Infect. Immun. *73*, 2828–2834.

Kannan, T.R., Musatovova, O., Balasubramanian, S., Cagle, M., Jordan, J.L., Krunkosky, T.M., Davis, A., Hardy, R.D., and Baseman, J.B. (2010). *Mycoplasma pneumoniae* Community Acquired Respiratory Distress Syndrome toxin expression reveals growth phase and infection-dependent regulation. Mol. Microbiol. *76*, 1127–1141.

Kornspan, J.D., Tarshis, M., and Rottem, S. (2010). Invasion of melanoma cells by *Mycoplasma hyorhinis*: enhancement by protease treatment. Infect. Immun. *78*, 611–617.

Krause, D.C., Leith, D.K., Wilson, R.M., and Baseman, J.B. (1982). Identification of *Mycoplasma pneumoniae* proteins associated with hemadsorption and virulence. Infect. Immun. *35*, 809–817.

Krause, D.C., Proft, T., Hedreyda, C.T., Hilbert, H., Plagens, H., and Herrmann, R. (1997). Transposon mutagenesis reinforces the correlation between *Mycoplasma pneumoniae* cytoskeletal protein HMW2 and cytadherence. J. Bacteriol. *179*, 2668–2677.

Kuwabara, S. (2004). Guillain-Barre syndrome: epidemiology, pathophysiology and management. Drugs *64*, 597–610.

Lipman, R.P., Clyde, W.A., Jr., and Denny, F.W. (1969). Characteristics of virulent, attenuated, and avirulent *Mycoplasma pneumoniae* strains. J. Bacteriol. *100*, 1037–1043.

Luo, W., Yu, H., Cao, Z., Schoeb, T.R., Marron, M., and Dybvig, K. (2008). Association of *Mycoplasma arthritidis* mitogen with lethal toxicity but not with arthritis in mice. Infect. Immun. *76*, 4989–4998.

Madsen, M.L., Nettleton, D., Thacker, E.L., and Minion, F.C. (2006). Transcriptional profiling of *Mycoplasma hyopneumoniae* during iron depletion using microarrays. Microbiology *152*, 937–944.

Madsen, M.L., Puttamreddy, S., Thacker, E.L., Carruthers, M.D., and Minion, F.C. (2008). Transcriptome changes in *Mycoplasma hyopneumoniae* during infection. Infect. Immun. *76*, 658–663.

Markham, J.F., Morrow, C.J., and Whithear, K.G. (1998). Efficacy of a temperature-sensitive *Mycoplasma synoviae* live vaccine. Avian Dis. *42*, 671–676.

Markham, P.F., Duffy, M.F., Glew, M.D., and Browning, G.F. (1999). A gene family in *Mycoplasma imitans* closely related to the pMGA family of *Mycoplasma gallisepticum*. Microbiology *145*, 2095–2103.

Marois, C., Gottschalk, M., Morvan, H., Fablet, C., Madec, F., and Kobisch, M. (2009). Experimental infection of SPF pigs with *Actinobacillus pleuropneumoniae* serotype 9 alone or in association with *Mycoplasma hyopneumoniae*. Vet. Microbiol. *135*, 283–291.

McGowin, C.L., Popov, V.L., and Pyles, R.B. (2009). Intracellular *Mycoplasma genitalium* infection of human vaginal and cervical epithelial cells elicits distinct patterns of inflammatory cytokine secretion and provides a possible survival niche against macrophage-mediated killing. BMC Microbiol. *9*, 139.

Medina, J.L., Coalson, J.J., Brooks, E.G., Winter, V.T., Chaparro, A., Principe, M.F., Kannan, T.R., Baseman, J.B., and Dube, P.H. (2012). *Mycoplasma pneumoniae* CARDS toxin induces pulmonary eosinophilic and lymphocytic inflammation. Am. J. Respir. Cell Mol. Biol. *46*, 815–822.

Metzger, J.W., Beck-Sickinger, A.G., Loleit, M., Eckert, M., Bessler, W.G., and Jung, G. (1995). Synthetic S-(2,3-dihydroxypropyl)-cysteinyl peptides derived from the N-terminus of the cytochrome subunit of the photoreaction centre of *Rhodopseudomonas viridis* enhance murine splenocyte proliferation. J. Pept. Sci. *1*, 184–190.

Mu, H.H., Pennock, N.D., Humphreys, J., Kirschning, C.J., and Cole, B.C. (2005). Engagement of Toll-like receptors by mycoplasmal superantigen: downregulation of TLR2 by MAM/TLR4 interaction. Cell. Microbiol. *7*, 789–797.

Mu, H.H., Humphreys, J., Chan, F.V., and Cole, B.C. (2006). TLR2 and TLR4 differentially regulate B7-1 resulting in distinct cytokine responses to the mycoplasma superantigen MAM as well as to disease induced by *Mycoplasma arthritidis*. Cell. Microbiol. *8*, 414–426.

Mu, H.H., Hasebe, A., Van Schelt, A., and Cole, B.C. (2011). Novel interactions of a microbial superantigen with TLR2 and TLR4 differentially regulate IL-17 and Th17-associated cytokines. Cell. Microbiol. *13*, 374–387.

Much, P., Winner, F., Stipkovits, L., Rosengarten, R., and Citti, C. (2002). *Mycoplasma gallisepticum*: influence of cell invasiveness on the outcome of experimental infection in chickens. FEMS Immunol. Med. Microbiol. *34*, 181–186.

Muneta, Y., Panicker, I.S., Kanci, A., Craick, D., Noormohammadi, A.H., Bean, A., Browning, G.F., and Markham, P.F. (2008). Development and immunogenicity of recombinant *Mycoplasma gallisepticum* vaccine strain ts-11 expressing chicken IFN-gamma. Vaccine *26*, 5449–5454.

Noormohammadi, A.H., Markham, P.F., Duffy, M.F., Whithear, K.G., and Browning, G.F. (1998). Multigene families encoding the major hemagglutinins in phylogenetically distinct mycoplasmas. Infect. Immun. *66*, 3470–3475.

Okusawa, T., Fujita, M., Nakamura, J., Into, T., Yasuda, M., Yoshimura, A., Hara, Y., Hasebe, A., Golenbock, D.T., Morita, M., *et al.* (2004). Relationship between structures and biological activities of mycoplasmal diacylated lipopeptides and their recognition by Toll-like receptors 2 and 6. Infect. Immun. *72*, 1657–1665.

Opriessnig, T., Thacker, E.L., Yu, S., Fenaux, M., Meng, X.J., and Halbur, P.G. (2004). Experimental reproduction of postweaning multisystemic wasting syndrome in pigs by dual infection with *Mycoplasma hyopneumoniae* and porcine circovirus type 2. Vet. Pathol. *41*, 624–640.

Papazisi, L., Frasca, S., Jr., Gladd, M., Liao, X., Yogev, D., and Geary, S.J. (2002). GapA and CrmA coexpression is essential for *Mycoplasma gallisepticum* cytadherence and virulence. Infect. Immun. *70*, 6839–6845.

Papazisi, L., Gorton, T.S., Kutish, G., Markham, P.F., Browning, G.F., Nguyen, D.K., Swartzell, S., Madan,

A., Mahairas, G., and Geary, S.J. (2003). The complete genome sequence of the avian pathogen *Mycoplasma gallisepticum* strain R(low). Microbiology *149*, 2307–2316.

Pilo, P., Vilei, E.M., Peterhans, E., Bonvin-Klotz, L., Stoffel, M.H., Dobbelaere, D., and Frey, J. (2005). A metabolic enzyme as a primary virulence factor of *Mycoplasma mycoides* subsp. *mycoides* small colony. J. Bacteriol. *187*, 6824–6831.

Sato, S., Takeuchi, O., Fujita, T., Tomizawa, H., Takeda, K., and Akira, S. (2002). A variety of microbial components induce tolerance to lipopolysaccharide by differentially affecting MyD88-dependent and -independent pathways. Int. Immunol. *14*, 783–791.

Schmidt, J.A., Browning, G.F., and Markham, P.F. (2007). *Mycoplasma hyopneumoniae* mhp379 is a Ca2+-dependent, sugar-nonspecific exonuclease exposed on the cell surface. J. Bacteriol. *189*, 3414–3424.

Schmidl, S.R., Gronau, K., Pietack, N., Hecker, M., Becher, D., and Stulke, J. (2010). The phosphoproteome of the minimal bacterium *Mycoplasma pneumoniae*: analysis of the complete known Ser/Thr kinome suggests the existence of novel kinases. Mol. Cell. Proteomics 9, 1228–1242.

Schmidl, S.R., Otto, A., Lluch-Senar, M., Pinol, J., Busse, J., Becher, D., and Stulke, J. (2011). A trigger enzyme in *Mycoplasma pneumoniae*: impact of the glycerophosphodiesterase GlpQ on virulence and gene expression. PLoS Pathog. 7, e1002263.

Shahzad, W., Ajuwape, A.T., and Rosenbusch, R.F. (2010). Global suppression of mitogen-activated ovine peripheral blood mononuclear cells by surface protein activity from *Mycoplasma ovipneumoniae*. Vet. Immunol. Immunopathol. *136*, 116–121.

Shil, P.K., Kanci, A., Browning, G.F., Marenda, M.S., Noormohammadi, A.H., and Markham, P.F. (2011). GapA+ *Mycoplasma gallisepticum* ts-11 has improved vaccine characteristics. Microbiology *157*, 1740–1749.

Sirand-Pugnet, P., Lartigue, C., Marenda, M., Jacob, D., Barre, A., Barbe, V., Schenowitz, C., Mangenot, S., Couloux, A., Segurens, B., *et al.* (2007). Being pathogenic, plastic, and sexual while living with a nearly minimal bacterial genome. PLoS Genet. 3, e75.

Skapski, A., Hygonenq, M.C., Sagne, E., Guiral, S., Citti, C., and Baranowski, E. (2011). Genome-scale analysis of *Mycoplasma agalactiae* loci involved in interaction with host cells. PLoS One 6, e25291.

Szczepanek, S.M., Frasca, S., Jr., Schumacher, V.L., Liao, X., Padula, M., Djordjevic, S.P., and Geary, S.J. (2010). Identification of lipoprotein MslA as a neoteric virulence factor of *Mycoplasma gallisepticum*. Infect. Immun. 78, 3475–3483.

Thacker, E.L., Halbur, P.G., Ross, R.F., Thanawongnuwech, R., and Thacker, B.J. (1999). *Mycoplasma hyopneumoniae* potentiation of porcine reproductive and respiratory syndrome virus-induced pneumonia. J. Clin. Microbiol. 37, 620–627.

Thanawongnuwech, R., Thacker, B., Halbur, P., and Thacker, E.L. (2004). Increased production of proinflammatory cytokines following infection with porcine reproductive and respiratory syndrome virus and *Mycoplasma hyopneumoniae*. Clin. Diagn. Lab. Immunol. *11*, 901–908.

van der Merwe, J., Prysliak, T., and Perez-Casal, J. (2010). Invasion of bovine peripheral blood mononuclear cells and erythrocytes by *Mycoplasma bovis*. Infect. Immun. *78*, 4570–4578.

van Noort, V., Seebacher, J., Bader, S., Mohammed, S., Vonkova, I., Betts, M.J., Kuhner, S., Kumar, R., Maier, T., O'Flaherty, M., *et al.* (2012). Cross-talk between phosphorylation and lysine acetylation in a genome-reduced bacterium. Mol. Syst. Biol. *8*, 571.

Vanden Bush, T.J., and Rosenbusch, R.F. (2004). Characterization of a lympho-inhibitory peptide produced by *Mycoplasma bovis*. Biochem. Biophys. Res. Commun. *315*, 336–341.

Vasconcelos, A.T., Ferreira, H.B., Bizarro, C.V., Bonatto, S.L., Carvalho, M.O., Pinto, P.M., Almeida, D.F., Almeida, L.G., Almeida, R., Alves-Filho, L., *et al.* (2005). Swine and poultry pathogens: the complete genome sequences of two strains of *Mycoplasma hyopneumoniae* and a strain of *Mycoplasma synoviae*. J. Bacteriol. *187*, 5568–5577.

Vilei, E.M., and Frey, J. (2001). Genetic and biochemical characterization of glycerol uptake in *Mycoplasma mycoides* subsp. *mycoides* SC: its impact on H(2)O(2) production and virulence. Clin. Diagn. Lab. Immunol. *8*, 85–92.

Vogl, G., Plaickner, A., Szathmary, S., Stipkovits, L., Rosengarten, R., and Szostak, M.P. (2008). *Mycoplasma gallisepticum* invades chicken erythrocytes during infection. Infect. Immun. *76*, 71–77.

Waites, K.B., and Talkington, D.F. (2004). *Mycoplasma pneumoniae* and its role as a human pathogen. Clin. Microbiol. Rev. *17*, 697–728, table of contents.

Waldo, R.H., 3rd, Popham, P.L., Romero-Arroyo, C.E., Mothershed, E.A., Lee, K.K., and Krause, D.C. (1999). Transcriptional analysis of the *hmw* gene cluster of *Mycoplasma pneumoniae*. J. Bacteriol. *181*, 4978–4985.

Waldo, R.H., 3rd, Jordan, J.L., and Krause, D.C. (2005). Identification and complementation of a mutation associated with loss of *Mycoplasma pneumoniae* virulence-specific proteins B and C. J. Bacteriol. *187*, 747–751.

Wawegama, N.K., Kanci, A., Marenda, M.S., Mansell, P.D., Browning, G.F., and Markham, P.F. (2012). Histochemical and morphometric characterization of broncho-pneumonia in calves caused by infection with *Mycoplasma bovis*. Vet. Microbiol. *158*, 220–224.

Whithear, K.G. (1996). Control of avian mycoplasmoses by vaccination. Rev. Sci. Tech. *15*, 1527–1553.

Whithear, K.G., Soeripto, Harrigan, K.E., and Ghiocas, E. (1990). Immunogenicity of a temperature sensitive mutant *Mycoplasma gallisepticum* vaccine. Aust. Vet. J. *67*, 168–174.

Winner, F., Rosengarten, R., and Citti, C. (2000). In vitro cell invasion of *Mycoplasma gallisepticum*. Infect. Immun. *68*, 4238–4244.

Yavlovich, A., Katzenell, A., Tarshis, M., Higazi, A.A., and Rottem, S. (2004a). *Mycoplasma fermentans* binds to and invades HeLa cells: involvement of plasminogen and urokinase. Infect. Immun. 72, 5004–5011.

Yavlovich, A., Tarshis, M., and Rottem, S. (2004b). Internalization and intracellular survival of *Mycoplasma pneumoniae* by non-phagocytic cells. FEMS Microbiol. Lett. 233, 241–246.

Zhao, Y., Li, Z., Drozd, S.J., Guo, Y., Mourad, W., and Li, H. (2004). Crystal structure of *Mycoplasma arthritidis* mitogen complexed with HLA-DR1 reveals a novel superantigen fold and a dimerized superantigen–MHC complex. Structure *12*, 277–288.

# Post-translational Modification of Proteins in the Mollicutes

5

Steven P. Djordjevic and Jessica L. Tacchi

## Abstract

The mycoplasmas are a genome-reduced, highly diverse group of bacteria. Although proteins constitute a significant proportion of mycoplasma membranes, genome sequence analyses reveal the presence of only a rudimentary general secretory pathway. Consequently, it is unclear how a wide array of proteins, including adherence and other virulence proteins, ABC transport proteins and key proteolytic enzymes, are translocated across their single bi-lipid membrane. Many species are host-specific and rely heavily on their hosts for the supply of essential metabolites. Nonetheless, different species employ remarkably diverse strategies to successfully colonize their respective hosts. Post-translational modifications (PTMs) of proteins profoundly influence the structure, and consequently the functions, of proteins and play fundamental roles in cellular physiology. PTMs are an integral component of the protein secretion apparatus and influence where proteins localize in cellular compartments and how they form functional complexes via interactions with other proteins. To date, post-translational modifications to mycoplasma proteins have been described in molecules that play key roles in host colonization, metabolism and immune evasion.

## Introduction

Although more than 300 post-translational modifications (PTMs) have been described to date (Witze *et al.*, 2007), many of these have only been described in eukaryotes. Post-translational modifications of proteins, specifically phosphorylation, glycosylation, acetylation, methylation and protein proteolysis, are now well described modifications on bacterial proteins (Kennelly, 2002; Scott *et al.*, 2009, 2010, 2011; Zhang *et al.*, 2009; Hu *et al.*, 2010; de Been *et al.*, 2011). Recent evidence suggests that the mycoplasmas also utilize some of these modifications to alter protein function (Djordjevic *et al.*, 2004; Burnett *et al.*, 2006; Wilton *et al.*, 2009; Schmidl *et al.*, 2010; van Noort *et al.*, 2012).

Despite undergoing reductive evolution from the low G+C Firmicutes (lactobacilli, streptococci and staphylococci) and losing genes encoding enzymes involved in cell wall biosynthesis, oxidative phosphorylation, and cholesterol and nucleic acid biosynthetic pathways (Woese *et al.*, 1980), the Mollicutes are highly successful pathogens and commensals of insects, plants and most vertebrates, including fish, reptiles, birds and mammals (Razin *et al.*, 1998).

Genetic manipulation methods, such as random transposon and site-specific mutagenic protocols, have been described for only a few *Mycoplasma* species. Despite these limitations, the mycoplasmas are attractive as model organisms that display a genetic repertoire that approaches that of a minimal cell. More than 75 Mollicute genomes, the majority of which are *Mycoplasma* species of human and animal origin, are now available in public databases. This wealth of genetic information provides opportunity to undertake detailed proteomic investigations. Unfortunately, only a few extensive proteome studies have been published and there is an urgent need to address this imbalance. Global proteome studies are important because they interrogate in silico predictions of protein coding sequences (proteogenomics), complement transcriptome studies, provide a systems approach to the study of global protein expression and provide one of the main avenues for elucidating how proteins are modified post-translationally. This chapter will examine what is known about how the mycoplasmas modify their proteins.

In an earlier study (Herrmann and Ruppert, 2006), four types of post-translational modifications were described:

- cleavage of signal peptides;
- acylation of lipoproteins;
- endoproteolytic cleavage of secreted proteins (e.g. virulence factors); and
- protein phosphorylation.

Since that report, a considerable body of work has been published that further underpins the importance of these protein modifications in the Mollicutes. In addition, lysine acetylation has recently been shown to be an important PTM in *Mycoplasma pneumoniae* (van Noort *et al.*, 2012).

## Post-translational protein processing plays a key role in protein secretion

Post-translational protein modification controls protein secretion and underpins many essential processes in biology, with profound effects on where proteins localize in cellular compartments and how they form functional complexes by interacting with other proteins. It is estimated that 30–45% of all proteins in the proteome are associated with cell membranes. Some of the largest gene families in Mollicute genomes encode ABC transporters, lipoproteins, adhesins and other secreted virulence factors (Razin *et al.*, 1998). All of these proteins are either integral to the cell membrane or must translocate across it. To perform their function(s) these proteins are often post-translationally modified. Because of the evolutionary links between mycoplasmas and low G + C Firmicutes, it is worthwhile examining key aspects of what is known of the general secretory pathway in the Mollicutes.

## Signal sequences are critical for trafficking proteins to secretion pathways

The primary feature of proteins that determines if they will remain localized in the cytosol or are destined for the cell membrane or secretion into the extracellular milieu, is the presence of a signal sequence. Two classes of signal peptides have been described; the general signal peptides (also known as type I signal peptides) and the type II peptides typically found associated with lipoproteins. The type 1 signal peptides comprise three domains known as the N, H and C domains (von Heijne, 1990). The N-domain is typically rich in positively charged K and R residues that are believed to interact both with the translocation machinery and the phospholipids in the lipid bilayer of the membrane. The H-domain immediately follows the N-region and comprises a central stretch of hydrophobic residues that is sufficient to span the cytoplasmic membrane. The H-domain adopts an α-helical conformation (Briggs *et al.*, 1986) and often contains several helix breaking residues (glycine and proline) that facilitate the formation of a hairpin-like structure. The hairpin structure inserts into the membrane and the process whereby this hairpin unloops within the bilipid layer facilitates insertion of the complete signal peptide across the membrane (de Vrije *et al.*, 1990). Helix breaking residues located at the end of the H-domain are believed to influence cleavage by signal peptidases (SPase) (Razin *et al.*, 1998). The C-terminal domain comprises a short run of hydrophilic amino acids that ends with a signal peptidase I cleavage motif. Signal peptidase I (SPase I) recognizes this motif and cleaves it from the nascent chain during or immediately following translocation. A signal peptide peptidase then degrades the cleaved signal peptide (Tjalsma *et al.*, 2004). Type II signal peptides have a similar structure, except they contain a lipoprotein box with a Leu-Ala-Gly-Cys consensus sequence. The cysteine residue in the consensus region is modified by covalent attachment of a lipid, anchoring the lipoprotein to the cell membrane. Signal peptidase II removes the signal sequence such that Cys is the first amino acid in mature lipoprotein sequences (van Wely *et al.*, 2001).

The general secretory pathway is responsible for the secretion of several hundred proteins in *Bacillus subtilis* (Tjalsma *et al.*, 2004). The signal sequence is recognized by cytosolic chaperones such as SecB and other chaperones such as DnaK and GroEL, which deliver the protein to receptors that comprise the export machinery (Schatz and Dobberstein, 1996). This process is known as post-translational translocation (Fink, 1999). Translation and secretion can also occur concurrently and this process is known as co-translational translocation (Phillips and Silhavy, 1992). The export machinery comprises SecA, a protein-stimulated translocation motor that drives translocation via the hydrolysis of ATP, and a complex of several integral membrane proteins (Sec YEG) that create an aqueous channel in the cytosolic membrane (van Wely *et al.*, 2001). SecB recognizes the carboxyl-terminus of SecA, a region that is highly conserved in most bacteria. Interestingly, this region is not conserved in mycoplasma SecA sequences (Fekkes *et al.*, 1997). Moreover, a homologue of SecB is typically not found in genomes of Gram-positive bacteria (van Wely *et al.*, 2001). The lack of a SecB homologue and the finding that most type I signal sequences are longer than their counterparts in Gram-negative bacteria are features that are more suggestive of co-translational translocation and most secreted proteins in the Gram-positive bacteria are believed to follow this route (Chen *et al.*, 1996).

The signal recognition particle (SRP) is a ribonucleotide–protein complex comprising the fifty-four homologue Ffh and small cytoplasmic RNA. The SRP recognizes the signal sequence as it exits the ribosome, traffics the nascent protein to membrane receptor FtsY and releases it to the SecYEG translocon (van Wely *et al.*, 2001; Luirink *et al.*, 2005). Removal of the signal sequence typically precedes release of the protein from the membrane and is considered a prerequisite for cell viability (Inada *et al.*, 1989; Phillips and Silhavy, 1992; Tjalsma *et al.*, 1997). However, recent studies in *Streptococcus mutans* indicate that mutants in which key genes encoding components of the SRP and its major membrane receptor FtsY are disrupted remain viable when grown on complex laboratory media (Hasona *et al.*, 2005; Zanen *et al.*, 2006). This conditional growth phenotype was mirrored in studies with Ffh mutants of *Streptococcus pyogenes*. However mutants with defects in the ability to construct a SRP were unable to process a subset of important virulence factors with cleavable signal sequences and were highly attenuated in animal models of infection (Rosch *et al.*, 2008).

Much remains to be learnt about how signal sequences are recognized and how they interact with secretion targeting pathways (Paetzel *et al.*, 2004; Carlsson *et al.*, 2006). The degree of hydrophobicity (Lee and Bernstein, 2001; Huber *et al.*, 2005) and the frequency of helix breaking amino acid residues in the H region of signal sequences (Adams *et al.*, 2002) have been implicated as mechanisms that influence trafficking. Nonetheless, recent studies indicate that protein trafficking pathways in Gram-positive bacteria are complex and highly organized. The picture is complicated further by observations that the protein export machinery may be influenced by cell wall synthesis machinery (Rafelski and Theriot, 2006; Rosch *et al.*, 2008) and that proteins which lack a classical N-terminal secretion signal but that carry a hydrophobic α–helix domain may traffic across the cell membrane (Yang *et al.*, 2011). This latter observation has profound implications for understanding how a subset of bacterial proteins with well-defined functions in the cytosol (moonlighting proteins) find their way onto the cell surface.

Analysis of mycoplasma genome sequences suggest that these genome-reduced bacteria contain genes that encode rudimentary protein secretion machinery. All species so far examined lack a SecB homologue, consistent with Gram-positive bacterial genomes (van Wely *et al.*, 2001). Notably absent are *secE*, *secG*, *secD* and *secF*. Further, the definitive identification of a *spase I* gene has been problematic in most completed mycoplasma genomes (Minion *et al.*, 2004), and a considerable number of completed mycoplasma genomes lack the *groEL-groES* operon (Minion *et al.*, 2004; Wong and Houry, 2004). Only members of the Pneumoniae clade (*Mycoplasma pneumoniae*, *Mycoplasma genitalium*, *Mycoplasma gallisepticum* and *Mycoplasma penetrans*) possess the groEL-groES operon (Minion *et al.*, 2004).

All *Mycoplasma* species characterized to date carry the *dnaK-dnaJ-grpE* operon, and *ffh* and *ftsY*, crucial genes encoding key components of the signal recognition pathway (SRP). The integral membrane protein encoded by the *prsA* gene is also notably absent from many mycoplasma genomes (Minion *et al.*, 2004). PrsA is thought to play a role as an extracellular chaperone that assists secreted proteins to fold. In support of this hypothesis, *prsA* mutants display excessive extracellular proteolytic degradation of secreted proteins (Jacobs *et al.*, 1993).

Many *Mycoplasma* species possess genes encoding putative proteases but little is known about what they do or how they function. Many of these are likely to play essential roles in recycling damaged proteins, degrading host immune proteins as a means to evade clearance and generating amino acids necessary for protein synthesis (Potempa and Pike, 2009).

## Are signal sequences removed from secreted mycoplasma proteins?

Many of most detailed studies describing post-translational processing events have been undertaken on *Mycoplasma pneumoniae* and *Mycoplasma hyopneumoniae*, pathogenic mycoplasma species that cause pneumonia in humans and swine, respectively. Although they inhabit similar anatomical sites, these two species have remarkably different genotypic and phenotypic attributes.

*M. pneumoniae*, the causative agent of primary atypical pneumonia in humans, is typically regarded as an extracellular pathogen that inhabits the respiratory tract of humans (Chanock *et al.*, 1963). The *M. pneumoniae* genome has a G+C content of 40% and comprises 816,394 base pairs encoding 693 open reading frames (Himmelreich *et al.*, 1996). *M. pneumoniae* is motile, grows well on abiotic surfaces, including glass, and attaches directly to respiratory epithelial cells via its adhesive tip by burrowing between cilia. *M. pneumoniae* has a complex cytoskeletal structure (Meng and Pfister, 1980; Gobel *et al.*, 1981) comprising a network of proteins that are resistant to extraction with Triton X-100 (Triton shell) (Krause and Balish, 2004). A key component of the cytoskeleton is the terminal organelle, which has a defined adhesive tip. The tip structure comprises an electron dense core that is capped by a terminal adhesive button (Krause and Balish, 2004; Henderson and Jensen, 2006). Because *M. pneumoniae* is a medically important pathogen that presents a small genome and the capability to be genetically manipulated, it has recently become the focus of a number of system-wide studies (Guell *et al.*, 2009; Kuhner *et al.*, 2009; Yus *et al.*, 2009; van Noort *et al.*, 2012). These studies have impacted significantly on our understanding of how this pathogen transcribes genes and the manner in which it modifies proteins posttranslationally. Of the 693 predicted open reading frames, 620 have been identified by proteome studies and functions for about 70% of the proteome have been assigned (Catrein and Herrmann, 2011).

*M. hyopneumoniae*, a member of the Hominis clade, is the causative agent of porcine enzootic pneumonia, one of the most economically significant diseases afflicting swine production. *M. hyopneumoniae* has a genome that is approximately the same size as *M. pneumoniae*, 892,758 base pairs encoding 692 putative ORFs, but with a G+C content of 28.6% (Minion *et al.*, 2004). The A+T bias of the *M. hyopneumoniae* genome is likely to profoundly influence promoter structure. A recent study of transcriptional start sites in this species showed that −10 regions that are recognized by $\sigma^{70}$ are well conserved but no −35 elements were detected, an observation consistent with other low G+C bacteria. Nearly half of the characterized promoters display a −16 element that is found in many Gram-positive and low G+C bacteria but not in *M. pneumoniae* (Weber Sde *et al.*, 2012). Unlike *M. pneumoniae*, *M. hyopneumoniae* does not display cellular asymmetry, but rather is ovoid with a pleiomorphic cell shape, and lacks evidence of a cytoskeletal structure or a defined adhesive tip. Electron micrographs show that the archetype cilium adhesin, P97 and its posttranslational cleavage fragments are distributed around the cell membrane (Zhang *et al.*, 1995; Djordjevic *et al.*, 2004). In contrast to *M. pneumoniae*, *M. hyopneumoniae* is not known to be motile, displays a clear affinity for epithelial cilia and has only rarely been reported in direct contact with epithelial cells (Mebus and Underdahl, 1977).

## Signal sequences are removed from secreted proteins in *Mycoplasma pneumoniae*

In silico analyses of the *M. pneumoniae* genome sequence failed to detect a gene that was likely to encode SPase I (Himmelreich *et al.*, 1996; Dandekar *et al.*, 2000). A gene encoding SPase I has been annotated in the genomes of *M. gallisepticum* and *M. pulmonis*, but not in most other completed mycoplasma genomes (Minion *et al.*, 2004). Indeed, there is a paucity of data confirming in silico predictions of the amino-termini of almost all proteins (proteogenomics). This is in part due to claims that many amino-termini are chemically blocked and cannot be determined by Edman sequencing (Wilton *et al.*, 2009; Bogema *et al.*, 2011) and the insensitive nature of the Edman sequencing reaction (Catrein *et al.*, 2005). Furthermore, reliable methodologies that enable global analyses of the amino and carboxyl- termini of proteins have only recently been developed (Butler and Overall, 2009). These global methodologies are yet to be applied to the Mollicutes.

Since the early 1990s it has been known that one important *M. pneumoniae* protein is subject to posttranslational processing. Mpn142 (formerly ORF6) is one of three genes in an operon that includes the major *M. pneumoniae* adhesin P1 (Mpn141) and Mpn140 (ORF4). According to TmPred (http://www.ch.embnet.org/software/TMPRED_form.html) Mpn142 encodes a cytadherence protein of 1218 amino acids (~130 kDa) and has three putative, yet significant (scores >500) transmembrane domains between amino acids 10–31 (score 1331), 292–308 (score 599) and 1125–1143 (score 2518). Mpn142 was shown to undergo endoproteolytic cleavage at amino acid 455 generating amino-terminal 40-kDa (P40) and carboxyl-terminal 90-kDa (P90) proteins (Sperker *et al.*, 1991; Layh-Schmitt and Herrmann, 1992). Both P40 and P90 were shown to reside on the cell surface of *M. pneumoniae* (Layh-Schmitt and Herrmann, 1992) and to complex with the P1 adhesin and other proteins (Feldner *et al.*, 1982; Layh-Schmitt *et al.*, 2000) in the adhesive tip. Edman sequencing determined that P90 has the amino acid sequence $^{455}$RAGNSSETDAL$^{465}$ (Layh-Schmitt and Herrmann, 1992). A 9-kDa discrepancy between the predicted mass of P40 of 44.873 kDa and the observed mass of ~36 kDa, as determined by migration during SDS-polyacrylamide gel electrophoresis, suggests that further endoproteolytic cleavage of products derived from Mpn142 may occur. Possible scenarios describing how these may occur have been summarized elsewhere (Catrein *et al.*, 2005).

Although attempts to determine the amino-terminal sequence of P40 by Edman sequencing were unsuccessful, Catrein and co-workers determined the amino-terminal sequence of P40 via an alternate strategy (Catrein *et al.*, 2005). P40 was enriched from cell extracts of *M. pneumoniae* by immunoprecipitation using anti-P40 antibodies and monitored by Western blotting, sliced from a polyacrylamide gel and chemically modified in-gel in a two-stage reaction: first, the ε-amino groups on lysine residues were converted to homoarginine with methylisourea hemisulfate followed by modification of the free amino terminal groups with the CAF (chemical assisted fragmentation) reagent. The modified P40 was then subjected to in-gel trypsin digestion and examined by ESI-QTOF-MS. The y- and b-fragment species unambiguously identified the amino-terminal residue as asparagine (residue 26). This cleavage event occurs at the carboxyl-terminal region of the transmembrane domain (residues 10–31 TmPred score 1331) and is predicted precisely by SignalP and EXPROT algorithms using the Gram-positive network (Catrein *et al.*, 2005). These data indicate that *M. pneumoniae* contains SPase I activity.

## Evidence that proteins are subject to endoproteolytic cleavage in *Mycoplasma pneumoniae*

Available evidence suggests that proteolytic processing plays an important role in maintaining the complex cytoskeletal structure. The adhesive tip comprises a complex of several adhesins, including P1 and P30, the cytadherence accessory protein

P90 and P40, several structural high molecular mass proteins, HMW1, HMW2 and HMW3, and the accessory proteins P41 and P65. Spontaneous mutants defective in the expression of P90, P40 or any of the HMW proteins display an ovoid morphology lacking a distinct terminal tip (Baseman *et al.*, 1982; Layh-Schmitt *et al.*, 1995). Moreover, the P1 adhesin is no longer focussed at the tip structure, but is randomly distributed around the cell membrane. Immunoaffinity chromatography of *M. pneumoniae* proteins using anti-P1 antibodies that were cross-linked with paraformaldehyde recovered a complex containing P1, P30, the accessory proteins P90 and P40, HMW1 and HMW3, DnaK and the E1-α subunit of pyruvate dehydrogenase (Layh-Schmitt *et al.*, 2000) suggesting that these proteins are essential components of the tip structure. Elongation factor Tu and the E1-β subunit of pyruvate dehydrogenase were also shown to be accessible on the surface of *M. pneumoniae* and function as fibronectin-binding proteins (Dallo *et al.*, 2002). Elongation factor Tu and the E1-β subunit of pyruvate dehydrogenase lack any evidence of a signal sequence and how they come to reside on the surface of *M. pneumoniae* has not been resolved.

HMW1-HMW3 proteins play important roles in the organization of the tip structure. Loss of function of HMW2 by transposon insertion alters the stability of HMW1 and HMW3. How this occurs is not well understood but recent evidence suggests that HMW2 is a key component of the core of the terminal organelle (Bose *et al.*, 2009). The mRNA transcripts of *hmw1* and *hmw3* are unaffected by mutation in *hmw2*, suggesting that post-transcriptional events play a key role in the turnover of HMW1 and HMW3 (Popham *et al.*, 1997). The carboxyl-terminal domains of HMW1 and HMW3 have similar motifs. A Pro-Xaa-Lys/Arg motif was reported in the carboxyl-terminal region of HMW1 and HMW3 (Popham *et al.*, 1997). A similar motif surrounds the endoproteolytic cleavage site in Mpn142. Downstream of this putative cleavage recognition site is a Ser-Ser motif similar to the cleavage recognition motif in *Caulobacter crescentus* (Alley *et al.*, 1993).

There is evidence that other proteins involved in the formation of the adhesive tip structure are targets of endoproteolytic cleavage events in *M. pneumoniae*. Amino-terminal sequence data for the P1 adhesin (Jacobs *et al.*, 1987; Su *et al.*, 1987) was reported more than 20 years ago and is the only other amino-terminal sequence derived experimentally for *M. pneumoniae.* Edman sequencing demonstrated that the amino-terminus of the mature form of the P1 adhesin commenced at amino acid 60, a considerable distance from the putative transmembrane domain identified between amino acids 14–31 (TmPred score 1736). Although it could not be determined if SPase I was required for processing the P1 adhesin, these data provided the first indication that proteins belonging to the P1 operon may be subject to endoproteolytic processing. Two-dimensional (2-D) gel electrophoresis coupled with mass spectrometric analyses of *M. pneumoniae* lysates identified numerous proteins belonging to different functional categories that migrated with a mass and/or isoelectric point (pI) that was different to the predicted values determined from the putative ORF. A 25-kDa carboxyl-terminal fragment of DnaK was identified. Amino-terminal sequence analysis showed that this protein started at amino acid 396 and commenced with an asparagine residue (Regula *et al.*, 2000).

## Signal sequences are not removed during secretion of large mass adhesins in *Mycoplasma hyopneumoniae*

Fourteen large mass proteins, six of which (Mhp493, Mhp684, Mhp271, Mhp385, Mhp107 and Mhp280) share sequence similarity with P97 (Mhp183), and another six (Mhp683, Mhp108, Mhp384, Mhp272, Mhp274 and Mhp275) which share sequence similarity with P102 (Mhp182), have been described in *M. hyopneumoniae* (Minion *et al.*, 2004). Many members of these families are prominently expressed during broth culture and mRNA transcripts encoding most of these adhesins have been detected in *M. hyopneumoniae* recovered from infected pigs (Adams *et al.*, 2005). Other than Mhp274 and Mhp275, these molecules have a large mass (100–220 kDa) and each have a single stretch of hydrophobic residues that comprises a putative transmembrane domain. Our studies have shown that most members of these two paralogous families undergo endoproteolytic cleavage. Cleavage fragments have been shown to reside on the surface of *M. hyopneumoniae* cells harvested from broth culture, but the mechanism enabling this to occur is not understood (Djordjevic *et al.*, 2004; Burnett *et al.*, 2006; Wilton *et al.*, 2009; Seymour *et al.*, 2010, 2011, 2012; Bogema *et al.*, 2011, 2012; Deutscher *et al.*, 2012). Fig. 5.1 depicts how proteolytic processing has a dramatic effect on how adhesin proteins migrate in 2-D gels compared with their predicted migration patterns based on genome sequence data. Cleavage fragments display multifunctional binding abilities for extracellular matrix components and other key host circulatory molecules, including various glycosaminoglycans (Zhang *et al.*, 1995; Burnett *et al.*, 2006; Jenkins *et al.*, 2006; Wilton *et al.*, 2009; Deutscher *et al.*, 2010, 2012; Bogema *et al.*, 2011, 2012; Seymour *et al.*, 2011), fibronectin (Deutscher *et al.*, 2010; Seymour *et al.*,

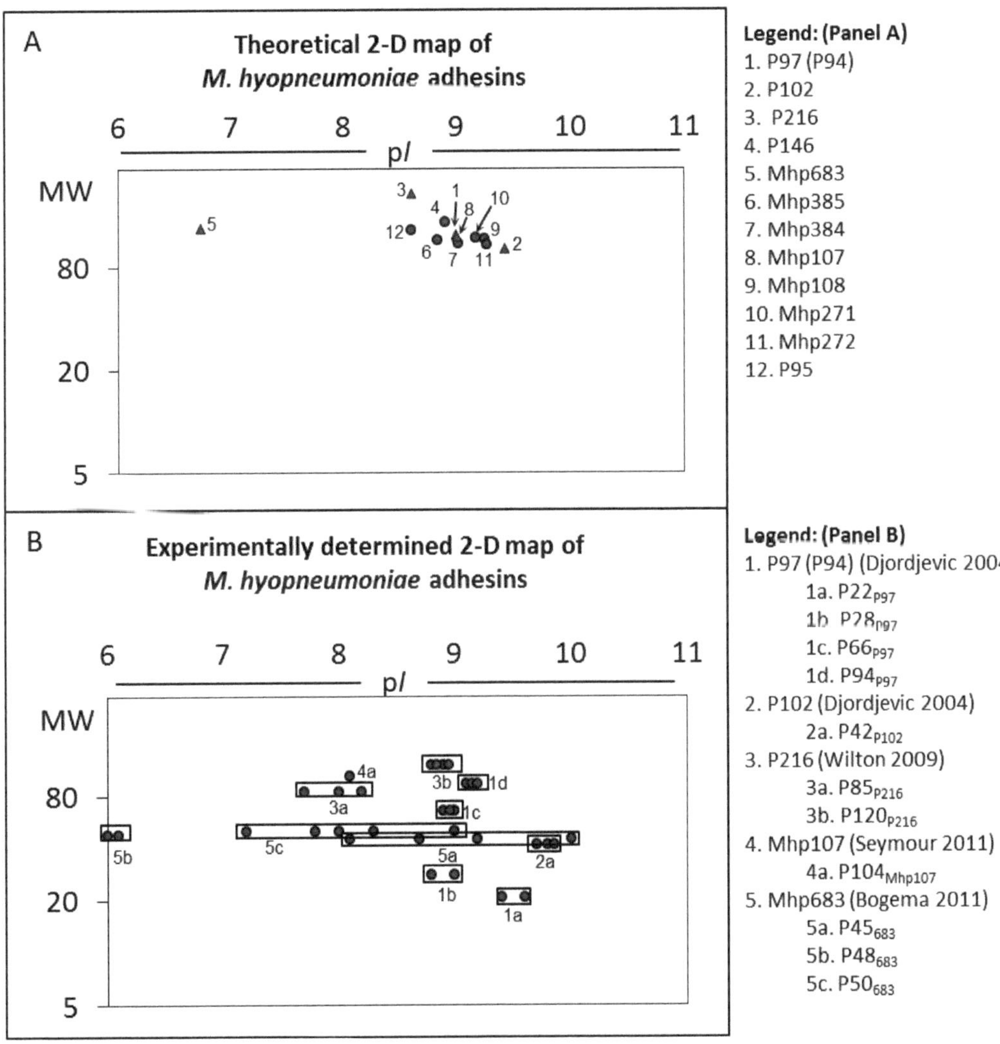


**Figure 5.1** Diagrammatical illustration of the effect of post-translational modifications of adhesin molecules of *Mycoplasma hyopneumoniae*. (A) Computationally composed theoretical 2D gel map of *Mycoplasma hyopneumoniae* adhesins based on in silico predicted open reading frames (ORFs). (B) Computationally composed 2D gel map of *Mycoplasma hyopneumoniae* adhesins based on experimentally determined isoelectric points and molecular mass of adhesins in panel A denoted by triangular markers. Molecular weights in kDa.

2010, 2011, 2012) and plasminogen (Seymour *et al.*, 2010, 2011, 2012; Bogema *et al.*, 2012). The amino acid sequences of P97 and P102 paralogues are highly conserved in the genomes of the four strains of *M. hyopneumoniae* so far sequenced. Although Mhp028 has a signal peptidase I signature motif, its remaining homology is considered too low to definitively annotate it as *lepB* (Minion *et al.*, 2004). Moitinho-Silva *et al.*, cloned and expressed the *sipS* gene (MHP0026) encoding a putative type I signal peptidase from *M. hyopneumoniae* strain 7448. SipS was recognized by swine serum collected from *M. hyopneumoniae*-infected pigs, however, signal peptidase activity was not demonstrated (Moitinho-Silva *et al.*, 2012).

Existing biochemical data from amino-terminal sequence analysis of two amino-terminal cleavage products derived from P97 (Mhp183) and P159 (Mhp494) indicate that *M. hyopneumoniae* lacks SPase I activity. The mhp183 gene, encoding the archetype cilium adhesin (P97), encodes a 124-kDa preprotein. Endoproteolytic cleavage at amino acid position 195 removes an amino-terminal fragment of approximately 22 kDa ($P22_{P97}$) and generates the mature cilium adhesin known as P97. Amino-terminal sequence analysis of P97 from strains J and 232 detected the sequence $^{195}$ADEKTSS$^{201}$ and an identical sequence is present in the cilium adhesin paralogue Mhp271 (Djordjevic *et al.*, 2004; Deutscher *et al.*, 2010). Consistent with this, P22 has been detected in cell lysates of *M. hyopneumoniae* strains J and 232 (Djordjevic *et al.*, 2004). The only putative transmembrane domain in P97 occurs between amino acids 8 and 22 (TMpred score 2202) and this region is likely to be critical for targeting the P124 preprotein to the general secretion machinery. Edman sequencing of the amino-terminal 22-kDa cleavage fragment (P22) detected the sequence $^{2}$SKKSKTF$^{8}$ (Djordjevic *et al.*, 2004). Clearly, Spase I activity is not required to remove the signal peptide during processing of Mhp183, but other processing events take place, presumably during or immediately after secretion. A carboxyl-terminal fragment of the cilium adhesin of approximately 28 kDa ($P28_{P97}$) is also removed by an endoproteolytic cleavage event. Edman analysis detected the amino-terminal sequence $^{863}$NTNTGFS$^{869}$. The lack of sequence similarity spanning the two endoproteolytic cleavage sites suggests that these cleavage events are executed by different proteases, but further studies are needed to confirm this hypothesis. Although not a paralogue of the adhesin families, Mhp494 (P159) is also subject to endoproteolytic cleavage, generating three fragments, P27, P110 and P52, that all reside on the external surface of the *M. hyopneumoniae* membrane (Burnett *et al.*, 2006). P110 and P52 bind heparin-like glycosaminoglycans and also recognize receptors displayed on the surface of porcine kidney epithelial-like cell (PK15) monolayers. Peptide mass mapping studies of P27, P110 and P52 show they map to the amino-terminal, central and carboxyl-terminal regions of P159, respectively (Burnett *et al.*, 2006). Edman sequencing of P27 detected the sequence $^{1}$MKKQIRN$^{7}$, consistent with the hypothesis that *M. hyopneumoniae* lacks SPase I activity. Edman sequencing and LC-MS/MS analyses have shown that surface accessible P97 paralogue Mhp385 (Deutscher *et al.*, 2012) and P102 paralogue Mhp683 (Bogema *et al.*, 2011) retain intact N-termini. Our data suggest that removal of the signal peptide is not required for efficient maturation of P97 and P102 paralogues.

Considerable progress has been made in deciphering proteolytic cleavage motifs that reside within the P97 and P102 paralogue families. In a landmark study, Bogema *et al.*, identified a motif TTKF↓QE in the P102 paralogue Mhp683 that resided in two regions of the molecule where cleavage was predicted to occur. A combination of Edman sequencing and LC-MS/MS was used to decipher the precise cleavage site at the phenylalanine residue in both these motifs that produced the three cleavage fragments $P45_{683}$, $P48_{683}$ and $P50_{683}$ (Bogema *et al.*, 2011). We were able to refine the cleavage site to S/T-X-F↓-X-D/E by examining cleavage data (Edman sequence and LC-MS/MS) for all members of the P97 and P102 paralogues and searching for atypical tryptic peptides (semitryptic peptides) that reside within regions of these molecules that span known cleavage sites (Djordjevic *et al.*, 2004; Burnett *et al.*, 2006; Wilton *et al.*, 2009; Bogema *et al.*, 2011, 2012; Deutscher *et al.*, 2012; Seymour *et al.*, 2012).

It is remarkable that *M. hyopneumoniae* can secrete and process paralogues of the P97 and P102 families without genetic evidence for the presence of genes with significant homology to *spaseI, secB, secD, secE, secF* and *secG* or the entire *groEL-groES* operon. Nonetheless, endoproteolytic cleavage fragments of the majority of the P97 and P102 paralogue families are prominent components comprising the surface topography of this species. Interestingly, our proteomic analyses indicate that the 53 lipoproteins in *M. hyopneumoniae* are not major targets for endoproteolytic cleavage because full length lipoproteins are readily identified. Minor lipoprotein cleavage fragments are detected but not to the same extend as we determined in the P97 and P102 paralogue families (Djordjevic, Tacchi and Padula unpublished results). Collectively, these observations suggest that *M. hyopneumoniae* maintains an 'adhesion-competent state' during broth culture and that expression of members of these two paralogue families are crucial during the infection process.

## Why are adhesins targets of endoproteolytic cleavage?

In other pathogenic bacteria, large extracellular bacterial proteins, particularly those that function as adhesins, are often processed by limited and timely endoproteolysis (Coutte *et al.*, 2001, 2003). Processing is considered essential for the maturation of large mass proteins to avoid aggregation on the external membrane surface (Coutte *et al.*, 2001). Like *M. hyopneumoniae*, various *Bordetella* species target the respiratory cilia as the preferred site of colonization. Adherence to cilia by *Bordetella bronchiseptica* is mediated by multiple adhesins to ensure maximal adherence and is a critical early step in pathogenesis (Edwards *et al.*, 2005). Mutants of the respiratory pathogen *Bordetella pertussis* deficient in the enzyme responsible for processing the major adhesin, filamentous haemagglutinin, are severely affected in their ability to colonize the murine respiratory tract. Processing involves chaperones and a protease(s) that has access to the proteolytic cleavage site(s), at least for a limited period of time before the molecule refolds on the extracellular side of the cell membrane. Secretion is likely to occur immediately following or concomitantly with translation. Once cleaved, the products either remain attached non-covalently to the external surface of the membrane or are released into the extracellular milieu. It has been suggested that adhesin processing plays a key role in respiratory tract colonization by assisting in the dispersal of *B. pertussis* from microcolonies (Coutte *et al.*, 2003).

## Processing of lipoproteins

Lipoproteins are prominent features of mycoplasma membranes and usually comprise one of the largest gene families in mycoplasma genomes (Razin *et al.*, 1998). Based on sequence similarity, lipoproteins in *M. pneumoniae* were subdivided into 6 classes (Himmelreich *et al.*, 1996; Hallamaa *et al.*, 2006). Reverse transcription-polymerase chain reaction (RT-PCR) assays detected transcripts for most (58/67) lipoproteins and mass spectrometric analyses concurred with RT-PCR data for 45 of the 67 ORFs (Hallamaa *et al.*, 2006). These data indicated, at least for *M. pneumoniae*, that most lipoprotein genes are expressed during broth culture. Processing of lipoproteins commences prior to membrane insertion, when the sulfhydryl moiety on a cysteine residue located immediately after the signal sequence is modified by the transfer of a diacylglycerol moiety from glycerophospholipid. Cleavage of the peptide bond that immediately precedes the modified cysteine by Spase II completes the processing event during or immediately following membrane translocation (Razin *et al.*, 1998). Although an *spase I* gene is rarely identifiable in genome sequence analyses, this is not the case for the *spase II* gene (Minion *et al.*, 2004), indicating that lipid modification is an important event in lipoprotein expression. Functions for most lipoproteins have not been elucidated.

Lipoproteins are also known to be targets of post-translational cleavage events. In *Mycoplasma fermentans*, the MALP-404 lipoprotein is a highly expressed surface accessible antigen that is targeted by endoproteolysis. A processing event removes 14 amino acids (2 kDa) from the amino-terminus (MALP-2) of the mature lipoprotein MALP-404 (P41) and MALP-2 remains attached to the *M.*

*fermentans* membrane via the diacyl-glycerol lipid anchor. The carboxyl-terminal region of MALP-404, referred to as the released fragment (RF), is secreted into the extracellular milieu. Efforts to detect RF on the cell surface have been unsuccessful (Davis and Wise, 2002). Edman sequencing of RF, purified by immunoprecipitation from culture supernatant, detected the unambiguous sequence DISKYTTTNA, corresponding to amino acids 15–24 in the deduced sequence of MALP-404. These data confirmed that a site-specific endoproteolytic cleavage event occurs between residues 14 and 15 of the mature lipoprotein (Davis and Wise, 2002). The MALP-2 fragment has potent immunomodulatory activity for macrophages and monocytes that is executed via Toll-like receptor pattern recognition (Muhlradt *et al.*, 1997; Kaufmann *et al.*, 1999; Nishiguchi *et al.*, 2001). Strains of *M. fermentans* produce different amounts of MALP-2 and MALP-404, suggesting that they carry out the endoproteolytic cleavage event with various degrees of efficiency (Calcutt *et al.*, 1999). Processing is likely to represent an efficient means of regulating the amount of MALP-2 on the surface of this organism and provides a mechanism to regulate the presentation of two important variants of this protein on the surface of *M. fermentans*. Although the precise function of the 41-kDa RF remains unknown, patient sera contains antibodies that recognize it (Davis and Wise, 2002). The RF also carries a conserved motif known as the selected lipoprotein associated (SLA) motif, which is conserved among lipoproteins from diverse bacterial species (Calcutt *et al.*, 1999). The protease that carries out this important processing event remains unidentified.

2-D gel electrophoresis of *M. pneumoniae* proteins identified four fragments spanning the entire sequence of lipoprotein 384 and other members of the lipoprotein family were also shown to be subject to endoproteolytic cleavage. Functions for these proteolytic cleavage fragments has not been reported (Regula *et al.*, 2000).

## Endoproteolytic processing in *Mycoplasma gallisepticum*

MGA0674 is a homologue of the Family 2 lipoproteins previously described in the phylogenetically related species *Mycoplasma pneumoniae* (Hallamaa *et al.*, 2006). In silico analyses predict that Family 2 lipoproteins carry a 'Mycoplasma lipoprotein X' central domain and a 'Mycoplasma lipoprotein 10' domain in the carboxyl-terminus of the molecule and are differentially expressed under various environmental conditions. The functions of these domains are poorly understood. Comparative transcriptomic analyses of the virulent $R_{low}$ strain and the attenuated vaccine strain F identified MGA1074 (*mlsA*) as one of a number of differentially expressed genes. Specifically, *mlsA* was expressed at 6-fold lower levels in F strain (Szczepanek *et al.*, 2010). Recently, we showed that the MGA0674-encoded lipoprotein, the only lipoprotein carrying a Family 2 signature in *M. gallisepticum*, is a highly expressed, proteolytically processed surface antigen. Two proteins of 22 kDa (P22) and 58 kDa (P57) were shown to represent the amino-terminal and carboxyl-terminal cleavage fragments of MGA0674, respectively. Edman sequencing found the amino-terminus of P58 to be $^{225}$AGEMAVTEN$^{233}$ (Szczepanek *et al.*, 2010), but the biological significance of the processing event is unknown. An $R_{low}$ ΔMGA1074 mutant appeared to colonize the tracheas of experimentally infected chickens with less efficiency than parental $R_{low}$ and resulted in reduced pathology (Szczepanek *et al.*, 2010). The function(s) of the MGA0674-encoded lipoprotein are unknown.

The lipoprotein designated pMGA_1.9 which encodes an 82-kDa protein was reported to be subject to posttranslational cleavage generating a 45-kDa C-terminal fragment (p45) (Markham *et al.*, 1998).

## Protein phosphorylation and acetylation

Van Noort *et al.*, characterized phosphorylation and acetylation sites in wild-type strains of *M. pneumoniae* and in isogenic mutant strains deficient in the protein kinases HprK and PknB and the phosphatase PrpC. 93 phosphorylation and 719 lysine acetylation sites were identified in 72 and 221 proteins respectively and represents the largest number of lysine acetylation sites reported for a prokaryote (van Noort *et al.*, 2012). Metabolic enzymes including those involved in glycolysis, components of the translational machinery and proteins with cytadherence and chaperone functions were phosphorylated and/or lysine acetylated (Schmidl *et al.*, 2010; van Noort *et al.*, 2012). Cytadherence proteins have been known to be phosphorylated for some time (Dirksen *et al.*, 1994). Phosphorylation and lysine acetylation sites were often conserved consistent with these being ancient modifications. Proteins confined to the Mycoplasmas including *M. pneumoniae*-specific proteins were also modified indicating that specific-specific regulation pathways have evolved. Phosphorylation was observed on serine, threonine and to a lesser extent tyrosine residues and 57% of all modified proteins display multiple modification sites. The level of modification targeted by these PTM varies from less than 10% of all serine, threonine, tyrosine and lysine residues for most proteins to approximately 40% for proteins undergoing extensive modification (e.g. ribosome recycling factor) (van Noort *et al.*, 2012). Phosphorylation was shown to profoundly alter cellular proteome primarily through altering the oligomeric state of proteins and influencing protein interaction pathways.

In conclusion, it is now recognized that many proteins are targets of phosphorylation and lysine acetylation and these PTM are likely to profoundly alter the functions of proteins and their interaction partners. Mycoplasmas translocate a significant proportion of their proteome to the membrane and many proteins are presented on the membrane surface. It is not known if PTM influence how proteins partition to different cellular locations nor is it known if PTM influence protein moonlighting behaviour in prokaryotes. Lipoproteins appear to be prominently expressed and constitute a considerable proportion of the protein component of the membrane and a proportion of these are subject to proteolytic cleavage. Gene families encoding adhesin proteins unrelated to lipoproteins (e.g. the P97/P102 paralogue families) are highly expressed, are targets of extensive endoproteolytic processing and are likely to play important roles in niche colonization. Currently it is not known if cleavage fragments are secreted into the extracellular milieu but many cleavage fragments remain attached to the mycoplasma membrane despite extensive washing and several rounds of centrifugation. The mechanism(s) enabling cleavage fragments to remain surface bound is poorly understood. Functions for a number of proteolytically cleaved surface proteins have been determined experimentally, but functions for the vast majority are yet to be defined. It is remarkable that these genome-reduced organisms are capable of presenting and modifying a complex repertoire of surface proteins with what appears to be rudimentary protein secretion capabilities. It is clear that the mycoplasmas utilize different proteolytic strategies to facilitate protein secretion and post-translationally modify protein function. The identification and characterization of proteases that perform many of these cleavage events remains an important challenge in future endeavours.

## References

Adams, C., Pitzer, J., and Minion, F.C. (2005). In vivo expression analysis of the P97 and P102 paralog families of *Mycoplasma hyopneumoniae*. Infect. Immun. 73, 7784–7787.

Adams, H., Scotti, P.A., De Cock, H., Luirink, J., and Tommassen, J. (2002). The presence of a helix breaker in the hydrophobic core of signal sequences of secretory proteins prevents recognition by the signal-recognition particle in *Escherichia coli*. Eur. J. Biochem. *269*, 5564–5571.

Alley, M.R., Maddock, J.R., and Shapiro, L. (1993). Requirement of the carboxyl terminus of a bacterial chemoreceptor for its targeted proteolysis. Science *259*, 1754–1757.

Baseman, J.B., Cole, R.M., Krause, D.C., and Leith, D.K. (1982). Molecular basis for cytadsorption of *Mycoplasma pneumoniae*. J. Bacteriol. *151*, 1514–1522.

Bogema, D.R., Scott, N.E., Padula, M.P., Tacchi, J.L., Raymond, B.B., Jenkins, C., Cordwell, S.J., Minion, F.C., Walker, M.J., and Djordjevic, S.P. (2011). Sequence TTKF downward arrow QE defines the site of proteolytic cleavage in Mhp683 protein, a novel glycosaminoglycan and cilium adhesin of *Mycoplasma hyopneumoniae*. J. Biol. Chem. *286*, 41217–41229.

Bogema, D.R., Deutscher, A.T., Woolley, L.K., Seymour, L.M., Raymond, B.B., Tacchi, J.L., Padula, M.P., Dixon, N.E., Minion, F.C., Jenkins, C., *et al.* (2012). Characterization of cleavage events in the multifunctional cilium adhesin Mhp684 (P146) reveals a mechanism by which *Mycoplasma hyopneumoniae* regulates surface topography. MBio. 3, pii: e00282–11.

Bose, S.R., Balish, M.F., and Krause, D.C. (2009). *Mycoplasma pneumoniae* cytoskeletal protein HMW2 and the architecture of the terminal organelle. J. Bacteriol. *191*, 6741–6748.

Briggs, M.S., Cornell, D.G., Dluhy, R.A., and Gierasch, L.M. (1986). Conformations of signal peptides induced by lipids suggest initial steps in protein export. Science *233*, 206–208.

Burnett, T.A., Dinkla, K., Rohde, M., Chhatwal, G.S., Uphoff, C., Srivastava, M., Cordwell, S.J., Geary, S., Liao, X., Minion, F.C., *et al.* (2006). P159 is a proteolytically processed, surface adhesin of *Mycoplasma hyopneumoniae*: defined domains of P159 bind heparin and promote adherence to eukaryote cells. Mol. Microbiol. *60*, 669–686.

Butler, G.S., and Overall, C.M. (2009). Proteomic identification of multitasking proteins in unexpected locations complicates drug targeting. Nat. Rev. Drug Discov. *8*, 935–948.

Calcutt, M.J., Kim, M.F., Karpas, A.B., Muhlradt, P.F., and Wise, K.S. (1999). Differential posttranslational processing confers intraspecies variation of a major surface lipoprotein and a macrophage-activating lipopeptide of *Mycoplasma fermentans*. Infect. Immun. *67*, 760–771.

Carlsson, F., Stalhammar-Carlemalm, M., Flardh, K., Sandin, C., Carlemalm, E., and Lindahl, G. (2006). Signal sequence directs localized secretion of bacterial surface proteins. Nature *442*, 943–946.

Catrein, I., and Herrmann, R. (2011). The proteome of *Mycoplasma pneumoniae*, a supposedly 'simple' cell. Proteomics *11*, 3614–3632.

Catrein, I., Herrmann, R., Bosserhoff, A., and Ruppert, T. (2005). Experimental proof for a signal peptidase I like activity in *Mycoplasma pneumoniae*, but absence of a gene encoding a conserved bacterial type I SPase. FEBS J. *272*, 2892–2900.

Chanock, R.M., Mufson, M.A., Somerson, N.L., and Couch, R.B. (1963). Role of Mycoplasma (Pplo) in human respiratory disease. Am. Rev. Respir. Dis. *88* (Suppl.), 218–239.

Chen, H., Kim, J., and Kendall, D.A. (1996). Competition between functional signal peptides demonstrates variation in affinity for the secretion pathway. J. Bacteriol. *178*, 6658–6664.

Coutte, L., Antoine, R., Drobecq, H., Locht, C., and Jacob-Dubuisson, F. (2001). Subtilisin-like autotransporter serves as maturation protease in a bacterial secretion pathway. EMBO J. *20*, 5040–5048.

Coutte, L., Alonso, S., Reveneau, N., Willery, E., Quatannens, B., Locht, C., and Jacob-Dubuisson, F. (2003). Role of adhesin release for mucosal colonization by a bacterial pathogen. J. Exp. Med. *197*, 735–742.

Dallo, S.F., Kannan, T.R., Blaylock, M.W., and Baseman, J.B. (2002). Elongation factor Tu and E1 beta subunit of pyruvate dehydrogenase complex act as fibronectin binding proteins in *Mycoplasma pneumoniae*. Mol. Microbiol. *46*, 1041–1051.

Dandekar, T., Huynen, M., Regula, J.T., Ueberle, B., Zimmermann, C.U., Andrade, M.A., Doerks, T., Sanchez-Pulido, L., Snel, B., Suyama, M., *et al.* (2000). Re-annotating the *Mycoplasma pneumoniae* genome sequence: adding value, function and reading frames. Nucleic Acids Res. *28*, 3278–3288.

Davis, K.L., and Wise, K.S. (2002). Site-specific proteolysis of the MALP-404 lipoprotein determines the release of a soluble selective lipoprotein-associated motif-containing fragment and alteration of the surface phenotype of *Mycoplasma fermentans*. Infect. Immun. *70*, 1129–1135.

de Been, M., Francke, C., Siezen, R.J., and Abee, T. (2011). Novel sigmaB regulation modules of Gram-positive bacteria involve the use of complex hybrid histidine kinases. Microbiology *157*, 3–12.

de Vrije, G.J., Batenburg, A.M., Killian, J.A., and de Kruijff, B. (1990). Lipid involvement in protein translocation in *Escherichia coli*. Mol. Microbiol. *4*, 143–150.

Deutscher, A.T., Jenkins, C., Minion, F.C., Seymour, L.M., Padula, M.P., Dixon, N.E., Walker, M.J., and Djordjevic, S.P. (2010). Repeat regions R1 and R2 in the P97 paralogue Mhp271 of *Mycoplasma hyopneumoniae* bind heparin, fibronectin and porcine cilia. Mol. Microbiol. *78*, 444–458.

Deutscher, A.T., Tacchi, J.L., Minion, F.C., Padula, M.P., Crossett, B., Bogema, D.R., Jenkins, C., Kuit, T.A., Walker, M.J., and Djordjevic, S.P. (2012). *Mycoplasma hyopneumoniae* surface proteins Mhp385 and Mhp384 bind host cilia and glycosaminoglycans and are endoproteolytically processed by proteases that recognize different cleavage motifs. J. Proteome Res. *11*, 1924–1936.

Dirksen, L.B., Krebes, K.A., and Krause, D.C. (1994). Phosphorylation of cytadherence-accessory proteins in *Mycoplasma pneumoniae*. J. Bacteriol. *176*, 7499–7505.

Djordjevic, S.P., Cordwell, S.J., Djordjevic, M.A., Wilton, J., and Minion, F.C. (2004). Proteolytic processing of the *Mycoplasma hyopneumoniae* cilium adhesin. Infect. Immun. *72*, 2791–2802.

Edwards, J.A., Groathouse, N.A., and Boitano, S. (2005). *Bordetella bronchiseptica* adherence to cilia is mediated by multiple adhesin factors and blocked by surfactant protein A. Infect. Immun. *73*, 3618–3626.

Fekkes, P., van der Does, C., and Driessen, A.J. (1997). The molecular chaperone SecB is released from the carboxy-terminus of SecA during initiation of precursor protein translocation. EMBO J. *16*, 6105–6113.

Feldner, J., Gobel, U., and Bredt, W. (1982). *Mycoplasma pneumoniae* adhesin localized to tip structure by monoclonal antibody. Nature *298*, 765–767.

Fink, A.L. (1999). Chaperone-mediated protein folding. Physiol. Rev. *79*, 425–449.

Gobel, U., Speth, V., and Bredt, W. (1981). Filamentous structures in adherent *Mycoplasma pneumoniae* cells treated with nonionic detergents. J. Cell. Biol. *91*, 537–543.

Guell, M., van Noort, V., Yus, E., Chen, W.H., Leigh-Bell, J., Michalodimitrakis, K., Yamada, T., Arumugam, M., Doerks, T., Kuhner, S., *et al.* (2009). Transcriptome complexity in a genome-reduced bacterium. Science *326*, 1268–1271.

Hallamaa, K.M., Browning, G.F., and Tang, S.L. (2006). Lipoprotein multigene families in *Mycoplasma pneumoniae*. J. Bacteriol. *188*, 5393–5399.

Hasona, A., Crowley, P.J., Levesque, C.M., Mair, R.W., Cvitkovitch, D.G., Bleiweis, A.S., and Brady, L.J. (2005). Streptococcal viability and diminished stress tolerance in mutants lacking the signal recognition particle pathway or YidC2. Proc. Natl. Acad. Sci. U.S.A. *102*, 17466–17471.

Henderson, G.P., and Jensen, G.J. (2006). Three-dimensional structure of *Mycoplasma pneumoniae's* attachment organelle and a model for its role in gliding motility. Mol. Microbiol. *60*, 376–385.

Herrmann, R., and Ruppert, T. (2006). Proteome of *Mycoplasma pneumoniae*. Methods Biochem. Anal. *49*, 39–56.

Himmelreich, R., Hilbert, H., Plagens, H., Pirkl, E., Li, B.C., and Herrmann, R. (1996). Complete sequence analysis of the genome of the bacterium *Mycoplasma pneumoniae*. Nucleic Acids Res. *24*, 4420–4449.

Hu, L.I., Lima, B.P., and Wolfe, A.J. (2010). Bacterial protein acetylation: the dawning of a new age. Mol. Microbiol. *77*, 15–21.

Huber, D., Boyd, D., Xia, Y., Olma, M.H., Gerstein, M., and Beckwith, J. (2005). Use of thioredoxin as a reporter to identify a subset of *Escherichia coli* signal sequences that promote signal recognition particle-dependent translocation. J. Bacteriol. *187*, 2983–2991.

Inada, T., Court, D.L., Ito, K., and Nakamura, Y. (1989). Conditionally lethal amber mutations in the leader peptidase gene of *Escherichia coli*. J. Bacteriol. *171*, 585–587.

Jacobs, E., Fuchte, K., and Bredt, W. (1987). Amino acid sequence and antigenicity of the amino-terminus of the 168 kDa adherence protein of *Mycoplasma pneumoniae*. J. Gen. Microbiol. *133*, 2233–2236.

Jacobs, M., Andersen, J.B., Kontinen, V., and Sarvas, M. (1993). *Bacillus subtilis* PrsA is required in vivo as an extracytoplasmic chaperone for secretion of active enzymes synthesized either with or without pro-sequences. Mol. Microbiol. *8*, 957–966.

Jenkins, C., Wilton, J.L., Minion, F.C., Falconer, L., Walker, M.J., and Djordjevic, S.P. (2006). Two domains within the *Mycoplasma hyopneumoniae* cilium adhesin bind heparin. Infect. Immun. *74*, 481–487.

Kaufmann, A., Muhlradt, P.F., Gemsa, D., and Sprenger, H. (1999). Induction of cytokines and chemokines in human monocytes by *Mycoplasma fermentans*-derived lipoprotein MALP-2. Infect. Immun. *67*, 6303–6308.

Kennelly, P.J. (2002). Protein kinases and protein phosphatases in prokaryotes: a genomic perspective. FEMS Microbiol. Lett. *206*, 1–8.

Krause, D.C., and Balish, M.F. (2004). Cellular engineering in a minimal microbe: structure and assembly of the terminal organelle of *Mycoplasma pneumoniae*. Mol. Microbiol. *51*, 917–924.

Kuhner, S., van Noort, V., Betts, M.J., Leo-Macias, A., Batisse, C., Rode, M., Yamada, T., Maier, T., Bader, S., Beltran-Alvarez, P., *et al.* (2009). Proteome organization in a genome-reduced bacterium. Science *326*, 1235–1240.

Layh-Schmitt, G., and Herrmann, R. (1992). Localization and biochemical characterization of the ORF6 gene product of the *Mycoplasma pneumoniae* P1 operon. Infect. Immun. *60*, 2906–2913.

Layh-Schmitt, G., Hilbert, H., and Pirkl, E. (1995). A spontaneous hemadsorption-negative mutant of *Mycoplasma pneumoniae* exhibits a truncated adhesin-related 30-kilodalton protein and lacks the cytadherence-accessory protein HMW1. J. Bacteriol. *177*, 843–846.

Layh-Schmitt, G., Podtelejnikov, A., and Mann, M. (2000). Proteins complexed to the P1 adhesin of *Mycoplasma pneumoniae*. Microbiology *146*, 741–747.

Lee, H.C., and Bernstein, H.D. (2001). The targeting pathway of *Escherichia coli* presecretory and integral membrane proteins is specified by the hydrophobicity of the targeting signal. Proc. Natl. Acad. Sci. U.S.A. *98*, 3471–3476.

Luirink, J., von Heijne, G., Houben, E., and de Gier, J.W. (2005). Biogenesis of inner membrane proteins in *Escherichia coli*. Annu. Rev. Microbiol. *59*, 329–355.

Markham, P.F., Glew, M.D., Browning, G.F., Whithear, K.G., and Walker, I.D. (1998). Expression of two members of the pMGA gene family of *Mycoplasma gallisepticum* oscillates and is influenced by pMGA-specific antibodies. Infect. Immun. *66*, 2845–2853.

Mebus, C.A., and Underdahl, N.R. (1977). Scanning electron microscopy of trachea and bronchi from gnotobiotic pigs inoculated with *Mycoplasma hyopneumoniae*. Am. J. Vet. Res. *38*, 1249–1254.

Meng, K.E., and Pfister, R.M. (1980). Intracellular structures of *Mycoplasma pneumoniae* revealed after membrane removal. J. Bacteriol. *144*, 390–399.

Minion, F.C., Lefkowitz, E.J., Madsen, M.L., Cleary, B.J., Swartzell, S.M., and Mahairas, G.G. (2004). The genome sequence of *Mycoplasma hyopneumoniae* strain 232, the agent of swine mycoplasmosis. J. Bacteriol. *186*, 7123–7133.

Moitinho-Silva, L., Heineck, B.L., Reolon, L.A., Paes, J.A., Klein, C.S., Rebelatto, R., Schrank, I.S., Zaha, A., and Ferreira, H.B. (2012). *Mycoplasma hyopneumoniae* type I signal peptidase: expression and evaluation of its diagnostic potential. Vet. Microbiol. *154*, 282–291.

Muhlradt, P.F., Kiess, M., Meyer, H., Sussmuth, R., and Jung, G. (1997). Isolation, structure elucidation, and synthesis of a macrophage stimulatory lipopeptide from *Mycoplasma fermentans* acting at picomolar concentration. J. Exp. Med. *185*, 1951–1958.

Nishiguchi, M., Matsumoto, M., Takao, T., Hoshino, M., Shimonishi, Y., Tsuji, S., Begum, N.A., Takeuchi, O., Akira, S., Toyoshima, K., *et al.* (2001). *Mycoplasma fermentans* lipoprotein M161Ag-induced cell activation is mediated by Toll-like receptor 2: role of N-terminal hydrophobic portion in its multiple functions. J. Immunol. *166*, 2610–2616.

Paetzel, M., Goodall, J.J., Kania, M., Dalbey, R.E., and Page, M.G. (2004). Crystallographic and biophysical analysis of a bacterial signal peptidase in complex with a lipopeptide-based inhibitor. J. Biol. Chem. *279*, 30781–30790.

Phillips, G.J., and Silhavy, T.J. (1992). The *E. coli ffh* gene is necessary for viability and efficient protein export. Nature *359*, 744–746.

Popham, P.L., Hahn, T.W., Krebes, K.A., and Krause, D.C. (1997). Loss of HMW1 and HMW3 in noncytadhering mutants of *Mycoplasma pneumoniae* occurs post-translationally. Proc. Natl. Acad. Sci. U.S.A. *94*, 13979–13984.

Potempa, J., and Pike, R.N. (2009). Corruption of innate immunity by bacterial proteases. J. Innate Immun. *1*, 70–87.

Rafelski, S.M., and Theriot, J.A. (2006). Mechanism of polarization of *Listeria monocytogenes* surface protein ActA. Mol. Microbiol. *59*, 1262–1279.

Razin, S., Yogev, D., and Naot, Y. (1998). Molecular biology and pathogenicity of mycoplasmas. Microbiol. Mol. Biol. Rev. *62*, 1094–1156.

Regula, J.T., Ueberle, B., Boguth, G., Gorg, A., Schnolzer, M., Herrmann, R., and Frank, R. (2000). Towards a two-dimensional proteome map of *Mycoplasma pneumoniae*. Electrophoresis *21*, 3765–3780.

Rosch, J.W., Vega, L.A., Beyer, J.M., Lin, A., and Caparon, M.G. (2008). The signal recognition particle pathway is required for virulence in *Streptococcus pyogenes*. Infect. Immun. *76*, 2612–2619.

Schatz, G., and Dobberstein, B. (1996). Common principles of protein translocation across membranes. Science *271*, 1519–1526.

Schmidl, S.R., Gronau, K., Pietack, N., Hecker, M., Becher, D., and Stulke, J. (2010). The phosphoproteome of the minimal bacterium *Mycoplasma pneumoniae*: analysis of the complete known Ser/Thr kinome suggests the existence of novel kinases. Mol. Cell. Proteomics *9*, 1228–1242.

Scott, N.E., Bogema, D.R., Connolly, A.M., Falconer, L., Djordjevic, S.P., and Cordwell, S.J. (2009). Mass spectrometric characterization of the surface-associated 42 kDa lipoprotein JlpA as a glycosylated antigen in strains of *Campylobacter jejuni*. J. Proteome Res. *8*, 4654–4664.

Scott, N.E., Marzook, N.B., Deutscher, A., Falconer, L., Crossett, B., Djordjevic, S.P., and Cordwell, S.J. (2010). Mass spectrometric characterization of the *Campylobacter jejuni* adherence factor CadF reveals post-translational processing that removes immunogenicity while retaining fibronectin binding. Proteomics *10*, 277–288.

Scott, N.E., Parker, B.L., Connolly, A.M., Paulech, J., Edwards, A.V., Crossett, B., Falconer, L., Kolarich, D., Djordjevic, S.P., Hojrup, P., *et al.* (2011). Simultaneous glycan-peptide characterization using hydrophilic interaction chromatography and parallel fragmentation by CID, higher energy collisional dissociation, and electron transfer dissociation MS applied to the N-linked glycoproteome of *Campylobacter jejuni*. Mol. Cell. Proteomics *10*, M000031-MCP000201.

Seymour, L.M., Deutscher, A.T., Jenkins, C., Kuit, T.A., Falconer, L., Minion, F.C., Crossett, B., Padula, M., Dixon, N.E., Djordjevic, S.P., *et al.* (2010). A processed multidomain *Mycoplasma hyopneumoniae* adhesin binds fibronectin, plasminogen, and swine respiratory cilia. J. Biol. Chem. *285*, 33971–33978.

Seymour, L.M., Falconer, L., Deutscher, A.T., Minion, F.C., Padula, M.P., Dixon, N.E., Djordjevic, S.P., and Walker, M.J. (2011). Mhp107 Is a member of the multifunctional adhesin family of *Mycoplasma hyopneumoniae*. J. Biol. Chem. *286*, 10097–10104.

Seymour, L.M., Jenkins, C., Deutscher, A.T., Raymond, B.B., Padula, M.P., Tacchi, J.L., Bogema, D.R., Eamens, G.J., Woolley, L.K., Dixon, N.E., *et al.* (2012). Mhp182 (P102) binds fibronectin and contributes to the recruitment of plasmin(ogen) to the *Mycoplasma hyopneumoniae* cell surface. Cell. Microbiol. *14*, 81–94.

Sperker, B., Hu, P., and Herrmann, R. (1991). Identification of gene products of the P1 operon of *Mycoplasma pneumoniae*. Mol. Microbiol. *5*, 299–306.

Su, C.J., Tryon, V.V., and Baseman, J.B. (1987). Cloning and sequence analysis of cytadhesin P1 gene from *Mycoplasma pneumoniae*. Infect. Immun. *55*, 3023–3029.

Szczepanek, S.M., Frasca, S., Jr., Schumacher, V.L., Liao, X., Padula, M., Djordjevic, S.P., and Geary, S.J. (2010). Identification of lipoprotein MslA as a neoteric virulence factor of *Mycoplasma gallisepticum*. Infect. Immun. *78*, 3475–3483.

Tjalsma, H., Noback, M.A., Bron, S., Venema, G., Yamane, K., and van Dijl, J.M. (1997). *Bacillus subtilis* contains four closely related type I signal peptidases with overlapping substrate specificities. Constitutive and temporally controlled expression of different sip genes. J. Biol. Chem. *272*, 25983–25992.

Tjalsma, H., Antelmann, H., Jongbloed, J.D., Braun, P.G., Darmon, E., Dorenbos, R., Dubois, J.Y., Westers, H., Zanen, G., Quax, W.J., *et al.* (2004). Proteomics of protein secretion by *Bacillus subtilis*: separating the 'secrets' of the secretome. Microbiol. Mol. Biol. Rev. *68*, 207–233.

van Noort, V., Seebacher, J., Bader, S., Mohammed, S., Vonkova, I., Betts, M.J., Kuhner, S., Kumar, R., Maier, T., O'Flaherty, M., *et al.* (2012). Cross-talk between phosphorylation and lysine acetylation in a genome-reduced bacterium. Mol. Syst. Biol. *8*, 571.

van Wely, K.H., Swaving, J., Freudl, R., and Driessen, A.J. (2001). Translocation of proteins across the cell envelope of Gram-positive bacteria. FEMS Microbiol. Rev. *25*, 437–454.

von Heijne, G. (1990). The signal peptide. J. Membr. Biol. *115*, 195–201.

Weber Sde, S., Sant'Anna, F.H., and Schrank, I.S. (2012). Unveiling *Mycoplasma hyopneumoniae* promoters: sequence definition and genomic distribution. DNA Res. *19*, 103–115.

Wilton, J., Jenkins, C., Cordwell, S.J., Falconer, L., Minion, F.C., Oneal, D.C., Djordjevic, M.A., Connolly, A., Barchia, I., Walker, M.J., *et al.* (2009). Mhp493 (P216) is a proteolytically processed, cilium and heparin binding protein of *Mycoplasma hyopneumoniae*. Mol. Microbiol. *71*, 566–582.

Witze, E.S., Old, W.M., Resing, K.A., and Ahn, N.G. (2007). Mapping protein post-translational modifications with mass spectrometry. Nat. Methods *4*, 798–806.

Woese, C.R., Maniloff, J., and Zablen, L.B. (1980). Phylogenetic analysis of the mycoplasmas. Proc. Natl. Acad. Sci. U.S.A. *77*, 494–498.

Wong, P., and Houry, W.A. (2004). Chaperone networks in bacteria: analysis of protein homeostasis in minimal cells. J. Struct. Biol. *146*, 79–89.

Yang, C.K., Ewis, H.E., Zhang, X., Lu, C.D., Hu, H.J., Pan, Y., Abdelal, A.T., and Tai, P.C. (2011). Nonclassical protein secretion by *Bacillus subtilis* in the stationary phase is not due to cell lysis. J. Bacteriol. *193*, 5607–5615.

Yus, E., Maier, T., Michalodimitrakis, K., van Noort, V., Yamada, T., Chen, W.H., Wodke, J.A., Guell, M., Martinez, S., Bourgeois, R., *et al.* (2009). Impact of genome reduction on bacterial metabolism and its regulation. Science *326*, 1263–1268.

Zanen, G., Antelmann, H., Meima, R., Jongbloed, J.D., Kolkman, M., Hecker, M., van Dijl, J.M., and Quax, W.J. (2006). Proteomic dissection of potential signal recognition particle dependence in protein secretion by *Bacillus subtilis*. Proteomics *6*, 3636–3648.

Zhang, J., Sprung, R., Pei, J., Tan, X., Kim, S., Zhu, H., Liu, C.F., Grishin, N.V., and Zhao, Y. (2009). Lysine acetylation is a highly abundant and evolutionarily conserved modification in *Escherichia coli*. Mol. Cell. Proteomics *8*, 215–225.

Zhang, Q.J., Young, T.F., and Ross, R.F. (1995). Identification and characterization of a *Mycoplasma hyopneumoniae* adhesin. Infect. Immun. *63*, 1013–1019.

# Multifunctional Cytoadherence Factors

6

Miriam Hopfe and Birgit Henrich

## Abstract

In the cell-wall-less Mollicutes cytoadherence factors that enable tight contact with the host often play additional roles in colonization, nutritional uptake and the host's immune response. Multifunctional cytoadhesins have been shown to interact with components of the extracellular matrix facilitating invasion of the host, and to hamper a successful immune response by through antigenic variation, mimicry or release of antigens into their surroundings. Cytoadhesins that condense at a tip structure in some species are also involved in gliding motility.

Detailed sequence analyses have enabled the detection of six mycoplasmal cytoadhesive moonlighters that are characterized by the independence of their different functions. Heparin binding of P97 and P110 of *M. hyopneumoniae* was shown to be mediated by regions distinct from their cytoadhesive domains. The α-enolase of *M. gallisepticum* acts at different locations, intracellularly as a cytoplasmic glycolytic enzyme and surface-exposed as an ECM-binding cytoadhesin. Antigenic mimicry of P1 and P30 are unique features of these proteins in *M. pneumoniae*, in contrast to their homologous in other mycoplasmas. The most outstanding moonlighter described thus far is OppA of *M. hominis*: it mediates cytoadherence, functions as peptide-binding domain for the oligopeptide permease and is the main ecto-ATPase, the action of which leads to apoptotic death in host cells. Phosphorylation appears to be one mechanism enabling coordinated cooperation between these distinct functions in a multifunctional cytoadhesin.

## Introduction

Mycoplasmas are the smallest self-replicating bacteria known. They have a reduced coding capacity and have lost several metabolic pathways, resulting in a need for parasitic behaviour just to be able to live. Like several other bacterial pathogens mycoplasmas engage in a daily game to struggle to survive:

- Keep me – feed me – but do not catch me!
- or, Colonization – nutrition – immune evasion.

Mycoplasmas have to be in close contact with their host to acquire their required nutrition. This is mediated, facilitated and modulated by cytoadherence factors. Initial adherence interactions between mycoplasmas and mammalian cells are important for host colonization and may also contribute to subsequent pathogenic processes, such as cytotoxicity and invasion. As reviewed by Razin in 1999, mycoplasmas have developed various strategies that enable modulated attachment to host tissues. They are able to generate a versatile surface coat through high-frequency phase, size and antigenic variation of surface-exposed proteins, which often function as cytoadhesins. Moreover, the flexible architecture of the surface provides a useful tool for immune evasion, allowing them to escape from attack by antibodies.

The reduced coding capacity of mycoplasmas also resulted in another survival mechanism: use of multifunctional proteins, which include adhesins that have additional functions besides cytoadhesion. Multifunctionality of proteins is not restricted to the Mollicutes, and occurs in more complex prokaryotic and eukaryotic microorganisms (Krulwich *et al.*, 2001; Kirn-Safran *et al.*, 2009). The aim of this chapter is to highlight the multifunctionality of mycoplasmal cytoadherence factors, but consideration of multifunctional proteins in other species may indicate how additional multifunctional homologues might be identified in the Mollicutes. Many protein functions can be inferred from the known functions of homologous proteins in other organisms, but in sequenced genomes of the Mollicutes up to 45% of the annotated proteins are of unknown function (Calderon-Copete *et al.*, 2009). The existence of multifunctional proteins complicates the interpretation of protein function using similarity searches, but also provides a fascinating view of the complex interactions among the various cell components of the minimized mycoplasma cell. We will focus on the description of protein-based studies of multifunctional cytoadhesins, including studies of the non-proteinaceous adhesins, in the context of homologous factors in other classes of bacteria and potential homologues detected by in silico analyses.

Before examining the mycoplasmal world of multifunctional cytoadherence factors, it is important to clarify the definitions used in this chapter. The term 'cytoadherence factor' is used to define proteins and non-proteinaceous structures that mediate close contact between the mycoplasma and the host cell. Attachment to plastic and binding to components of the host's extracellular matrix are not sufficient to establish a role in *cyto*adherence. The demonstration of attachment of a factor to the host cell *and* of the decrease or inhibition of mycoplasmal cytoadherence by masking or deletion of this molecule are useful tools for the identification of a true cytoadherence factor.

Our definition of the term 'multifunctional' is much more ambitious. We wish to discriminate between 'multifunctional' and 'moonlighting' factors. 'Multifunctional' is be used to describe cytoadherence factors with effects or features other than cytoadherence that are not common to all other cytoadherence factors (e.g. in antigenic variation or immune modulation, motility, invasion, cell fusion and virulence). The term 'moonlighting' is used in the style of Constance Jeffrey, who characterized moonlighting proteins predominantly in higher organisms (Jeffrey, 1999) and defined them as a subset of multifunctional proteins in which more than one function is found in one polypeptide chain (Jeffrey, 2009). This *ex*cludes eukaryotic proteins that are the result of gene fusions, proteins derived from a gene that generates multiple translational products or that vary as a result of post-translational modifications, and proteins that have a single function but can operate in different locations or utilize different substrates. These molecules are rather defined as multifunctional. Features suggestive of a mycoplasma moonlighter can therefore include: different locations of the factor within a cell, or even presence inside and outside the cell; binding to different cofactors; different binding sites for distinct factors; or a different oligomerization state. An important criterion of moonlighting factors is the independency of their distinct functions – inactivation of one function, for example by mutation, should not affect the other functions (Huberts and van der Klei, 2010).

As a result of our experience in characterizing the multifunctionality of cytoadhesins of *Mycoplasma hominis*, we have taken closely examined the literature and searched mycoplasma surface-exposed structures for cytoadhesive and accessory features and functions, and will describe typical examples of mycoplasmal cytoadherence factors with multiple roles and highlight the moonlighting factors.

## Cytoadherence factors

Many animal mycoplasmas depend on adhesion to host tissues for infection. Cytoadhesion is a prerequisite for successful colonization, preventing the mycoplasma cell from being flushed away and this securing its survival, and thus acts as a major virulence factor, as adherence-deficient mutants have been shown to be avirulent (Razin and Jacobs, 1992; Ben-Menachem *et al.*, 2001).

All surface-exposed structures of the cell wall-less Mollicutes have to be considered as potential cytoadherence factors, including lipids, intrinsic membrane proteins and peripherical bound proteins, and particularly the wide range of lipoproteins and (ABC) transporters used for nutritional uptake and export of noxious substances, extra-cellular toxins and targeting membrane components. In a few species, such as *M. mycoides* subspecies *mycoides*, capsules are composed of lipoglycans and saccharides (Rosenbusch and Minion, 1992).

## Non-proteinaceous cytoadherence factors with multiple functions

In the cell membrane of *Mycoplasma fermentans*, choline-containing lipids have been shown to participate in cytoadhesion to the surface of eukaryotic cells (Rottem, 2002). Analysis of the phosphocholine-containing glycoglycerolipids (GGPLs), the main lipid antigens of *M. fermentans*, showed that GGPL-III (=MfGL-II) constitutes up to 20% of the total phospholipids of this organism. GGPL-III is responsible for attachment of *M. fermentans* to host cells (Yavlovich *et al.*, 2004), as well as for stimulation of glial cells to secret nitric oxide and prostaglandin E2 (Ben-Menachem *et al.*, 1998), mainly through its phosphocholine moiety. The earlier hypothesis that GGPL-III stimulates fusion between mycoplasmas and eukaryotic cells (Salman *et al.*, 1994) has been disproved, as GGPL-III has been shown to have no fusogenic properties, and thus this function is more likely performed by a newly identified lysolipid in the membrane of *M. fermentans* (Ben-Menachem *et al.*, 2001). It has been hypothesized that GGPL-III plays a role in rheumatoid arthritis (Kawahito *et al.*, 2008), and it has recently been shown to promote rheumatoid arthritis and metal allergy in a mouse model (Sato *et al.*, 2010).

Taken together, these findings suggest that GGPL-III is an important pathophysiological mediator during infection with *M. fermentans*. As it functions as a cytoadherence factor and also triggers an inflammatory response in the host, it fulfils the requirements of a multifunctional cytoadherence factor.

## Membrane proteins as cytoadhesins

Several mycoplasmal cytoadhesins have been described. The best studied adherence systems are those of mycoplasmas that attach using a specialized attachment organelle, with *M. pneumoniae* the most studied example (Chaudhry *et al.*, 2005; Hatchel and Balish, 2008). Three distinct subcellular protein complexes, HMW1–3, P1-P90-P40 and P30-P65, have been shown to be essential for assembling the tip structure that is involved in adhesion, gliding and cell division (Seto and Miyata, 2003). Nevertheless, identification of cytoadherence proteins is somewhat hindered by the fact that tip formation is a complex outcome of interactions between cytoadhesive and accessory proteins. Loss of HMW2, which lacks cytoadherence properties, is accompanied by reduced levels of the terminal organelle proteins HMW1, HMW3, P24, P28, P41 and P65, failure to localize the major adhesin P1 to the terminal organelle, and loss of cytoadherence (Bose *et al.*, 2009). Adhesion and inhibition assays are useful tools to confirm that surface-exposed (lipo-) proteins are cytoadherence factors. Thus, P116, an immunogenic lipoprotein of *M. pneumoniae* was shown to be surface exposed and to participate in cytoadhesion, independent of the function of the main adhesin P1 (Svenstrup *et al.*, 2002), and P64, a 64-kDa lipoprotein of *M. gallisepticum*, was shown to function as cytoadhesin in addition to the main cytoadhesin GapA (Kheyar *et al.*, 1995).

The structure of P1 and the P1 homologues of *M. gallisepticum* (GapA) and *M. genitalium* (MgPa) is in some respects highly conserved. The main cytoadhesins carry an amino-terminal signal peptide (with a more or less conserved signal peptidase (SPase) I cleavage site) and a further transmembrane region in the carboxyl-terminal part. This is also seen in P89 (SARP1) of *Spiroplasma citri*, which also contains five Reg-Prop repeats (PFAM PF07494), which are common in two-component regulator proteins towards its amino-terminal end (Berg *et al.*, 2001). Gli349, the cytoadhesin of *M. mobile*, which belongs to the hominis phylogenetic group, shares low sequence similarities with the P1 homologues. The carboxyl-terminal transmembrane (TM) region is missing, and two repeat regions of about 660 AA occur within the first and second thirds of the section of the polypeptide missing from the P1 homologues. However, the amino-terminal signal peptide with a predicted SPase I cleavage site, its location, clustered at in a tip-structure-like attachment organelle, and the structures targeting the host look similar to those seen associated with P1 and its homologues. Binding is mediated by a highly conserved region within the carboxyl-terminal part of P1 and attachment to sialylated oligosaccharides (N-acetylneuraminic acid α-2,3 linked to galactose) effected by the sequence motif [G-I-V-R-T-P-L-A-E-L-L-D] (Dallo *et al.*, 1988). Similarly, Gli349 preferentially binds to a slightly modified sialylgalactose (N-acetylneuraminic acid α-2,6 linked to galactose) (Miyata, 2010). Repeats are a common feature of surface exposed proteins in the Mollicutes and are thought to play an important role in immune evasion by modulating the surface architecture of the cell. Thus it is not surprising that even the conserved P1 cytoadhesin can occur as size variants as a result of recombination within its repetitive domains and homopolymeric repeats (Su *et al.*, 1993). The deletion of 8 of 13 proline-rich repeats of P30 renders *M. pneumoniae* haemadsorption negative (Dallo *et al.*, 1996) and, in addition, acidic proline-rich (APR) domains are found in HMW1, HMW3 and P65, although their function is as yet unknown (Balish *et al.*, 2001).

In Mollicutes lacking a specialized attachment organelle, cytoadhesive membrane proteins seem to have evolved independently or divergently in different species, rendering the detection of cytoadhesins more difficult because of the low level of inter-species homology. Cytoadhesins of Mollicutes lacking a tip structure are often lipoproteins or membrane-bound proteins with amino-terminal signal peptides, as summarized in Table 6.1. In these Mollicutes, variation and modulation of surface-exposed structures is common, including cytoadhesins such as P50/Vaa of *M. hominis* (Zhang and Wise, 1996; Henrich *et al.*, 1998a) and its homologues P89 of *Spiroplasma citri* and P40 in *M. agalactiae* (Fleury *et al.*, 2002). The function of the P40 lipoprotein of *M. agalactiae* as cytoadhesin has been confirmed using P40-specific antibodies, which inhibit adherence to lamb joint synovial cells. Using an analogous experimental setting, an abundant lipoprotein of *M. fermentans*, P29, has been shown to act as cytoadhesin, whereas the highly phase variable lipoprotein, P78, which is encoded in an ABC transporter operon, did not have this function (Theiss and Wise, 1997; Leigh and Wise, 2002).

It should be remembered that such a model system can have flaws when used to identify cytoadherence factors. A decrease in mycoplasmal adherence to the host cell may not be caused by masking of the putative cytoadhesin by the antibody, but rather a consequence of steric hindrance as a result of binding to a neighbouring, but unrecognised, cytoadherence factor.

Washburn and coworkers used transposon mutagenesis to obtain definitive evidence that the MAA1 lipoprotein of *Mycoplasma arthritidis* acts as a cytoadhesin (Washburn *et al.*, 2003). Further analyses revealed that the MAA2 lipoprotein, formerly thought to also function in cytoadhesion, was more likely to be a negative regulator of MAA1, probably serving to release organisms from microcolonies at specific stages of infection (Bird *et al.*, 2008).

Attachment-induced host–pathogen interactions, comprising inflammatory responses of the host, including cytokine secretion, antibody production and apoptosis, and reactions by the pathogen, including immune evasion by secretion and shedding of antigens, mimicry of host cell structures, biofilm formation, motility, invasion of and fusion with the host cell, will then determine the further fortunes of the mycoplasma cell. In some, but not always all cases, cytoadherence factors themselves lead to these host–pathogen interactions, thus earning the description 'multifunctional'. In this context the following cytoadhesins can be considered multifunctional:

## Cytoadhesins that are variable in phase, size and antigenicity

In Mollicutes most of the surface-exposed proteins are immunogenic, and especially regions of cytoadhesins that mediate close contact with the host, as seen, for example, for the carboxyl-terminal regions of P1 and P30 of *M. pneumoniae*, which induce particularly prominent humoral responses *in vivo* (Chaudry *et al.*, 2005; Varshney *et al.*, 2008; Schurwanz *et al*, 2009). Although most cytoadhesins are well recognized as antigens, antigenicity and immunogenicity are

**Table 6.1** Mycoplasma cytoadhesins

| Organism | Protein | Reference |
|---|---|---|
| *M. genitalium* | MgPa (mg191) P110 (mg192) | Burgos *et al.* (2006), Pich *et al.* (2009) |
| *M. gallisepticum* | CrmA | Goh *et al.* (1998) |
| | _-enolase | Chen *et al.* (2011) |
| | GapA | Mudahi-Orenstein *et al.* (2003) |
| | P64 | Kheyar *et al.* (1995) |
| | PvpA | Jiang *et al.* (2009) |
| *M. penetrans* | P50 precursor | Bendjennat *et al.* (1997) |
| *M. pneumoniae* | CARDS toxin | Kannan and Baseman (2006) |
| | EF-Tu | Balasubramanian *et al.* (2008, 2009) |
| | GAPDH | Alvarez *et al.* (2003) |
| | HMW1+3 | Schmidl *et al.* (2010) |
| | P1 | Su *et al.* (1993) |
| | P116 | Svestrup *et al.* (2002) |
| | P30 | Romero-Arroyo *et al.* (1999) |
| | PDH | Dallo *et al.* (2002) |
| *M. mobile* | Gli349 | Adan-Kubo *et al.* (2006) |
| *S. citri* | P32 | Killiny *et al.* (2006) |
| | P89 (SARP1) | Berg *et al.* (2001) |
| | Spiralin | Killiny *et al.* (2005) |
| *M. agalactiae* | P40 | Fleury *et al.* (2002) |
| *M. arthritidis* | MAA1 | Washburn *et al.* (2003) |
| *M. bovis* | Vsp | Sachse *et al.* (2000) |
| *M. conjunctivae* | LppS LppT | Belloy *et al.* (2003), Zimmermann *et al.* (2010) |
| *M. fermentans* | GGPL III | Bon-Menachem *et al.* (1998) |
| | P29 | Leigh and Wise (2002) |
| *M. hominis* | OppA | Hopfe *et al.* (2004) |
| | P50/Vaa | Henrich *et al.* (1993) |
| | P80+P60 | Henrich *et al.* (1993) |
| *M. hyopneumoniae* | P126 derived P97 | Djordjevic *et al.* (2004); Okamba *et al.* (2010) |
| | P135 derived P45+P48+P50 | Bogema *et al.* (2011) |
| | P159 derived P27+P110+P52 | Burnett *et al.* (2006) |
| | P216 derived P120+P85 | Wilton *et al.* (2008) |
| *M. hyorhinis* | P37 | Dudler (1988) |
| *M. pulmonis* | Vsa | Simmons and Dybvig (2003) |
| *U. urealyticum SV8* | P96 | Smith *et al.* (1994) |

more likely benefits to the host rather additional functions of the cytoadherence factors. However the situation differs for membrane proteins that exhibit phase-variable expression, or size and/or antigenic variation. This is the simplest approach of the mycoplasma cell to evasion of the adaptive immune system – rapidly changing non-essential epitopes on its surface while keeping essential epitopes concealed.

Lipoproteins are the main structures affected by phase, size and antigenic variations. As reviewed by Christine Citti and colleagues, these gene families are predominantly found in Mollicutes that lack an attachment organelle (e.g. *vaa*, *vlp*, *vmc*, *vsa*, *vsp*, *vpma* and *mba*), but are also found in a few that have a tip structure for cytoadherence (*vlhA* and *mpl*) (Citti *et al.*, 2010). In most cases variable lipoproteins do not function as cytoadhesins. Nevertheless, in searching for multifunctional cytoadherence factors we found four exceptions: the *mgp* system of *M. genitalium*, the *vsp* system of *M. bovis*, the *vsa* system of *M. pulmonis*, and the *vaa* system of *M. hominis*.

The *mgp* system of *M. genitalium* effects variation of the main cytoadhesin, MgPa. In *M. genitalium* nine MgPar regions 0.4 kb to 2.4 kb in length have been described containing sequences homologous to three regions within the MgPa-encoding *mgp*B gene and to two regions within the adjacent *mgp*C gene, which encodes P110, thus leading to the generation of a diversity of variants by recombination (Iverson-Cabral *et al.*, 2006, 2007).

The *vsp* system of *M. bovis* expresses a family of phase-variable Vsp lipoproteins that carry different immunodominant Vsp repeats, some of which have been shown to mediate cytoadhesion of *M. bovis* cells to embryonic bovine lung cells. The highly immunogenic nature of distinct Vsp variants suggests a specific function in pathogenesis (Sachse *et al.*, 2000). In the murine respiratory pathogen *Mycoplasma pulmonis*, attachment is affected by the lengths of the variable surface antigens (Vsa). Strains producing Vsa variants with about 40–60 tandem repeats in the carboxyl-terminal region do not (cyto-) adhere, whereas strains with Vsa proteins with 0–5 tandem repeats do adhere to red blood cells and pulmonary epithelial cells, and became the target of complement-induced killing (Simmons and Dybvig, 2003; Bolland and Dybvig, 2012). In addition, *M. pulmonis* cells that are encased within a biofilm produce a short Vsa protein with a few tandem repeats. As variation in the number of Vsa tandem repeats occurs by slipped-strand mispairing, the ability of mycoplasma cells to attach to or release from a biofilm switches stochastically (Simmons and Dybvig, 2007). Thus, biofilm formation may be one of the mechanisms that protect mycoplasmas from host immunity but, due to the phase-variable ON/OFF switch, it isn't a one-way trip.

The main cytoadhesin of *M. hominis*, P50/Vaa, was first described in the early 1990s. Immunoelectron microscopy revealed that this protein is distributed evenly over the cell surface (Feldmann *et al.*, 1992) and adhesion and inhibition assays provided evidence of its cytoadherence function (Henrich *et al.*, 1993). The single copy gene encodes a 441 amino acid lipoprotein, with an extended, predominantly helical structure and three large repeats in the carboxyl-terminal part, which are highly homologous, but not identical, allowing differentiation by monoclonal antibodies (Henrich *et al.*, 1996). Thus, P50 shares structural similarities with variable proteins of mycoplasmas, such as Vlp in *M. hyorhinis*. They are all composed of three domains: domain I, consisting of an amino-terminal signal peptide, a variable domain II, and domain III, comprising the carboxyl-terminal part, with its repetitive elements. Genetic analyses revealed that inter-strain variability is attributable to specific truncations as well as duplications of the P50 repeats (Henrich *et al.*, 1998a). Subsequent cytoadherence studies provided evidence that each repeat region of P50 binds to HeLa cells and that this binding is inhibited by high molecular weight dextran sulphate. This indicates that P50 will bind to sulfatides on the host, which are highly concentrated in the endometrium, the usual site of infection for *M. hominis* (Kitzerow *et al.*, 1999).

Although this inter-strain variation did not justify classifying P50 as a multifunctional cytoadherence factor, the studies of Zhang and Wise do. They discovered that P50 varies in size and antigenicity, leading to it being renamed Vaa (Zhang and Wise, 1996). Additionally, they demonstrated that Vaa undergoes high-frequency phase variation in expression, and that this phase variation correlates with cytoadherence. An oscillating single nucleotide deletion/insertion in a short polyadenine repeat near the amino-terminus of the mature Vaa was shown to result in an ON/OFF switch in expression of Vaa that was also seen during *M. hominis* infection in the natural host (Zhang and Wise, 1996, 1997).

These findings clearly demonstrate that repetitive sequence elements play an essential role in the phase variable expression and size or antigenic variation of mycoplasma surface molecules.

Cytoadhesins like P30 of *M. pneumoniae* and its homologues in other mycoplasma, as well as Vaa, Vsa and Vsp that display such variation confer a plasticity of surface architecture to the cell that facilitates modulation of colonization, motility and immune evasion, and therefore they warrant classification as multifunctional cytoadhesins.

## Cytoadhesins that help to move

Gliding motility is an important feature of some mycoplasmas, facilitating successful colonization of the host. Spread from the initial site of infection provides an advantage, allowing escape from local host defence mechanisms. The attachment organelle was considered to be essential for gliding motility and only Mollicutes carrying this organelle were thought to be able to move (Seto and Miyata, 2003, Balish and Krause, 2006). However, the detection of *Mycoplasma insons*, which glides but does not have a differentiated tip structure, this dogma has had to be re-evaluated (Relich *et al.*, 2009). The rod-like shape of this species may be indicative of a polarized architecture, in which cytoadhesins may be embedded in an as yet undefined, but well-organized, protein complex.

The modulation of cytoadherence from attachment to release and gliding motility has been best studied in the fish pathogen *Mycoplasma mobile* (reviewed by Miyata, 2010, and elsewhere in this book). The domains of the gliding machine have been identified by comparing the proteomes of non-gliding and gliding strains and using monoclonal antibodies that inhibit cytoadherence. Four operon-encoded proteins Gli349, Gli521, Gli123 and P42 have been identified. Using sophisticated analyses, including electron microscopy, Gli349 was identified as main cytoadhesin of *M. mobile*. In the centipede model, Miyata has postulated that Gli349 forms multiple legs at the neck of *M. mobile*, the feet of which bind to sialylgalactose. Periodic repetitions of [Stroke – Return – Release – Binding]$^{n}$ have been proposed as necessary steps in gliding. Binding and release of the Gli349 feet require the ATPase activity of P42 for supply of energy and the additional action of Gli521 as a motor (called the gear by Miyata) that induces conformational changes in Gli349 (Miyata, 2010). Gliding motility of *Mycoplasma pneumoniae*, a species phylogenetically distant from *M. mobile*, shares some similarities with this gliding model. Conformational changes of the P1 cytoadhesin resulting in displacement from the host cell have been suggested to be involved in the gliding mechanism in *M. pneumoniae* (Seto *et al.*, 2005). The P1 adhesin, which, by analogy with Gli349, is proposed to function as a 'leg' in gliding, has been shown to form a heterodimer complex with the accessory protein P90 (Nakane *et al.*, 2010). Detachment of the tip structure seems to require the help of further tip structure embedded cytoadhesive proteins (e.g. P30), as well as cytoskeletal proteins (e.g. P41), enabling gliding of the cell (Hasselbring at al., 2005; Hasselbring and Krause, 2007a,b). Cytoadherence-associated proteins like P30 may play a central role as it appears to have multiple modulating functions in adhesion and gliding (Chang *et al.*, 2011).

Although the main cytoadhesins of motile mycoplasmas seems to need assistance when detaching from the host surface, gliding motility depends on their attachment. As they are capable of both attachment and gliding, the main cytoadhesins Gli349 and P1 and its homologues warrant annotation as multifunctional cytoadhesins.

## Immune evasion

Mycoplasmas have evolved some further strategies to circumvent an effective host defence. Invasion of the host cell is one of them, as there mycoplasmas are shielded from direct contact with immune cells, antibodies and complement. Surface molecules, such as proteins and lipids that facilitate the processes of adhesion and motility, may also have an effect on invasion. Mycoplasmas with an attachment organelle (such as *M. penetrans*, *M. gallisepticum* and *M. genitalium*), as well as those without a tip structure (such as *M. fermentans* and *M. hominis*) have been found to invade host cells (Razin *et al.*, 1998; Winner *et al.*, 2000). This ability has been linked to systemic dissemination of the organism *in vivo*, resulting in formation of multiple loci of infection, rendering an effective host defence more difficult (Much *et al.*, 2002). Unfortunately, the mechanism used by mycoplasmas to gain entry into host cells remains unresolved. In other organisms binding to components of the extracellular matrix has been shown to play a role in invasion. HBHA, a heparin-binding haemagglutinin, is needed for extrapulmonary dissemination of *Mycobacterium tuberculosis* (Pethe *et al.*, 2001). In other cases bacterial invasion has been shown to be based on the ability of the pathogen to bond plasminogen and to bind to fibronectin or sulfated polysaccharides, like dextran sulfates or heparin (Marques *et al.*, 2010). This leads to formation of a bridge between the pathogen and different host cell surface proteins and finally enables invasion (Rottem, 2003). For this reason, mycoplasma proteins that bind these extracellular matrix molecules of the host, the so-called microbial surface components recognizing adhesive matrix molecules (MSCRAMMs), seem to be a useful target for future work to elucidate the molecular mechanisms involved in invasion and to clarify their role in immune evasion.

## MSCRAMMs

The structures of the host's extracellular matrix (ECM) are formed from of the interstitial matrix, which is composed of glycosylated proteins, glycosaminoglycans and individual molecules, such as heparin sulfate, fibronectin, proteoglycan, plasminogen and laminin, and the basement membrane, upon which the epithelial cells rest (Dunsmore and Rannels, 1996). The extracellular matrix provides structural support for cells. In intact tissue the ECM is covered by epithelial or endothelial cells and is therefore not accessible for bacterial binding and colonization. However any trauma or damage to host cells leads to exposure of the ECM, enabling bacterial colonization (Patti and Hook, 1994). ECM-embedded plasminogen will be activated by secreted eukaryotic and prokaryotic activators to the proteolytic enzyme plasmin (Coleman and Benach, 1999), which degrades fibrin clots and other extracellular matrix proteins during cell migration (Markus, 1996). Plasminogen has been shown to bind to exposed lysine residues on cell surfaces or in fibrin structures through its amino-terminal binding domain, while its carboxyl-terminal domain is responsible for the protease activity (Ponting *et al.*, 1992).

A number of bacterial pathogens are known to acquire host plasminogen on their cell surface to assist in invasion and colonisation of host tissue (Pollanen *et al.*, 1991). ECM-targeting structures of a pathogen, the MSCRAMMs, allow the infectious agent to colonize and subsequently invade the host cell through proteolytic degradation of the ECM, in the worst case for the host resulting in translocation of the pathogen into the blood stream (Duensing *et al.*, 1999).

As cytoadherence factors may not be restricted to mediating mycoplasma binding to the host cells, but may also function in binding the ECM, we examined mycoplasma species that cytoadhere by an attachment organelle (Giron *et al.*, 1996; Dallo *et al.*, 2002; Alvarez *et al.*, 2003; Kannan *et al.*, 2005; May *et al.*, 2006), as well as those lacking a specialized structure for cytoadherence (Burnett *et al.*, 2006; Jenkins *et al.*, 2006; Yavlovich and Rottem, 2007; Zimmermann *et al.*, 2010), for MSRCAMM proteins. In recent years many mycoplasma MSCRAMM proteins have been characterized (see Table 6.2), most comprehensively in *M. hyopneumoniae*, but only a subset have also been shown to act as cytoadherence factors (depicted in bold letters).

*M. hyopneumoniae*, the aetiological agent of porcine enzootic pneumonia, adheres to swine cilia and damages respiratory epithelial cells. *M. hyopneumoniae* cells have been shown to bind plasminogen and the ECM components fibronectin and heparin (a heparan sulfate analogue commonly used to identify glycosaminoglycan (GAG)-binding proteins). A number of surface-exposed proteins of *M. hyopneumoniae* have been characterized that are derived from a handful of genes by post-translational processing of high-molecular weight precursors (reviewed by Djordjevic and Tacchi in this book). Beside the plasminogen- and/or fibronectin-binding MSCRAMM proteins, the cytoadhesin function of which remains to be verified, seven of the heparin-binding surface-exposed proteins of *M. hyopneumoniae* have been shown to act as cytoadhesins: the *mhp*493-encoded proteins, P120 (amino-terminal part of P216) and P85 [carboxyl-terminal part of P216 the *mhp*494-encoded proteins, P110 and P52 (Burnett *et al.*, 2006)], as well as the *mhp*683-encoded proteins, P48 and P50 (Bogema *et al.*, 2011) and the main

**Table 6.2** Mycoplasma MSCRAMMs

| Organism | Surface exposed protein[a] | SP(I/II)[b] | Gene | ECM component[c] | Reference |
|---|---|---|---|---|---|
| *M. gallisepticum* | _-enolase | – | MGA-0209 | Plasminogen | Chen *et al.* (2011) |
| | HMW3 homologue | – | MGA-0928 | Fibronectin (Fn) | May *et al.* (2006) |
| | OsmC | – | MGA-1142 | Heparin | Jenkins *et al.* (2007) |
| | PlpA | – | MGA-1199 | Fibronectin | May *et al.* (2006) |
| *M. genitalium* | GAPDH | – | MG_301 | Mucin | Alvarez *et al.* (2003) |
| *M. pneumoniae* | CARDS toxin | – | MPN372 | Hsp-A | Kannan *et al.* (2005, 2006) |
| | EF-Tu | – | MPN665 | Fn | Balasubramanian *et al.* (2008, 2009) |
| | GAPDH | – | MPN430 | Fn | Dumke *et al.* (2011) |
| | PDH-B | – | MPN392 | Fn | Dallo *et al.* (2002) |
| *M. hyopneumoniae* | P97 (R1+R2) | SpI | Mhp183 | Heparin | Jenkins *et al.* (2006) |
| | P97 paralogue (R1+R2) | SpI | Mhp271 | Heparin, Plg, Fn | Deutscher *et al.* (2010) |
| | P102 ($P60^N$+$P42^C$) | $SpI^N$ | Mhp182 | Plg, Fn | Seymour *et al.* (2012) |
| | P102 paralogue ($P60^N$+$P42^C$) | $SpI^N$ | Mhp384 | Heparin | Deutscher *et al.* (2012) |
| | P104 | SpI | Mhp107 | Heparin, Plg, Fn | Seymour *et al.* (2011) |
| | P115 (P88) | $SpI^N$ | Mhp385 | Heparin | Deutscher *et al.* (2012) |
| | P116 | SpI | Mhp108 | Plg, Fn | Seymour *et al.* (2010) |
| | P135 ($P45^N$+$P48^M$+$P50^C$) | $SpI^N$ | Mhp683 | Heparin | Bogema *et al.* (2011) |
| | P146 ($P50^N$+$P40^M$+$P85^C$) | $SpI^N$ | Mhp684 | Heparin, Plg | Bogema *et al.* (2011) |
| | P159 ($P110^M$+$P52^C$) | $SpI^N$ | Mhp494 | Heparin | Burnett *et al.* (2006) |
| | P216 ($P120^N$+$P85^C$) | $SpI^N$ | Mhp493 | Heparin | Wilton *et al.* (2008) |
| *M. conjunctivae* | LppT | SpI | *lppT* | Fibronectin | Zimmermann *et al.* (2010) |
| *M. fermentans* | _-enolase | – | MFE_01090 | Plasminogen | Yavlovich *et al.* (2004) |
| *M. hyorhinis* | P37 | SpII | MHR_0625 | MMP2 | Gong *et al.* (2008) |

[a]Cytoadhesins are indicated in bold; [N]N-terminal; [M]middle or [C]C-terminal region of the precursor. [b]SP, signal peptide; SpI, cleavage by SPaseI; SpII, cleavage by SPaseII and lipid anchoring as predicted by LipoP (http://www.cbs.dtu.dk/services/LipoP/). [c]Fn, Fibronectin; Plg, Plasminogen.

cytoadhesin P97 (Jenkins *et al.*, 2006; Hsu and Minion, 1998a,b). These findings are in accordance with the observation that *M. hyopneumoniae* binds specifically to ciliary sulfated glycolipids and that this binding can be partially inhibited by heparin (Zhang *et al.*, 1994). Interestingly, regions of the amino and carboxyl terminal half of P110 are essential for cytoadhesion, whereas binding to heparin is mediated by the carboxyl -terminal half and invasion is mediated by the amino-terminal portion of P110 (for a detailed description see the review by Djordjevic and Tacchi in this book).

The MSCRAMM roles of mycoplasma cytoadhesins are not restricted to heparin-binding in other mycoplasmas. In addition to MG1142 of *M. gallisepticum*, an OsmC-like protein that belongs to a class of osmotically inducible proteins that has been shown to reside on the cell surface and to have heparin-binding capacity (Jenkins *et al.*, 2007), two fibronectin-binding components of the attachment organelle have been identified in *M. gallisepticum*: PlpA, the P65 homologue of *M. pneumoniae*, and MGA-0928, a HMW3 homologue (May *et al.*, 2006). In *M. pneumoniae* the amino-terminal half of P65 have been shown to contain several penta- and hexapeptides (DPNAY and DPNQAY) that form a proline-rich acidic domain that is suggestive of an extended and rigid structure. An [Arg-Gly-Asp] (RGD) motif, which is known to play an important role in binding of extracellular matrix proteins, has been identified near the carboxyl-terminus of P65 (Proft *et al.*, 1995). Definitive evidence of the cytoadherence-mediating properties of these *M. gallisepticum* proteins is still not available. Nevertheless, the likely function of the RGD motif in fibronectin binding by mycoplasma proteins has been supported by the recent finding that the cytoadhesive LppT protein of *M. conjunctivae* has an RGD motif that has been shown to be essential for both binding to fibronectin and binding to beta integrins of the host and, LppT has been proposed to be the first mycoplasmal RGD lectin to be identified (Zimmermann *et al.*, 2010). Thus it should be anticipated that the *mhp*384 encoded P102 paralogue of *M. hyopneumoniae* that shows similarity to LppT of *M. conjunctivae* (Deutscher *et al.*, 2012) but lacks the RGD motif is not capable to bind to fibronectin. Plasminogen binding by P116 of *M. hyopneumoniae* has been shown to depend on the carboxyl-terminal lysine, while heparin binding by the cytoadhesin P97 of *M. hyopneumoniae* depends on repeat regions R1 and R2 (Jenkins *et al.*, 2006). These data have provided evidence that cytoadherence factors of Mollicutes with and without an attachment organelle have additional roles in binding to components of the extracellular matrix. As P97-mediated cytoadhesion is independent of repeat region R2, the cytoadhesin P97 of *M. hyopneumoniae* is a cytoadherence moonlighter.

The last group of MSCRAMM proteins is the most controversial as the function of their members as surface-exposed structures is still contentious. Elongation factor EF-Tu, pyruvate dehydrogenase (PDH), glyceraldehyde-3-phosphate dehydrogenase (GAPDH) and α-enolase, all mycoplasmal enzymes that normally play important roles in the cytoplasm, have been detected as surface-exposed and attached to the mycoplasma membrane. The mediators responsible for transporting these molecules through the mycoplasmal membrane and facilitating their attachment to the mycoplasmal surface are still unknown. Nevertheless, recent analyses have revealed that these proteins have the capacity to bind to the ECM. EF-Tu of *M. genitalium* and *M. pneumoniae* and PDH-B of *M. pneumoniae* have been shown to bind fibronectin (Dallo *et al.*, 2002; Balasubramanian *et al.*, 2008, 2009). Adherence of viable *M. pneumoniae* to immobilized fibronectin is inhibited by antibodies directed against EF-Tu and the amino-terminal part (amino acids 1–244) of PDH-B (Dallo *et al.*, 2002). Further analyses have revealed that the surface-exposed carboxyl-terminal region of *Mycoplasma pneumoniae* EF-Tu interacts with fibronectin (Balasubramanian *et al.*, 2008). The fibronectin-binding region has been mapped to amino acids 340–358, with serine 343, proline 345 and threonine 357 found to be essential. Modification within this region led to loss of fibronectin binding, as seen for EF-Tu of *M. genitalium* (Balasubramanian *et al.*, 2009). We believe EF-Tu and PDH of *M. pneumoniae* have to be considered cytoadherence factors, as their masking by antibody-binding inhibits cytoadherence to the host's extracellular matrix protein fibronectin. Given their additional, major function in the cytoplasm, they are clearly multifunctional MSCRAMMs.

GAPDH of *M. genitalium* has been shown to attach to mucin, a heavily glycosylated high molecular weight protein that is produced by epithelial tissues (Alvarez *et al.*, 2003), GAPDH of *M. pneumoniae* to bind to fibrinogen (Dumke *et al.*, 2011) and the α-enolase of *M. fermentans* and *M. gallisepticum* to bind to plasminogen (Yavlovich *et al.*, 2004; Chen *et al.*, 2011). In lactobacilli and streptococci, α-enolase has been shown to be moonlighting: it plays a role in glycolysis inside the cell and, additionally, serves as a plasmin(ogen) receptor on the cell surface, enhancing cytoadhesion to and invasion of host cells (Antikainen *et al.*, 2007). Moreover, α-enolase has been shown to function as binding partner for attachment to extracellular matrix proteins (e.g. fibronectin) and to play an important role as potential autoantigen (Nahm *et al.*, 2006). In *M. fermentans*, binding of plasminogen (Plg) to the cell surface markedly increases mycoplasmal cytoadherence to HeLa cells. As the addition of α-enolase results in competitive reduction in plasminogen-binding to *M. fermentans*, it has been suggested that α-enolase also functions as a cell-surface plasminogen receptor (Yavlovich *et al.*, 2007). The α-enolases of *M.*

*agalactiae*, *M. synoviae*, *M. mobile*, *M. capricolum* and *M. pneumoniae* have lysine-rich Plg binding motifs, the carboxyl-terminal Plg1 [F-Y-N-I-K], with the lysine indispensable for Plg-binding, and Plg2 [I-Y-D-E-K-S-K-K-Y-V], which was firstly described in *S. pneumoniae* (Bergmann *et al.*, 2003). It has been suggested that the surface-localized enolase binds through the lysine-rich binding motif Plg1 to the amino-terminal lysine binding Kringle domain of Plg, thus promoting mycoplasmal tissue invasion (Yavlovich *et al.*, 2007). With the demonstration of the cytoadhesive function of α-enolase of *M. gallisepticum* (Chen *et al.*, 2011), mycoplasmal α-enolases turn to good candidates as moonlighting cytoadhesins.

A further candidate for the group of surface-exposed enzymes, the 'community acquired respiratory distress syndrome'(CARDS) toxin of *M. pneumoniae*, warrants closer attention. This mpn372-encoded toxin possesses ADP-ribosyltransferase activity, is surface exposed and mediates attachment of mycoplasma cells to human surfactant protein A (hSP-A) (Kannan *et al.*, 2005), thus fulfilling all requirements of an adhesive MSCRAMM protein. As the respiratory mucosa is enriched with extracellular matrix components, including surfactant proteins, it represents a significant *in vivo* target for mycoplasma parasitism. Although the definitive proof of a cytoadherence role for these group members is still required to demonstrate their moonlighting in cytoadherence, they are still moonlighters with respect to their intracellular activities and their pathophysiological role as MSCRAMMs.

From the mycoplasma's point of view, tight contact of mycoplasma with the host is a prerequisite for successful colonization, but this step is also the initial point of recognition by the immune system of the host. Direct contact between *M. pneumoniae* and sialyated residues on a mast cell surface has been demonstrated to mediate activation and secretion of anti-inflammatory cytokines such as IL-4. That IL-4 induction depends on the P1 cytoadhesion has been shown by analysis of a P1-deficient mutant of *M. pneumoniae*, which was unable to efficiently induce an IL-4 release (Hoek *et al.*, 2005). However mycoplasmas have evolved still further strategies to circumvent elimination from their hosts.

## Mimicry

Berg and colleagues showed in 2009 that mycoplasma surface antigens such as pyruvate dehydrogenase (PDH) may act as a trigger for the induction of anti-mitochondrial antibodies in primary biliary cirrhosis (PBC). Cross-reactive antibodies against PDH of human (hPDH) and *M. pneumoniae* (mPDH) were detected in sera from patients with PBC (Berg *et al.*, 2009). This epitope mimicry leads to a misdirected immune response of the colonized host that, in extreme cases, will lead to autoimmunity. In the mid 1990s Jacobs and colleagues analysed the immunogenicity of regions of the P1 adhesin of *Mycoplasma pneumoniae* that mediate adherence and identified epitopes that induced antibodies against intracellular antigens of eukaryotic cell lines, which were identified as human glyceraldehyde-3-phosphate dehydrogenase (hGAPDH) and 2-phospho-D-glycerate hydrolase (Jacobs *et al.*, 1995). Subsequently, regions homologous to parts of P1 and P30 were identified in ECM proteins, including fibrinogen, keratin, myosin and actin. Further analysis of P30 revealed that its amino-terminal region shares cross-reactive epitopes with keratin, myosin, actin and troponin, while antibodies against the proline-rich A and B domains of the carboxyl-terminal region cross-react with keratin, myosin and fibrinogen (Dallo *et al.*, 1996). This suggests that antigenic mimicry of eukaryotic structures by functional sites of cytoadhesins may influence pathogenesis. Antigenic mimicry may assist in hiding mycoplasmas from B-cell mediated immune response, but post-infectious autoimmunity may be a consequence and one explanation chronic diseases mycoplasmas are responsible for.

## Secreted cytoadhesins

A major challenge for the mycoplasmas is camouflaging of their surface-exposed structures from the immune system of the host. In addition to using antigenic mimicry, many microbial pathogens secrete virulence factors, such as toxins and immune modulators, into the external milieu, or even into host cells, to attenuate or deflect the host's immune response (Finley and McFadden, 2006). Mycoplasma lipoproteins (Into *et al.*,

2002; Hopfe *et al.*, 2008), as well as secreted proteins (Paddenberg *et al.*, 1996; Bendjennat *et al*, 1999), have been shown to cause apoptotic cell death in the host. Although elucidation of the molecular mechanisms that lead to apoptosis require further study (Dedieu *et al.*, 2005), we will here focus on one of the first steps and examine secreted proteins of mycoplasmas that may have a function in cytoadherence.

In bacteria export of immune modulators or toxins can be affected through the Sec pathway, which transports protein precursors that are hallmarked for secretion by an amino-terminal signal peptide across the membrane. Signal peptidases then process the protein, leading to membrane anchoring, through acylation to generate lipoproteins (SPase II) or release of the mature protein into the surrounding milieu (SPase I). Although the main domains of the Sec pathway are encoded in Mollicutes, even by the minimal genome of *M. genitalium*, in silico analyses suggest that signal peptidase I is missing in some Mollicutes, including *M. genitalium*, *M. pneumoniae*, *M. penetrans*, *M. mobile* (and *M. hominis* (Staats *et al.*, 2007; Pereyre *et al.*, 2009). In these species precursor proteins with an amino-terminal SPase I recognition site would be expected to be surface-exposed membrane proteins or to be processed by an enzyme encoded by an NOD gene (one that has evolved by non-orthologous gene displacement). As there are many mycoplasma genes of unknown function and there is evidence presented below that these SPase I-precursor proteins are processed, it is likely that another protein might have evolved to serve a similar function to SPase I.

In *Mycoplasma hominis*, a facultative pathogen of the human urogenital tract, a cytoadhesive membrane protein complex has been shown to be composed of two antigenically and size invariant proteins, the lipoprotein P60 and P80, which has an uncleaved amino-terminal signal peptide mediating membrane anchoring (Henrich *et al.*, 1993, 1998b). Both domains are concomitantly expressed with HinT, a cytoplasmic, nucleotide-binding protein, by the three genes comprising the *hit* locus. Homologous *hit* loci are found predominantly in Mollicutes lacking an attachment organelle, but the genes encoding the P60/P80 membrane complex have been lost by members of the pneumoniae phylogenetic group. The sole exception so far known at present is *Mycoplasma mobile*, which has both, an attachment organelle and the *hit* locus genes encoding P80, P60 and HinT.

Interestingly, the signal sequences of the P80 precursors of *M. mycoides* subspecies *mycoides* SC and *Mesoplasma florum* contain recognition sites for signal peptidase II (SPase II) rather than SPase I, which would be expected to result in lipid anchoring (Hopfe *et al.*, 2005). Further analysis led to the proposal that the uncleaved amino-terminus of the membrane-embedded P80 precursor extends into the cytoplasm, there interacting with the cytosolic HinT (Kitzerow and Henrich, 2001). Loss of the 4.7-kDa signal peptide was shown to be a prerequisite for P80 secretion, and this is followed by a proteolytic process resulting in a helical 74-kDa product that can be detected in the supernatant of (late-)logarithmic phase cultures (Hopfe *et al.*, 2004). Although the mechanisms and additional factors involved in processing are still unknown, P80 may represent a new prototype of mycoplasma secretins: with initial stable insertion in the membrane as antigenic cytoadhesin and subsequent release from the membrane into the supernatant after cleavage. In other mycoplasmas several membrane proteins have been shown to possess amino-terminal signal sequences for anchorage with poorly conserved SPase I cleavage sites (Table 6.2). Thus, it may be a common phenomenon for mycoplasmas to use secreted proteins in a dual role, as surface localised proteins anchored in the membrane for cell surface architecture, and, in response to a change in the environment, as a soluble protein released from the cell surface. Interestingly, several mycoplasmal MRSCAMMs have amino-terminal signal sequences, the processing of which is yet to be determined.

HinT proteins are also considered to play a role in metabolic processes such as cell cycle regulation and secretion (CD01277). An increase in calcium ions has been shown to prevent release of P80 from the membrane, and a local decrease of $Ca^{2+}$ may activate its processing. This results in a soluble antigenic P80 helix and a membrane anchored signal peptide that subsequently may enter the cytoplasm. Its interaction with cytoplasmic

HinT may then promote cellular processes, such as growth or activation of gene expression. P80 has been shown to contribute to different steps in infection, in cytoadhesion and secretion of an immunogenic antigen, which provides evidence of its role as multifunctional cytoadherence factor of *M. hominis*. Definitive proof of its predicted effect in modulation of mycoplasma growth would demonstrate that it fulfils the requirements of a moonlighter.

Although final evidence of the patho-physiological properties of the secreted part of P80 is yet to be obtained, a few studies have examined the affect of other secreted mycoplasma proteins on their host. In characterizing the cytotoxicity of *M. penetrans*, Bendjennat and colleagues identified a 40-kDa protein effector that was derived from a membrane-anchored 50-kDa precursor protein of *M. penetrans*. The secreted P40 adheres to host cells, functions as $Ca^{2+}/Mg^{2+}$-dependent endonuclease with a cytotoxic effect, and when internalized it causes apoptosis (Bendjennat *et al.*, 1997, 1999). Unfortunately, the gene that encodes it has not yet been identified, and its capacity to mediate cytoadherence has not been assessed.

It took a long time to discover that the final example of a secreted mycoplasma protein did not originate from the murine cells to which it adheres. Its presence on the surface of many tumour cells correlates highly with increased neoplastic invasiveness and metastatic activity. After the recognition that P37 originated from *Mycoplasma hyorhinis*, its characterization made progress. P37 was shown to be expressed as lipoprotein, concomitantly with two additional proteins that are very similar to components of the periplasmic binding-protein-dependent transporter of Gram-negative bacteria. Dudler and colleagues suggested that P37 was part of a homologous, high-affinity ABC transport system (Dudler *et al.*, 1988). It was predicted to be outer membrane protein, and was shown to cytoadhere to many tumour cells, enabling close contact between these eukaryotic cells (Reutzel *et al.*, 2002), thus promoting invasion and metastasis (Schmidhauser *et al.*, 1990; Ketcham *et al.*, 2005). P37 was also shown to inhibit mammalian cell adhesion (Liu *et al.*, 2006, 2007) by activation of MMP-2 and subsequent phosphorylation of EGFR (Gong *et al.*, 2008), and to alter gene expression, growth and morphology of prostate cancer cells (Goodison *et al.*, 2007). Because of its capacity to bind thiamine, P37 was renamed CypI (extracytoplasmic thiamine-binding lipoprotein). Sequence homologues have been identified in the ruminant pathogens, *M. mycoides*, *M. capricolum* and *M. agalactiae*, and two human pathogens, *M. genitalium* and *M. pneumoniae* (Sippel *et al.*, 2009). There is no doubt about the function of CypI/P37 as mycoplasmal virulence factor. However, lack of an experimental evidence that CypI/P37 is not only an potent adhesin when attached to host cells, but also mediates mycoplasma adhesion to host cells is the sole obstacle to defining P37 as a multifunctional cytoadhesin, or as cytoadherence moonlighter.

## Moonlighting cytoadhesins

Thus, it is clear that not only gene products of higher organisms moonlight (Jeffery, 1999) and that mycoplasma cytoadhesive factors can act as moonlighters.

Mpn133 of *M. pneumoniae* is a good candidate moonlighter that is a surface-exposed protein with enzymatic function. Mpn133 was initially characterized as 301 amino acid cytotoxic lipoprotein, with $Ca^{2+}$-dependent nucleolytic activity conferred by the region between amino acids 114 and 168. Binding to and internalization by human A549 cells depends on a glutamate, lysine and serine rich domain, called the EKS region, at the amino-terminal end (amino acids 72–110) (Somarajan *et al.*, 2010). Homologous proteins have been found in *M. genitalium* (MG186; Glass *et al.*, 2006) and in *M. hyopneumoniae* (Mhp379), but both lack the EKS region. In all three species homologues of an ATP-binding cassette (ABC) transport system have been identified immediately downstream of this gene, suggesting a polycistronic organization and the participation of Mpn133 and its homologues in their respective ABC transporters. A biochemical approach has also proven the $Ca^{2+}$-dependent exonuclease activity of Mhp379 (Schmidt *et al.*, 2007), leading to the suggestion that the nucleolytic activity of these surface-localized lipoproteins may indicate that the function of the ABC transport system is in import of nucleic acid precursors. The *M. genitalium* protein MG185, which is encoded by a gene upstream of mg186, has been shown to have cytoadhesive functions, whereas in *M. pneumoniae* the corresponding gene is missing. Thus, the EKS region of Mpn133 enables it to bind to and enter host cells and to function as a moonlighting adhesive factor whose capability to mediate cytoadhesion has yet to be tested to be certified as cytoadhesive moonlighter.

One of the best-studied multifunctional cytoadhesin in Mollicutes is OppA, the substrate-binding domain of the oligopeptide permease of *Mycoplasma hominis*. In 1993, using monoclonal antibodies (mAbs) it was shown to be one of four surface-localized antigens of *M. hominis* (the lipoproteins P50,which was later renamed Vaa, P60 and P100, which was later renamed OppA, as well as the P80 precursor mentioned above). Their function as cytoadhesins of *M. hominis* was clearly demonstrated by their capacity to bind to HeLa cells in a quantifiable adhesion and inhibition assay (Henrich *et al.*, 1993). In contrast to P50/Vaa, which was later characterized as a size and phase variable cytoadhesin (Zhang and Wise, 1996, 1997), P100 is highly conserved in sequence and size within and between strains. The P100 proteins of all of the more than 200 *M. hominis* isolates tested were detected by the mAb DC10 and 36% of them by the mAb BG11, indicating some variation in the epitope for that mAb.

Further characterization of OppA revealed that its gene is polycistronically organized with four core domains, *opp*BCDF, of an oligopeptide permease that belongs to the ATP-binding cassette (ABC) family of transporters. In silico analysis of the deduced 961 amino acid polypeptide revealed that P100 is expressed as a pre-lipoprotein with a structure in the amino-terminal region common to peptide-binding proteins (amino acids 176–243) and an ATP or GTP binding P-loop structure in the carboxyl-terminal region (amino acids 874–881). The peptide-binding capacity of P100 was confirmed by fluorescence spectroscopy, strongly suggesting that the cytoadherence-mediating lipoprotein P100 is OppA, the substrate-binding domain of an oligopeptide transport system in *M. hominis* (Henrich *et al.*, 1999). As the oligopeptide-binding proteins are not surface-exposed in Gram-negative bacteria, but rather restricted to the periplasmic space, cytoadherence is not a feature shared by all bacterial oligopeptide-binding domains (Brass *et al.*, 1986). Thus, the multifunctionality of a mycoplasma OppA has been demonstrated, with it having roles in both, cytoadhesion and peptide-binding. As peptide binding has no influence on cytoadhesion (Hopfe, personal communication), OppA fulfils the criteria for classification as a moonlighter.

For a long time a further role of OppA *of Mycoplasma hominis* that had never been described before for a substrate-binding protein remained unrecognised. Following the in silico detection of a putative ATPase site in the carboxyl-terminal region of OppA, ATP binding and hydrolysis was demonstrated using wild type OppA and recombinant OppA variants differing in the Walker A (or P-loop) structure (amino acids 869–876: G-K-D-S-S-G-K-S) or the less conserved Walker B region (amino acids 737–752: R-F-D-G-V-T-G-E-N-L-L-A-W-S-A-D). ATP-binding by OppA was confirmed using ATP affinity chromatography and a tryptic digestion pattern assay. Dose dependent ATPase activity of OppA was demonstrated by measuring the release of phosphate from ATP ($K_m$ of 0.18 ± 0.04 mM). Mutation of the Walker A motif at amino acid 875 from K to R preserved ATP binding, but led to an 85% decrease in ATP hydrolysis. The OppA$^{\Delta P\text{-loop}}$ variant, in which the whole P-loop motif (GKDSSGKS) was changed to THASSSAH, was no longer able to bind ATP, demonstrating that the Walker A motif is essential for binding ATP (Hopfe *et al.*, 2004). In tryptic digestion pattern assay the membrane-anchored OppA of intact mycoplasma cells was protected from proteolysis by ATP, but was rapidly digested in the absence of ATP. These studies clearly showed that the ATP-binding site of OppA is located outside the mycoplasma cell and disclosed that OppA can also act as an ecto-ATPase of *Mycoplasma hominis*.

A current BLAST search for OppA homologues proteins in other Mollicutes detects more than 100 putative OppA proteins, but none of them has Walker A or B motifs in their carboxyl-termini suggestive of ATPase activity. Only Mypu_2820, a putative OppA homologue of *M. pulmonis*, has a Walker A motif (G-L-Q-S-Y-G-K-T), but in the amino-terminal half of the predicted sequence. However, replacement of the P-loop region in OppA of *M. hominis* (G-K-D-S-S-G-K-S) with the P-loop region of Mypu_2820 did not preserve the ATPase activity (Miriam Hopfe, personal communication). These findings strongly suggest that the ecto-ATPase activity of OppA is a moonlighting function that, to date, is exclusively to *Mycoplasma hominis*.

The substrate for the ecto-ATPase activity of OppA could only be extra-cellular ATP. The possibility that this ATP could be provided by colonized host cells was assessed by measuring ATP release from HeLa cells after incubation with different preparations of *Mycoplasma hominis* (with or without OppA) and recombinant variants of OppA. Release of ATP into the supernatant of the HeLa cells was primarily observed with *M. hominis* or OppA that lacked the ecto-ATPase activity or in the presence of the ATPase inhibitor 4 ,4′-diisothiocyanostilbene 2′,2′-disulfonic acid (DIDS). In addition to revealing that OppA was involved in induction of release of ATP from the host cell, real-time PCR and flow cytometry experiments revealed that proliferation of the infected cell is reduced and apoptosis induced by the ecto-ATPase activity of OppA (Hopfe *et al.*, 2008). *M. hominis* utilizes the arginine dihydrolase pathway for generating ATP and rapidly depletes the host's arginine reserves, affecting host cell division and growth (Rottem and Barile, 1993). As the ecto-ATPase activity of OppA of *M. hominis* results in an ATP-deficient environment, the subsequent induction of apoptosis is reminiscent of the creation of a choline-deficient environment and the subsequent induction of apoptosis by infection with *M. fermentans* (Ben-Menachem *et al.*, 2001).

The cytoadhesive sites of OppA were identified using OppA variants (Hopfe *et al.*, 2011). Mutation within the mAb BG11-binding region leads to a 50% reduction in adherence of *M. hominis* to HeLa cells. Deletion of a larger region ($OppA^{\Delta 176-243}$) that is conserved in all bacterial extra-cellular solute-binding proteins or destruction of the domain for ATP hydrolysis ($OppA^{\Delta P\text{-loop}}$ variant) reduces cytoadherence down to 18%. As ATPase inhibitors also reduce the attachment of OppA to HeLa cells, and 5′-fluorosulfonylbenzoyl-5′-adenosine (FSBA), a non-hydrolysing derivative of adenosine, inhibits ATP hydrolysis as well as adhesion, it appears that OppA-mediated cytoadherence requires energy.

Because the ecto-ATPase domain is required for cytoadherence using OppA of *M. hominis*, it may not meet the criteria set for classification as a moonlighter, as inactivation of one function destroys the other. However, we feel it should be regarded as a moonlighter because, in addition to its function as peptide-binding domain for the oligopeptide permease, it mediates cytoadhesion, a role not shared by other bacterial OppA homologues, and is thus far the only one thus far recognized to also function as an ecto-ATPase.

## Vice versa: cytoplasmic factors with cytoadhesive functions

A comprehensive search for multifunctional cytoadherence factors is somewhat hindered by the fact that cytoadhesion is not restricted to membrane-embedded structures, such as lipoproteins or secreted precursor proteins, with or without signal peptidase I recognition sites. In *M. pneumoniae*, Mpn 474 has been shown to cover the mycoplasma cell surface, even though it lacks an export signal sequence (Hegermann *et al.*, 2008). As discussed above, several cytoplasmic enzymes have been shown to also function as surface-localized adhesins binding to the extracellular matrix. We will conclude with consideration of the role of heat shock proteins as they are traditionally regarded to be intracellular chaperones. Nevertheless, some members of the heat shock family appear to have evolved to cytoadherence-associated functions.

Hsp70 (or DnaK) of *Haemophilus influenzae* (Hartmann *et al.*, 2001) and Hsp70 of *Helicobacter pylori* have been shown to reside on the bacterial surface and to have sulfatide binding functions (Huesca *et al.*, 1996). Lingwood and colleagues demonstrated in 1995 that cell surface-exposed mycoplasma heat shock proteins, especially those belonging to the Hsp70 family, also bind sulfatide (sulfated glycolipids). Cross-linking studies have revealed the presence of Hsp70 in complexes with cytadherence-associated membrane proteins of *M. pneumoniae* (Layh-Schmitt *et al.*, 2000).

The diverse functions of intracellular Hsp70 proteins as chaperones are influenced by its amino-terminal ATPase activity. Their function as cell surface localized adhesins that attach sulfogalactolipids (SGL) has been shown using monoclonal antibodies (Boulanger *et al.*, 1995). The facts that members of the Hsp70 family in prokaryotes and eukaryotes function as SGL-specific adhesins and SGL are concentrated in specific sites of the mammalian body suggest a physiological role in mycoplasma–host interactions. Interestingly, conserved residues within the ATPase domain of Hsp70 have been shown to mediate SGL binding (Mamelak and Lingwood, 2001), evoking the moonlighting functions of OppA in *M. hominis*, as its ecto-ATPase domain is also necessary for adhesion.

Hsp60 (GroEL) was, for a long time, predicted to be essential for bacterial life because of its function as a chaperone. However transposon mutagenesis has shown that this is not the case for *M. pneumoniae* or *M. genitalium* (Hutchison *et al.*, 1999). That GroEL is an optional in some mycoplasma species was confirmed by the Williams and Fares, who demonstrated the absence of GroEL in 11 species (Williams and Fares, 2010). In some bacteria, including *Clostridium difficile* (Hennequin *et al.*, 2001), *Legionella pneumophila* and *Helicobacter pylori* (Hoffman and Garduno, 1999), GroEL has been demonstrated to mediate cytoadherence. Phylogenetic analyses have suggested lateral transfer of the GroEL gene from the proteobacteria to some of the mycoplasmas and GroEL is proposed to act as an adhesin-invasin,

**Table 6.3** Multifunctionality of mycoplasmal cytoadhesins

| Molecule[a,b] | Membrane anchoring[c] | Adhesion | Nutrition | Immune modulation | Repeats | Tip | Organism |
|---|---|---|---|---|---|---|---|
| Vsp | Lipoprotein | Cytoadhesin | | Antigenic/phase variation | + | – | *M. bovis* |
| LppT | Spl precursor | Cytoadhesin | | MSCRAMM | – | – | *M. conjunctivae* |
| OppA | Lipoprotein | Cytoadhesin | Peptide importer | ecto-ATPase | – | – | *M. hominis* |
| P50/Vaa | Lipoprotein | Cytoadhesin | | Antigenic/phase variation | + | | |
| P80 | Spl precursor | Cytoadhesin | | Secretion | – | | |
| P37 | Lipoprotein | Adhesive | ABC transporter | Invasion | – | – | *M. hyorhinis* |
| $P48^M$+$P50^C$ | – | Cytoadhesins | | MSCRAMMs | – | -- | *M. hyopneumoniae* |
| P97 | Spl-precursor | Cytoadhesin | | MSCRAMM | + | | |
| $P110^M$+$P52^C$ | – | Cytoadhesins | | MSCRAMMs+Invasins | – | | |
| $P120^N$+$P85^C$ | Spl-precursor$^N$ | Cytoadhesins | | MSCRAMMs | – | | |
| GGPLIII | Glycolipid | Cytoadherent | | Immune stimulator | – | - | *M. fermentans* |
| Vsa | Lipoprotein | Cytoadhesin | | Antigenic/phase variation | + | - | *M. pulmonis* |
| α-enolase | – | Cytoadhesin | Glycolysis | MSCRAMM | – | + | *M. gallisepticum* |
| MgPa | Spl precursor+1TM | Cytoadhesin | | Antigenic variation | | | |
| Gli349 | Spl precursor | Cytoadhesin | | Gliding | – | + | *M. mobile* |
| P1 | Spl precursor+1TM | Cytoadhesin | | Gliding+Mimicry | + | + | *M. pneumoniae* |
| P30 | Spl precursor+1TM | Cytoadhesin | | Mimicry | Proline-rich | | |

[a]Moonlighter, the different functions of which are conferred by distinct sequence regions. [b]mhp493 encoded N-terminal P120 ($P120^N$) and C-terminal P85 ($P85^C$); mhp494 encoded $P110^M$ (middle section) and $P52^C$ (C-terminal region). [c]SPase I (Spl) or SPase II (lipoprotein) sites predicted by lipoP 1.0 (Juncker, A.S., Willenbrock, H., von Heijne, G., Nielsen, H., Brunak, S., Krogh, A. (2003) Prediction of lipoprotein signal peptides in Gram-negative bacteria. Protein Sci. *12*, 1652–1662). TM, transmembrane region.

rather than as chaperone in some of the mycoplasmas (Clark and Tillier, 2010). Taken together, these data suggest that GroEL may have cytoadhesive features in some Mollicutes, but, as its essential function as chaperone remains uncertain, it does not appear to be a multifunctional cytoadherence factor.

Interestingly, mice infected with *M. genitalium* develop specific antibodies against the major outer membrane protein MgPa, but also against EF-Tu, PDH and Hsp70 (McGowin *et al.*, 2010), all of which with cytoadhesive functions (see above). A study by Su and colleagues suggests that these cytoadherence proteins may be regulated by phosphorylation. In *M. pneumoniae* and *M. genitalium* 24 phospho-proteins were identified, including PDH-E1α and PDH-E1β, enolase, Hsp70 and Hsp60, EF-Tu, cytadherence accessory proteins and hypothetical proteins (Su *et al.*, 2007). As moonlighters such as OppA can exist in a phosphorylated and unphosphorylated state (Hopfe *et al.*, 2004), it is possible that some of these cytoadhesive moonlighters are regulated by phosphorylation.

Returning to our initial proposition that the survival of mycoplasmas depends on the successful management of the three pillars of life – colonization, nutrition, immune modulation and evasion – we can complete our search for multifunctional cytoadherence factors with a summary of those pillars that they contribute to (Table 6.3). In one respect it is surprising how few multifunctional cytoadhesins have been characterized, given the minimal coding capacity of the Mollicutes. However, it is also surprising how varied and complex the systems are that the Mollicutes have developed to generate cell surface diversity, to obtain the nutrition they need and to modulate and evade their hosts' immune systems.

## References


Adan-Kubo, J., Uenoyama, A., Arata, T., and Miyata, M. (2006). Morphology of isolated Gli349, a leg protein responsible for *Mycoplasma mobile* gliding via glass binding, revealed by rotary shadowing electron microscopy. J. Bacteriol. *188*, 2821–2828.

Alvarez, R.A., Blaylock, M.W., and Baseman, J.B. (2003). Surface localized glyceraldehyde-3-phosphate dehydrogenase of *Mycoplasma genitalium* binds mucin. Mol. Microbiol. *48*, 1417–1425.

Antelmann, H., Tjalsma, H., Voigt, B., Ohlmeier, S., Bron, S., van Dijl, J.M., and Hecker, M. (2001). A proteomic view on genome-based signal peptide predictions. Genome Res. *11*, 1484–1502.

Antikainen, J., Kuparinen, V., Lähteenmäki, K., and Korhonen, T.K. (2007). Enolases from Gram-positive bacterial pathogens and commensal lactobacilli share functional similarity in virulence-associated traits. FEMS Immunol. Med. Microbiol. *51*, 526–534.

Balasubramanian, S., Kannan, T.R., and Baseman, J.B. (2008). The surface-exposed carboxyl region of *Mycoplasma pneumoniae* elongation factor Tu interacts with fibronectin. Infect. Immun. *76*, 3116–3123.

Balasubramanian, S., Kannan, T.R., Hart, P.J., and Baseman, J.B. (2009). Amino acid changes in elongation factor Tu of *Mycoplasma pneumoniae* and *Mycoplasma genitalium* influence fibronectin binding. Infect. Immun. *77*, 3533–3541.

Balish, M.F., and Krause, D.C. (2006). Mycoplasmas: a distinct cytoskeleton for wall-less bacteria. J. Mol. Microbiol. Biotechnol. *11*, 244–255.

Balish, M.F., Hahn, T.W., Popham, P.L., and Krause, D.C. (2001). Stability of *Mycoplasma pneumoniae* cytadherence-accessory protein HMW1 correlates with its association with the triton shell. J. Bacteriol. *183*, 3680–3688.

Belloy, L., Vilei, E.M., Giacometti, M., and Frey, J. (2003). Characterization of LppS, an adhesion of *Mycoplasma conjunctivae*. Microbiology *149*, 185–193.

Bendjennat, M., Blanchard, A., Loutfi, M., Montagnier, L., and Bahraoui, E. (1997). Purification and characterization of *Mycoplasma penetrans* $Ca^{2+}/Mg^{2+}$-dependent endonuclease. J. Bacteriol. *179*, 2210–2220.

Bendjennat, M., Blanchard, A., Loutfi, M., Montagnier, L., and Bahraoui, E. (1999). Role of *Mycoplasma penetrans* endonuclease P40 as a potential pathogenic determinant. Infect. Immun. *67*, 4456–4462.

Ben-Menachem, G., Rottem, S., Tarshis, M., Barash, V., and Brenner, T. (1998). *Mycoplasma fermentans* glycolipid triggers inflammatory response in rat astrocytes. Brain Res. *803*, 34–38.

Ben-Menachem, G., Mousa, A., Brenner, T., Pinto, F., Zähringer, U., and Rottem, S. (2001). Choline deficiency induced by *Mycoplasma fermentans* enhances apoptosis of rat astrocytes. FEMS Microbiol. Lett. *201*, 157–162.

Berg, M., Melcher, U., and Fletcher, J. (2001). Characterization of *Spiroplasma citri* adhesion related protein SARP1, which contains a domain of a novel family designated sarpin. Gene *275*, 57–64.

Berg, C.P., Kannan, T.R., Klein, R., Gregor, M., Baseman, J.B., Wesselborg, S., Lauber, K., and Stein, G.M. (2009). Mycoplasma antigens as a possible trigger for the induction of antimitochondrial antibodies in primary biliary cirrhosis. Liver Int. *29*, 797–809.

Bergmann, S., Wild, D., Diekmann, O., Frank, R., Bracht, D., Chhatwal, G.S., and Hammerschmidt, S. (2003). Identification of a novel plasmin(ogen)-binding motif in surface displayed alpha-enolase of *Streptococcus pneumoniae*. Mol. Microbiol. *49*, 411–423.

Bird, D.W., Graber, K., Knutson, A., and Washburn, L.R. (2008). Mutation of two *Mycoplasma arthritidis* surface lipoproteins with divergent functions in cytadherence. Infect. Immun. *76*, 5768–5776.

Bogema, D.R., Scott, N.E., Padula, M.P., Tacchi, J.L., Raymond, B.B., Jenkins, C., Cordwell, S.J., Minion, F.C., Walker, M.J., and Djordjevic, S.P. (2011). Sequence TTKF ↓ QE defines the site of proteolytic cleavage in Mhp683 protein, a novel glycosaminoglycan and cilium adhesin of *Mycoplasma hyopneumoniae*. J. Biol. Chem. *286*, 41217–41229.

Bogema, D.R., Deutscher, A.T., Woolley, L.K., Seymour, L.M., Raymond, B.B., Tacchi, J.L., Padula, M.P., Dixon, N.E., Minion, F.C., Jenkins, C., *et al.* (2012). Characterization of cleavage events in the multifunctional cilium adhesin Mhp684 (P146) reveals a mechanism by which *Mycoplasma hyopneumoniae* regulates surface topography. MBio. 3, e00282–11.

Boguslavsky, S., Menaker, D., Lysnyansky, I., Liu, T., Levisohn, S., Rosengarten, R., García, M., and Yogev, D. (2000). Molecular characterization of the *Mycoplasma gallisepticum pvpA* gene which encodes a putative variable cytadhesin protein. Infect. Immun. *68*, 3956–3964.

Bolland, J.R., and Dybvig, K. (2012). *Mycoplasma pulmonis* Vsa proteins and polysaccharide modulate adherence to pulmonary epithelial cells. FEMS Microbiol. Lett. *331*, 25–30.

Bose, S.R., Balish, M.F., and Krause, D.C. (2009). *Mycoplasma pneumoniae* cytoskeletal protein HMW2 and the architecture of the terminal organelle. J. Bacteriol. *191*, 6741–6748.

Boulanger, J., Faulds, D., Eddy, E.M., and Lingwood, C.A. (1995). Members of the 70 kDa heat shock protein family specifically recognize sulfoglycolipids: role in gamete recognition and mycoplasma-related infertility. J. Cell Physiol. *165*, 7–17.

Brass, J.M., Higgins, C.F., Foley, M., Rugman, P.A., Birgham, J., and Garland, P.J. (1986). Lateral diffusion of proteins in the periplasm of *Escherichia coli*. J. Bacteriol. 7, 3971–3974.

Burgos, R., Pich, O.Q., Ferrer-Navarro, M., Baseman, J.B., Querol, E., and Piñol, J. (2006). *Mycoplasma genitalium* P140 and P110 cytadhesins are reciprocally stabilized and required for cell adhesion and terminal-organelle development. J. Bacteriol. *188*, 8627–8637.

Burnett, T.A., Dinkla, K., Rohde, M., Chhatwal, G.S., Uphoff, C., Srivastava, M., Cordwell, S.J., Geary, S., Liao, X., Minion, F.C., *et al.* (2006). P159 is a proteolytically processed, surface adhesin of *Mycoplasma hyopneumoniae*: defined domains of P159 bind heparin and promote adherence to eukaryote cells. Mol. Microbiol. *60*, 669–686.

Calderon-Copete, S.P., Wigger, G., Wunderlin, C., Schmidheini, T., Frey, J., Quail, M.A., and Falquet, L. (2009). The *Mycoplasma conjunctivae* genome sequencing, annotation and analysis. BMC Bioinformatics *10*, S7.

Chang, H.Y., Jordan, J.L., and Krause, D.C. (2011). Domain analysis of protein P30 in *Mycoplasma pneumoniae* cytadherence and gliding motility. J. Bacteriol. *193*, 1726–1733.

Chaudhry, R., Nisar, N., Hora, B., Chirasani, S.R., and Malhotra, P. (2005). Expression and immunological characterization of the carboxy-terminal region of the P1 adhesin protein of *Mycoplasma pneumoniae*. J. Clin. Microbiol. *43*, 321–325.

Chen, A.Y., Fry, S.R., Forbes-Faulkner, J., Daggard, G., and Mukkur, T.K. (2006). Evaluation of the immunogenicity of the P97R1 adhesin of *Mycoplasma hyopneumoniae* as a mucosal vaccine in mice. J. Med. Microbiol. *55*, 923–929.

Chen, H., Yu, S., Shen, X., Chen, D., Qiu, X., Song, C., and Ding, C. (2011). The *Mycoplasma gallisepticum* α-enolase is cell surface-exposed and mediates adherence by binding to chicken plasminogen. Microb. Pathog. *51*, 285–290.

Citti, C., Nouvel, L.X., and Baranowski, E. (2010). Phase and antigenic variation in mycoplasmas. Future Microbiol. *5*, 1073–1085.

Clark, G.W., and Tillier, E.R. (2010). Loss and gain of GroEL in the Mollicutes. Biochem. Cell. Biol. *88*, 185–194.

Coleman, J.L., and Benach, J.L. (1999). Use of the plasminogen activation system by microorganisms. J. Lab. Clin. Med. *134*, 567–576.

Dallo, S.F., Su, C.J., Horton, J.R., and Baseman, J.B. (1988). Identification of P1 gene domain containing epitope(s) mediating *Mycoplasma pneumoniae* cytoadherence. J. Exp. Med. *167*, 718–723.

Dallo, S.F., Lazzell, A.L., Chavoya, A., Reddy, S.P., and Baseman, J.B. (1996). Biofunctional domains of the *Mycoplasma pneumoniae* P30 adhesin. Infect. Immun. *64*, 2595–2601.

Dallo, S.F., Kannan, T.R., Blaylock, M.W., and Baseman, J.B. (2002). Elongation factor Tu and E1 beta subunit of pyruvate dehydrogenase complex act as fibronectin binding proteins in *Mycoplasma pneumoniae*. Mol. Microbiol. *46*, 1041–1051.

Davis, K.L., and Wise, K.S. (2002). Site-specific proteolysis of the MALP-404 lipoprotein determines the release of a soluble selective lipoprotein-associated motif-containing fragment and alteration of the surface phenotype of *Mycoplasma fermentans*. Infect. Immun. *70*, 1129–1135.

Dedieu, L., Chapey, E., and Balcer-Rodrigues, V. (2005). *Mycoplasma mycoides* ssp. *mycoides* biotype small colony-secreted components induce apoptotic cell death in bovine leucocytes. Scand. J. Immunol. *62*, 528–538.

Deutscher, A.T., Jenkins, C., Minion, F.C., Seymour, L.M., Padula, M.P., Dixon, N.E., Walker, M.J., and Djordjevic, S.P. (2010). Repeat regions R1 and R2 in the P97 paralogue Mhp271 of *Mycoplasma hyopneumoniae* bind heparin, fibronectin and porcine cilia. Mol. Microbiol. *78*, 444–458.

Deutscher, A.T., Tacchi, J.L., Minion, F.C., Padula, M.P., Crossett, B., Bogema, D.R., Jenkins, C., Kuit, T.A., Walker, M.J., and Djordjevic, S.P. (2012). *Mycoplasma hyopneumoniae* surface proteins Mhp385 and Mhp384 bind host cilia and glycosaminoglycans and are endoproteolytically processed by proteases that recognize different cleavage motifs. J. Proteome Res. *11*, 1924–1936.

Djordjevic, S.P., Cordwell, S.J., Djordjevic, M.A., Wilton, J., and Minion, F.C. (2004). Proteolytic processing of the *Mycoplasma hyopneumoniae* cilium adhesin. Infect. Immun. *72*, 2791–2802.

Dudler, R., Schmidhauser, C., Parish, R.W., Wettenhall, R.E., and Schmidt, T. (1988). A mycoplasma high-affinity transport system and the in vitro invasiveness of mouse sarcoma cells. EMBO J. *7*, 3963–3970.

Duensing, T.D., Wing, J.S., and van Putten, J.P. (1999). Sulfated polysaccharide-directed recruitment of mammalian host proteins: a novel strategy in microbial pathogenesis. Infect. Immun. *67*, 4463–4468.

Dumke, R., Hausner, M., and Jacobs, E. (2011). Role of *Mycoplasma pneumoniae* glyceraldehyde-3-phosphate dehydrogenase (GAPDH) in mediating interactions with the human extracellular matrix. Microbiology *157*, 2328–2338.

Dunsmore, S.E., and Rannels, D.E. (1996). Extracellular matrix biology in the lung. Am. J. Physiol. *270*, L3–27.

Feldmann, R.C., Henrich, B., Kolb-Bachofen, V., and Hadding, U. (1992). Decreased metabolism and viability of *Mycoplasma hominis* induced by monoclonal antibody-mediated agglutination. Infect. Immun. *60*, 166–174.

Finlay, B.B., and McFadden, G. (2006). Anti-immunology: evasion of the host immune system by bacterial and viral pathogens. Cell *124*, 767–782.

Fleury, B., Bergonier, D., Berthelot, X., Peterhans, E., Frey, J., and Vilei, E.M. (2002). Characterization of P40, a cytadhesin of *Mycoplasma agalactiae*. Infect. Immun. *70*, 5612–5621.

Girón, J.A., Lange, M., and Baseman, J.B. (1996). Adherence, fibronectin binding, and induction of cytoskeleton reorganization in cultured human cells by *Mycoplasma penetrans*. Infect. Immun. *64*, 197–208.

Glass, J.I., Assad-Garcia, N., Alperovich, N., Yooseph, S., Lewis, M.R., Maruf, M., Hutchison, C.A., 3rd, Smith, H.O., and Venter, J.C. (2006). Essential genes of a minimal bacterium. Proc. Natl. Acad. Sci. U.S.A. *103*, 425–430.

Goh, M.S., Gorton, T.S., Forsyth, M.H., Troy, K.E., and Geary, S.J. (1998). Molecular and biochemical analysis of a 105 kDa *Mycoplasma gallisepticum* cytadhesin (GapA). Microbiology *144*, 2971–2978.

Gong, M., Meng, L., Jiang, B., Zhang, J., Yang, H., Wu, J., and Shou, C. (2008). p37 from *Mycoplasma hyorhinis* promotes cancer cell invasiveness and metastasis through activation of MMP-2 and followed by phosphorylation of EGFR. Mol. Cancer Ther. *7*, 530–537.

Goodison, S., Nakamura, K., Iczkowski, K.A., Anai, S., Boehlein, S.K., and Rosser, C.J. (2007). Exogenous mycoplasmal p37 protein alters gene expression, growth and morphology of prostate cancer cells. Cytogenet. Genome Res. *118*, 204–213.

Hardy, R.D., Coalson, J.J., Peters, J., Chaparro, A., Techasaensiri, C., Cantwell, A.M., Kannan, T.R., Baseman, J.B., and Dube, P.H. (2009). Analysis of pulmonary inflammation and function in the mouse and baboon after exposure to *Mycoplasma pneumoniae* CARDS toxin. PLoS One *4*, e7562.

Hartmann, E., Lingwood, C.A., and Reidl, J. (2001). Heat-inducible surface stress protein (Hsp70) mediates sulfatide recognition of the respiratory pathogen *Haemophilus influenzae*. Infect. Immun. *69*, 3438–3441.

Hasselbring, B.M., Jordan, J.L., and Krause, D.C. (2005). Mutant analysis reveals a specific requirement for protein P30 in *Mycoplasma pneumoniae* gliding motility. J. Bacteriol. *187*, 6281–6289.

Hasselbring, B.M., and Krause, D.C. (2007a). Cytoskeletal protein P41 is required to anchor the terminal organelle of the wall-less prokaryote *Mycoplasma pneumoniae*. Mol. Microbiol. *63*, 44–53.

Hasselbring, B.M., and Krause, D.C. (2007b). Proteins P24 and P41 function in the regulation of terminal-organelle development and gliding motility in *Mycoplasma pneumoniae*. J. Bacteriol. *189*, 7442–7449.

Hatchel, J.M., and Balish, M.F. (2008). Attachment organelle ultrastructure correlates with phylogeny, not gliding motility properties, in *Mycoplasma pneumoniae* relatives. Microbiology *154*, 286–295.

Hegermann, J., Halbedel, S., Dumke, R., Regula, J., Gabdoulline, R.R., Mayer, F., Stülke, J., and Herrmann, R. (2008). The acidic, glutamine-rich Mpn474 protein of *Mycoplasma pneumoniae* is surface exposed and covers the complete cell. Microbiology *154*, 1185–1192.

Hennequin, C., Porcheray, F., Waligora-Dupriet, A., Collignon, A., Barc, M., Bourlioux, P., and Karjalainen, T. (2001). GroEL (Hsp60) of *Clostridium difficile* is involved in celladherence. Microbiology *147*, 87–96.

Henrich, B., Feldmann, R.C., and Hadding, U. (1993). Cytoadhesins of *Mycoplasma hominis*. Infect. Immun. *61*, 2945–2951.

Henrich, B., Kitzerow, A., Feldmann, R.C., Schaal, H., and Hadding, U. (1996). Repetitive elements of the *Mycoplasma hominis* adhesin p50 can be differentiated by monoclonal antibodies. Infect. Immun. *64*, 4027–4034.

Henrich, B., Lang, K., Kitzerow, A., MacKenzie, C., and Hadding, U. (1998a). Truncation as a novel form of variation of the p50 gene in *Mycoplasma hominis*. Microbiology *144*, 2979–2985.

Henrich, B., Berns, G., Weinhold, M., Kitzerow, A., Schaal, H., and Hadding, U. (1998b). Cloning and expression of P60, a conserved surface-localized protein of *Mycoplasma hominis*, in *Escherichia coli*. Biol. Chem. *379*, 1143–1150.

Henrich, B., Hopfe, M., Kitzerow, A., and Hadding, U. (1999). The adherence-associated lipoprotein P100, encoded by an opp operon structure, functions as the oligopeptide-binding domain OppA of a putative oligopeptide transport system in *Mycoplasma hominis*. J. Bacteriol. *181*, 4873–4878.

Hoek, K.L., Duffy, L.B., Cassell, G.H., Dai, Y., and Atkinson, T.P. (2005). A role for the *Mycoplasma pneumoniae* adhesin P1 in interleukin (IL)-4 synthesis and release from rodent mast cells. Microb. Pathog. *39*, 149–158.

Hoffman, P.S., and Garduno, R.A. (1999). Surface-associated heat shock proteins of *Legionella pneumophila* and *Helicobacter pylori*: roles in pathogenesis and immunity. Infect. Dis. Obstet. Gynecol. *7*, 58–63.

Hopfe, M., and Henrich, B. (2004). OppA, the substrate-binding subunit of the oligopeptide permease, is the major Ecto-ATPase of *Mycoplasma hominis*. J. Bacteriol. *186*, 1021–1028.

Hopfe, M., and Henrich, B. (2008). OppA, the ecto-ATPase of *Mycoplasma hominis* induces ATP release and cell death in HeLa cells. BMC Microbiol. *8*, 55.

Hopfe, M., Hoffmann, R., and Henrich, B. (2004). P80, the HinT interacting membrane protein, is a secreted antigen of *Mycoplasma hominis*. BMC Microbiol. *4*, 46.

Hopfe, M., Hegemann, J.H., and Henrich, B. (2005). HinT proteins and their putative interaction partners in Mollicutes and Chlamydiaceae. BMC Microbiol. *5*, 27.

Hopfe, M., Dahlmanns, T., and Henrich, B. (2011). In *Mycoplasma hominis* the OppA-mediated cytoadhesion depends on its ATPase activity. BMC Microbiol. *11*,185.

Hsu, T., and Minion, F.C. (1998a). Molecular analysis of the P97 cilium adhesin operon of *Mycoplasma hyopneumoniae*. Gene *214*, 13–23.

Hsu, T., and Minion, F.C. (1998b). Identification of the cilium binding epitope of the *Mycoplasma hyopneumoniae* P97 adhesin. Infect. Immun. *66*, 4762–4766.

Huberts, D.H., and van der Klei, I.J. (2010). Moonlighting proteins: an intriguing mode of multitasking. Biochim. Biophys. Acta *1803*, 520–525.

Huesca, M., Borgia, S., Hoffman, P., and Lingwood, C.A. (1996). Acidic pH changes receptor binding specificity of *Helicobacter pylori*: a binary adhesion model in which surface heat shock (stress) proteins mediate sulfatide recognition in gastric colonization. Infect. Immun. *64*, 2643–2648.

Hutchison, C.A., Peterson, S.N., Gill, S.R., Cline, R.T., White, O., Frase, C.M., Smith, H.O., and Venter, J.C. (1999). Global transposon mutagenesis and a minimal mycoplasma genome. Science *286*, 2165–2169.

Into, T., Nodasaka, Y., Hasebe, A., Okuzawa, T., Nakamura, J., Ohata, N., and Shibata, K. (2002). Mycoplasmal lipoproteins induce Toll-like receptor 2- and caspases-mediated cell death in lymphocytes and monocytes. Microbiol. Immunol. *46*, 265–276.

Iverson-Cabral, S.L., Astete, S.G., Cohen, C.R., Rocha, E.P., and Totten, P.A. (2006). Intrastrain heterogeneity of the mgpB gene in *Mycoplasma genitalium* is extensive in vitro and in vivo and suggests that variation is generated via recombination with repetitive chromosomal sequences. Infect. Immun. *74*, 3715–3726.

Iverson-Cabral, S.L., Astete, S.G., Cohen, C.R., and Totten, P.A. (2007). mgpB and mgpC sequence diversity in *Mycoplasma genitalium* is generated by segmental reciprocal recombination with repetitive chromosomal sequences. Mol. Microbiol. *66*, 55–73.

Jacobs, E., Bartl, A., Oberle, K., and Schiltz, E. (1995). Molecular mimicry by *Mycoplasma pneumoniae* to evade the induction of adherence inhibiting antibodies. J. Med. Microbiol. *43*, 422–429.

Jarrell, K.F., and McBride, M.J. (2008). The surprisingly diverse ways that prokaryotes move. Nat. Rev. Microbiol. *6*, 466–476.

Jeffery, C.J. (1999). Moonlighting proteins. Trends Biochem. Sci. *24*, 8–11.

Jeffery, C.J. (2009). Moonlighting proteins-an update. Mol. Biosyst. *5*, 345–350.

Jenkins, C., Wilton, J.L., Minion, F.C., Falconer, L., Walker, M.J., and Djordjevic, S.P. (2006). Two domains within the *Mycoplasma hyopneumoniae* cilium adhesin bind heparin. Infect. Immun. *74*, 481–487.

Jenkins, C., Geary, S.J., Gladd, M., and Djordjevic, S.P. (2007). The *Mycoplasma gallisepticum* OsmC-like protein MG1142 resides on the cell surface and binds heparin. Microbiology *153*, 1455–1463.

Jiang, H.X., Chen, J.R., Yan, H.L., Li, X.N., Chen, Z.L., and Zeng, Z.L. (2009). Molecular variability of DR-1 and DR-2 within the *pvpA* gene in *Mycoplasma gallisepticum* isolates. Avian Dis. *53*, 124–128.

Jönsson, K., Signäs, C., Müller, H.P., and Lindberg, M. (1991). Two different genes encode fibronectin binding proteins in *Staphylococcus aureus*. The complete nucleotide sequence and characterization of the second gene. Eur. J. Biochem. *202*, 1041–1048.

Jordan, J.L., Chang, H.Y., Balish, M.F., Holt, L.S., Bose, S.R., Hasselbring, B.M., Waldo, R.H. 3rd, Krunkosky, T.M., and Krause, D.C. (2007). Protein P200 is dispensable for *Mycoplasma pneumoniae* hemadsorption but not gliding motility or colonization of differentiated bronchial epithelium. Infect. Immun. *75*, 518–522.

Kannan, T.R., Provenzano, D., Wright, J.R., and Baseman, J.B. (2005). Identification and characterization of human surfactant protein A binding protein of *Mycoplasma pneumoniae*. Infect. Immun. *73*, 2828–2834.

Kannan, T.R., and Baseman, J.B. (2006). ADP-ribosylating and vacuolating cytotoxin of *Mycoplasma pneumoniae* represents unique virulence determinant among bacterial pathogens. Proc. Natl. Acad. Sci. U.S.A. *103*, 6724–6729.

Kawahito, Y., Ichinose, S., Sano, H., Tsubouchi, Y., Kohno, M., Yoshikawa, T., Tokunaga, D., Hojo, T., Harasawa, R., Nakano, T., *et al.* (2008). *Mycoplasma fermentans* glycolipid-antigen as a pathogen of rheumatoid arthritis. Biochem. Biophys. Res. Commun. *369*, 561–566.

Ketcham, C.M., Anai, S., Reutzel, R., Sheng, S., Schuster, S.M., Brenes, R.B., Agbandje-McKenna, M., McKenna, R., Rosser, C.J., and Boehlein, S.K. (2005). p37 Induces tumor invasiveness. Mol. Cancer Ther. *4*, 1031–1038.

Kheyar, A., Reddy, S.K., and Silim, A. (1995). The 64 kDa lipoprotein of *Mycoplasma gallisepticum* has two distinct epitopes responsible for haemagglutination and growth inhibition. Avian Pathol. *24*, 55–68.

Killiny, N., Castroviejo, M., and Saillard, C. (2005). *Spiroplasma citri* spiralin acts in vitro as a lectin binding to glycoproteins from its insect vector *Circulifer haematoceps*. Phytopathology 95, 541–548.

Killiny, N., Batailler, B., Foissac, X., and Saillard, C. (2006). Identification of a *Spiroplasma citri* hydrophilic protein associated with insect transmissibility. Microbiology *152*, 1221–1230.

Kirn-Safran, C., Farach-Carson, M.C., and Carson, D.D. (2009). Multifunctionality of extracellular and cell surface heparan sulfate proteoglycans. Cell. Mol. Life Sci. *66*, 3421–3434.

Kitzerow, A., and Henrich, B. (2001). The cytosolic HinT protein of *Mycoplasma hominis* interacts with two membrane proteins. Mol. Microbiol. *41*, 279–287.

Kitzerow, A., Hadding, U., and Henrich, B. (1999). Cytoadherence studies of the adhesin P50 of *Mycoplasma hominis*. J. Med. Microbiol. *48*, 485–493.

Kornspan, J.D., Tarshis, M., and Rottem, S. (2010). Invasion of melanoma cells by *Mycoplasma hyorhinis*: enhancement by protease treatment. Infect. Immun. *78*, 611–617.

Krulwich, T.A., Jin, J., Guffanti, A.A., and Bechhofer, H. (2001). Functions of tetracycline efflux proteins that do not involve tetracycline. J. Mol. Microbiol. Biotechnol. *3*, 237–246.

Layh-Schmitt, G., Podtelejnikov, A., and Mann, M. (2000). Proteins complexed to the P1 adhesin of *Mycoplasma pneumoniae*. Microbiology *146*, 741–747.

Leigh, S.A., and Wise, K.S. (2002). Identification and functional mapping of the *Mycoplasma fermentans* P29 adhesin. Infect. Immun. *70*, 4925–4935.

Li, Y.Z., Ho, Y.P., Chen, S.T., Chiou, T.W., Li, Z.S., and Shiuan, D. (2009). Proteomic comparative analysis of pathogenic strain 232 and avirulent strain J of *Mycoplasma hyopneumoniae*. Biochemistry *74*, 215–220.

Liu, W.B., Zhang, J.Z., Jiang, B.H., Ren, T.T., Gong, M.M., Meng, L., and Shou, C.C. (2006). Lipoprotein p37 from *Mycoplasma hyorhinis* inhibiting mammalian cell adhesion. J. Biomed. Sci. *13*, 323–331.

Liu, W., Ren, T., Jiang, B., Gong, M., and Shou, C. (2007). Mycoplasmal membrane protein p37 promotes malignant changes in mammalian cells. Can. J. Microbiol. *53*, 270–276.

Mamelak, D., and Lingwood, C.A. (2001). The ATPase domain of Hsp70 possesses a unique binding specificity for 3′ sulfated galactolipids. J. Biol. Chem. *276*, 449–456.

Markus, G. (1996). Conformational changes in plasminogen, their effect on activation, and the agents that modulate activation rates: a review. Fibrinolysis *10*, 75–85.

Marques, L.M., Ueno, P.M., Buzinhani, M., Cortez, B.A., Neto, R.L., Yamaguti, M., Oliveira, R.C., Guimarães, A.M., Monezi, T.A., Braga, A.C. Jr, *et al.* (2010). Invasion of *Ureaplasma diversum* in Hep-2 cells. BMC Microbiol. *10*, 83.

May, M., Papazisi, L., Gorton, T.S., and Geary, S.J. (2006). Identification of fibronectin-binding proteins in *Mycoplasma gallisepticum* strain R. Infect. Immun. *74*, 1777–1785.

McCarthy, A.J., and Lindsay, J.A. (2010). Genetic variation in *Staphylococcus aureus* surface and immune evasion genes is lineage associated: implications for vaccine design and host–pathogen interactions. BMC Microbiol. *10*, 173.

McGowin, C.L., Spagnuolo, R.A., and Pyles, R.B. (2010). *Mycoplasma genitalium* rapidly disseminates to the upper reproductive tracts and knees of female mice following vaginal inoculation. Infect. Immun. *78*, 726–736.

Minion, F.C., Adams, C., and Hsu, T. (2000). R1 region of P97 mediates adherence of *Mycoplasma hyopneumoniae* to swine cilia. Infect. Immun. *68*, 3056–3060.

Miyata, M. (2010). Unique centipede mechanism of Mycoplasma gliding. Annu. Rev. Microbiol. *64*, 519–537.

Monnet, V. (2003). Bacterial oligopeptide-binding proteins. Cell. Mol. Life Sci. *60*, 2100–2114.

Much, P., Winner, F., Stipkovits, L., Rosengarten, R., and Citti, C. (2002). *Mycoplasma gallisepticum*: influence of cell invasiveness on the outcome of experimental infection in chickens. FEMS Immunol. Med. Microbiol. *34*, 181–186.

Mudahi-Orenstein, S., Levisohn, S., Geary, S.J., and Yogev, D. (2003). Cytadherence-deficient mutants of *Mycoplasma gallisepticum* generated by transposon mutagenesis. Infect. Immun. *71*, 3812–3820.

Nahm, D.H., Lee, K.H., Shin, J.Y., Ye, Y.M., Kang, Y., and Park, H.S. (2006). Identification of alpha-enolase as an autoantigen associated with severe asthma. J. Allergy Clin. Immunol. *118*, 376–381.

Nakane, D., Adan-Kubo, J., Kenri, T., and Miyata, M. (2010). Isolation and characterization of P1 adhesin, leg protein of gliding bacterium, *Mycoplasma pneumoniae*. J. Bacteriol. *193*, 715–722.

Ogawa, Y., Oishi, E., Muneta, Y., Sano, A., Hikono, H., Shibahara, T., Yagi, Y., and Shimoji, Y. (2009). Oral vaccination against mycoplasmal pneumonia of swine using a live *Erysipelothrix rhusiopathiae* vaccine strain as a vector. Vaccine 27, 4543–4550.

Okamba, F.R., Arella, M., Music, N., Jia, J.J., Gottschalk, M., and Gagnon, C.A. (2010). Potential use of a recombinant replication-defective adenovirus vector carrying the C-terminal portion of the P97 adhesin protein as a vaccine against *Mycoplasma hyopneumoniae* in swine. Vaccine *28*, 4802–4809.

Ozeri, V., Tovi, A., Burstein, I., Natanson-Yaron, S., Caparon, M.G., Yamada, K.M., Akiyama, S.K., Vlodavsky, I., and Hanski, E. (1996). A two-domain mechanism for group A streptococcal adherence through protein F to the extracellular matrix. EMBO J. *15*, 989–998.

Paddenberg, R., Wulf, S., Weber, A., Heimann, P., Beck, L.A., and Mannherz, H. (1996). Internucleosomal DNA fragmentation in cultured cells under conditions reported to induce apoptosis may be caused by mycoplasma endonucleases. Eur. J. Cell. Biol. *71*: 105–119.

Patti, J.M., and Höök, M. (1994). Microbial adhesins recognizing extracellular matrix macromolecules. Curr. Opin. Cell. Biol. *6*, 752–758.

Pereyre, S., Sirand-Pugnet, P., Beven, L., Charron, A., Renaudin, H., Barré, A., Avenaud, P., Jacob, D., Couloux, A., Barbe, V., *et al.* (2009). Life on arginine for *Mycoplasma hominis*: clues from its minimal genome and comparison with other human urogenital mycoplasmas. PLoS Genet. 5, e1000677.

Pethe, K., Alonso, S., Biet, F., Delogu, G., Brennan, M.J., Locht, C., and Menozzi, F.D. (2001). The heparin-binding haemagglutinin of *M. tuberculosis* is required for extrapulmonary dissemination. Nature *412*, 190–194.

Pich, O.Q., Burgos, R., Ferrer-Navarro, M., Querol, E., and Piñol, J. (2008). Role of *Mycoplasma genitalium* MG218 and MG317 cytoskeletal proteins in terminal organelle organization, gliding motility and cytadherence. Microbiology *154*, 3188–3198.

Pich, O.Q., Burgos, R., Querol, E., and Piñol, J. (2009). P110 and P140 cytadherence-related proteins are negative effectors of terminal organelle duplication in *Mycoplasma genitalium*. PLoS One *4*, e7452.

Pollanen, J., R., Stephens, W., and Vaheri, A. (1991). Directed plasminogen activation at the surface of normal and malignant cells. Adv. Cancer Res. *57*, 273–328.

Ponting, C.P., Marshall, J.M., and Cederholm-Williams, S.A. (1992). Plasminogen: a structural review. Blood Coag. Fibrinol. *3*, 605–614.

Proft, T., Hilbert, H., Layh-Schmitt, G., and Herrmann, R. (1995). The proline-rich P65 protein of *Mycoplasma pneumoniae* is a component of the Triton X-100-insoluble fraction and exhibits size polymorphism in the strains M129 and FH. J. Bacteriol. *177*, 3370–3378.

Raulston, J.E., Davis, C.H., Paul, T.R., and Wayrick, P.B. (1998). Heat shock protein 70 kDa and the chlamydial envelope: an entry level position? In Chlamydial infections, Stephens, R.S., Byrne, G.I., Christiansen, M.D., Clarke, I.N., and Graystone, J.T., eds. (University of California – Berkeley Press, San Francisco), pp. 83–86.

Razin, S. (1999). Adherence of pathogenic mycoplasmas to host cells. BioSci. Rep. *19*, 367–372.

Razin, S., and Jacobs, E. (1992). Mycoplasma adhesion. J. Gen. Microbiol. *138*, 407–422.

Razin, S., Yogev, D., and Naot, Y. (1998). Molecular biology and pathogenicity of mycoplasmas. Microbiol. Mol. Biol. Rev. *62*, 1094–1156.

Relich, R.F., Friedberg, A.J., and Balish, M.F. (2009). Novel cellular organization in a gliding mycoplasma, *Mycoplasma insons*. J. Bacteriol. *191*, 5312–5314.

Reutzel, R., Boehlein, S.K., Govindasamy, L., Brenes, R.B., Agbandje-McKenna, M., Schuster, S.M., and McKenna, R. (2002). Crystallization and preliminary X-ray analysis of the tumor metastasis factor p37. Acta Crystallogr. D Biol. Crystallogr. *58*, 2141–2144.

Romero-Arroyo, C.E., Jordan, J., Peacock, S.J., Willby, M.J., Farmer, M.A., and Krause, D.C. (1999). *Mycoplasma pneumoniae* protein P30 is required for cytadherence and associated with proper cell development. J. Bacteriol. *181*, 1079–1087.

Rosenbusch, R.F., and Minion, F.C. (1992). Cell envelcope: morphology and biochemistry. In Mycoplasmas, Maniloff, J., McElhaney, R.N., Finch, L.R., and Baseman, J.B., eds. (ASM Press, Washington, DC), pp.73–76.

Rottem, S. (2002). Choline-containing lipids in mycoplasmas. Microbes Infect. *4*, 963–968.

Rottem, S., and Barile, M.F. (1993). Beware of mycoplasmas. Trends Biotechnol. *11*, 143–151.

Sachse, K., Helbig, J.H., Lysnyansky, I., Grajetzki, C., Müller, W., Jacobs, E., and Yogev, D. (2000). Epitope mapping of immunogenic and adhesive structures in repetitive domains of *Mycoplasma bovis* variable surface lipoproteins. Infect. Immun. *68*, 680–687.

Salman, M., Deutsch, I., Tarshis, M., Naot, Y., and Rottem, S. (1994). Membrane lipids of *Mycoplasma fermentans*. FEMS Microbiol. Lett. *123*, 255–260.

Sato, N., Oizumi, T., Kinbara, M., Sato, T., Funayama, H., Sato, S., Matsuda, K., Takada, H., Sugawara, S., and Endo, Y. (2010). Promotion of arthritis and allergy in mice by aminoglycoglycerophospholipid, a membrane antigen specific to *Mycoplasma fermentans*. FEMS Immunol. Med. Microbiol. *59*, 33–41.

Schmidhauser, C., Dudler, R., Schmidt, T., and Parish, R.W. (1990). A mycoplasmal protein influences tumour cell invasiveness and contact inhibition in vitro. J. Cell. Sci. *95*, 499–506.

Schmidl, S.R., Gronau, K., Hames, C., Busse, J., Becher, D., Hecker, M., and Stülke, J. (2010). The stability of cytadherence proteins in *Mycoplasma pneumoniae* requires activity of the protein kinase PrkC. Infect. Immun. *78*, 184–192.

Schmidt, J.A., Browning, G.F., and Markham, P.F. (2007). *Mycoplasma hyopneumoniae* mhp379 is a $Ca^{2+}$-dependent, sugar-nonspecific exonuclease exposed on the cell surface. J. Bacteriol. *189*, 3414–3424.

Schurwanz, N., Jacobs, E., and Dumke, R. (2009). Strategy to create chimeric proteins derived from functional adhesin regions of *Mycoplasma pneumoniae* for vaccine development. Infect. Immun. *77*, 5007–5015.

Seto, S., and Miyata, M. (2003). Attachment organelle formation represented by localization of cytadherence proteins and formation of the electron-dense core in wild-type and mutant strains of *Mycoplasma pneumoniae*. J. Bacteriol. *185*, 1082–1091.

Seymour, L.M., Deutscher, A.T., Jenkins, C., Kuit, T.A., Falconer, L., Minion, F.C., Crossett, B., Padula, M., Dixon, N.E., Djordjevic, S.P., *et al.* (2010). A processed multidomain *Mycoplasma hyopneumoniae* adhesin binds fibronectin, plasminogen, and swine respiratory cilia. J. Biol. Chem. *285*, 33971–33978.

Seymour, L.M., Falconer, L., Deutscher, A.T., Minion, F.C., Padula, M.P., Dixon, N.E., Djordjevic, S.P., and Walker, M.J. (2011). Mhp107 is a member of the multifunctional adhesin family of *Mycoplasma hyopneumoniae*. J. Biol. Chem. *286*, 10097–10104.

Seymour, L.M., Jenkins, C., Deutscher, A.T., Raymond, B.B., Padula, M.P., Tacchi, J.L., Bogema, D.R., Eamens, G.J., Woolley, L.K., Dixon, N.E., *et al.* (2012). Mhp182 (P102) binds fibronectin and contributes to the recruitment of plasmin(ogen) to the *Mycoplasma hyopneumoniae* cell surface. Cell. Microbiol. *14*, 81–94.

Simmons, W.L., and Dybvig, K. (2003). The Vsa proteins modulate susceptibility of *Mycoplasma pulmonis* to complement killing, hemadsorption, and adherence to polystyrene. Infect. Immun. *71*, 5733–5738.

Simmons, W.L., and Dybvig, K. (2007). Biofilms protect *Mycoplasma pulmonis* cells from lytic effects of complement and gramicidin. Infect. Immun. *75*, 3696–3699.

Sippel, K.H., Robbins, A.H., Reutzel, R., Boehlein, S.K., Namiki, K., Goodison, S., Agbandje-McKenna, M., Rosser, C.J., and McKenna, R. (2009). Structural insights into the extracytoplasmic thiamine-binding lipoprotein p37 of *Mycoplasma hyorhinis*. J. Bacteriol. *191*, 2585–2592.

Smith, D.G., Russell, W.C., and Thirkell, D. (1994). Adherence of *Ureaplasma urealyticum* to human epithelial cells. Microbiology *140*, 2893–2898.

Somarajan, S.R., Kannan, T.R., and Baseman, J.B. (2010). *Mycoplasma pneumoniae* Mpn133 is a cytotoxic nuclease with a glutamic acid-, lysine- and serine-rich region essential for binding and internalization but not enzymatic activity. Cell. Microbiol. *12*, 1821–1831.

Staats, C.C., Boldo, J., Broetto, L., Vainstein, M., and Schrank, A. (2007). Comparative genome analysis of proteases, oligopeptide uptake and secretion systems in *Mycoplasma* spp. Genet. Mol. Biol. *30*, 225–229.

Su, C.J., Dallo, S.F., Chavoya, A., and Baseman, J.B. (1993). Possible origin of sequence divergence in the P1 cytadhesin gene of *Mycoplasma pneumoniae*. Infect. Immun. *61*, 816–822.

Su, H.C., Hutchison, C.A. 3rd, and Giddings, M.C. (2007). Mapping phosphoproteins in *Mycoplasma genitalium* and *Mycoplasma pneumoniae*. BMC Microbiol. *7*, 63.

Svenstrup, H.F., Nielsen, P.K., Drasbek, M., Birkelund, S., and Christiansen, G. (2002). Adhesion and inhibition assay of *Mycoplasma genitalium* and *M. pneumoniae* by immunofluorescence microscopy. J. Med. Microbiol. *51*, 361–373.

Tabassum, I., Chaudhry, R., Chourasia, B.K., and Malhotra, P. (2010). Identification of an N-terminal 27 kDa fragment of *Mycoplasma pneumoniae* P116 protein as specific immunogen in *M. pneumoniae* infections. BMC Infect. Dis. *10*, 350.

Theiss, P., and Wise, K.S. (1997). Localized frameshift mutation generates selective, high-frequency phase variation of a surface lipoprotein encoded by a mycoplasma ABC transporter operon. J. Bacteriol. *179*, 4013–4022.

Varshney, A.K., Chaudhry, R., Kabra, S.K., and Malhotra, P. (2008). Cloning, expression, and immunological characterization of the P30 protein of *Mycoplasma pneumoniae*. Clin. Vaccine Immunol. *15*, 215–220.

Washburn, L.R., Bird, D.W., and Dybvig, K. (2003). Restoration of cytoadherence to an adherence-deficient mutant of *Mycoplasma arthritidis* by genetic complementation. Infect. Immun. *71*, 671–675.

Williams, T.A., and Fares, M.A. (2010). The effect of chaperonin buffering on protein evolution. Genome Biol. Evol *2*, 609–619.

Wilton, J., Jenkins, C., Cordwell, S.J., Falconer, L., Minion, F.C., Oneal, D.C., Djordjevic, M.A., Connolly, A., Barchia, I., Walker, M.J., *et al.* (2008). Mhp493 (P216) is a proteolytically processed, cilium and heparin binding protein of *Mycoplasma hyopneumoniae*. Mol. Microbiol. *71*, 566–582.

Winner, F., Rosengarten, R., and Citti, C. (2000). In vitro cell invasion of *Mycoplasma gallisepticum*. Infect. Immun. *68*, 4238–4244.

Yavlovich, A., and Rottem, S. (2007). Binding of host extracellular matrix proteins to *Mycoplasma fermentans* and its effect on adherence to, and invasion of HeLa cells. FEMS Microbiol. Lett. *266*, 158–162.

Yavlovich, A., Katzenell, A., Tarshis, M., Higazi, A.A., and Rottem, S. (2004). *Mycoplasma fermentans* binds to and invades HeLa cells: involvement of plasminogen and urokinase. Infect. Immun. *72*, 5004–5011.

Yavlovich, A., Rechnitzer, H., and Rottem, S. (2007). Alpha-enolase resides on the cell surface of *Mycoplasma fermentans* and binds plasminogen. Infect. Immun. *75*, 5716–5719.

Zhang, Q., and Wise, K.S. (1996). Molecular basis of size and antigenic variation of a *Mycoplasma hominis* adhesin encoded by divergent vaa genes. Infect. Immun. *64*, 2737–2744.

Zhang, Q., and Wise, K.S. (1997). Localized reversible frameshift mutation in an adhesin gene confers a phase-variable adherence phenotype in mycoplasma. Mol. Microbiol. *25*, 859–869.

Zhang, Q., Young, T.F., and Ross, R.F. (1994). Glycolipid receptors for attachment of *Mycoplasma hyopneumoniae* to porcine respiratory ciliated cells. Infect. Immun. *62*, 4367–4373.

Zimmermann, L., Peterhans, E., and Frey, J. (2010). RGD motif of lipoprotein T, involved in adhesion of *Mycoplasma conjunctivae* to lamb synovial tissue cells. J. Bacteriol. *192*, 3773–3779.

# The Glycocalyx of Mollicutes

# 7

James M. Daubenspeck, David S. Jordan and Kevin Dybvig

## Abstract

Virtually all bacteria have a glycocalyx, a sugar shell, and probably all bacterial pulmonary pathogens produce a capsule under appropriate conditions. The glycomoieties produced by these bacteria are often critical for immune evasion and survival in the animal host. The mycoplasmas are no exception. Most, if not all, mycoplasmas that are animal pathogens produce polysaccharides, glycolipids and glycoproteins. In addition, the mycoplasmas can and do adsorb glycoconjugates from their environment that are incorporated into the glycocalyx and serve to further camouflage the organism from host immunity. The adsorption of host molecules contributes to the difficulty in determining which glycoconjugates are produced by the Mollicutes. The machinery for glycoconjugate synthesis in mycoplasmas is for the most part unknown. Some mycoplasmal glycosyltransferases closely resemble those of other bacteria and are implicated in the synthesis of glycolipids. The machinery for polysaccharide synthesis in mycoplasmas is currently obscure and potentially novel.

## Introduction

The word 'glycocalyx', meaning 'sweet husk' or 'sugar shell', was coined to describe the polysaccharide-rich coating of cells (Bennett, 1963). The glycocalyx can be defined as the set of macromolecules exposed on the surface of the organism that contain monosaccharide as a key component. The glycocalyx includes the polysaccharide capsule or slime layer, lipoglycans, glycolipids, adhesive polysaccharides associated with biofilms, and any other glycoconjugates that might be produced by Mollicutes or adsorbed onto the cell surface from the environment. In this chapter we will discuss these molecules and the enzymatic machinery used to synthesize them, including the annotated glycosyltransferase (GT) and nucleotidyltransferase genes identified in the published genomes of Mollicutes.

Evidence for the importance of the glycocalyx to the survival of bacteria began to accumulate in the 1930s. It was hypothesized in 1933 that acetylated polysaccharide components isolated from *Streptococcus pneumoniae* were immunogenic in mice (Avery and Goebol, 1933). Although the ground-breaking work in 1944 of Avery and colleagues is primarily remembered for demonstrating the transformative effects of the DNA isolated from a smooth phenotype of *S. pneumoniae* on the rough variant, and thus for proving that DNA was the genetic material of the cell, their findings also showed that the smooth phenotype was associated with virulence (Avery *et al.*, 1944). The smooth phenotype is the result of capsule production. The first evidence for a polysaccharide in Mollicutes came as early as 1958, with data suggesting that *Mycoplasma mycoides* produced a galactan (Plackett and Buttery, 1958). As we progress in our understanding of the roles of glycoconjugates, our appreciation of the importance of the glycocalyx will grow.

## Capsule

Polysaccharide capsules or slime layers are generally critical for the survival of a bacterium in one or more niches. These complex macromolecules exist at the interface of the organism and its environment. The distinction between capsule and slime layer is not necessarily obvious, but a capsule is firmly attached to the surface of the bacterium and often structurally complex, while a slime layer is loosely associated and usually structurally simple. We refer here to polysaccharides exported to the surface of the bacterium or secreted as exopolysaccharides without regard to function, such as incorporation into a capsule or slime layer or providing adhesiveness within a biofilm. These polysaccharides in pathogenic organisms are the first and major defence against the host immune system. The list of activities modulated by the polysaccharide capsule continues to grow and includes antibacteriolytic activity, antiphagocytosis, modulation and evasion of immune responses, adhesion and biofilm formation (Comstock and Kasper, 2006). One potentially important role for the capsule in Mollicutes is host mimicry, whereby bacterial polymers effectively camouflage the bacterium by mimicking the exposed surface of host epithelial cells. Here we will look at examples of bacterial exopolysaccharides that are essential for an organism's success in their environmental niche and that interact with the host immune system.

Perhaps the most intensely studied polysaccharide capsule is that of *S. pneumoniae*, a Gram-positive coccoid human pathogen and the causative agent of pneumococcal pneumonia, sinusitis, meningitis, bacteraemia and otitis media, among others (Bryce *et al.*, 2005). Its capsule is up-regulated in the lung and one of the first to be shown to be a virulence factor (Avery *et al.*, 1944). Over 90 antigenically distinct pneumococcal capsular serotypes have been described that vary by order and type of the monosaccharide units composing the polysaccharide chains and that also may contain side branches (Miyachi *et al.*, 2009). The *S. pneumoniae* capsule has been shown to inhibit complement activity and complement-dependent phagocytosis (Hyams *et al.*, 2010). Effective vaccines have been developed for *S. pneumoniae* against the most common of the capsular serotypes, including a 23-valent polysaccharide vaccine recommended for individuals over 65 and a 13-valent conjugate vaccine licensed in 2010 for use in children (Nuorti and Whitney, 2010). For this bacterium, the most effective vaccines are derived from the polysaccharide capsule.

The polysaccharide capsule is the major virulence factor of the Gram-negative human pulmonary pathogen *Haemophilus influenzae* (Hallström and Riesbeck, 2010). *H. influenzae* is an invasive pathogen that affects infants and children and causes bacteraemia, influenza, and acute meningitis (Watt *et al.*, 2009). Non-typeable (unencapsulated) *H. influenzae* is a commensal of the respiratory tract and generally does not cause disease (Erwin and Smith, 2007). There are six distinct serotypes of encapsulated invasive strains, with type b being the most virulent (Adderson *et al.*, 2001). The pathogenesis associated with encapsulated serotypes of *H. influenzae* is dependent on polysaccharide. The introduction of a capsular polysaccharide vaccine against serotype b in the early 1990s reduced the incidence of invasive disease caused by *H. influenzae* (Morris *et al.*, 2008).

Studies with *S. pneumoniae* and *H. influenzae* are a small representation of the cumulative data that establishes the importance and the functionality of the capsule in pulmonary pathogens. In each of 14 species of Mollicutes we have tested, gas chromatographic (GC) analysis supports production of at least one polysaccharide (unpublished). It is anticipated that these polysaccharides will prove to be virulence factors.

The term capsule is used here in a general sense without differentiating a true capsule from a slime layer. Many studies conducted on Mollicutes have relied on electron microscopy to observe what appears to be a capsule. Fixed samples were negatively stained with a compound such as uranyl acetate and, in many studies, also stained with ruthenium red, which binds to anionic molecules (Luft, 1971). It seems that most of the polysaccharides produced by Mollicutes contain neutral sugars and the contribution of the ruthenium red to the staining of the capsule is unclear. The common fixation techniques for electron microscopy use paraformaldehyde, glutaraldehyde, or both, which would react with amino sugars such as N-acetylglucosamine (GlcNAc) and the free reducing ends of neutral polysaccharides that would produce negatively charged molecules capable of binding cationic stains such as ruthenium red.

Images suggesting the presence of a capsule have been obtained for most *Mycoplasma* species that have been examined, including *M. dispar* (Almeida and Rosenbusch, 1991), *M. gallisepticum* (Howard and Gourlay, 1974), *M. hominis* (Furness *et al.*, 1976), *M. hyopneumoniae* (Tajima *et al.*, 1982), *M. meleagridis* (Green and Hanson, 1973), *M. mycoides* subsp. *mycoides* (Howard and Gourlay, 1974), *M. ovipneumoniae* (Niang *et al.*, 1998), *M. penetrans* (Neyrolles *et al.*, 1998), *M. pulmonis* (Taylor-Robinson *et al.*, 1981) and *M. synoviae* (Ajufo and Whithear, 1978). The possible presence of a capsule in ureaplasmas is strengthened by experiments using the lectin concanavalin A (ConA), which preferentially binds mannose and glucose residues. ConA conjugated to iron particles was observed bound to the surface of *Ureaplasma urealyticum*, revealing a capsule-like structure (Robertson and Smook, 1976). Similarly, isolates of *M. ovipneumoniae* react strongly with the lectin wheatgerm agglutinin) (WGA), which binds to GlcNAc and, to a lesser degree, other sugars, and have a capsule observable by electron microscopy (Niang *et al.*, 1998). Capsule-like material purified by protease digestion of whole cells of *M. dispar* was used to produce rabbit antibodies that were utilized in immunogold labelling of the organism (Almeida and Rosenbusch, 1991). The immunogold-labelled antibodies colocalized with the ruthenium red-stained material in electron micrographs. In one study, electron microscopy did not reveal an obvious capsule on *Acholeplasma laidlawii*, *M. bovis*, *M. bovigenitalium*, *M. bovirhinis* or *M. pneumoniae* strain NCTCIOII (Howard and Gourlay, 1974). However, another study did observe a ruthenium red-staining layer enveloping cells of *M. pneumoniae* strain M129 (Wilson and Collier, 1976), and we believe that most pathogenic *Mycoplasma* species have a glycocalyx containing one or more polysaccharides. The observed difference between *M. pneumoniae* strains NCTCIOII and M129 in the two studies may be due to strain variation or differences in technique.

One feature of polysaccharides is their ability to retain water within their structure. When they are dehydrated, such as by treatment with alcohol during sample preparation for electron microscopy, this water is lost, causing the entire structure to shrink. The shrinkage of capsular polysaccharides in bacteria such as *Escherichia coli* (Bayer and Thurow, 1977) and group B streptococci (Mackie *et al.*, 1979) can be prevented by the addition of specific antibodies. Similarly, this collapse was effectively prevented by treatment of *M. hyopneumoniae* and *M. gallisepticum* with specific antisera before fixation and staining (Tajima *et al.*, 1985). When stabilized the capsule of *M. hyopneumoniae* extended approximately 125 nm from the membrane and was three times thicker, and less dense, than those of controls. The stabilized capsule of *M. gallisepticum* was 40 nm in width, two times thicker than those of controls. These data strongly suggest the presence of a polysaccharide capsule in both of these species and the formation of antibodies against those molecules by the host.

Despite considerable electron microscopic evidence for polysaccharide capsules in Mollicutes, it is generally not known whether the capsular material is synthesized by the mycoplasma or adsorbed from the medium. There are a number of examples of serum glycoproteins, such as fibronectin, that adsorb to the surface of the mycoplasma and contribute to the glycocalyx (Girón *et al.*, 1996; May *et al.*, 2006; Balasubramanian *et al.*, 2009; Deutscher *et al.*, 2010; Seymour *et al.*, 2010). The capsule of *M. dispar* is thin when the mycoplasma is grown in pure culture, but becomes more pronounced when the mycoplasma is cultured with host cells (Almeida and Rosenbusch, 1991). The more pronounced capsule might result from the induction of capsule synthesis when host cells are present, or might be due to the adsorption of host material onto the mycoplasma cell surface. A ruthenium red-staining capsule-like material surrounding *M. penetrans* was removed by treatment with neuraminidase (Neyrolles *et al.*, 1998). Neuraminidases, also known as sialidases, are glycosidases that catalyse the removal of terminal sialic acid (neuraminic acid) residues. Sialic acid refers to a family of nine-carbon monosaccharides primarily found in eukaryotic glycoproteins. This family of sugars is rare in prokaryotic organisms, and it is unlikely that any member of this family of carbohydrates is synthesized by Mollicutes. Treatment with neuraminidase may have removed serum glycoproteins that had adsorbed onto the surface of *M. penetrans*.

## Polysaccharides

### Filamentous threads

The history of polysaccharide research in mycoplasmology can be traced to a point before the genus *Mycoplasma* was recognized. As early as 1935, the causal agent of bovine pleuropneumonia (*M. mycoides* subsp. *mycoides*) was described as forming 'filamentous threads' during growth (Tang *et al.*, 1935). A serologically active fraction from these threads yielded positive Molisch and negative biuret reactions (Kurotchkin, 1937). The Molisch test is a chemical assay for the presence of carbohydrates and the biuret test is for the presence of proteins. A direct analysis of these filamentous threads in 1966 showed that they stained with periodic-acid Schiff (PAS) and that the addition of glucose to the medium caused the threads to increase in size (Gourlay and MacLeod, 1966). The periodic acid oxidizes vicinal hydroxyl groups of carbohydrates, generating aldehydes that react with the Schiff reagent to produce a distinctive pink colour. The threads have been described as a 'mucinous homogeneous matrix' (Gourlay and Thrower, 1968). We have observed similar threads in cryo-electron microscopy of *M. pulmonis* (Fig. 7.1).

### Polysaccharides and biofilms

The filamentous threads described for some species may prove to be an integral part of the adhesive matrix encasing the biofilm. Biofilms have been described for several species of mycoplasma (McAuliffe *et al.*, 2006, 2008; Simmons *et al.*, 2007; García-Castillo *et al.*, 2008; Simmons and Dybvig, 2009; Kornspan *et al.*, 2011). Lectin binding studies indicate the presence of polysaccharides, which may be produced by the mycoplasma or adsorbed from the medium or host. *M. pulmonis* produces the polysaccharide EPS-I, which contains equimolar amounts of glucose and galactose. Mutants that lack EPS-I have

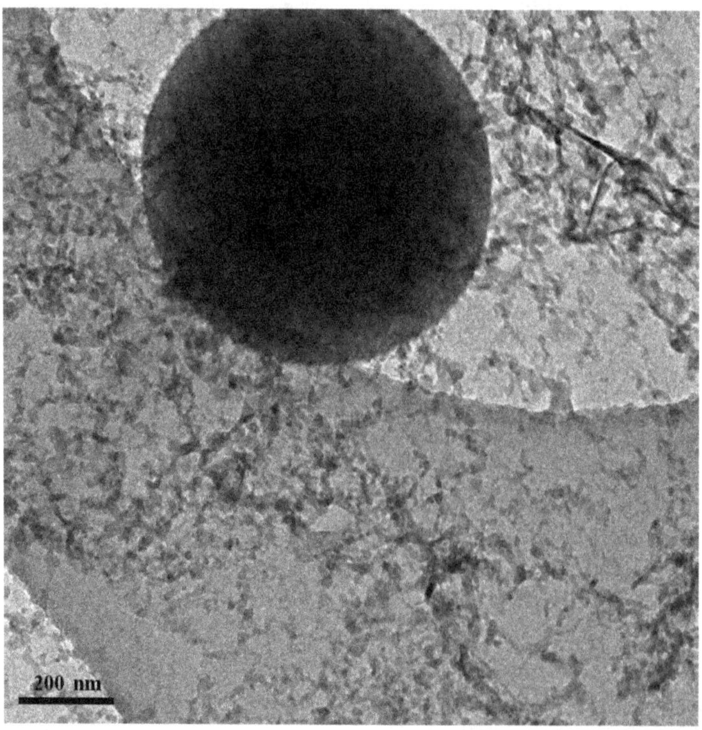


**Figure 7.1** Cryo-electron micrograph of *M. pulmonis* showing filamentous threads.

an enhanced ability to form a biofilm and have a 8-fold increase in what appears to be a second polysaccharide (EPS-II) that contains GlcNAc (Daubenspeck *et al.*, 2009). Many bacteria produce an adhesive polysaccharide composed of poly-GlcNAc that has a critical role in biofilm formation (Jefferson, 2009), and it is likely that a similar molecule is produced by *M. pulmonis* and other Mollicutes that can form biofilms. It has recently been shown that the quality of the biofilm produced by strains of *M. pneumoniae* differs. Attached to the cells of strains that produce a more robust biofilm is a GlcNAc-containing polysaccharide that might have a role in tethering the cells in the biofilm. A GlcNAc-containing polysaccharide is also produced by strains that form only a modest biofilm, but this polysaccharide is found in culture supernatants and is unattached to the cells (Simmons *et al.*, 2013).

## Polysaccharide pathogenesis and immune responses

Polysaccharides are major virulence factors of many, if not most, bacterial pathogens. They are often components of capsules or slime layers and serve to protect the bacteria from various aspects of host immunity. Some polysaccharides are adhesive and required for the formation of bacterial biofilms. Perhaps all bacterial polysaccharides have a role in survival of the bacterium in at least one of its niches. However, the possible association between polysaccharides and the virulence of mycoplasmas has not yet been proven. The galactan of *M. mycoides* subsp. *mycoides* reportedly has toxic effects (Lloyd *et al.*, 1971). However, we view the toxicity of the galactan as unlikely. The toxicity of mycoplasma lipid moieties, such as those of lipoproteins is well known, and lipid impurities within the galactan preparations or a potential lipid anchor were likely responsible for the reported toxicity. The galactan has been described as a slime layer, but whether it protects from host defences or has some other role in pathogenesis has not been studied. The polysaccharides of *M. pulmonis* have a role in biofilm formation (Daubenspeck *et al.*, 2009). When encased in a biofilm, the mycoplasma resists some aspects of innate immunity (Simmons and Dybvig, 2007). Structures resembling a mycoplasma biofilm form *ex vivo* and *in vivo* on the tracheal mucosa of infected mice (Simmons and Dybvig, 2009). Polysaccharide production in *M. pulmonis* can also modulate adherence of the mycoplasma to epithelial cells and the ability of the mycoplasma to resist killing by complement and macrophages (Daubenspeck *et al.*, 2009; Bolland and Dybvig, 2012; Bolland *et al.*, 2012; Shaw *et al.*, 2013). The thickness of the ruthenium red-stained matrix around *M. pulmonis* has been correlated with strain virulence (Taylor-Robinson *et al.*, 1981). Collectively, these studies strongly suggest a role for the polysaccharides of *M. pulmonis* in pathogenesis. The capsule of *M. dispar* may exert antiphagocytic activity (Almeida *et al.*, 1992), as has been found for many other bacterial capsules (Wood and Smith, 1949). A ruthenium red-staining capsule was observed on two pathogenic strains of the avian pathogen *M. gallisepticum*, but not on a non-pathogenic strain (Tajima *et al.*, 1982). They reported a correlation between the presence of the ruthenium red-binding material and cytadsorption to chicken tracheal epithelia. The virulent Gladysdale and T2 vaccine strains of *M. mycoides* subsp. *mycoides* produced filamentous threads while the avirulent strain KH3J and the T1 vaccine strain did not (Gourlay and MacLeod, 1966). Little is known about the host immune responses to polysaccharides of Mollicutes, but a polysaccharide fraction from *M. pneumoniae* reacted with sera from rabbits that had been immunized with *M. pneumoniae* (Sobeslavsky *et al.*, 1966, 1967). This fraction was somewhat protective when used as an immunogen in a hamster model (Brunner, 1981).

## Biochemical analysis of polysaccharides

One of the major complications in the isolation and characterization of capsules from mycoplasma species is the requirement for a complex medium for growth. From the point of view of polysaccharide research, the two most detrimental medium components are yeast extract and serum. Yeast extract contains mannosylated glycoproteins and serum contains many glycoconjugates that adsorb to the surface of mycoplasmas and contribute to the glycocalyx. The lot of serum used can affect results because not all sera contain the same glycoconjugates and variation can even be found when blood is drawn from the same animal on different days. Glycoconjugates from the medium are thus present in variable amounts in preparations of polysaccharide material isolated from Mollicutes.

The early observations of filamentous threads led researchers to attempt to purify polysaccharides from Mollicutes. An antigenic water-soluble fraction was isolated from the contagious bovine pleuropneumonia organism and described as a crude polysaccharide fraction (Dafaalla, 1957). The galactan molecule from *M. mycoides* subsp. *mycoides* was purified using centrifugation or ion exchange chromatography (Plackett and Buttery, 1958). Paper chromatographic analysis found that 89% of the material was the monosaccharide galactose. They also determined, using the anthrone procedure (Trevelyan *et al.*, 1952), that 10% of the cell by weight was composed of carbohydrate. A lipid fraction was identified that accounted for up to 5% of the sample (Plackett and Buttery, 1960). It is feasible that the polysaccharide isolated from *M. mycoides* had a lipid anchor or that there may have been lipid contaminants in the polysaccharide preparation. Using limited acid hydrolysis and periodate oxidation, which cleaves the ring structure at vicinal hydroxyls, the structure of the capsular polysaccharide in the attenuated vaccine strain V5 of *M. mycoides* subsp. *mycoides* was proposed to be β-1→6-linked galactose with a putative lipid anchor (Plackett and Buttery, 1964). The early biochemical work on *M. mycoides* was a significant step in proving that mycoplasmas produce a polysaccharide capsule. A neutral polysaccharide composed of glucose has been isolated from *M. mycoides* subsp. *capri* (Jones *et al.*, 1965). The capsule of *M. dispar* is reportedly a polymer of galacturonic acid, but the experimental evidence to support this claim is unpublished (Rosenbusch and Minion, 1992). The EPS-I polysaccharide of *M. pulmonis* is composed of the neutral sugars glucose and galactose with undermined linkage(s) (Daubenspeck *et al.*, 2009).

GC analysis of a polysaccharide fraction from *M. pneumoniae* strain FH revealed the monosaccharides glucose, galactose, rhamnose and mannose (Allen and Prescott, 1978). Similar results from our laboratory for strains M129 and UAB PO1 reveal differences in the cellularly associated polysaccharide composition, suggesting strain variation (Simmons *et al.*, 2013). The data we have accumulated would suggest that most mycoplasma species produce a polysaccharide composed of neutrally charged sugar residues.

Bacterial polysaccharide synthesis often occurs on the membrane and is initiated on a lipid anchor (Cartee *et al.*, 2005). Depending on the bacterium, the synthesized polysaccharide may remain anchored or become disassociated from the lipid. If the polysaccharide remains attached to the lipid, it might be considered a lipoglycan. There are numerous reports of lipoglycans in Mollicutes (Smith, 1984, 1985, 1992). It is unknown whether the EPS-I and II polysaccharides of *M. pulmonis* are attached to lipid because of the method used for sample preparation. To isolate these molecules, samples were sonicated using conditions that should shear the polysaccharide from any anchor that may be present (Daubenspeck *et al.*, 2009). However, it has been reported that *M. pulmonis* lacks lipoglycan (Smith, 1984).

## Glycoproteins

Glycoproteins are common in eukaryotes but until recently were thought to be less prevalent in bacteria. There are striking structural differences between the common linkages and glycomoieties of eukaryotic glycoproteins and those of prokaryotes (Messner, 2004). Data that have accumulated over the intervening years lead to the conclusion that prokaryotes show a greater diversity of glycan structures and compositions than do eukaryotes (Abu-Qarn *et al.*, 2008). There are five distinct glycosidic bonds that covalently link the glycan to the peptide backbone in eukaryotes: *C*-glycosides, *N*-glycosides, *O*-glycosides, *P*-glycosides, and *S*-glycosides. Only *N*- and *O*-linkages have been characterized in bacterial species. In bacteria, glycans are usually linked to proteins through an N-linkage at an asparagine residue (consensus sequence Asn-X-Ser/Thr, where X represents any amino acid except proline) or through an *O*-linkage at a serine/threonine residue (Szymanski and Wren, 2005).

Glycosylation will alter the function of a protein and can affect its conformation, stability, proteolytic susceptibility, adhesiveness and physiochemical properties such as viscosity, solubility and surface charge (Upreti *et al.*, 2003). The first convincing bacterial glycoprotein was purified from the cell envelope of *Halobacterium salinarum* (Mescher and Strominger, 1976). This glycoprotein was a cell surface layer (S-layer) protein. S-layer proteins are often glycosylated and are widespread in bacteria. They contribute to the organism's potential to diversify its cell surface (Schäffer and Messner, 2004). In pathogenic bacteria, such as *Neisseria gonorrhoeae*, *Neisseria meningitidis* and *Pseudomonas aeruginosa*, pili are also commonly composed of glycosylated proteins (Banerjee and Ghosh, 2003). Nevertheless, bacterial glycoproteins are rare and have not been identified in *E. coli*, *Salmonella* sp. or *Bacillus subtilis* (Messner, 2004).

Examples of *Mycoplasma* species that have been reported to contain a glycoprotein include *M. bovis*, *M. gallisepticum*, *M. genitalium*, *M. hyopneumoniae*, *M. hyorhinis* and *M. pneumoniae* (Goel and Lemcke, 1975; Kahane and Brunner, 1977; Nicolet *et al.*, 1980; Geary *et al.*, 1981; Kobayashi and Watanabe, 1991). These reports detected glycoproteins through PAS staining of gels or lectin binding. These early reports were generally unconvincing, as there was no evidence that the glycoproteins were synthesized by the mycoplasma, rather than adsorbed from the medium. However, a proteomic study of *M. gallisepticum* identified two candidate glycoproteins (Demina *et al.*, 2009). More recently, unpublished data from our laboratory indicate that *O*-glycosylation at serine and threonine residues is widespread in Mollicutes.

## Glycolipids

Glycolipids are a ubiquitous and complex family of molecules found in most bacterial species, including Mollicutes. These amphipathic molecules contain a monosaccharide, disaccharide, or a more complex glycomoiety attached to a lipid derivative that sometimes includes a charged head group. The structures have been elucidated for numerous glycolipids. However, the function of these molecules is less well defined. It is agreed that glycolipids are a major structural building block in biological membranes (Hölzl and Dörmann, 2007). It has been proposed that glycolipids also function as biosurfactants, surface-active compounds that lower surface and interfacial tensions (Soberón-Chávez *et al.*, 2005; Cameotra *et al.*, 2010). Several groups have reported toxicity associated with glycolipids. The cord factor of *Corynebacterium diphtheriae* is a trehalose-6,6′-dicorynomycolate glycolipid shown to have lethal toxicity in a murine model (Kato, 1970). In *Mycobacterium tuberculosis*, the glycolipid lipoarabinomannan has been described as a virulence factor that can bind to leucocytes and modulate immune responses (Strohmeier and Fenton, 1999). Glycolipids are as varied in structure and putative function as the species in which they are found.

One of the earliest reports of glycolipids in Mollicutes was an unknown glycolipid in *Mycoplasma* sp. avian strain J (*M. gallisepticum*) (Smith and Koostra, 1967), followed closely by the identification of a galactolipid synthesized by *M. mycoides* strain V5 (Plackett, 1967) and a putative glucolipid in *Acholeplasma laidlawii* identified based on the incorporation of $^{14}C$ glucose into non-acidic lipids (Plackett and Shaw, 1967). Over 50% of the total lipids of *A. laidlawii* are glycolipids, with two main glycosides with the structures *O*-α-D-glucopyranosyl-(1→1)-D-glycerol and *O*-α-D-glucopyranosyl-(1→2)-*O*-α-D-glucopyranosyl-(1→1)-D-glycerol (Shaw *et al.*, 1968). As for polysaccharide research, little research on the glycolipids of Mollicutes was performed during the 20-year period following these early reports.

A major finding came in *Mycoplasma fermentans*, in which novel phosphocholine-containing glycophospholipids (GGPL-I and GGPL-III) were identified. The structures were determined by NMR to be 6′-*O*-phosphocholine-α-glucopyranosyl-(1′→3)–1,2-diacyl-*sn*-glycerol and 1″-phosphocholine,2″-amino dihydroxypropane-3″-phospho-6′-α-glucopyranosyl-(1′→3)–1,2-diacyl-glycerol for GGPL-I and III, respectively (Matsuda *et al.*, 1994, 1997a). GGPL-I and GGPL-III are the main lipid components and species-specific major lipid antigens of *M. fermentans* (Matsuda *et al.*, 1997b). A third glycolipid isolated from *M. fermentans*, designated MfGL-II (*M. fermentans* glycolipid-II) has a structure of 6′-O-(3′-phosphocholine-2′-amino-1′-phospho-1′,3′-propanediol)-α-D-glucopyranosyl-(1→3)-1,2-diacyl-glycerol (Zähringer *et al.*, 1997). A putative glycolipid fraction (GLF) isolated from *Mycoplasma penetrans* has a monosaccharide composition of mannose: glucose: N-acetyl-galactosamine: GlcNAc in a ratio of 1: 6: 2: 1 (Neyrolles *et al.*, 1998). It is possible that the complexity of the sugar mixture found in the GLF preparations is due, in part, to medium contaminants. We have found that glycoconjugates from the medium adsorb readily to Mollicutes and give rise to highly variable, complex monosaccharide compositions of glycolipid fractions. Glucose and galactose seem to be the primary monosaccharides incorporated into glycolipid structures produced by Mollicutes. In *M. pneumoniae* a pair of β-glycolipids, a β-D-glucopyranoside and a β-D-galactopyranoside, were isolated that are the major immunodeterminants of this pathogenic species (Miyachi *et al.*, 2009). In *M. pulmonis* and *M. arthritidis*, we have identified a lipid fraction with a strong glucose signal when analysed by GC (unpublished data).

There have also been reports of Mollicutes that do not produce glycolipids. A survey of the total lipid fractions of *Mycoplasma hominis* did not detect any glycolipids (Rottem and Razin, 1973). Whether these negative results are due to a lack of sensitivity in the assays or due to variability in the lipid fraction of the bacteria is not known. *M. mycoides* subsp. *capri* varies in the amount of glycolipids produced when the amount of cholesterol in the medium is altered (Archer, 1975). We have observed that glycolipid fractions isolated from *M. pulmonis*, *M. arthritidis* and *M. pneumoniae* are highly variable and dependent on the age of the cultures and the formulation of the growth medium.

The functions of glycolipids in Mollicutes remain to be elucidated. Glycolipid-rich isolates from *M. pneumoniae* are serologically active when tested against human and rabbit antisera (Razin *et al.*, 1970). Glycolipids have been shown to be major immunodeterminants in several mycoplasma species, including *M. pneumoniae* and *M. fermentans* (Li *et al.*, 1997; Miyachi *et al.*, 2009). The primary function of glycolipids in Mollicutes should not be related to their antigenicity. Logically, an organism that exists without a cell wall has other means to stabilize their membrane, and glycolipids may have such a role.

## Glycoconjugate synthesis machinery

Most of the mycoplasma species we have discussed have limited machinery for the synthesis of glycoconjugates based on the bioinformatic analysis of the genome sequences. Using *M. pulmonis* as an example, the MYPU_7700 gene codes for the only annotated GT in this species that might support glycoconjugate synthesis. We believe that this putative enzyme is involved in the production of a glycolipid and not a polysaccharide. How do *M. pulmonis* and other species of mycoplasma synthesize polysaccharides? In most bacteria, glycoconjugate synthesis occurs at the membrane. Often a nucleotide sugar such as UDP-glucose is the substrate for synthesis, with the dinucleotide providing the energy to drive the reaction. The information at hand suggests that the mycoplasmas produce simple, linear polysaccharides such as polymers of glucose, galactose, or other hexoses. Because of the absence of a cell wall and periplasm, it is anticipated that polysaccharide synthesis and export will be simultaneous. A model for polysaccharide production in mycoplasmas may be type III capsule synthesis in *S. pneumoniae*. A single GT, the synthase, is a membrane protein that synthesizes and exports the growing polysaccharide chain. The type III capsular polysaccharide consists of alternating residues of glucuronic acid and glucose and is synthesized from substrates of UDP-glucuronic acid and UDP-glucose on a lipid anchor (Cartee *et al.*, 2005).

Glucolipids in *A. laidlawii* are synthesized from the precursors UDP-glucose and 1,2-diglyceride, and the reaction is specific for UDP (Smith, 1969). *Mycoplasma gallinarum* strain J transfers glucose from UDP-glucose to membrane-bound sterol and only membrane-bound sterol (Smith, 1971). Gas chromatograms of a *M. pulmonis* mutant with a disrupted MYPU_3090 gene, the only potential nucleotidyltransferase gene that we have identified in the genome, show a reduction in a lipid signal associated with glycolipids (unpublished data), suggesting that glycolipids in this species are synthesized by using a nucleotide sugar precursor. Glycolipid production in most mycoplasma species may occur through the standard nucleotide sugar pathway.

## Glycosyltransferases

Glycosyltransferases catalyse reactions that result in the covalent linkage of carbohydrate residues to lipids, proteins, nucleic acids or carbohydrates. The CAZy (Carbohydrate-Active enZYmes) database (www.cazy.org) currently lists 94 families of GT and nearly 1900 unclassified GT sequences amongst all forms of life. GT families are grouped by amino acid sequence similarity. Some families are small and described as 'monospecific', meaning that all members of the family are thought to have one specific function. Some families have thousands of members and are called 'polyspecific' because several different experimentally established GT reactions are catalysed by them.

The protein structure of at least one enzyme of each family member exists for 35 of the families and it has been noted that many of these enzymes form one of two types of Rossmann-like folds, type A or type B. A number of amino acids sequences of low similarity among the GT families are thought to have evolutionarily converged to form the GT-A or GT-B folds. GT enzymes are described as either retaining or inverting, referring to whether the anomeric configuration of the donor molecule is preserved as it transferred to the acceptor molecule. There are examples of retaining and inverting GT within both the GT-A and GT-B groups. A useful review describing the structural differences between the two types of folds is available (Lairson *et al.*, 2008).

Six mycoplasma genomes and the CAZy website were searched for genes with annotations describing potential glycosyltransferases and grouped by similarity based on BLAST analysis, resulting in Table 7.1. All but two of the mycoplasmal GT proteins belong to the polyspecific GT-A family 2 (inverting) enzymes, the exceptions being *Mycoplasma fermentans* JER GT-B family 4 (retaining) protein and *Mycoplasma mobile* 163K GT-B family 5 (retaining) protein. GT family 2

**Table 7.1** Potential glycosyltransferases in mycoplasmas

| Homologue cluster | Species searched (ORF) | Gene | GT family | Fold | Mycoplasma homologues |
|---|---|---|---|---|---|
| 1 | *Mycoplasma mycoides* PG1 (MSC0108, MSC0980, MSC0987 and MSC0993) | *epsG* | 2 | A | *Mycoplasma mycoides* subsp. *capri*<br>*Mycoplasma alligatoris*<br>*Mycoplasma crocodyli* |
| 2 | *Mycoplasma genitalium* PG37 (MG335.2)<br>*Mycoplasma pneumonia* M129 (MPN483) | *yibD* | 2 | A | *Mycoplasma conjunctivae* |
| 3 | *Mycoplasma arthritidis* 158L3–1 (MARTH849) | | 2 | A | *Mycoplasma agalactiae*<br>*Mycoplasma mobile*<br>*Mycoplasma fermentans* |
| | *Mycoplasma mycoides* PG1 (MSC771) | | | | *Mycoplasma alligatoris*<br>*Mycoplasma crocodyli*<br>*Mycoplasma synoviae* |
| | *Mycoplasma pulmonis* UAB CTIP (MYPU7700) | | | | *Mycoplasma hyorhinis*<br>*Mycoplasma hominis*<br>*Mycoplasma capricolum* |
| 4 | *Mycoplasma genitalium* PG37 (MG025 and MG060)<br>*Mycoplasma mycoides* PG1 (MSC0109, MSC0974, MSC0982 and MSC0988)<br>*Mycoplasma pneumonia* (MPN028 and MPN075) | | 2 | A | *Mycoplasma mobile*<br>*Mycoplasma alligatoris* |

proteins constitute the largest family of all the GT families, with almost 27,000 members. The structure of only four of these enzymes has been solved. The known substrates and products of the GT family 2 members are diverse. Specific functions may be proposed, but cannot be assigned based upon homology alone (Henrissat *et al.*, 2008).

The *M. mycoides* PG1 *epsG* gene is homologous to GT enzymes, the function of which has been studied in *Streptococcus thermophilus* Sfi6, *Lactococcus lactis* NIZO B40 and *Lactobacillus helveticus* NCC2745. The *eps* gene cluster was first defined in *S. thermophilus*. Although no clear functional assessment of *epsG* GT activity was characterized, the enzyme was noted to have homologues in *S. thermophilus*, *S. pneumoniae*, *Klebsiella pneumoniae* and *Streptococcus suis* (Stingele *et al.*, 1996; Germond *et al.*, 2001).

*E. coli* DH5α (with no GT activity) harbouring plasmids containing *L. lactis epsDEFG* are able to use UDP-galactose to produce a lipid-linked glucose-glucose-galactose trisaccharide while *E. coli* harbouring plasmids containing only the *L. lactis epsDEF* genes only produce lipid-linked glucose-glucose disaccharides (van Kranenburg *et al.*, 1999). GST-tagged EpsG from *L. helveticus* purified from *E. coli* BL21-SI can produce the lipid-linked glucose-glucose-galactose molecule in the presence of UDP-galactose, as well as lipid-linked glucose-glucose-glucose molecules (using UDP-glucose), although this activity is 10-fold less than for the UDP-galactose reaction (Jolly *et al.*, 2002). The galactosyltransferase activity attributed to *epsG* in these studies would be consistent with reports demonstrating that *M. mycoides* produces a galactose polymer (Plackett and Buttery, 1958; Buttery and Plackett, 1960).

*Mycoplasma pneumoniae* M129 possesses a GT (MPN483) that has been shown to use UDP-galactose or UDP-glucose as substrate with lipid or terminal galactose-containing glycolipid precursors, resulting in various combinations of galactose and glucose-containing glycolipids with up to three hexose residues per molecule (Klement *et al.*, 2007). No experimental data are available for the homologous protein (MG355.2) of the closely related species *M. genitalium*.

One group of mycoplasma GT proteins sharing high amino acid sequence similarity are MPN028, MPN075, MG025, MG060, MSC_0109, MSC_0974, MSC_0982 and MSC_0988. Many of these genes are thought to be essential for growth based on transposon mutagenesis studies and no experimental data pertaining to function is available.

A smaller group of enzymes with sequence similarity consists of MYPU_7700, MARTH_orf849 and MSC_0771. The MYPU_7700 protein shares 24% amino acid sequence identity with the *S. suis* Cps2J GT, which is hypothesized to be a galactosyltransferase (Smith *et al.*, 1999). *M. pulmonis* does produce a galactose-containing polysaccharide (EPS-I) that is part of the glycocalyx. However, the MYPU_7700 protein may not be a galactosyltransferase, as the homologous enzyme in *M. arthritidis* (the MARTH_orf849 protein) is not a galactosyltransferase because we have found that *M. arthritidis* produces no galactose-containing glycoconjugates (unpublished data).

The GT-A and GT-B families require UDP-hexose substrates. A pronounced existence of other GT families devoid of GT-A or GT-B folds is emerging, but little is known about these enzymes, as structural studies are hindered by the presence of several membrane-spanning domains. Some of these enzymes have structural features that resemble a glycohydrolase fold. It is noted that many of the families with uncharacterized structures use carbohydrate-lipid donors rather than NDP-hexoses (Henrissat *et al.*, 2008).

There are two lines of evidence that support the hypothesis that some mycoplasma species have GT enzymes other than GT-A and GT-B folded proteins. The first is the observation that most bacterial GT genes comprise 1–2% of each species' genome, with mycoplasmas being the 'exception' to the rule because of the low number of GT genes found among them (Lairson *et al.*, 2008). We predict that many of the mycoplasma GT enzymes have yet to be identified and that there is a significant representation in uncharacterized structures and mechanisms. GT activity in an environment lacking NDP-hexose, discussed below, strongly suggests the presence of alternatively folded enzymes.

Mutants of *M. pulmonis* that do not produce the EPS-I polysaccharide have been obtained by screening a transposon library using PAS staining of polyacrylamide gels. The mutants were further characterized using GC and fluorescence of bound *Griffonia simplicifolia* I (GS-I) lectin (Daubenspeck *et al.*, 2009). All of the EPS-I mutants had transposon disruptions in the overlapping gene pair MYPU_7410 and 7420. Complementation of the mutants with the two-gene operon of MYPU_7410 and 7420 restored function.

Structurally, the MYPU_7410 and 7420 proteins resemble closely one another and the multidrug exporter Sav1866 from *Staphylococcus aureus* (Dawson and Locher, 2006). Each protein has a conserved domain suggestive of a glucan exporter. It is likely that these proteins are required for export of EPS-I, but are they also required for synthesis? Do these proteins combine to form the GT responsible for EPS-I synthesis. Bioinformatics does not suggest that these proteins would have GT activity, but if they do not, what gene does code for the GT and why were mutants with this gene knocked out not identified in the transposon library? We suggest that these proteins do indeed have GT activity and form a complex that both synthesizes and exports EPS-I. Protein pairs highly similar to MYPU_7410 and 7420 have been identified in *M. hyopneumoniae* and *M. penetrans*, suggesting that the putative polysaccharide synthesis machinery of *M. pulmonis* may be widespread in Mollicutes (Daubenspeck *et al.*, 2009).

If MYPU_7410 and 7420 are novel GT proteins, they may use something other than nucleotide sugars as substrates. There are examples of glycoconjugates that are synthesized from substrates other than nucleotide sugars. Glycansucrases use sucrose, or other polymers containing glucose such as starch or maltose, as a substrate for polysaccharide synthesis (Plou *et al.*, 2002). Such enzymes are common in bacteria that colonize the oral cavity, where sucrose is often available. Cyclodextrin glucanotransferases use cyclodextrins, cyclic oligosaccharides, as substrates. These enzymes are primarily found in, but are not limited to, *Bacillus* species (Biwer *et al.*, 2002). Various phosphorylases can synthesize polysaccharides in the presence of excess levels of sugar-1-phosphate (Illingworth *et al.*, 1961; Luley-Goedl and Nidetzky, 2010). MYPU_7410 and 7420 might use one of these alternative substrates or something entirely new for EPS-I synthesis.

## Nucleotidyltransferases

The GT enzymes described above that are readily identifiable through bioinformatics may use nucleotide sugars as substrates. Since the structure of UDP-glucose was first proposed (Caputto *et al.*, 1950), the role of nucleotide sugars in the production of bacterial glycoconjugates and the importance of nucleotidyltransferase enzymes in capsule, biofilm, and glycolipid production has been documented for a variety of bacterial species (Bernheimer *et al.*, 1968; Bronner *et al.*, 1993; Rocchetta *et al.*, 1998; Gründling and Schneewind, 2007). In general, a hexose-1-phosphate must first be covalently linked to a nucleotide by a nucleotidyltransferase before the sugar residue can be polymerized. Different nucleotides with different levels of phosphorylation have been found to be substrates for nucleotidyltransferase reactions, but there are indications that each specific enzyme generally uses only one type of nucleotide as a substrate (Kim *et al.*, 2010).

Nucleotidyltransferases are essential for production of the most studied glycoconjugates in eukaryotics and prokaryotics. The eukaryotic nucleotidyltransferases are vastly different from the prokaryotic enzymes. Thus, the bacterial nucleotidyltransferases are potential targets for engineering new antimicrobial compounds (Mollerach *et al.*, 1998). The identification of new and potentially novel nucleotidyltransferases (Geerlof *et al.*, 1999) in Mollicutes would be a significant finding.

The sequenced genomes of six Mollicutes were surveyed for genes homologous to previously characterized nucleotidyltransferases and resulted in the identification of UTP-glucose-1-phosphate uridylyltransferase in *M. mycoides* subsp. *mycoides* (MSC_0110 and MSC_0990), *M. pneumoniae* (MPN667) and *M. genitalium* (MG453). This enzyme is referred to as GalU in some bacteria and GtaB in others. The enzyme catalyses the reversible reaction that produces UDP-glucose and pyrophosphate from UTP and glucose-1-phosphate with $Mg^{2+}$ as a cofactor (Thoden and Holden, 2007a,b; Kim *et al.*, 2010).

The MSC_0110 and MSC_0990 genes of *M.*

*mycoides* subsp. *mycoides* PG1 code for UTP-glucose-1-phosphate uridylyltransferases of identical amino acid sequence. The enzymes share 46% amino acid sequence identity with *M. pneumoniae* MPN667 and 35% identity with GalU of *E. coli*. GalU mutants in *E. coli*, *P. aeruginosa* and *Vibrio cholerae* have an altered lipopolysaccharide (LPS) structure (Genevaux *et al.*, 1999; Nesper *et al.*, 2001; Choudhury *et al.*, 2005). *S. pneumoniae* serotypes 1 and 3 and *L. lactis* harbouring mutations in their glucose-1-phosphate uridylyltransferase homologues produce no capsule or exopolysaccharide, respectively, and the enzyme is also required for glycolipid synthesis in *S. aureus* (Mollerach *et al.*, 1998; Boels *et al.*, 2001). BLAST analysis of the mycoplasmal glucose-1-phosphate uridylyltransferase sequences revealed highly homologous proteins in several *Clostridium* and *Bacillus* species and various other genera. Although there are regions of sequence similarity throughout the length of the protein sequence, the highest similarity in all these organisms is observed within the first 60 amino acid residues of the amino-terminus. Two short sequences within these 60 residues are identical in 93% of these species: GTRFLP at residues 12–17 and KEMLPI at residues 24–29. The *E. coli* enzyme has been noted to contain GTRMLP residues close to its amino-terminus and fits within the broader nucleotidyltransferase motif of GXGTR($X_9$)K, as does the *M. mycoides* protein. The proline within the GTRMLP sequence of *E. coli* is estimated to be 15 Å from the active site and may be important for function. The arginine within the GTRFLP motif is within 6 Å of the active site and mutating this residue to alanine in the *Helicobacter pylori* homologue (38% identity) results in an 86% decrease in function, measured as a decrease in UDP-glucose pyrophosphorylase activity. The *H. pylori* study also indicated that the lysine within the KEMLPI motif binds to a phosphate group of the uridine residue and is a crucial component of the active site (Kim *et al.*, 2010). The structural studies conclude that this enzyme is a tetramer that multimerizes as two subunits in tight association, which in turn associate less tightly with another tightly associated dimer. This region of weaker association between the two dimers consists of a β-strand that includes many of the amino-terminal residues (Thoden and Holden, 2007a,b; Kim *et al.*, 2010). Notably, *M. mycoides* subsp. *mycoides* SC strain Gladysdale is missing the first 113 residues of this protein, but it does possess a second full-length copy of the gene. As the Gladysdale strain is pathogenic and strain PG1 is not, the truncated nucleotidyltransferase has no dominant-negative effect on its pathogenicity. Perhaps this is not too surprising since the complete active site is present within a single subunit (Thoden and Holden, 2007a) and subunits may have transferase activity without multimerization (Kim *et al.*, 2010).

UTP-glucose-1-phosphate uridylyltransferase may not be required for production of some glycoconjugates in Mollicutes. *M. pulmonis* strain UAB CT lacks an obvious UTP-glucose-1-phosphate uridylyltransferase gene, but produces a polysaccharide capsule (Daubenspeck *et al.*, 2009). A possible candidate for a *M. pulmonis* nucleotidyltransferase gene is MYPU_3090, but our group has shown that transposon mutants with this gene disrupted retain the ability to produce EPS-I (unpublished data). Some mycoplasma genomes encode histidine triad proteins, which are sometimes associated with nucleotidyltransferase activity. Mycoplasmal HIT proteins, however, are of unknown function and apparently associate with membrane proteins, representing a different localization from the *Chlamydiae* HIT proteins that are predicted to interact with cytoplasmic proteins (Hopfe *et al.*, 2005).

## Summary

The available information including unpublished gas chromatography analysis from our group suggests that linear neutral polysaccharides, simple polymers of glucose, galactose or other hexoses, are produced by all species of *Mycoplasma* and transported to the cell surface. In some cases, the machinery for synthesis of these polysaccharides may be novel and use substrates other than nucleotide sugars. Some of the polysaccharides produced by the mycoplasmas contribute to the capsule or slime layer with other, adhesive polysaccharides contributing to the matrix of biofilms. Probably all species of Mollicutes produce glycolipids, which contribute to the glycocalyx. It is evident that glycoconjugates adsorbed from the environment also contribute to the glycocalyx. As for other bacteria, mycoplasma capsules likely serve to shield the organism from host immunity and modulate adherence properties. Molecular mimicry and camouflage using bacterially synthesized polysaccharides in conjunction with adsorbed host glycoconjugates would be an effective means of evading immune surveillance.

## References

Abu-Qarn, M., Eichler, J., and Sharon, N. (2008). Not just for Eukarya anymore: protein glycosylation in Bacteria and Archaea. Curr. Opin. Struct. Biol. *18*, 544–550.

Adderson, E.E., Byington, C.L., Spencer, L., Kimball, A., Hindiyeh, M., Carroll, K., Mottice, S., Korgenski, S.E.K., Christenson, J.C., and Pavia, A.T. (2001). Invasive serotype a *Haemophilus influenzae* infections with a virulence genotype resembling *Haemophilus influenzae* type b: emerging pathogen in the vaccine era? Pediatrics *108*, E18.

Ajufo, J.C., and Whithear, K.G. (1978). Evidence for a ruthenium red-staining extracellular layer as the haemagglutinin of the WVU 1853 strain of *Mycoplasma synoviae*. Aust. Vet. J. *54*, 502–504.

Allen, P.Z., and Prescott, B. (1978). Immunochemical studies on a *Mycoplasma pneumoniae* polysaccharide fraction: cross-reactions with type 23 and 32 antipneumococcal rabbit sera. Infect. Immun. *20*, 421–429.

Almeida, R.A., and Rosenbusch, R.F. (1991). Capsulelike surface material of *Mycoplasma dispar* induced by in vitro growth in culture with bovine cells is antigenically related to similar structures expressed in vivo. Infect. Immun. *59*, 3119–3125.

Almeida, R.A., Wannemuehler, M.J., and Rosenbusch, R.F. (1992). Interaction of *Mycoplasma dispar* with bovine alveolar macrophages. Infect. Immun. *60*, 2914–2919.

Archer, D.B. (1975). Modification of the membrane composition of *Mycoplasma mycoides* subsp. *capri* by the growth medium. J. Gen. Microbiol. *88*, 329–338.

Avery, O.T., and Goebol, W.F. (1933). Chemoimmunological studies on the soluble specific substances of pneumococcus. I. The isolation and properties of the acetyl polysaccharide of pneumococcus Type I. J. Exp. Med. *58*, 731–755.

Avery, O.T., Macleod, C.M., and McCarty, M. (1944). Studies on the chemical nature of the substance inducing transformation of pneumococcal types: I. Induction of transformation by a desoxyribonucleic acid fraction isolated from pneumococcus type III. J. Exp. Med. *79*, 137–158.

Balasubramanian, S., Kannan, T.R., Hart, P.J., and Baseman, J.B. (2009). Amino acid changes in elongation factor-Tu of *Mycoplasma pneumoniae* and *Mycoplasma genitalium* influence fibronectin binding. Infect. Immun. *77*, 3533–3541.

Banerjee, A., and Ghosh, S.K. (2003). The role of pilin glycan in neisserial pathogenesis. Mol. Cell. Biochem. *253*, 179–190.

Bayer, M.E., and Thurow, H. (1977). Polysaccharide capsule of *Escherichia coli*: microscopic study of its size, structure, and sites of synthesis. J. Bacteriol. *130*, 911–936.

Bennett, H.S. (1963). Morphological aspects of extracellular polysaccharides. J. Histochem. Cytochem. *11*, 14–23.

Bernheimer, H.P., Wermundsen, I.E., and Austrian, R. (1968). Mutation in pneumococcus type III affecting multiple cistrons concerned with the synthesis of capsular polysaccharide. J. Bacteriol. *96*, 1099–1102.

Biwer, A., Antranikian, G., and Heinzle, E. (2002). Enzymatic production of cyclodextrins. Appl. Microbiol. Biotechnol. *59*, 609–617.

Boels, I.C., Ramos, A., Kleerebezem, M., and de Vos, W.M. (2001). Functional analysis of the *Lactococcus lactis galU* and *galE* genes and their impact on sugar nucleotide and exopolysaccharide biosynthesis. Appl. Environ. Microbiol. *67*, 3033–3040.

Bolland, J.R., and Dybvig, K. (2012). *Mycoplasma pulmonis* Vsa proteins and polysaccharide modulate adherence to pulmonary epithelial cells. FEMS Microbiol. Lett. *331*, 25–30.

Bolland, J.R., Simmons, W.L., Daubenspeck, J.M., and Dybvig, K. (2012). Mycoplasma polysaccharide protects against complement. Microbiology *158*, 1867–1873.

Bronner, D., Sieberth, V., Pazzani, C., Roberts, I.S., Boulnois, G.J., Jann, B., and Jann, K. (1993). Expression of the capsular K5 polysaccharide of *Escherichia coli*: biochemical and electron microscopic analyses of mutants with defects in region 1 of the K5 gene cluster. J. Bacteriol. *175*, 5984–5992.

Brunner, H. (1981). Protective efficacy of *Mycoplasma pneumoniae* polysaccharides. Isr. J. Med. Sci. *17*, 678–681.

Bryce, J., Boschi-Pinto, C., Shibuya, K., and Black, R.E. (2005). WHO estimates of the causes of death in children. Lancet *365*, 147–1152.

Buttery, S.H., and Plackett, P. (1960). A specific polysaccharide from *Mycoplasma mycoides*. J. Gen. Microbiol. *23*, 357–368.

Cameotra, S.S., Makkar, R.S., Kaur, J., and Mehta, S.K. (2010). Synthesis of biosurfactants and their advantages to microorganisms and mankind. Adv. Exp. Med. Biol. *672*, 261–280.

Caputto, R., Leloir, L.F., Cardini, C.E., and Paladini, A.C. (1950). Isolation of the coenzyme of the galactose phosphate-glucose phosphate transformation. J. Biol. Chem. *184*, 333–350.

Cartee, R.T., Forsee, W.T., and Yother, J. (2005). Initiation and synthesis of the *Streptococcus pneumoniae* type 3 capsule on a phosphatidylglycerol membrane anchor. J. Bacteriol. *187*, 4470–4479.

Choudhury, B., Carlson, R.W., and Goldberg, J.B. (2005). The structure of the lipopolysaccharide from a *galU* mutant of *Pseudomonas aeruginosa* serogroup-O11. Carbohydr. Res. *340*, 2761–2772.

Comstock, L.E., and Kasper, D.L. (2006). Bacterial glycans: key mediators of diverse host immune responses. Cell *126*, 847–850.

Dafaalla, E.N. (1957). A study of the antigenic structure of the contagious bovine pleuropneumonia organism. Bull. Epizoot. Dis. Afr. *5*, 135–145.

Daubenspeck, J.M., Bolland, J.R., Luo, W., Simmons, W.L., and Dybvig, K. (2009). Identification of exopolysaccharide-deficient mutants of *Mycoplasma pulmonis*. Mol. Microbiol. 72, 1235–1245.

Dawson, R.J.P., and Locher, K.P. (2006). Structure of a bacterial multidrug ABC transporter. Nature *443*, 180–185.

Demina, I.A., Serebryakova, M.V., Ladygina, V.G., Rogers, M.A., Zgoda, V.G., Korzhenevskyi, D.A., and Govorun, V.M. (2009). Proteome of the bacterium *Mycoplasma gallisepticum*. Biochemistry 74, 165–174.

Deutscher, A.T., Jenkins, C., Minion, F.C., Seymour, L.M., Padula, M.P., Dixon, N.E., Walker, M.J., and Djordjevic, S.P. (2010). Repeat regions R1 and R2 in the P97 paralogue Mhp271 of *Mycoplasma hyopneumoniae* bind heparin, fibronectin and porcine cilia. Mol. Microbiol. *78*, 444–458.

Erwin, A.L., and Smith, A.L. (2007). Nontypeable *Haemophilus influenzae*: understanding virulence and commensal behavior. Trends Microbiol. *15*, 355–362.

Furness, G., Whitescarver, J., Trocola, M., and DeMaggio, M. (1976). Morphology, ultrastructure, and mode of division of *Mycoplasma fermentans, Mycoplasma hominis, Mycoplasma orale,* and *Mycoplasma salivarium*. J. Infect. Dis. *134*, 224–229.

García-Castillo, M., Morosini, M.I., Gálvez, M., Baquero, F., del Campo, R., and Meseguer, M.A. (2008). Differences in biofilm development and antibiotic susceptibility among clinical *Ureaplasma urealyticum* and *Ureaplasma parvum* isolates. J. Antimicrob. Chemother. *62*, 1027–1030.

Geary, S.J., Tourtelotte, M.E., and Cameron, J.A. (1981). Inflammatory toxin from *Mycoplasma bovis*: isolation and characterization. Science *212*, 1032–1033.

Geerlof, A., Lewendon, A., and Shaw, W.V. (1999). Purification and characterization of phosphopantetheine adenylyltransferase from *Escherichia coli*. J. Biol. Chem. *274*, 27105–27111.

Genevaux, P., Bauda, P., Dubow, M.S., and Oudega, B. (1999). Identification of Tn*10* insertions in the *rfaG, rfaP,* and *galU* genes involved in lipopolysaccharide core biosynthesis that affect *Escherichia coli* adhesion. Arch. Microbiol. *172*, 1–8.

Germond, J.E., Delley, M., D'Amico, N., and Vincent, S.J.F. (2001). Heterologous expression and characterization of the exopolysaccharide from *Streptococcus thermophilus* Sfi39. Eur. J. Biochem. *268*, 5149–5156.

Girón, J.A., Lange, M., and Baseman, J.B. (1996). Adherence, fibronectin binding, and induction of cytoskeleton reorganization in cultured human cells by *Mycoplasma penetrans*. Infect. Immun. *64*, 197–208.

Goel, M.C., and Lemcke, R.M. (1975). Dissociation of *Mycoplasma gallisepticum* membranes with lithium diiodosalicylate and isolation of glycoprotein. Ann. Microbiol. (Paris) *126*, 299–312.

Gourlay, R.N., and MacLeod, A.K. (1966). Fermentation of glucose by *Mycoplasma mycoides* and its effect on viability. Bull. Epizoot. Dis. Afr. *14*, 373–381.

Gourlay, R.N., and Thrower, K.J. (1968). Morphology of *Mycoplasma mycoides* thread-phase growth. J. Gen. Microbiol. *54*, 155–159.

Green, F., and Hanson, R.P. (1973). Ultrastructure and capsule of *Mycoplasma meleagridis*. J. Bacteriol. *116*, 1011–1018.

Gründling, A., and Schneewind, O. (2007). Genes required for glycolipid synthesis and lipoteichoic acid anchoring in *Staphylococcus aureus*. J. Bacteriol. *189*, 2521–2530.

Hallström, T., and Riesbeck, K. (2010). *Haemophilus influenzae* and the complement system. Trends Microbiol. *18*, 258–265.

Henrissat, B., Sulzenbacher, G., and Bourne, Y. (2008). Glycosyltransferases, glycoside hydrolases: surprise, surprise! Curr. Opin. Struct. Biol. *18*, 527–533.

Hölzl, G., and Dörmann, P. (2007). Structure and function of glycoglycerolipids in plants and bacteria. Prog. Lipid Res. *46*, 225–243.

Hopfe, M., Hegemann, J.H., and Henrich, B. (2005). HinT proteins and their putative interaction partners in Mollicutes and Chlamydiaceae. BMC Microbiol. *5*, 27.

Howard, C.J., and Gourlay, R.N. (1974). An electron-microscopic examination of certain bovine mycoplasmas stained with ruthenium red and the demonstration of a capsule on *Mycoplasma dispar*. J. Gen. Microbiol. *83*, 393–398.

Hyams, C., Camberlein, E., Cohen, J.M., Bax, K., and Brown, J.S. (2010). The *Streptococcus pneumoniae* capsule inhibits complement activity and neutrophil phagocytosis by multiple mechanisms. Infect. Immun. *78*, 704–715.

Illingworth, B., Brown, D.H., and Cori, C.F. (1961). The *de novo* synthesis of polysaccharide by phosphorylase. Proc. Nat. Acad. Sci. U.S.A. *47*, 469–478.

Jefferson, K.K. (2009). Poly-*N*-acetyl-glucosamine as a mediator of bacterial biofilm formation. In Bacterial Polysaccharides, Ullrich, M., ed. (Caister Academic Press, Norfolk, UK), pp. 175–186.

Jolly, L., Newell, J., Porcelli, I., Vincente, S.J.F., and Stingele, F. (2002). *Lactobacillus helveticus* glycosyltransferases: from genes to carbohydrate synthesis. Glycobiology *12*, 319–327.

Jones, A.S., Tittensor, J.R., and Walker, R.T. (1965). The chemical composition of the nucleic acids and other macromolecular constituents of *Mycoplasma mycoides* var. *capri*. J. Gen. Microbiol. *40*, 405–411.

Kahane, I., and Brunner, H. (1977). Isolation of a glycoprotein from *Mycoplasma pneumoniae* membranes. Infect. Immun. *18*, 273–277.

Kato, M. (1970). Action of a toxic glycolipid of *Corynebacterium diphtheriae* on mitochondrial structure and function. J. Bacteriol. *101*, 709–716.

Kim, H., Choi, J., Kim, T., Lokanath, N.K., Ha, S.C., Suh, S.W., Hwang, H.Y., and Kim, K.K. (2010). Structural basis for the reaction mechanism of UDP-glucose pyrophosphorylase. Mol. Cells *29*, 397–405.

Klement, M.L., Ojemyr, L., Tagscherer, K.E., Widmalm, G., and Wieslander, A. (2007). A processive lipid glycosyltransferase in the small human pathogen *Mycoplasma pneumoniae*: involvement in host immune response. Mol. Microbiol. *65*, 1444–1457.

Kobayashi, K., and Watanabe, T. (1991). Comparison of glycoproteins from ten human *Mycoplasma* species. Microbios *67*, 23–33.

Kornspan, J.D., Tarshis, M., and Rottem, S. (2011). Adhesion and biofilm formation of *Mycoplasma pneumoniae* on an abiotic surface. Arch. Microbiol. *193*, 833–836.

Kurotchkin, T.J. (1937). Specific carbohydrate from *Asterococcus mycoides* for serological tests of bovine pleuropneumonia. Proc. Soc. Exp. Biol. Med. *37*, 21.

Lairson, L.L., Henrissat, B., Davies, G.J., and Withers, S.G. (2008). Glycosyltransferases: structures, functions and mechanisms. Annu. Rev. Biochem. *77*, 521–555.

Li, J.L., Matsuda, K., Takagi, M., and Yamamoto, N. (1997). Detection of serum antibodies against phosphocholine-containing aminoglycoglycerolipid specific to *Mycoplasma fermentans* in HIV-1 infected individuals. J. Immunol. Methods *208*, 103–113.

Lloyd, L.C., Buttery, S.H., and Hudson, J.R. (1971). The effect of the galactan and other antigens of *Mycoplasma mycoides* var. *mycoides* on experimental infection with that organism in cattle. Med. Microbiol. *4*, 425–439.

Luft, J.H. (1971). Ruthenium red and violet I. Chemistry, purification, methods of use for electron microscopy and mechanism of action. Anat. Rec. *171*, 347–368.

Luley-Goedl, C., and Nidetzky, B. (2010). Carbohydrate synthesis by disaccharide phophorylases: reactions, catalytic mechanisms and application in the glycosciences. Biotechnol. J. *5*, 1324–1338.

Mackie, E.B., Brown, K.N., Lam, J., and Costerton, J.W. (1979). Morphological stabilization of capsules of group B streptococci, types Ia, Ib, II, and III, with specific antibody. J. Bacteriol. *138*, 609–617.

Matsuda, K., Kasama, T., Ishizuka, I., Handa, S., Yamamoto, N., and Taki, T. (1994). Structure of a novel phosphocholine-containing glycoglycerolipid from *Mycoplasma fermentans*. J. Biol. Chem. *269*, 33123–33128.

Matsuda, K., Ishizuka, I., Kasama, T., Handa, S., Yamamoto, N., and Taki, T. (1997a). Structure of a novel phosphocholine-containing aminoglycoglycerolipid of *Mycoplasma fermentans*. Biochim. Biophys. Acta *1349*, 1–12.

Matsuda, K., Li, J.L., Harasawa, R., and Yamamoto, N. (1997b). Phosphocholine-containing glycoglycerolipids (GGPL-I and GGPL-III) are species-specific major immunodeterminants of *Mycoplasma fermentans*. Biochem. Biophys. Res. Comm. *233*, 644–649.

May, M., Papazisi, L., Gorton, T.S., and Geary, S.J. (2006). Identification of fibronectin-binding proteins in *Mycoplasma gallisepticum* strain R. Infect. Immun. *74*, 1777–1785.

McAuliffe, L., Ellis, R.J., Miles, K., Ayling, R.D., and Nicholas, R.A.J. (2006). Biofilm formation by mycoplasma species and its role in environmental persistence and survival. Microbiology *152*, 913–922.

McAuliffe, L., Ayling, R.D., Ellis, R.J., and Nicholas, R.A. (2008). Biofilm-grown *Mycoplasma mycoides* subsp. *mycoides* SC exhibit both phenotypic and genotypic variation compared with planktonic cells. Vet. Microbiol. *129*, 315–324.

Mescher, M.F., and Strominger, J.L. (1976). Purification and characterization of a prokaryotic glucoprotein from the cell envelope of *Halobacterium salinarium*. J. Biol. Chem. *251*, 2005–2014.

Messner, P. (2004). Prokaryotic glycoproteins: unexplored but important. J. Bacteriol. *186*, 2517–2519.

Miyachi, A., Miyazaki, A., Shingu, Y., Matsuda, K., Dohi, H., and Nishida, Y. (2009). Synthesis and absolute structures of *Mycoplasma pneumoniae* β-glyceroglycolipid antigens. Carbohydr. Res. *344*, 36–43.

Mollerach, M., López, R., and García, E. (1998). Characterization of the *galU* gene of *Streptococcus pneumoniae* encoding a uridine diphosphoglucose pyrophosphorylase: a gene essential for capsular polysaccharide biosynthesis. J. Exp. Med. *188*, 2047–2056.

Morris, S.K., Moss, W.J., and Halsey, N. (2008). *Haemophilus influenzae* type b conjugate vaccine use and effectiveness. Lancet Infect. Dis. *8*, 435–443.

Nesper, J., Lauriano, C.M., Klose, K.E., Kapfhammer, D., Kraiss, A., and Reidl, J. (2001). Characterization of *Vibrio cholerae* O1 El Tor *galU* and *galE* mutants: influence on lipopolysaccharide structure, colonization, and biofilm formation. Infect. Immun. *69*, 435–445.

Neyrolles, O., Brenner, C., Prevost, M.C., Fontaine, T., Montagnier, L., and Blanchard, A. (1998). Identification of two glycosylated components of *Mycoplasma penetrans*: a surface-exposed capsular polysaccharide and a glycolipid fraction. Microbiology *144*, 1247–1255.

Niang, M., Rosenbusch, R.F., Andrews, J.J., and Kaeberle, M.L. (1998). Demonstration of a capsule on *Mycoplasma ovipneumoniae*. Am. J. Vet. Res. *59*, 557–562.

Nicolet, J., Paroz, P., and Kristensen, B. (1980). Growth medium constituents contaminating mycoplasma preparations and their role in the study of membrane glycoproteins in porcine mycoplasmas. J. Gen. Microbiol. *119*, 17–26.

Nuorti, J.P., and Whitney, C.G. (2010). Prevention of pneumococcal disease among infants and children – use of 13-valent pneumococcal conjugate vaccine and 23-valent pneumococcal polysaccharide vaccine – recommendations of the Advisory Committee on Immunization Practices (ACIP). MMWR Recomm. Rep. *59*, 1–18.

Plackett, P. (1967). The synthesis of polar lipids by mycoplasma. Ann. N. Y. Acad. Sci. *143*, 158–164.

Plackett, P., and Buttery, S.H. (1958). A galactan from *Mycoplasma mycoides*. Nature *182*, 1236–1237.

Plackett, P., and Buttery, S.H. (1960). A specific polysaccharide from *Mycoplasma mycoides*. J. Gen. Microbiol. *23*, 357–368.

Plackett, P., and Buttery, S.H. (1964). A galactofuranose disaccharide from the galactan of *Mycoplasma mycoides*. Biochem. J. *90*, 201–205.

Plackett, P., and Shaw, E.J. (1967). Glucolipids from *Mycoplasma laidlawii* and *Streptococcus* MG. Biochem. J. *104*, 61C-62C.

Plou, F.J., Martin, M.T., Segura, A.G.d., Alcalde, M., and Ballestros, A. (2002). Glucosyltransferases acting on starch or sucrose for the synthesis of oligosaccharides. Can. J. Chem. *80*, 743–752.

Razin, S., Prescott, B., Caldes, G., James, W.D., and Chanock, R.M. (1970). Role of glycolipids and phosphatidylglycerol in the serological activity of *Mycoplasma pneumoniae*. Infect. Immun. *1*, 408–416.

Robertson, J., and Smook, E. (1976). Cytochemical evidence of extramembranous carbohydrates on *Ureaplasma urealyticum* (T-strain mycoplasma). J. Bacteriol. *128*, 658–660.

Rocchetta, H.L., Pacan, J.C., and Lam, J.S. (1998). Synthesis of the A-band polysaccharide sugar D-rhamnose requires Rmd and WbpW: identification of multiple AlgA homologues, WbpW and ORF488, in *Pseudomonas aeruginosa*. Mol. Microbiol. *29*, 1419–1434.

Rosenbusch, R.F., and Minion, F.C. (1992). Cell envelope: morphology and biochemistry. In Mycoplasmas: Molecular Biology and Pathogenesis, Maniloff, J., McElhaney, R.N., Finch, L.R., and Baseman, J.B., eds. (American Society for Microbiology, Washington, D.C.), pp. 73–77.

Rottem, S., and Razin, S. (1973). Membrane lipids of *Mycoplasma hominis*. J. Bacteriol. *113*, 565–571.

Schäffer, C., and Messner, P. (2004). Surface-layer glycoproteins: an example for the diversity of bacterial glycosylation with promising impacts on nanobiotechnology. Glycobiology *14*, 31R-42R.

Seymour, L.M., Deutscher, A.T., Jenkins, C., Kuit, T.A., Falconer, L., Minion, F.C., Crossett, B., Padula, M., Dixon, N.E., Djordjevic, S.P., *et al.* (2010). A processed multidomain *Mycoplasma hyopneumoniae* adhesin binds fibronectin, plasminogen, and swine respiratory cilia. J. Biol. Chem. *285*, 33971–33978.

Shaw, B.M., Daubenspeck, J.M., Simmons, W.L., and Dybvig, K. (2013). EPS-I polysaccharide protects *Mycoplasma pulmonis* from phagocytosis. FEMS Microbiol. Lett. 338, 155–160.

Shaw, N., Smith, P.F., and Koostra., W.L. (1968). The lipid composition of *Mycoplasma laidlawii* strain B. Biochem. J. *107*, 329–333.

Simmons, W.L., and Dybvig, K. (2007). Biofilms protect *Mycoplasma pulmonis* cells from lytic effects of complement and gramicidin. Infect. Immun. *75*, 3696–3699.

Simmons, W.L., and Dybvig, K. (2009). Mycoplasma biofilms *ex vivo* and *in vivo*. FEMS Microbiol. Lett. *295*, 77–81.

Simmons, W.L., Bolland, J.R., Daubenspeck, J.M., and Dybvig, K. (2007). A stochastic mechanism for biofilm formation by *Mycoplasma pulmonis*. J. Bacteriol. *189*, 1905–1913.

Simmons, W.L., Daubenspeck, J.M., Osborne, J.D., Balish, M.F., Waites, K.B., and Dybvig, K. (2013). Type 1 and type 2 strains of *Mycoplasma pneumoniae* form different biofilms. Microbiology *159*, 737–747.

Smith, H.E., Damman, M., van der Velde, J., Wagenaar, F., Wisselink, H.J., Stockhofe-Zurwieden, N., and Smits, M.A. (1999). Identification and characterization of the *cps* locus of *Streptococcus suis* serotype 2: the capsule protects against phagocytosis and is an important virulence factor. Infect. Immun. *67*, 1750–1756.

Smith, P.F. (1969). Biosynthesis of glucosyl diglycerides by *Mycoplasma laidlawii* strain B. J. Bacteriol. *99*, 480–486.

Smith, P.F. (1971). Biosynthesis of cholesteryl glucoside by *Mycoplasma gallinarum*. J. Bacteriol. *108*, 986–991.

Smith, P.F. (1984). Lipoglycans from mycoplasmas. Crit. Rev. Microbiol. *11*, 157–186.

Smith, P.F. (1985). Detection of lipoglycans in ureaplasmas. J. Bacteriol. *162*, 611–614.

Smith, P.F. (1992). Membrane lipid and lipopolysaccharide structures. In Mycoplasmas: Molecular Biology and Pathogenesis, Maniloff, J., McElhaney, R.N., Finch, L.R., and Baseman, J.B., eds. (American Society for Microbiology, Washington, D.C.), pp. 79–91.

Smith, P.F., and Koostra, W.L. (1967). Phospholipids and glycolipids of sterol-requiring *Mycoplasma*. J. Bacteriol. *93*, 1853–1862.

Soberón-Chávez, G., Lépine, F., and Déziel, E. (2005). Production of rhamnolipids by *Pseudomonas aeruginosa*. Appl. Microbiol. Biotechnol. *68*, 718–725.

Sobeslavsky, O., Prescott, B., James, W.D., and Chanock, R.M. (1966). Isolation and characterization of fractions of *Mycoplasma pneumoniae*. II. Antigenicity and immunogenicity. J. Bacteriol. *91*, 2126–2138.

Sobeslavsky, O., Prescott, B., James, W.D., and Chanock, R.M. (1967). Serological and immunogenic activities of different fractions of *Mycoplasma pneumoniae*. Ann. N. Y. Acad. Sci. *143*, 682–690.

Stingele, F., Neeser, J.-R., and Mollet, B. (1996). Identification and characterization of the *eps* (exopolysaccharide) gene cluster from *Streptococcus thermophilus* Sfi6. J. Bacteriol. *178*, 1680–1690.

Strohmeier, G.R., and Fenton, M.J. (1999). Roles of lipoarabinomannan in the pathogenesis of tuberculosis. Microbes Infect. *1*, 709–717.

Szymanski, C.M., and Wren, B.W. (2005). Protein glycosylation in bacterial mucosal pathogens. Nat. Rev. Microbiol. 3, 225–237.

Tajima, M., Yagihashi, T., and Miki, Y. (1982). Capsular material of *Mycoplasma gallisepticum* and its possible relevance to the pathogenic process. Infect. Immun. *36*, 830–833.

Tajima, M., Yagihashi, T., and Nunoya, T. (1985). Ultrastructure of mycoplasmal capsules as revealed by stabilization with antiserum and staining with ruthenium red. Jpn. J. Vet. Sci. *47*, 217–223.

Tang, F.F., Wei, H., McWhirter, D., and Edgar, J. (1935). An investigation of the causal agent of bovine pleuropneumonia. J. Pathol. Bacteriol. *40*, 391–406.

Taylor-Robinson, D., Furr, P.M., Davies, H.A., Manchee, R.J., Mouches, C., and Bove, J.M. (1981). Mycoplasmal adherence with particular reference to the pathogenicity of *Mycoplasma pulmonis*. Isr. J. Med. Sci. *17*, 599–603.

Thoden, J.B., and Holden, H.M. (2007a). Active site geometry of glucose-1-phosphate uridylyltransferase. Protein Sci. *16*, 1379–1388.

Thoden, J.B., and Holden, H.M. (2007b). The molecular architecture of glucose-1-phosphate uridylyltransferase. Protein Sci. *16*, 432–440.

Trevelyan, W.E., Forrest, R.S., and Harrison, J.S. (1952). Determination of yeast carbohydrates with the anthrone reagent. Nature *170*, 626–627.

Upreti, R.K., Kumar, M., and Shankar, V. (2003). Bacterial glycoproteins: functions, biosynthesis and applications. Proteomics *3*, 363–379.

van Kranenburg, R., van Swam, I.I., Marugg, J.D., Kleerebezem, M., and de Vos, W.M. (1999). Exopolysaccharide biosynthesis in *Lactococcus lactis* NIZO B40: functional analysis of the glycosyltransferase genes involved in synthesis of the polysaccharide backbone. J. Bacteriol. *181*, 338–340.

Watt, J.P., Wolfson, L.J., O'Brien, K.L., Henkle, E., Deloria-Knoll, M., McCall, N., Lee, E., Levine, O.S., Hajjeh, R., Mulholland, K., *et al.* (2009). Burden of disease caused by *Haemophilus influenzae* type b in children younger than 5 years: global estimates. Lancet *374*, 903–911.

Wilson, W.H., and Collier, A.M. (1976). Ultrastructural study of *Mycoplasma pneumoniae* in organ culture. J. Bacteriol. *125*, 332–339.

Wood, W.B., Jr., and Smith, M.R. (1949). Inhibition of surface phagocytosis by the capsular 'slime layer' of pneumococcus type III. J. Exp. Med. *90*, 85–96.

Zähringer, U., Wagner, F., Reitschel, E.T., Ben-Menachem, G., Deutsch, J., and Rottem, S. (1997). Primary structure of a new phosphocholine-containing glycoglycerolipid of *Mycoplasma fermentans*. J. Biol. Chem. *272*, 26262–26270.

# Glycosidase Activity in Mollicutes

# 8

Meghan May and Daniel R. Brown

## Abstract

Non-metabolic manipulation of carbohydrates is associated with pathogenicity in many bacteria, but was historically thought to be absent or very rare among the Mollicutes. The sequencing of numerous mycoplasma genomes has allowed the recognition of several features previously unrecognised or under-reported in mycoplasmas. As a result, glycosidase genes that can be phenotypically validated have been identified in numerous species. The characterization of these enzymes and the genes encoding them has led to a greater understanding of the ecology of the organisms and of horizontal gene transfer between Mycoplasma species. Such studies have also highlighted an area with potential for novel intervention strategies in the treatment and/or prevention of mycoplasmosis. Here we present a review of the known mycoplasmal glycosidases describing their biological characteristics, putative evolutionary origins and potential roles in pathogenicity.

## Introduction

Glycosidases are enzymes that hydrolyse glycosidic linkages in carbohydrate polymers. Their principal function is the scavenging of carbohydrate monomers from large, complex sugars for use in energy generation. The universal nature of this function makes glycosidases, as a superfamily of enzymes, very common in nature. More specialized functions include roles in cellular turnover, cell signal regulation, innate antibacterial defences, synthesis of extracellular structures such as capsules or biofilms, and in the virulence of many pathogens (reviewed by Henrissat and Romeu, 1995).

Glycosidase activity in mycoplasmas is not universal, but is not as uncommon as once thought. The specific activities sialidase (neuraminidase), hyaluronidase, β-galactosidase, *N*-β-hexosaminidase and β-glucosidase are associated with virulence in certain pathogens, and have been reported in various *Mycoplasma* species. Sialidase activity is involved in bacterial colonization and dissemination, extracellular matrix (ECM) degradation and induced host-cell death (Corfield, 1992; Vimr and Lichtensteiger, 2004; King *et al.*, 2006; Hunt and Brown, 2007). The role of bacterial hyaluronidases in pathogenicity and sterile inflammation has been documented (Horton *et al.*, 1998, 1999; Knudson *et al.*, 2000 Termeer *et al.*, 2002; Starr *et al.*, 2006). *N*-β-hexosaminidase can potentially modulate attachment and dispersion in biofilms produced by several Gram-positive and Gram-negative species (Manuel *et al.*, 2007). The β-glucosidases are associated with nutritional fitness and prolonged bacteraemia, and with invasion of host cells (Vilei and Frey, 2007; Epel, 2009), while β-galactosidase activity has been shown to play a role in bacterial adherence (Limoli *et al.*, 2007). The relevance of α-mannosidase to bacterial virulence is unclear, but genes encoding this enzyme are almost exclusively found in pathogenic bacteria, rather than in commensals (Suits, 2010). Finally, it has also been proposed that the bacterial deglycosylation of host glycoconjugates that can be achieved through the cooperative activity of these enzymes can lead to highly invasive disease associated with marked tissue destruction (i.e. flesh-eating disease), or could expose or form new host antigens that may play a role in autoimmune complications of infection (Biberfeld, 1979; Matsushita and Okabe, 2001; King *et al.*, 2006).

## Virulence-associated glycosidases in Mycoplasmas

### Sialidase

Exo-α-sialidases (EC 3.2.1.18) catalyse hydrolysis of α-(2–3)-, α-(2–6)-, and/or α-(2–8)-glycosidic linkages of terminal sialic acid residues on oligosaccharides, glycoproteins, glycolipids, colominic acid (a homopolymer of N-acetylneuraminic acid) and synthetic substrates. Synonyms for sialidase include neuraminidase, α-neuraminidase and N-acetylneuraminate glycohydrolase. The term sialic acid is the family name covering all derivatives of neuraminic acid (Blix *et al.*, 1957), the aldol condensation product of D-mannosamine and pyruvic acid, which are potential bacterial nutrients. In vertebrates, diverse sialic acid derivatives are involved in recognition processes, cellular connections with extracellular matrix (ECM) components and intercellular interactions (Achyuthan and Achyuthan, 2001). They protect against hydrolysis of the glycosidic or peptide bonds of oligosaccharides, glycoproteins, glycolipids and gangliosides located on eukaryotic cell surfaces, and against degradation of the ECM. In addition, sialylated lipopolysaccharide and polysialic acid capsules are surface features of certain Gram-negative and Gram-positive bacteria (Severi *et al.*, 2007). Most bacterial sialidases preferentially cleave α-(2–3)-linked sialic acids, and are found in species that live in close contact with vertebrate host cells as commensals or facultative pathogens. In bacterial genomes sialidase genes are often part of a locus encoding additional enzymes that enable the import and intracellular catabolism of free sialic acid. This pathway culminates with the production of fructose-6-phosphate for entry into glycolysis (Vimr and Lichtensteiger, 2004), constituting a bacterial nutrient stream that might be essential to offset any selective disadvantage of increased virulence attributable to desialylation of host glycoconjugates (Corfield, 1992). A canonical sialic acid scavenging and degradation pathway typically includes the enzymes N-acetylneuraminate lyase (NanA), N-acetylmannosamine kinase (NagC), N-acetylmannosamine-6-phosphate epimerase (NanE), N-acetylglucosamine-6-phosphate deacetylase (NagA), and glucosamine-6-phosphate deaminase (NagB) (Fig. 8.1). Certain species also produce specialized transporters (e.g. NanT) to import liberated sialic acid.

The first description of sialidase activity in a mycoplasma was published in 1967, and two detailed characterizations of this activity were described almost simultaneously in 1972 (Roberts, 1967; Mueller and Sethi, 1972; Sethi and Mueller, 1972). These reports described activity in two strains of the avian pathogen *Mycoplasma gallisepticum*, while subsequent reports indicated a lack of sialidase activity in other strains (Glasgow and Hill, 1980; Kahane *et al.*, 1990). Sequencing of the *M. gallisepticum* Strain R genome revealed the presence of a putative sialidase gene, MGA_0329 (Papazisi *et al.*, 2003). The functionality of this gene was substantiated by the loss of a sialidase-positive phenotype following its transposon-mediated disruption, and the restoration of activity by complementation of MGA_0329 (May *et al.*, 2012). The presence of this gene in *M. gallisepticum* occurs in isolation; that is, a sialic acid degradation pathway does not seem to be encoded by the organism. The complete genome of the poultry pathogen *Mycoplasma synoviae* (Strain 53) indicated that this species has a homologue of MGA_0329, MS53_0199. Unlike *M. gallisepticum*, *M. synoviae* encodes a canonical sialic acid degradation pathway that would enable the use of free sialic acid as an energy source (Vasconcelos *et al.*, 2005). Quantitative measurements of sialidase activity among strains of *M. synoviae* indicate that there are statistically significant level differences between strains, and that the level of activity correlates significantly with strain virulence. The nucleotide and of the genes encoding these enzymes and the predicted protein sequences of the sialidases they produce are highly variable across strains (May *et al.*, 2007, May and Brown, 2008).

Until recently, the hypervirulent species Mycoplasma alligatoris A21JP2$^{T}$ was the only

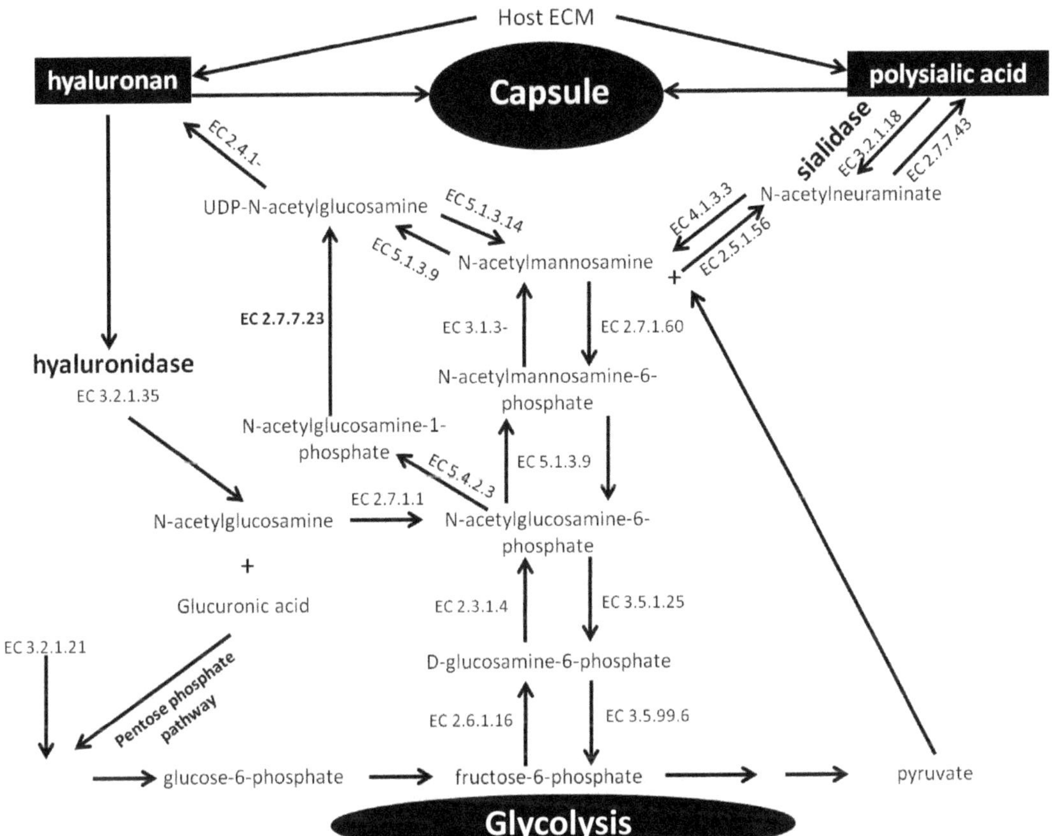


**Figure 8.1** Glycan degradation and capsule biosynthetic pathways. Common catabolic pathways for the degradation and entry into glycolysis of sialic acid, hyaluronic acid, and β-D-glycosides and pathways for the synthesis of hyaluronan and polysialic acid (i.e. capsular polysaccharides) are depicted. The majority of the biochemical intermediates are common to the two processes. The enzymes to carry out these pathways can be found in some, but not all, Mollicutes species.

extant isolate in which sialidase activity had been characterized. The activity is congruous with the invasiveness of *M. alligatoris*, which causes rapidly fatal disease in susceptible hosts, in contrast to the chronic epithelial lesions observed during typical mycoplasmal infections (Baseman and Tully, 1997). Sequencing of the *M. alligatoris* A21JP2 genome revealed the presence of two orthologous sialidase genes. Transposon mutagenesis studies indicated that disrupting a single sialidase gene does not abolish the sialidase-positive phenotype, suggesting that both genes produce functional sialidase proteins under standard conditions. Deglycosylation of digoxigenin-labelled lectins indicated that these sialidases are specific for α-(2 3)-linkages, and do not cleave α-(2–6)-linkages (our unpublished data). Sequencing of the closely related species *Mycoplasma crocodyli* MP145 genome indicated that a key difference between these species is the presence of sialidase in *M. alligatoris* and its absence from *M. crocodyli* (Brown *et al.*, 2011). This finding is potentially very meaningful, as *M. crocodyli* tends to cause disease with a classical course for mycoplasmosis, whereas *M. alligatoris* has a degree of virulence that is arguably unprecedented in the genus (Brown *et al.*, 2004, 2011).

Phenotypic surveys for sialidase activity have been conducted in other avian mycoplasmas. Sialidase was detected in some strains of the turkey pathogens *Mycoplasma meleagridis* and *Mycoplasma iowae*, but was completely absent in others. The single strains of the goose commensal *Mycoplasma anseris* and of the galliform pathogens

*Mycoplasma pullorum* and *Mycoplasma cloacale* that were tested had weak sialidase activity. Conversely, the single strain of the vulture pathogen *Mycoplasma corogypsi* examined had very potent sialidase activity (Bercic *et al*, 2008). The genes encoding the sialidases of these species have yet to be identified. Similarly, a phenotypic survey of eleven *Mycoplasma* species associated with infection of canids revealed secreted sialidase activity in *Mycoplasma canis, Mycoplasma cynos* and *Mycoplasma molare*. As observed with *M. synoviae*, significant differences in sialidase activity were detected between strains of *M. canis* (May and Brown, 2009).

Cell-associated sialidases have been documented in *M. alligatoris* strain A21JP2$^T$, virulent strains of *M. gallisepticum* and *M. synoviae*, certain strains of *M. meleagridis, M. iowae* and *M. pullorum, M. anseris, M. cloacale* and *M. corogypsi* (Brown *et al.*, 2004; Berčič *et al.*, 2008; May and Brown, 2008). Secreted sialidases have been observed in *M. canis, M. cynos* and *M. molare*. Like the sialidase of *M. alligatoris*, the secreted sialidase of *M. canis* cleaves α-(2–3)- but not α-(2–6)-linkages (our unpublished data). Though the activity has been described for eleven species, this represents a minority of known mycoplasmas and, as such sialidase activity is still considered anomalous for the genus at this time. The distribution of sialidase in the genus Mycoplasma has not been fully elucidated. In addition, the demonstration of strains lacking detectable sialidase activity underscores the importance of assessing multiple strains of each species to thoroughly investigate the phenomenon.

## β-galactosidase

β-Galactosidase (EC 3.2.1.23) hydrolyses terminal β-D-galactosides attached to glycoproteins, glycosaminoglycans and synthetic substrates to D-galactose and alcohol, and degrades lactose into free galactose and glucose. Synonyms include exo-(1,4)- β-D-galactanase and lactase, and the commonly used gene symbol is *lacZ*. Most bacteria producing β-galactosidase utilize it for energy generation via lactose catabolism. In bacterial pathogens β-galactosidase tends to be associated with a capacity for host cell deglycosylation (King *et al.*, 2006; Burnaugh *et al*, 2008).

Five mycoplasma species have been found to have β-galactosidase activity. Kahane *et al.* measured moderate levels of enzymatic activity in *Mycoplasma pneumoniae* strain M129 and *Mycoplasma hominis* ATCC15056, and low levels of activity in *M. gallisepticum* strain A5969 and *Mycoplasma capricolum* California kid. Curiously, sequencing of the *M. pneumoniae* M129 and the *M. capricolum* California kid genomes did not reveal the presence of a *lacZ* gene in the primary annotation or the reannotation (Himmelreich *et al*, 1996; Dandekar *et al*, 2000; Glass *et al*, 2005). The source of these activities remains cryptic. Similarly, the genomes of *M. gallisepticum* strain R and *M. hominis* strain PG21 do not encode recognizable *lacZ* genes. However, the activity described by Kahane *et al.* could be the result of interstrain diversity. A phenotypic screen for β-galactosidase activity including in these four species did not detect any activity (our unpublished data), although this too could be attributed to differences between strains.

Phenotypic and genomic data are in complete agreement for *M. alligatoris* A21JP2. Washed cells or whole-cell antigen both cleave the synthetic substrate 4-methylumbelliferyl N-acetyl-β-D-galactosaminide (MUG), strongly indicating endogenous β-galactosidase activity (Ortiz *et al.*, 2007). However the chromogenic substrate 5-bromo-4-chloro-3-indolyl-beta-D-galactopyranoside (X-Gal) is not cleaved. The A21JP2 genome encodes two putative *lacZ* genes, supporting the phenotypic observation of cleavage of MUG. Conversely, the *M. crocodyli* MP145 genome encodes a putative *lacZ*, yet washed cells do not exhibit β-galactosidase activity as measured by cleavage of MUG or X-gal (Brown *et al.*, 2011). These data indicate that the aglycone linkage may affect the ability to detect endogenous LacZ, and thus negative phenotypes could be considered presumptive in the absence of genotypic data. The distribution of β-galactosidase activity in mycoplasmas is somewhat unclear, but it is clear that this enzyme occurs in a very small minority of species. Further characterization of sequence motifs responsible for cryptic β-galactosidase activity or the inclusion of additional synthetic substrates in phenotypic screens may alter this perception.

## Hyaluronidase

Bacterial hyaluronidases (EC 3.2.1.35) are endo-*N*-acetylhexosaminidases that hydrolyse glycosidic linkages between *N*-acetyl-β-D-glucosamine and D-glucuronate in hyaluronan or linkages between N-acetyl-galactosamine (or N-acetylgalactosamine sulfate) and glucuronic acid in chondroitin (or chondroitin sulfate). Synonyms include hyaluronoglucosaminidase, hyaluronoglucosidase, chondroitinase and chondroitinase I. The substrate hyaluronan is a polymer of *N*-acetylglucosamine and glucuronic acid disaccharide repeats, which can serve as a source of bacterial nutrients. Hyaluronidase genes (*nagH*) are often accompanied by genes encoding glucuronyl hydrolase (*ugl*). Ugl enzymes catalyse the release of glucuronic acid from the disaccharides cleaved from hyaluronan, liberating *N*-acetylglucosamine for conversion to fructose-6-phosphate and entry into glycolysis. Hyaluronan is found in many animal tissues and fluids, and is a major component of the vertebrate ECM. The chemical composition of glycosaminoglycan polymers such as hyaluronan and chondroitin is highly variable, even between tissues within a vertebrate species, with these heterogeneities including differences in isomerization, acetylation, phosphorylation and sulfation (Ernst *et al*., 1995).

The lack of a quantitative and consistently detectable assay for hyaluronidase activity has severely limited characterization of this enzyme in mycoplasmas. Turbidity-based assays used in other bacteria are principally qualitative, and are of such low sensitivity that they are not practical for thorough investigations. Turbidity-based assays were used to confirm the presence of hyaluronidase in *M. alligatoris* A21JP2 and *M. crocodyli* MP145, though the activity bordered on the limits of detection of the assay (Brown *et al*., 2004). Early genome surveys indicated that *nagH* genes were present in both species, and complete genome sequencing confirmed the presence of a single copy of *nagH* in *M. crocodyli* and five orthologous copies in *M. alligatoris* (Brown *et al*., 2011). Though functional analysis has not been performed, the genome of the tortoise commensal *Mycoplasma testudinis* 01008-A1 appears to encode four orthologous copies of *nagH* (Balish *et al*., 2010).

## *N*-acetyl-β-hexosaminidase

*N*-acetyl-β-hexosaminidases (EC 3.2.1.52) hydrolyse terminal *N*-acetyl-D-hexosamine residues in *N*-acetyl-β-D-hexosaminides. This allows for the liberation of, most notably, *N*-acetylglucosamine (GlcNAc) and *N*-acetylgalactosamine (GalNAc), which can be used as nutritional substrates. The common gene name is *nagZ*, and synonymous names are hexosaminidase, β-acetylaminodeoxyhexosidase, *N*-acetyl-β-D-hexosaminidase, β-hexosaminidase, β-D-*N*-acetylhexosaminidase, β-*N*-acetyl-D-hexosaminidase, β-*N*-acetylglucosaminidase and *N*-acetylhexosaminidase, β-D-hexosaminidase. Hypothesized nutritional use is often supported by the presence of GlcNAc-specific transporters (e.g. *nagE*).

Phenotypic screening for hexosaminidase indicated that the enzyme was produced by *M. pneumoniae* strain M129, *M. hominis* ATCC15056, *M. gallisepticum* strain A5969, *M. capricolum* California kid and *Acholeplasma laidlawii* oral strain (Kahane *et al*., 1990). As was the case with β-galactosidase, there was no obvious *nagZ* gene in genomes of *M. pneumoniae* strain M129, *M. hominis* PG21, *M. gallisepticum* strain R or *M. capricolum* California kid. Additional phenotypic screening of *M. pneumoniae* strain PI1428, *M. hominis* ATCC15056, *M. gallisepticum* strains R, S6, A5969, JR-67 and F, *M. capricolum* California kid and *Acholeplasma laidlawii* oral strain utilizing the synthetic substrate 4-methylumbelliferyl N-acetyl-β-D-glucosaminide (MUGlcNAc) did not reveal any hexosaminidase activity.

A more recent phenotypic screen of several mycoplasma species using washed whole cells and MUGlcNAc demonstrated hexosaminidase activity in *M. alligatoris* A21JP2, *M. crocodyli* MP145, *Mycoplasma mycoides* subspecies *capri* strain GM684–13, *Mycoplasma anatis* strain 1340 and several strains of *M. canis*. The amount of hexosaminidase varied significantly across species, and varied significantly between strains of *M. canis*. Two of the ten isolates examined had no detectable activity, underscoring the need to assess multiple strains during phenotypic glycosidase screens. *NagZ* genes have yet to be identified in *M. mycoides* subspecies *capri*, or *M. anatis* due to a lack of available genome data, but were readily identified in *M. alligatoris, M. crocodyli*, and *M. canis*. Two tandem, paralogous copies of *nagZ* sit downstream of a putative N-acetylhexosamine-binding protein in *M. crocodyli*, composing a putative hexosamine degradation locus. The genome of *M. alligatoris* has only a single copy of *nagZ*. The gene seems to be a transposon 'hot spot' (Brown *et al.*, 2011; Brown *et al.*, 2012).

## β-Glucosidase

β-Glucosidases (EC 3.2.1.21) hydrolyse terminal β-D-glycosyl groups and result in the release of glucose or other monosaccharides. A synonym is β-glucoside glucohydrolase, and the standard bacterial gene name is *bglX*. The enzyme is typically used to liberate glucose for use in glycolysis.

Only a few *Mycoplasma* species have been shown to have β-glucosidase activity, but, to our knowledge, there has not been the extensive screening of isolates required to comment on its rarity in the genus. Bradbury demonstrated the presence of qualitatively varying levels of activity in several strains of *Acholeplasma axanthum* and *A. laidlawii*, the latter being independently confirmed by Kahane *et al.* Phenotypic screening also indicated the presence of β-glucosidase in *M. gallisepticum, M. hominis* and *M. capricolum* (Bradbury, 1977, Kahane *et al.*, 1990, Henrikson and Smith, 1964), but, no identifiable *bglX* gene can be found in their genomes.

Recent studies focussed on β-glucosidase have described nucleotide sequence polymorphisms in the *bgl* gene of *Mycoplasma mycoides* subspecies *mycoides* (formerly known as 'small colony' variant), the agent of contagious bovine pleuropneumonia (CBPP). The impact of this sequence diversity on β-glucosidase activity is not known. *Bgl* genes are present in the goat pathogen *M. mycoides* subsp. *capri* GM12 (formerly known as *Mycoplasma mycoides* subspecies *mycoides* 'large colony' variant), the bovine pathogen *M. leachii*, the putative human pathogens *Mycoplasma penetrans* HF-2 and *Mycoplasma fermentans* PG18, and the floral commensal *Mesoplasma florum* L1. While all of these genes contain a complete copy of the functional motif for β-glucosidase, no phenotypic confirmation of enzymatic activity has been performed. A *bglX* gene is also present in 'Mollicutes bacterium D7', an uncharacterized isolate from an intestinal biopsy of a patient with Crohn's disease, but it will not be discussed further because the isolate is not a valid member of the Mollicutes, further analysis of the rRNA sequence indicating that it is a member of the *Firmicutes* (genus *Coprobacillus*).

## α-mannosidase

The enzyme α-mannosidase (EC 3.2.1.24) hydrolyses terminal α-D-mannose in α-D-mannosides and results in the release of free mannan. Synonyms include α-D-mannoside mannohydrolase, α-D-mannosidase, α-D-mannopyranosidase and exo-α-mannosidase. At this time, there is no consensus on a standard gene name for bacterial α-mannosidases.

The confirmed presence of α-mannosidases in the genus *Mycoplasma* has yet to be reported in the published literature. Hypothetical open reading frames in *M. leachii* PG50, *M. mycoides* subspecies *mycoides* PG1 and Gladysdale, and *M. mycoides capri* GM112 encode proteins with the appropriate functional motifs to act as α-mannosidases, but phenotypic confirmation of the activity has not been performed in these species. In the context of examining specific lectin-binding activities and glycosidic linkage specificities for *M. alligatoris* and *M. canis*, potent N-linked α-mannosidase activity was observed for *M. canis* PG14, but not *M. alligatoris* (our unpublished data). This is the first and only phenotypically confirmed α-mannosidase activity in the genus, but an extensive phenotypic analysis has not been carried out across species.

# Putative roles in cellular biology or disease processes of mycoplasmas

While most bacteria utilize glycosidases to liberate nutritional substrates, there is growing evidence that this is not necessarily true of mycoplasmas. The presence of a sialidase gene in *M. gallispeticum* in the absence of any other genes mediating sialic acid catabolism suggests that the organism has retained the gene for an alternative, non-nutritive purpose. Though *M. synoviae* has retained these genes, at least two strains have been described in which the *nanA* gene contains large, frame-shifting deletions, ultimately resulting in premature truncation of the protein. Because NanA is required for the continuation of sialic acid catabolism (Fig. 8.1), its loss without a corresponding effect on growth rate is suggestive of the redundancy of the pathway (May *et al.*, 2007; May and Brown, 2008). The utilization of any glycosidase for nutritional fitness *in vivo* is unknown.

The use of sialidase as a virulence mechanism by *M. alligatoris* and *M. gallisepticum* have been examined most extensively. The association with virulence was first suspected for *M. alligatoris* following a genome-wide survey to identify genes uncommon or unique among mycoplasmas in an effort to explain the organism's remarkable pathogenicity. Serial deglycosylation of host cells initiated by sialidase is very consistent with the highly invasive disease caused by infection with *M. alligatoris*. Infection of alligator fibroblasts with *M. alligatoris* induces apoptosis and addition of the sialidase inhibitor 2-deoxy-2,3-didehydro-*N*-acetylneuraminic acid inhibits this effect, implicating the enzyme in the induction of cell death (Hunt and Brown, 2005, 2007). In contrast to *M. synoviae, in vitro* growth studies with *M. alligatoris* indicate that its glycosidases likely play a minor role in energy generation. Growth on sialic acid as a primary carbon source does not significantly alter *in vitro* growth rates and *nanI* knockout mutants do not have a significantly different growth rate to that of wild-type A21JP2. Isogenic mutants lacking *N*-acetylneuriminate lyase (NanA) and gluconoryl hydrolase (Ugl), which represent the first essential steps in sialic acid and hyaluronic acid scavenging, respectively, do not have a reduced growth rate compared to wild type when free glucose is provided. When glucose is depleted, but *N*-acetylneuraminate is provided, both wild type and NanA knockout cells grow poorly, but NanA knockouts are more growth-deficient than wild-type *M. alligatoris* or irrelevant mutants. Ugl mutants and wild-type cells grew at similar rates in glucose-depleted medium, indicating that free glucoronic acid and N-acetylglucosamine are not important growth substrates. Paradoxically, Ugl mutant cells and wild-type cells cultured in glucose-depleted medium supplemented with free glucuronic acid and *N*-acetylglucosamine both demonstrate increased nutritional fitness. These findings indicate that the ability of *M. alligatoris* to scavenge sugars via the actions of extracellular sialidase and hyaluronidase likely makes a modest contribution to intracellular carbon and energy flux when free glucose is limiting, in addition to non-nutritional roles of these enzymes in the overall fitness of the organism (our unpublished data).

The majority of research on the virulence of *M. synoviae* has focused on cytadherence. An analysis by Lockaby *et al.* (1999) examined the relationship between virulence of *M. synoviae* isolates and their cytadherence capabilities and concluded that there are likely to be additional virulence factors, because cytoadherence did not completely correlate with pathogenicity. However, sialidase activity in *M. synoviae* is correlated with strain virulence, suggesting a biological relationship between sialidase and pathogenicity. Furthermore, quantitative measurements of haemagglutination vary between strains of *M. synoviae* and have a significant ($P < 0.001$) correlation with sialidase activity. This indicates that there is a proportional relationship between attachment to sialic acid residues, as measured by haemagglutination, and detachment from host cell surfaces, as measured by sialic acid cleavage (May and Brown, 2011).

Because a reliable method for gene disruption has not been developed for *M. synoviae*, the effect of sialidase on virulence was assessed for *M. gallisepticum*. Variants of strain R with a disrupted copy of MGA_0329 were recovered less frequently from infected chickens, and caused tracheal lesions that were significantly less severe than those caused by wild-type strain R (May *et al.*, 2012; May and Brown, 2008). The role of sialidase in the virulence of canine or other avian mycoplasmas is unknown. The sialidase-positive *M. cynos* and *M. canis* are the species most consistently associated with canine mycoplasmosis (Chalker, 2004), but to date no specific virulence factor had been proposed as a basis of their pathogenicity. Though *M. molare* is not a definitive canine pathogen, its ability to secrete sialidase in the urogenital tract may allow it also to act as a primary pathogen or as a co-factor in disease.

Hyaluronan degradation increases the permeability of connective tissues and decreases the viscosity of body fluids, facilitating the spread of bacteria and their toxins, and may promote direct contact between bacteria and host cell surfaces (Brown *et al.*, 2004). Depolymerization of hyaluronan disrupts the role of the ECM as a reservoir of cytokines and enzymes involved in signal transduction (Brown *et al.*, 2004). These effects are all consistent with the pathological effects of infection with *M. alligatoris* and *M. crocodyli*.

The role of bacterial hyaluronidases in sterile inflammation has been documented. This immunopathological response is often precipitated by induction of the transcriptional regulator NFκB. Expression of the *NFκB1* gene can be induced by myeloid differentiation factor 88 (MyD88), and often results in the downstream induction of proinflammatory cytokines. Signalling through Toll-like receptors 2 or 4 (TLR2, TLR4) can drive MyD88 translocation to the nucleus, where it can mediate up-regulation of *NFκB1*. Infection of alligator fibroblasts with *M. alligatoris* resulted in the translocation of MyD88 from the cytoplasm to the nucleus and the production of phosphorylated NFκB. Treatment of uninfected fibroblasts with synthetic hyaluronan fragments induced the nuclear translocation of MyD88 at levels near that of *M. alligatoris*-infected fibroblasts, suggesting that fragmented hyaluronan generated by mycoplasmal hyaluronidase may be the TLR agonist associated with this strong, proinflammatory signal (our unpublished data).

The roles of β-galactosidase and α-mannosidase in the virulence of mycoplasmas are uncharacterized. Because the hypervirulent *M. alligatoris* is the only species to have solidly demonstrated activity and a corresponding *lacZ* gene, it is tempting to speculate that β-galactosidase has a unique role that contributes to the unusual virulence phenotype of the organism. The complete destruction of host ECM that can be accomplished by serial deglycosylation to produce flesh-eating disease would require the presence of β-galactosidase (King *et al.*, 2006). Thus, the pathology associated with *M. alligatoris* infection is consistent with the utilization of β-galactosidase, in collaboration with all of its other glycosidases, for virulence (Pye *et al.*, 2001).

The role of β-hexosaminidases in the biology and the virulence of mycoplasmas is similarly uncharacterised. Biofilm formation by mycoplasmas has been described and characterized in recent years, and seems a plausible factor to consider, because β-hexosaminidases are involved in biofilm escape in other bacteria. An alternative hypothesis is that β-hexosaminidase serves as a bacteriocin, because it could theoretically target peptidoglycan linkages in walled bacteria, and thus its presence would not be detrimental to mycoplasmas.

Sequence diversity in the *bgl* (β-glucosidase) gene of *M. mycoides* subspecies *mycoides* has been observed between strains with different degrees of pathogenicity. A single substitution of alanine for valine at amino acid residue 204 is characteristic of attenuated strains, indicating the potential for the $Val_{204}$ allele to contribute to the disease process. The β-glucosidase activity was not associated with the release of $H_2O_2$ or any defined toxic metabolite. Co-incubation of embryonic bovine lung cells with virulent strains of *M. mycoides* subspecies *mycoides* possessing $Val_{204}$ and exogenous disaccharides resulted in rapid cell death, whereas co-incubation of host cells with attenuated strains of *M. mycoides* subspecies *mycoides* possessing $Ala_{204}$ and the same sugars, or with $Val_{204}$ strains in the absence of exogenous sugars, did not. Paradoxically, strains possessing the $Val_{204}$ allele had lower β-glucosidase activity, suggesting that additional virulence factors may be under catabolite repression. The cytotoxic effect of exogenous sugars may be the result of toxic metabolites that are produced downstream of β-glucosidase activity that attenuated strains are defective in, of the induction of unrecognized virulence factors in *M. mycoides* subspecies *mycoides* that are similarly defective in attenuated strains, or of a currently unrecognized glycosidase activity. Strains with the $Val_{204}$ allele were also found to have higher rates of survivability and persistence. At present, it appears that β-glucosidase may play a role in the disease process by prolonging bacteraemia (Vilei *et al.*, 2004; Vilei and Frey, 2007).

Though specific linkages between the activities have not been experimentally demonstrated in mycoplasmas, it is appropriate and important to discuss the connection between glycosidase

**Table 8.1** Glycosidase activity and capsule formation of Mycoplasmas

| Mycoplasma Species[a] | Glycosidase activity | Genomic evidence for capsule synthesis and carbohydrate catabolism overlap |
|---|---|---|
| *M. mycoides* subspecies *mycoides* | β-Glucosidase | *wcaA, nanE, nagC, nagB, nagA* |
| *M. mycoides* subspecies *capri* | N-β-Hexosaminidase | *wcaA, nanE, nagC, nagB, nagA* |
| *M. bovigenitalium* | ND[c] | NS[c] |
| *M. pulmonis* | ND[c] | *nanE, nagC, nagB, nagA* |
| *M. meleagridis* | Sialidase | NS[c] |
| *M. synoviae*[b] | Sialidase | *nanA, nanE, nagC, nagB, nagA* |

[a]*Mycoplasma dispar, Mycoplasma glycophilum* and *Mycoplasma ovipneumoniae* are also reported to form capsules or extracellular layers, but were excluded for lack of glycosidase screening and genomic data. [b]*M. synoviae* forms an 'extracellular layer'. [c]ND: none detected; NS: no sequence data available.

activity and the formation of polysaccharide capsules. Some bacterial capsules are composed of glycosaminoglycans (e.g. polysialic acid or hyaluronic acid capsules), which can be synthesized by bacterial cells or are pillaged from the ECM of host cells by glycosaminoglycan-liberating enzymes such as hyaluronidases (DeAngelis, 2002). More importantly, many of the same intermediates of carbohydrate metabolism initiated by glycosidases are employed in capsule biosynthesis, such that the two pathways intersect on numerous occasions (Fig. 8.1). A comparison between *Mycoplasma* species with known glycosidase activity and known capsule formation demonstrates considerable overlap (Table 8.1). Further analysis of the capsular polysaccharide composition of each species is required in order to determine if the connection to glycosidase activity is essential, but there is great potential for synergy between these activities.

## Acquisition, diversity and evolutionary fitness

The sporadic presence of glycosidase genes throughout the genus indicates their acquisition by lateral transfer, rather than vertical inheritance, and examination of sialidase, hyaluronidase and β-glucosidase genes provides evidence for this (Fig. 8.2A). Closely related species with the same glycosidase phenotype, in some cases, have highly diverse alleles that show greater similarity to the glycosidase genes of different species, and homologous glycosidase genes can be found in some distantly related species that share a common host habitat. In addition, certain species have glycosidase genes that are not present in their closest relatives. Taken together, these findings clearly indicate that the evolutionary origin of glycosidase genes is not from a common ancestor, but rather from contemporary, unrelated species.

This concept is most evident with sialidase genes (Fig. 8.2B). Six of the eleven known sialidase-positive species (*M. alligatoris, M. synoviae, M. cynos, M. canis, M. meleagridis* and *M. corogypsi*) are close phylogenetic relatives, and all fall within the *M. synoviae* cluster. Of the remaining species, *M. molare, M. gallisepticum, M. cloacale* and *M. pullorum* each share a common host habitat (domestic dogs and galliform birds, respectively) with a sialidase-positive species from the *M. synoviae* cluster. The remaining unrelated species, *M. anseris*, has only been found in geese, but *M. synoviae* and *M. pullorum* have both been isolated from geese, even though their primary habitat is galliform birds (Brown *et al.*, 2010). All known sialidase-positive species thus share a commonality of phylogeny or of host habitat.

Remarkably, those species from the *M. synoviae* cluster with extensively characterized sialidase activity (namely, *M. alligatoris, M. synoviae, M. canis* and *M. cynos*) do not necessarily share sialidase homologues. The *nanI* gene of *M. alligatoris* is highly homologous to *nanI* from *Clostridium perfringens*, and it is clear that the transfer of *nanI* to *M. alligatoris* was a relatively recent evolutionary event (Brown *et al.*, 2004). In contrast, the *nanI* allele(s) of *M. synoviae* and *M. gallisepticum* are

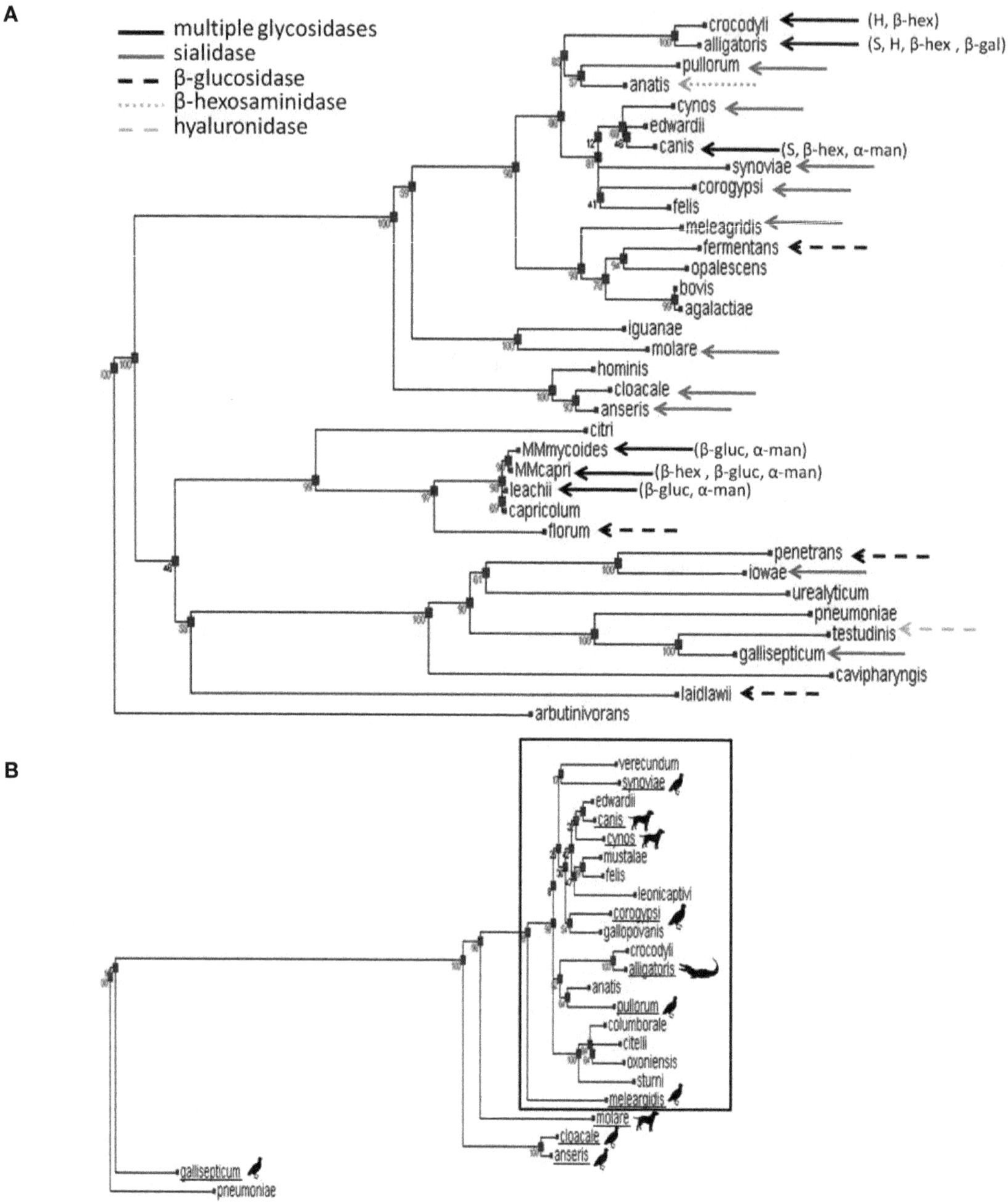


**Figure 8.2** Glycosidase distribution among Mollicutes species. (A) Arrows pointing to species names indicate the presence of glycosidase activity. The distribution of glycosidases (genotype and/or phenotype) does not follow phylogeny as depicted by neighbour-joining trees generated by the 16S rRNA gene. (B) The sialidase phenotype (underlined species) is heavily associated with the *M. synoviae* cluster (box) regardless of genotype. Sialidase activity in additional species is associated with common host habitats (indicated as avian, canine, or crocodilian) with species from the *M. synoviae* cluster.

so different from *M. alligatoris* and *C. perfringens* that either extensive evolutionary distancing has occurred, or, more likely, the *nanI* donor was a different, distantly related species that has yet to be sequenced. Regardless, it is very clear that these alleles have a distinct evolutionary origin from one another, despite the close phylogenetic relationship between their source species. Though a *nanI* gene has not been identified in *M. canis* or *M. cynos*, it is likely that these species share a common gene that is distinct from that of *M. alligatoris* and *M. synoviae*, because their sialidase is secreted rather than cell associated (May and Brown, 2009).

The importance of shared host habitat is evident from an examination of the sialidase gene and activity of *M. gallisepticum* and *M. molare*. The *nanI* gene from *M. gallisepticum* is highly homologous to that of *M. synoviae*, and clearly derives from the same evolutionary source. We and others have reported evidence that one species (probably *M. gallisepticum*) acquired the gene and then horizontally transferred it to the other (probably *M. synoviae*) during co-infection of gallinaceous birds (Vasconcelos *et al.*, 2005; May and Brown, 2008). It is likely that the sialidase gene expressed by *M. molare* was acquired in a similar fashion during mixed infections of domestic dogs because it has the unique aspect of secretion coupled with the shared host habitat. Although *M. molare* is an inhabitant of the urogenital tract and *M. cynos* is a pathogen of the respiratory tract, *M. canis* readily infects both body sites (Chalker, 2004). A scenario wherein either *M. molare* or *M. cynos* acquired a gene encoding the secreted sialidase and transferred the gene to *M. canis*, which subsequently transferred the gene to the other species in the distant body site, seems plausible.

The origins of the known mycoplasmal hyaluronidases are well understood. The five hyaluronidase genes (three *nagH* paralogues and two *nagK* paralogues) of *M. alligatoris* are highly homologous to those of *C. perfringens*, suggesting that the gene suite is likely to have been a recent acquisition from a *Clostridium* species. The *nagH* gene of *M. crocodyli* is the sole hyaluronidase gene in that species (Brown *et al.*, 2004). Three possibilities exist for its acquisition: a single lateral transfer event of only *nagH* from *M. alligatoris* to *M. crocodyli* during mixed infection of a crocodilian host; a single lateral transfer of only *nagH* from the donor *Clostridium* species directly to *M. crocodyli*; or acquisition of *nagH* before the two sibling species diverged from their common ancestor, with *M. alligatoris* subsequently acquiring the remaining *nag* genes. Similarly, a common *nagZ* (β-heximanidase) gene of clostridial origin is also shared by the two species (Brown *et al.*, 2011).

Despite sharing a common human host, the *bglH* gene of *M. penetrans* and the unannotated β-glucosidase gene of *M. fermentans* appear to have arisen from two independent transfer events. The *bglH* of *M. penetrans* appears to be derived from a *Bacillus* species, while the two paralogous copies of the β-glucosidase gene carried by *M. fermentans* share greatest similarity with that of *Listeria monocytogenes*. These genes are thus analogous rather than orthologous, and lateral transfer from one species to the other is unlikely to have occurred.

The *bgl* and α-mannosidase genes of *M. mycoides* subspecies *mycoides*, *M. mycoides* subspecies *capri* and *M. leachii* are linked both by ancestry and genomic location. The two genes are part of a seven gene locus in all three species that is not present in any other members of the class Mollicutes. The organization of the locus is conserved between *M. mycoides* subspecies *capri* and *M. leachii*, and all seven genes are part of a single operon. The operon is disrupted by a transposase in *M. mycoides* subspecies *mycoides* PG1 and Gladysdale (Fig. 8.3). Genes flanking this region are distinct in all three species.

In addition to β-glucosidase and α-mannosidase, the locus consists of a conserved ROK family sugar kinase and two fructose-specific phosphotransferase (PTS) system components, indicating a potential role in carbohydrate processing. The remaining two genes include a conserved hypothetical protein and a predicted lipoprotein unique to these organisms that has been predicted in *M. mycoides* subspecies *mycoides* to be a member of the *lpp* gene family involved in antigenic variation (Westberg *et al.*, 2004). This prediction was based on sequence similarity and on the presence of a polyadenosine (poly A) tract upstream of the promoter in strain PG1. Poly-A tracts are also present (but of differing length) in strain Gladysdale, *M. mycoides* subspecies *capri* and *M. leachii* (Lartigue *et al.*, 2009; Wise *et al.*, 2010a,b; our unpublished observations). Because the lipoprotein is the first gene in the operon in all three species, potential phase variation of β-glucosidase and α-mannosidase activity is predictable.

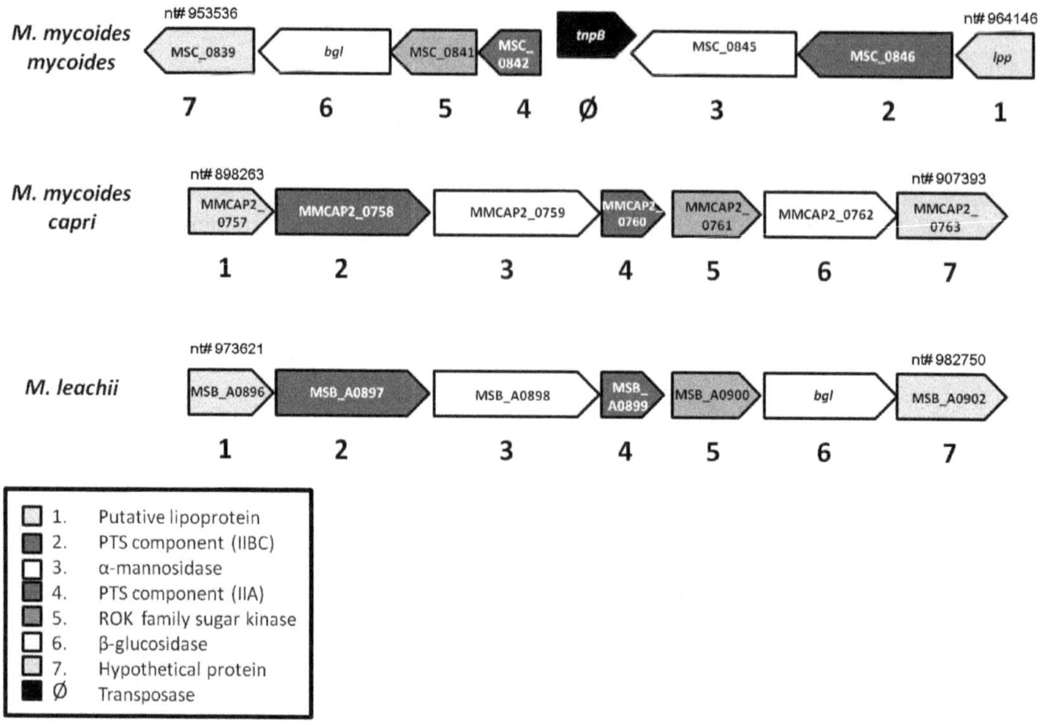


**Figure 8.3** The glycosidase locus of *M. mycoides mycoides* PG1, *M. mycoides capri* GM112, and *M. leachii* PG50. The β-glucosidase and α-mannosidase genes of these three species are part of a locus putatively associated with carbohydrate processing. All seven genes are homologous across species, and all display >95% identity with their counterparts. The gene order and orientation is strictly conserved between *M. mycoides capri*, and *M. leachii*, while the locus is interrupted by mobile element in *M. mycoides mycoides* PG1. The organization of the locus in *M. mycoides mycoides* strain Gladysdale (not shown) is identical to strain PG1.

The orthologous genes composing this locus show >95% similarity among the three species, and are clearly derived from the same source. Sequence identity indicates that the donor was almost certainly a low G+C Gram-positive (i.e. *Firmicutes*) species, but these specific genes in this arrangement are not currently found in the NCBI database. This may be reflective of the absence of the true donor's sequence, or subsequent sequence drift and/or genomic rearrangement following the lateral transfer. The sequences with the highest identity across the locus are *Clostridium* species, but significant similarity is also found in *Lactobacillus* and *Streptococcus* species. Interestingly, one of the PTS system protein genes and the ROK family kinase gene, which are located in tandem, have the greatest similarity to genes in members of the *Erysipelothrichaceae*, while none of the remaining genes have similarity to genes in those species. This is suggestive of three possible scenarios: the original *Mycoplasma* recipient obtained two distinct sets of genes, one from a *Clostridiaceae* species and one from an *Erysipelothrichaceae* species, and assembled the current operon; the original *Clostridiaceae* donor transferred only the PTS and ROK kinase genes to *Erysipelothrichaceae*, in addition to transferring the entire operon to a *Mycoplasma* species; or the original *Clostridiaceae* donor transferred the entire operon to both an ancestral *Mycoplasma* species and *Erysipelothrichaceae* species, and the *Erysipelotrichaceae* eliminated all but the PTS and ROK kinase genes.

*M. leachii* and *M. mycoides* subspecies *mycoides* share a host habitat (cattle) and *M. mycoides* subspecies *capri* has a distinct habitat (goats), yet the arrangement of this locus indicates less evolutionary distance between *M. mycoides* subspecies *capri* and *M. leachii*. It seems most plausible that the genes were acquired by the ancestral *M. mycoides* prior to the subspecies divergence between *capri* and *mycoides*, and that one subspecies subsequently transferred it to *M. leachii*. The probability of coincident infection would suggest the donor was *M. mycoides* subspecies *mycoides*, but the presence of the transposase would thus reflect a disruption following the transfer to *M. leachii*. Alternatively, it is probable that the capacity of these species to infect both small and large ruminants is greater than appreciated, because historically definitive distinction between members of the *mycoides* cluster has been difficult (Brown, 2010). Numerous reports have indicated the possible identification of *M. leachii* and *M. mycoides* subspecies *mycoides* in goats or sheep (Sharew, 2005, Atalaia, 1987, Smith, 1981) and the infection of cattle with *M. mycoides* subspecies *capri* (Sylla *et al.*, 2005; Stradaioli *et al.*, 1999), making the direction of transfer difficult to establish without extensive analysis of the sequence. Regardless, it is clear that this locus is xenogenic and unique to these three species within the *mycoides* clade and the genus at large. Because β-glucosidase, α-mannosidase and antigenic variation of lipoproteins are potentially associated with virulence, this locus can be considered a putative pathogenicity island, the second ever described for the genus (Jechlinger, 2004).

It is important to note the reported strains of *M. synoviae*, *M. canis*, *M. gallisepticum*, *M. meleagridis* and *M. iowae* with no measurable sialidase activity (Bercic *et al*, 2007; May and Brown, 2009; Kahane *et al.*, 1990) in the context of lateral gene transfer. These strains highlight the extreme plasticity of the glycosidase phenotype and the considerable limitation of examining a single strain. Glycosidase genes and sialidase genes, in particular, are extremely mobile (Roggentin *et al*, 1993), and the presence or absence of a glycosidase-positive phenotype in a single strain is not necessarily reflective of the species as a whole. This has potential relevance to strains that appear unusually invasive in an immunocompetent host, because the elevated virulence may reflect acquisition of a glycosidase gene that is not common across the species. Accordingly, an examination of over one hundred respiratory tract (i.e. non-invasive) isolates of *M. pneumoniae* indicated that all were sialidase-negative (May and Brown, 2011b). Examination of isolates from disseminated infection would be illuminating in this regard.

Extensive characterization of the diversity of glycosidase alleles and the impact of this diversity on evolutionary fitness has only been performed on the sialidase gene of *M. synoviae*. We explored the genetic basis for quantitative variation between strains in sialidase-specific activity. Marked heterogeneity (single-nucleotide polymorphisms) exists within the locus of six contiguous *M. synoviae* genes associated with sialic acid catabolism, but is remarkably high in the sialidase gene itself (May and Brown, 2008, 2009). We analysed the polymorphisms in detail by estimating ω values (ratio of synonymous to non-synonymous mutations; $K_a/K_s$) globally for each of these enzymes, and then for each amino acid residue, to interpret the nature of selection on this locus. We found significant diversifying selection ($\omega > 1$) acting on the sialidase gene of *M. synoviae*, but neutral to purifying selection acting on the remainder of the catabolic pathway (May and Brown, 2009). This is suggestive of the organism being at its greatest level of fitness when diversity is maintained in the sialidase gene, and further supports the notion that it serves a non-nutritive purpose in the biology of the organism. The striking correlation between the ranked haemagglutination and endogenous sialidase activities of these strains is evidence that host-induced *vlhA* allele switching indirectly drives sequence diversity in the passenger sialidase gene of *M. synoviae* (May and Brown, 2011a).

## Therapeutic potential

The substrate specificities of glycosidases make them extremely attractive therapeutic targets in the treatment of infection (Ponnuraj and Jedrzejas, 2000), and management of tumour metastasis (Delpech *et al.*, 2002) and wound healing (Bos *et al.*, 2002, Bradbury *et al.*, 2002). The ability to specifically block glycosidase activities is well-known, and is often associated with minimal toxicity to host cells. Many inhibitors are substrate analogues that function by competitive inhibition, and no effect is seen on the damaged tissue. The β-glucosidase inhibitors, such as deoxynojirimycin, have antimicrobial and antiretroviral activity *in vitro* (Gruters *et al.*, 1987). The clinical successes of oseltamivir (TamiFlu®) and zanamivir (Relenza®) in treating influenza infection by inhibiting neuraminidase highlights the outstanding therapeutic potential of targeting glycosidases to alleviate infection. Successful strategies could include their ancillary use for rapid relief of clinical signs when used in conjunction with antibiotics to alleviate infection.

## References

Achyuthan, K.E., and Achyuthan, A.M. (2001). Comparative enzymology, biochemistry and pathophysiology of human exo-alpha-sialidases (neuraminidases). Comp. Biochem. Physiol. B Biochem. Mol. Biol. *129*, 29–64.

Atalaia, V., Machado, M., and Frazão, F.F. (1987). Patologia dos pequenos ruminantes infecções em ovinos e caprinos, originadas pelo micoplasma do grupo 7, Leach. Repos. Trab. Lab. Nac. Invest. Vet. *19*, 55–60.

Balish, M.F., Friedberg, I., Fulton, L., Pritchard, R.E., Hatchel, J.M., and Wilson, R.K. (2010). Preliminary Comparative Analysis of the Genomes of Four Species of the *Mycoplasma pneumoniae* Cluster. In International Organization for Mycoplasmology 18th Biennial Congress (Chianciano Terme, Siena, Italy).

Baseman, J.B., and Tully, J.G. (1997). Mycoplasmas: sophisticated, reemerging, and burdened by their notoriety. Emerg. Infect. Dis. *3*, 21–32.

Bercic, R.L., Slavec, B., Lavric, M., Narat, M., Zorman-Rojs, O., Dovc, P., and Bencina, D. (2008). A survey of avian Mycoplasma species for neuraminidase enzymatic activity. Vet. Microbiol. *130*, 391–397.

Biberfeld, G. (1979). Autoimmune reactions associated with *Mycoplasma pneumoniae* infection. Zentralbl. Bakteriol. Orig. A *245*, 144–149.

Blix, F.G., Gottschalk, A., and Klenk, E. (1957). Proposed nomenclature in the field of neuraminic and sialic acids. Nature *179*, 1088.

Bos, P.K., DeGroot, J., Budde, M., Verhaar, J.A., and van Osch, G.J. (2002). Specific enzymatic treatment of bovine and human articular cartilage: implications for integrative cartilage repair. Arthritis Rheum. *46*, 976–985.

Bradbury, E.J., Moon, L.D., Popat, R.J., King, V.R., Bennett, G.S., Patel, P.N., Fawcett, J.W., and McMahon, S.B. (2002). Chondroitinase ABC promotes functional recovery after spinal cord injury. Nature *416*, 636–640.

Bradbury, J.M. (1977). Rapid biochemical tests for characterization of the Mycoplasmatales. J. Clin. Microbiol. *5*, 531–534.

Brown, D., Zacher, L., and Farmerie, W. (2004). Spreading factors of *Mycoplasma alligatoris*, a flesh-eating mycoplasma. J. Bacteriol. *186*, 3922–3927.

Brown, D.R., May, M., and Glass, J.I. (2008). Annotation of the *Mycoplasma crocodyli* Genome. In International Organization for Mycoplasmology 17th Biennial Congress (Tianjin, People's Republic of China).

Brown, D.R., May, M., Bradbury, J.M., Balish, M.F., Calcutt, M.J., Glass, J.I., Tasker, S., Messick, J.B., Johansson, K.-E., and Neimark, H. (2010). Genus I: *Mycoplasma*. In Bergey's Manual of Systematic Bacteriology, Krieg, N.R., Ludwig, W., Whitman, W.B., Hedlund, B., Paster, B.J., Staley, J.T., Ward, N., Brown, D.R., and Parte, A., eds. (Springer, New York), pp. 575–612.

Brown, D.R., May, M., Michaels, D.L., and Barbet, A.F. (2012). Genome annotation of five *Mycoplasma canis* strains. J. Bacteriol. *194*(15), 4138–4139.

Burnaugh, A.M., Frantz, L.J., and King, S.J. (2008). Growth of *Streptococcus pneumoniae* on human glycoconjugates is dependent upon the sequential activity of bacterial exoglycosidases. J. Bacteriol. *190*, 221–230.

Chalker, V. (2005). Canine mycoplasmas. Res. Vet. Sci. *79*, 1–8.

Corfield, T. (1992). Bacterial sialidases--roles in pathogenicity and nutrition. Glycobiology *2*, 509–521.

Dandekar, T., Huynen, M., Regula, J.T., Ueberle, B., Zimmermann, C.U., Andrade, M.A., Doerks, T., Sánchez-Pulido, L., Snel, B., Suyama, M., *et al.* (2000). Re-annotating the *Mycoplasma pneumoniae* genome sequence: adding value, function and reading frames. Nucleic Acids Res. *28*, 3278–3288.

DeAngelis, P.L. (2002). Evolution of glycosaminoglycans and their glycosyltransferases: Implications for the extracellular matrices of animals and the capsules of pathogenic bacteria. Anat. Rec. *268*, 317–326.

Delpech, B., Laquerriere, A., Maingonnat, C., Bertrand, P., and Freger, P. (2002). Hyaluronidase is more elevated in human brain metastases than in primary brain tumours. Anticancer Res. *22*, 2423–2427.

Epel, B.L. (2009). Plant viruses spread by diffusion on ER-associated movement-protein-rafts through plasmodesmata gated by viral induced host beta-1,3-glucanases. Semin. Cell. Dev. Biol. *20*, 1074–1081.

Ernst, S., Langer, R., Cooney, C.L., and Sasisekharan, R. (1995). Enzymatic degradation of glycosaminoglycans. Crit. Rev. Biochem. Mol. Biol. *30*, 387–444.

Feizi, T., and Loveless, R.W. (1996). Carbohydrate recognition by *Mycoplasma pneumoniae* and pathologic consequences. Am. J. Respir. Crit. Care Med. *154*, S133–136.

Glasgow, L.R., and Hill, R.L. (1980). Interaction of *Mycoplasma gallisepticum* with sialyl glycoproteins. Infect. Immun. *30*, 353–361.

Gruters, R.A., Neefjes, J.J., Tersmette, M., de Goede, R.E., Tulp, A., Huisman, H.G., Miedema, F., and Ploegh, H.L. (1987). Interference with HIV-induced syncytium formation and viral infectivity by inhibitors of trimming glucosidase. Nature *330*, 74–77.

Henrikson, C.V., and Smith, P.F. (1964). Beta-glucosidase activity in mycoplasma. J. Gen. Microbiol. *37*, 73–80.

Henrissat, B., and Romeu, A. (1995). Families, superfamilies and subfamilies of glycosyl hydrolases. Biochem. J. *311*, 350–351.

Himmelreich, R., Hilbert, H., Plagens, H., Pirkl, E., Li, B.C., and Herrmann, R. (1996). Complete sequence analysis of the genome of the bacterium *Mycoplasma pneumoniae*. Nucleic Acids Res. *24*, 4420–4449.

Horton, M.R., Burdick, M.D., Strieter, R.M., Bao, C., and Noble, P.W. (1998). Regulation of hyaluronan-induced chemokine gene expression by IL-10 and IFN-gamma in mouse macrophages. J. Immunol. *160*, 3023–3030.

Horton, M.R., Shapiro, S., Bao, C., Lowenstein, C.J., and Noble, P.W. (1999). Induction and regulation of macrophage metalloelastase by hyaluronan fragments in mouse macrophages. J. Immunol. *162*, 4171–4176.

Hunt, M., and Brown, D. (2007). Role of sialidase in *Mycoplasma alligatoris*-induced pulmonary fibroblast apoptosis. Vet. Microbiol. *121*, 73–82.

Hunt, M.E., and Brown, D.R. (2005). *Mycoplasma alligatoris* infection promotes CD95 (FasR) expression and apoptosis of primary cardiac fibroblasts. Clin. Diagn. Lab. Immunol. *12*, 1370–1377.

Kahane, I., Banai, M., Razin, S., and Feldner, J. (1982). Attachment of mycoplasmas to host cell membranes. Rev. Infect. Dis. *4* (Suppl.), S185–192.

Kahane, I., Reisch-Saada, A., Almagor, M., Abeliuck, P., and Yatziv, S. (1990). Glycosidase activities of mycoplasmas. Zentralbl. Bakteriol. *273*, 300–305.

King, S.J., Hippe, K.R., and Weiser, J.N. (2006). Deglycosylation of human glycoconjugates by the sequential activities of exoglycosidases expressed by *Streptococcus pneumoniae*. Mol. Microbiol. *59*, 961–974.

Knudson, W., Casey, B., Nishida, Y., Eger, W., Kuettner, K.E., and Knudson, C.B. (2000). Hyaluronan oligosaccharides perturb cartilage matrix homeostasis and induce chondrocytic chondrolysis. Arthritis Rheum. *43*, 1165–1174.

Limoli, D., Sladek, J., and King, S.J. (2007). *Streptococcus pneumoniae* beta-galactosidase, BgaA is an adhesin. In The 3rd Midwest Carbohydrate Meeting (Columbus, OH).

Lockaby, S., Hoerr, F., Lauerman, L., Smith, B., Samoylov, A., Toivio-Kinnucan, M., and Kleven, S. Factors associated with virulence of *Mycoplasma synoviae*. Avian Dis. *43*, 251–261.

Manuel, S.G., Ragunath, C., Sait, H.B., Izano, E.A., Kaplan, J.B., and Ramasubbu, N. (2007). Role of active-site residues of dispersin B, a biofilm-releasing beta-hexosaminidase from a periodontal pathogen, in substrate hydrolysis. FEBS J. *274*, 5987–5999.

Matsushita, O., and Okabe, A. (2001). Clostridial hydrolytic enzymes degrading extracellular components. Toxicon *39*, 1769–1780.

May, M., and Brown, D. (2008). Genetic variation in sialidase and linkage to N-acetylneuraminate catabolism in *Mycoplasma synoviae*. Microb. Pathog. *45*, 38–44.

May, M., and Brown, D.R. (2009). Secreted sialidase activity of canine mycoplasmas. Vet. Microbiol. *137*, 380–383.

May, M., and Brown, D.R. (2011a). Diversity of expressed *ulhA* adhesin sequences and intermediate hemagglutination phenotypes in *Mycoplasma synoviae*. J. Bacteriol. *193*(9), 2116–2121.

May, M., and Brown, D.R. (2011b). Retrospective survey for sialidase activity in *Mycoplasma pneumoniae* isolates from cases of community-acquired pneumonia. BMC Res. Notes *4*, 195.

May, M., Kleven, S., and Brown, D. (2007). Sialidase activity in *Mycoplasma synoviae*. Avian Dis. *51*, 829–833.

May, M., Szczepanek, S.M., Frasca, S. Jr., Gates, A.E., Demcovitz, D.L., Moneypenny, C.G., Brown, D.R., and Geary, S.J. (2012). Effects of sialidase knockout and complementation on virulence of *Mycoplasma gallisepticum*. Vet. Microbiol. *157*(1–2), 91–95.

Müller, H.E., and Sethi, K.K. (1972). [The occurrence of neuraminidase in *Mycoplasma gallisepticum*]. Med. Microbiol. Immunol. *157*, 160–168.

Ortiz, G.J., Hunt, M.E., and Brown, D.R. (2007). Beta-galactosidase Activity in *Mycoplasma alligatoris* A21JP2$^{T}$ (American Society for Microbiology 107th General Meeting, Toronto, Canada).

Papazisi, L., Gorton, T.S., Kutish, G., Markham, P.F., Browning, G.F., Nguyen, D.K., Swartzell, S., Madan, A., Mahairas, G., and Geary, S.J. (2003). The complete genome sequence of the avian pathogen *Mycoplasma gallisepticum* strain R(low). Microbiology *149*, 2307–2316.

Ponnuraj, K., and Jedrzejas, M.J. (2000). Mechanism of hyaluronan binding and degradation: structure of *Streptococcus pneumoniae* hyaluronate lyase in complex with hyaluronic acid disaccharide at 1.7 A resolution. J. Mol. Biol. *299*, 885–895.

Pye, G., Brown, D., Nogueira, M., Vliet, K., Schoeb, T., Jacobson, E., and Bennett, R. (2001). Experimental inoculation of broad-nosed caimans (*Caiman latirostris*) and Siamese crocodiles (*Crocodylus siamensis*) with *Mycoplasma alligatoris*. J. Zoo. Wildl. Med. *32*, 196–201.

Roberts, D.H. (1967). Neuramindase-Like Enzyme Present in *Mycoplasma gallisepticum*. Nature *213*, 87–88.

Roggentin, P., Schauer, R., Hoyer, L., and Vimr, E. (1993). The sialidase superfamily and its spread by horizontal gene transfer. Mol. Microbiol. *9*, 915–921.

Sethi, K.K., and Müller, H.E. (1972). Neuraminidase activity in *Mycoplasma gallisepticum*. Infect. Immun. *5*, 260–262.

Severi, E., Hood, D.W., and Thomas, G.H. (2007). Sialic acid utilization by bacterial pathogens. Microbiology *153*, 2817–2822.

Starr, C.R., and Engleberg, N.C. (2006). Role of hyaluronidase in subcutaneous spread and growth of group A streptococcus. Infect. Immun. *74*, 40–48.

Termeer, C., Benedix, F., Sleeman, J., Fieber, C., Voith, U., Ahrens, T., Miyake, K., Freudenberg, M., Galanos, C., and Simon, J.C. (2002). Oligosaccharides of hyaluronan activate dendritic cells via Toll-like receptor 4. J. Exp. Med. *195*, 99–111.

Vasconcelos, A., Ferreira, H., Bizarro, C., Bonatto, S., Carvalho, M., Pinto, P., Almeida, D., Almeida, L., Almeida, R., Alves-Filho, L., *et al.* (2005). Swine and poultry pathogens: the complete genome sequences of two strains of *Mycoplasma hyopneumoniae* and a strain of Mycoplasma synoviae. J. Bacteriol. *187*, 5568–5577.

Vilei, E.M., Correia, I., Ferronha, M.H., Bischof, D.F., and Frey, J. (2007). Beta-D-glucoside utilization by *Mycoplasma mycoides* subsp. *mycoides* SC: possible involvement in the control of cytotoxicity towards bovine lung cells. BMC Microbiol. 7, 31.

Vilei, E.M., and Frey, J. (2004). Differential clustering of *Mycoplasma mycoides* subsp. *mycoides* SC strains by PCR-REA of the *bgl* locus. Vet. Microbiol. *100*, 283–288.

Vimr, E., and Lichtensteiger, C. (2002). To sialylate, or not to sialylate: that is the question. Trends Microbiol. *10*, 254–257.

# Current Insights into Phase and Antigenic Variation in Mycoplasmas

9

Carl-Ulrich Zimmerman

## Abstract

Despite their small genomes and the lack of a protecting cell wall, many *Mycoplasma* species successfully establish chronic infections in diverse hosts, even in the presence of a specific immune response. This success is, in part, attributable to genetic switches that rapidly alter the expression, size or structure of their surface exposed proteins. These stochastic events create highly versatile and dynamic membranes, enabling these bacteria to escape the host immune response and enhance bacterial–host interactions essential for survival. This chapter provides an overview of the current knowledge of phase- and antigenic-variation in Mycoplasmas and documents on the diverse genetic mechanisms, such as reversible point mutation, slipped-strand mispairing and DNA rearrangement via site-specific or homologous recombination that lead to antigenic variation of distinct proteins in individual *Mycoplasma* species. The chapter contains one diagram illustrating the causes and effects of antigenic variation, three figures illustrating diverse mechanisms of phase variation and one table describing gene loci from *Mycoplasma* species that are affected by phase variation.

## Introduction

Pathogenic microbes all face similar challenges when infecting a susceptible host. Firstly, they must avoid mechanical clearance to successfully colonize their preferred tissue or niche, a process that frequently involves the use of specific adhesive molecules that utilize various host ligands as anchors. Secondly, they must avoid recognition by the host immune system, a challenge that is circumvented by the use of hypervariable surface molecules. Both approaches give microbial pathogens the potential to successfully colonize their host environments. This common need to evade the host immune system has resulted in the concomitant development of remarkably similar survival strategies among evolutionary distant pathogens. One such strategy is antigenic variation, the capacity of an organism to systematically alter the proteins displayed on the cell surface (Fig. 9.1). The term antigenic variation generally encompasses both phase variation, the ON/OFF expression of a particular antigen, and antigenic variation (domain variation), the expression of alternative forms of a particular antigen. Both cases can contribute to yet another form of antigenic switching, namely 'epitope masking'. Epitope masking is a phenomenon in which the epitopes of a constitutively expressed surface protein are subject to variable surface exposure, either due to a secondary protein that sterically blocks accessibility of the surface epitopes or as a consequence of size variation. Antigenic variation has been extensively studied in a number of microbial systems, leading to the generation of several models of the mechanisms underlying this phenomenon (van der and Baumler, 2004; Wisniewski-Dye and Vial, 2008; Deitsch *et al.*, 2009).

The availability of extensive genome sequence data and improvements in the tools available to study non-model pathogenic microorganisms made it possible to obtain a deeper insight into how microorganisms alter their genomic make-up to yield antigenic variability. Moreover, it enables one to predict the mechanisms that a

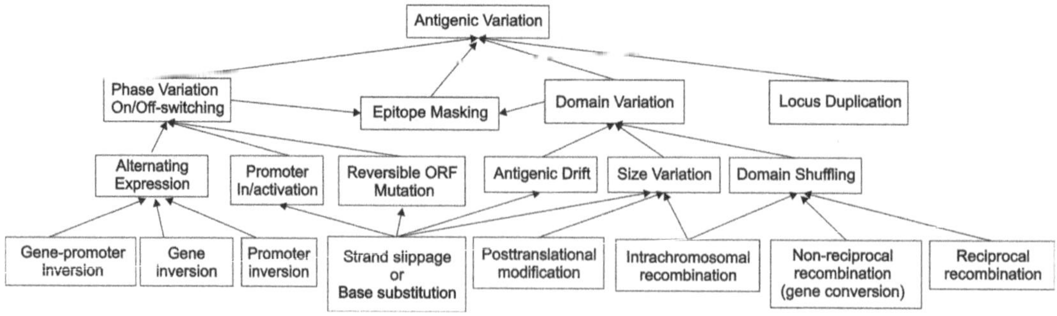


**Figure 9.1** Causes and effects of antigenic variation described in mycoplasmas.

microorganism may apply to achieve this variation. Such predictions can be based on various types of sequence repeats (e.g. simple sequence repeats, large repeats, close repeats, direct repeats and inverse repeats) found within coding and non-coding sequences (Rocha and Blanchard, 2002).

Surface antigenic variation in bacterial microbes including the wall-less mycoplasmas is achieved by three general mechanisms acting on three genetic levels. These are: genetic mechanisms, epigenetic mechanisms and post-genetic mechanisms. Genetic events like point mutations, recombinations, excisions and duplications all may change the DNA sequence of either an antigen-encoding gene or its regulatory elements, thereby altering either the level of expression of a gene or the amino acid sequence of a gene product. Among these, antigenic variation via point mutations, strand-slippage and recombination events are by far the most frequently found and described for many pathogenic *Mycoplasma* species (Citti *et al.*, 2010). Epigenetic mechanisms affect the expression of a gene without altering the primary nucleotide sequence, for example through differential methylation of DNA. This mechanistic event has so far received little attention in mycoplasma research, was however suggested to occur as a consequence of the ON/OFF regulated methyltransferases located in the type I restriction modification systems in *M. pulmonis* (Sitaraman and Dybvig, 1997). Post-genetic mechanisms include modifications of transcripts and translational products. Post-translational cleavage has been described for some variable mycoplasma surface proteins, such as the macrophage activating lipoprotein MALP-404 of *M. fermentans* which is cleaved into a membrane anchored N-terminal lipid-modified region and an external hydrophilic region (Calcutt *et al.*, 1999; Davis and Wise, 2002).

Phase variation in expression and size variation of mycoplasma surface protein antigens have been shown to rapidly occur *in vivo* and *in vitro* with a high frequency of $10^{-2}$ to $10^{-5}$ events/cell/generation and were first discovered in the swine pathogen *M. hyorhinis* (Rosengarten and Wise, 1990). Since these events are reversible with the same frequency, subpopulations resulting from a single cell are never pure. Phase variation of mycoplasma surface proteins can easily be recognized by observing switching between two alternative phenotypes (phases) among the cells in a clonal population and is best illustrated by colony immunoblotting using specific antibodies directed against the variable surface protein. A qualitative estimation of the switching frequency can already be made through the number of revertants and the size of sectors displayed within the immunostained colonies. The presence of microsectors, for example, is indicative of high-frequency phase variation, while the presence of uniformly immunostained colonies expressing switched antigens is indicative of a slower or prolonged (through multiple steps) phase variation.

In many *Mycoplasma* species, more than one gene within a family of genes is often regulated by phase variation, in which case each gene may switch ON and OFF independently. The possible number of phenotypes that can be created by n phase-variable genes is then $2^n$. For the seven *vlp* genes of *M. hyorhinis* strain HUB-1 (Liu *et al.*,

2010), which can all be expressed in a non-co-ordinate manner, this would mean that a total of 128 Vlp combinations are theoretically possible. Some systems, however, display a phenomenon called 'mutually exclusive expression' in which any member of a gene family can be expressed but only one member of the gene family is expressed at one time. In this case, the number of different phenotypes is at least as high as the number of family members. For the six *vpma* genes of the *M. agalactiae* strain PG2, which share one promoter, this would mean that only one of six variations can possibly be expressed at a time. In another *M. agalactiae* strain (strain 5632), however, in which duplication and sequence variation of the *vpma* locus had occurred, at least 91 Vpma configurations can be predicted from its 23 *vpma* genes that are distributed among two separate clusters, each cluster bearing one promoter (Nouvel *et al.*, 2009, 2010). The actual number of possible phenotypes can, of course, be much greater than the number of individual genes of a gene family in either case if other factors contribute to antigenic variation, such as mutations in coding regions or recombinations occurring among genes within gene families. The latter can produce virtually unlimited diversity through the production of chimeric sequences, as was predicted for the MgPa protein of *M. genitalium*, whose gene utilizes a pool of repeated but non-identical sequences through homologous recombination to generate sequence variations (Ma *et al.*, 2007).

Often governed by spontaneous insertion/deletion mutations or through DNA inversions driven by a single recombinase, the mechanisms of phase variation in protein expression in some mycoplasmas may appear rather rudimentary, however, the potential of these strategies should not be underestimated. Complexity is often achieved by combining different strategies such as a combined phase and size variation or through gene and/or locus duplication that continuously increases the pool of available donor antigens.

While phase variation is most often described for lipoproteins, it is not limited to surface proteins and has been reported for other systems, such as the type I restriction and modification system of *M. pulmonis* (Dybvig *et al.*, 1998; Gumulak-Smith *et al.*, 2001; Sitaraman *et al.*, 2002). Moreover, in silico predictions for phase variation have been made for genes encoding ribosomal proteins in *M. pulmonis* and *U. urealyticum* and for genes encoding subunits of RNA polymerase in *M. genitalium* (Rocha and Blanchard, 2002).

## Genetic mechanisms of Mycoplasma phase and antigenic variation at glance

At least four molecular strategies have been described in mycoplasmas that affect the expression status of antigens. Two of these are molecular switches, one based on spontaneous mutations in regions prone to DNA slippage and the other based on DNA rearrangements such as promoter or gene inversions. While the latter mechanism only affects the expression of the antigen, with the promoter usually remaining in an ON state, the first may affect either the activity of the promoter, and thereby the expression of an antigen, or the structure of the antigen, when the mutation lies within the reading frame.

A third mechanism involves homologous recombination. Non-reciprocal or unidirectional (gene conversion) and reciprocal recombination are two prominent and effective systems which affect the structure of a particular gene rather than its expression, as the event usually occurs within the coding region. Other intrachromosomal homologous recombination events that were described, lead to the development of chimeric genes and gene loss.

The fourth mechanism entails the interplay between a phase variable and a constitutively expressed partner, leading to high frequency masking and unmasking of epitopes or functional domains on the constitutively expressed protein.

Many mycoplasmas proved to be equipped with more than just one of these four general strategies and are able to achieve both phase and size variation in the same locus using different mechanisms (Table 9.1). The outcome of simultaneously occurring, high-frequency alterations of multiple factors is a highly dynamic and plastic membrane by which mycoplasmas may rapidly adapt to their habitats.

**Table 9.1** Main genetic mechanisms governing surface variation in mycoplasmas

| Type of variation | Genetic event | Mycoplasma species (gene locus) | Reference |
|---|---|---|---|
| ON/OFF switching | DNA slippage involving SSR in promoter regions | *M. hyorhinis* (*vlp*) | Yogev *et al.* (1991), Citti and Wise (1995) |
| | | *M. gallisepticum* (*vlh*A)* | Markham *et al.* (1994), Glew *et al.* (1998), Liu *et al.* (2002) |
| | | *M. arthritidis* (*maa2*) | Washburn *et al.* (1998) |
| | | *M. mycoides* subsp. *mycoides* SC (*vmm*) | Persson *et al.* (2002) |
| | | *M. capricolum* subsp. *capricolum* (*vmc*) | Wise *et al.* (2006) |
| | Site-specific recombination (gene inversion) | *M. pulmonis* (*hsd*) | Dybvig *et al.* (1998) |
| | | *M. bovis* (*vsp*) | Lysnyansky *et al.* (1999, 2001b) |
| | | *M. agalactiae* (*vpma*)‡ | Glew *et al.* (2002), Chopra-Dewasthaly *et al.* (2008) |
| | Site-specific recombination (gene-promoter inversion) | *M. pulmonis* (*vsa*) | Bhugra *et al.* (1995), Shen *et al.* (2000) |
| | | *U. parvum* (*mba*, UU172) | Zimmerman *et al.* (2009, 2011) |
| | Site-specific recombination (promoter inversion) | *M. penetrans* (*mpl*) | Roske *et al.* (2001), Horino *et al.* (2003, 2009) |
| Size variation | DNA slippage involving close repeats within CDSs | *M. hyorhinis* (*vlp*) | Yogev *et al.* (1991), Citti and Wise (1995), Citti *et al.* (1997) |
| | | *M. pulmonis* (*vsa*) | Simmons *et al.* (1996) |
| | | *M. bovis* (vsp) | Behrens *et al.* (1994) |
| | | *M. arthritidis* (*maa2*) | Washburn *et al.* (1998) |
| | | *M. hyopneumoniae* (adhesion related CDS products) | Wilton *et al.* (1998), de Castro *et al.* (2006) |
| | | *U. urealyticum* (*mba*) | Teng *et al.* (1994), Zheng *et al.* (1994, 1995) |
| | DNA slippage involving SSR within CDSs | *M. fermentans* (*p78*) | Theiss and Wise (1997) |
| | | *M. hominis* (*vaa*) | Zhang and Wise (1996, 1997) |
| | Base substitution | *M. mycoides* subsp. *mycoides* SC (*pts*G)§ | Gaurivaud *et al.* (2004) |
| | | *M. gallisepticum* (*gap*A) | Winner *et al.* (2003) |
| | posttranslational modification | *M. fermentans* (*malp*) | Calcutt *et al.* (1999) |
| Domain shuffling | Non-reciprocal (unidirectional) recombination (gene conversion) | *M. synoviae* (*vlh*A) | Noormohammadi *et al.* (2000) |
| | | *M. pneumoniae* (RepMP2/3, RepMP4, RepMP5) | Kenri *et al.* (1999), Spuesens *et al.* (2009, 2010) |
| | | *M. genitalium* (MgPar) | Iverson-Cabral *et al.* (2006, 2007), Ma *et al.* (2007) |
| | Reciprocal recombination | *M. genitalium* (MgPar) | Iverson-Cabral *et al.* (2006, 2007), Ma *et al.* (2007) |
| | Intrachromosomal recombination with deletion | *M. bovis* (*vsp*; between *vspO* and *vspA*) | Lysnyansky *et al.* (2001a) |
| | | *M. agalactiae* (*vpma* at inversion sites) | Czurda *et al.* (2010) |

| Type of variation | Genetic event | Mycoplasma species (gene locus) | Reference |
|---|---|---|---|
| Locus duplication§ | Unknown | *M. pulmonis* (*vsaC*, *-D*, *-E*) | Shen *et al.* (2000) |
| | Unknown | *M. agalactiae* (*vpma*) | Nouvel *et al.* (2009) |
| Epitope masking | Unknown | *M. fermentans* (P29 by ND) | Theiss *et al.* (1996) |
| | Unknown | *M. hominis* (P56 by P120) | Zhang and Wise (2001) |
| Size and phase variation | Unknown | *M. equigenitalium*T34 (protein: pET45) | Tortschanoff *et al.* (2005) |
| Phase variation | DNA slippage suggested | *M. mycoides* subsp. *mycoides* SC (Vmm-type proteins) | Hamsten *et al.* (2008) |
| Size and phase variation | Unknown | *M. meleagridis* (locus ND) | Dufour-Gesbert *et al.* (2001) |
| Domain variation | Intrachromosomal recombination with deletion | *M. pneumoniae* (RepMP1 between MPN137 and MPN138) | Musatovova *et al.* (2008) |

*The *vlh*A gene family was previously designated as pMGA. ‡Some of the *vpma* genes were also designated as *avg*. §Expected to occur at low frequency (rare events). SSR: short sequence repeat; CDS: coding sequence; ND, not yet defined.

## Surface antigen ON/OFF switching via DNA slippage and point mutations

Molecular switches involving DNA slippage have been described for single genes and also gene families for a number of *Mycoplasma* species. The insertion/deletion of nucleotides can occur in either the non-coding or coding regions, usually at favoured target sites termed 'hot spots'. Slipped-strand mispairing or illegitimate recombination is a RecA-independent process that can occur during chromosomal replication, DNA repair or recombination and requires DNA synthesis. Mispairing on the synthesis strand generates addition events, whereas slipped-strand mispairing on the template strand results in deletions. Although this type of reversible mutation often occurs at poly-nucleotide repeats or close repeats, it can likewise happen at non-repeated sites. Also, not all poly-nucleotide repeats are hot-spots and similar sites on the chromosome must not be equally affected.

Among mycoplasmas, the first mechanism described regulating the expression of a family of genes encoding surface exposed lipoproteins, was the molecular switch involving DNA slippage in the promoter regions of the *vlp* genes of the swine pathogen *M. hyorhinis* (Yogev *et al.*, 1991, 1993; Citti and Wise, 1995). This species possesses a genetic system composed of three to eight genes, encoding the 'variant surface antigens A-G', that are clustered on the chromosome (Yogev *et al.*, 1995; Citti *et al.*, 2000). Each gene represents a single transcription unit that contains a poly-A tract lying in the promoter region between the −35 and the −10 sequences. When the poly-A tract is 17 bp long, the downstream Vlp is expressed (Fig. 9.2a). Elongation or contraction by insertion or deletion of single nucleotides is sufficient to abolish *vlp* transcription, presumably due to structural alterations in the DNA curvature that is disadvantageous for transcription initiation (Citti and Wise, 1995).

A similar transcription regulation mechanism has been proposed for the phase-variable *maa2* gene in *M. arthritidis,* which encodes a major cytadhesin (Washburn *et al.*, 1998). Its promoter region includes a −10 sequence that lies approximately 100 nucleotides away from the start codon with an immediate upstream poly-T tract. In clones where the poly-T tract was made up of 14 thymidines, Maa2 expression was off. When the poly-T tract contained 16 thymidines, Maa2 was expressed.

Phase variation of the *vmm* gene of *M. mycoides* subsp. *mycoides* SC and of the *vmc* genes of the

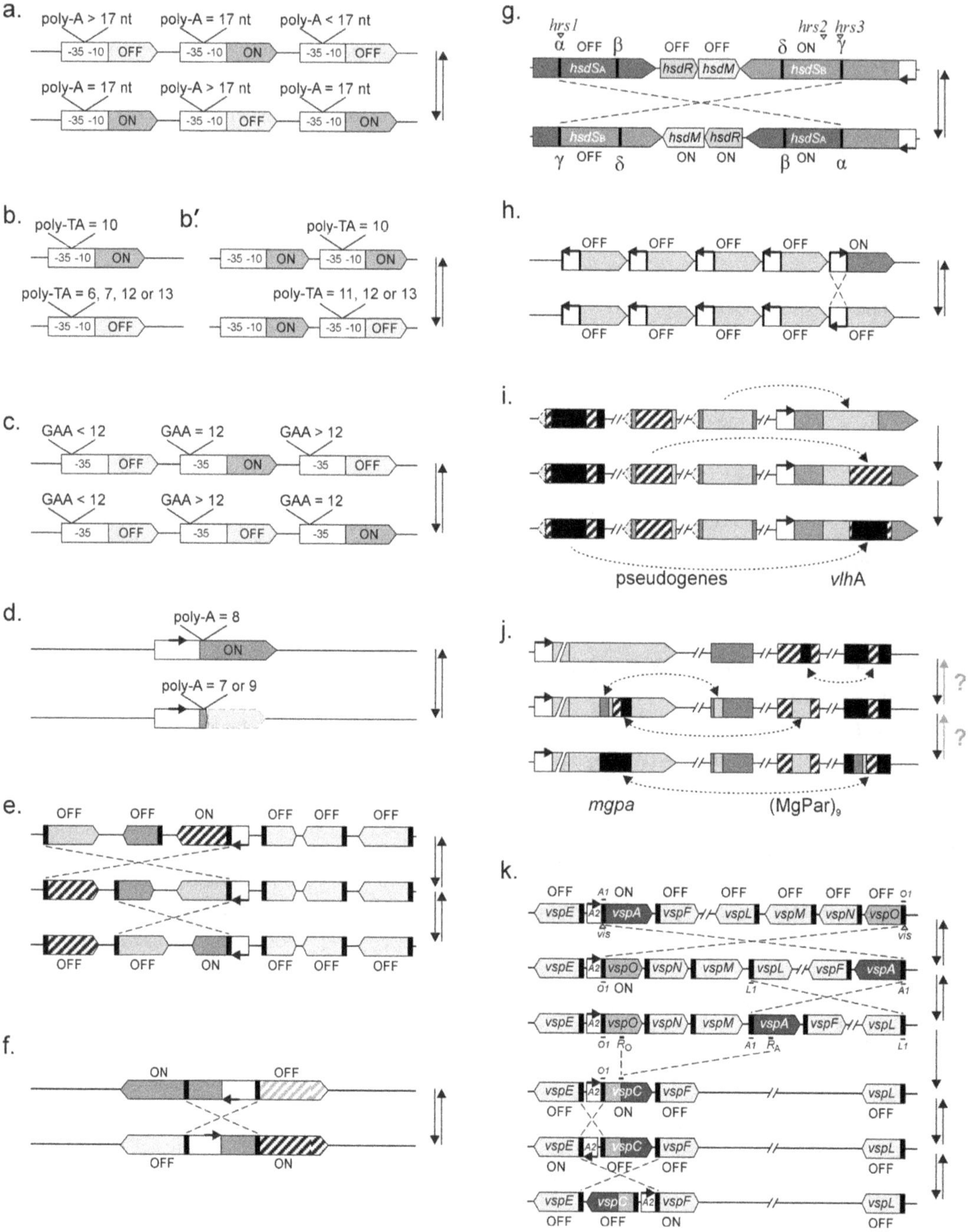


**Figure 9.2** Schematic illustrations of genetic mechanisms governing high-frequency phase and antigenic variation of mycoplasma surface proteins. (a) ON/OFF switching by DNA slippage in the promoter of the *vlp* genes of *M. hyorhinis*. Spontaneous, high-frequency mutations occur in the poly-A tract located between the −35 and the −10 promoter region of each *vlp* gene. Transcription of any *vlp* gene only takes place when the poly-A is 17nt long. (b) ON/OFF switching by DNA slippage in the promoter of the *vmm* gene of *M. mycoides* and (b′) of the *vmcE* and *vmcF* genes of *M. capricolum*. Spontaneous, high-frequency mutations occur in a poly-TA tract located between the −35 and the −10 promoter regions of each gene. Proteins are expressed when the poly-TA is made up of 10 TA repeats. (c) ON/OFF switching by DNA slippage in the promoter region of the *vlh*A genes of *M. gallisepticum*. Spontaneous, high-frequency mutations occur in a poly-GAA tract located upstream of the −35 promoter region of each *vlh*A gene. Transcription takes place when the poly-

phylogenetically related *M. capricolum* subsp. *capricolum* is also governed by reversible promoter mutations between the −35 and -10 box 5′ to each of these structural genes (Persson *et al.*, 2002; Wise *et al.*, 2006). The region prone to strand-slippage is a poly-TA tract that is ideally made up of 10 reiterated di-nucleotides (Fig. 9.2b and 9.2b'). Mutants containing 12 TA repeats did not express the lipoproteins.

The *vlh*A genes (previously pMGAs) of the avian pathogen *M. gallisepticum* underlie a similar regulatory mechanism. The gene family comprises approximately 40 related lipoprotein genes distributed within five clusters on the genome (Papazisi *et al.*, 2003). Most *vlh*A gene family members are not transcribed and only a single member of the gene family is expressed at any time (Glew *et al.*, 1995; Liu *et al.*, 2002). Phase variation in VlhA expression is also governed at the transcription level by alterations in the length of a trinucleotide repeat (GAA) located upstream of the −35 box of the promoter of each individual *vlhA* gene (Glew *et al.*, 1998, 2000a; Liu *et al.*, 2000, 2002). Optimal VlhA expression is achieved when the GAA region comprises 12 repeats, while the genes are transcriptionally silent when the region is composed of 5, 6, 8, 9, 10, 11, 13, 14 18, 20 or 24 reiterated GAA tri-nucleotides (Fig. 9.2c). The mechanism by which this element controls transcription is not clear, and although length of the promoter region seems to be a critical factor, regulatory proteins might also be involved. Interestingly, an insertion of either a 10-bp or a 12-bp linker at the 3′ region of the GAA element could compensate the loss of four GAA repeats, resulting, however, in a net gain of 1 and 3 bp, respectively, in the promoter region downstream of the GAA element. Clones bearing these constructs were also VlhA positive (Liu *et al.*, 2002).

Overall, these genetic systems offer an efficient ON/OFF switch that operates, in many cases, independently for each gene in a non-coordinate manner. For *M. hyorhinis* and *M. capricolum* this means that one cell can express either a single or a combination of their phase-variable proteins at one time (Rosengarten and Wise, 1991; Wise *et al.*, 2006). Unlike for the *vlp* and *vmc* systems, the *vlh*A system of *M. gallisepticum* seems to underlie

---

GAA tract has 12 reiterated GAA tri-nucleotides. (d) Phase variation by DNA slippage in a poly-A tract within the *vaa* gene of *M. hominis*. Addition or deletion of one nucleotide in the tract causes a frameshift, resulting in the expression of a short protein. (e) ON/OFF switching by site-specific recombination as in the *vpma* cluster of *M. agalactiae*; similar events occur in the *vsp* cluster of *M. bovis* (j). Recombination/inversion sites (▮) are located 5′ to the start ATG of each coding sequence. Inversion is processed by a single recombinase. (f) ON/OFF switching by site-specific recombination as in the *mba* locus of *U. parvum* serovar 3 and the *vsa* cluster of *M. pulmonis*. Recombination/inversion sites (▮) are located within the coding sequence of one gene and at the 5'end of another putative gene that lacks the N-terminal encoding sequence for membrane anchorage. (g) Site-specific α-γ DNA inversion as in the *hsd1* and *hsd2* loci of *M. pulmonis*; the *hsd2* locus of strain KD735–15 is illustrated. Four recombination/inversion sites (▮) designated *vip*α, -β, -γ, -δ are located in two potential $hsdS_{(A\ and\ B)}$ genes. DNA inversions at all four sites can produce variable HsdS products and cause ON/OFF switching of *hsdM* and *hsdR*. (h) ON/OFF switching by site-specific recombination as in the *mpl* system of *M. penetrans*. Two short recombination/inversion sites are located at the 3′ and 5′ end of individual promoters. (i) Antigenic variations by unidirectional recombination (gene conversion), as in the *vlh*A cluster of *M. synoviae*, between an expressed locus and donor sequences (pseudogenes) located in close proximity to the gene. (j) Antigenic variations by reciprocal recombination, as occur in the *mgp* operon of *M. genitalium* between an expressed locus and archived donor sequences distributed throughout the chromosome. Similar events occur between genes of the P1 operon and RepMP elements of *M. pneumoniae*. (k) Reversible DNA inversion events as in the *vsp* locus of *M. bovis* PG45 and a potentially irreversible intrachromosomal recombination event between *vspO* and *vspA* leading to the generation of a chimeric *vspC* and loss of *vspM* and *vspN*. Inversions occur at oppositely oriented *vis* sites that are located within cassette 1 (▮; e.g. *A1*, *O1* and *L1*) positioned 5′ to each coding sequence. One active promoter lies in a cassette 2, designated *A2*. Each *vsp* gene contains a cassette 2, which are not shown in this illustration. Intrachromosomal recombination is believed to occur between the homologous repetitive domains $R_O$ and $R_A$ of the *vspO* and the *vspA* genes, generating the chimeric gene *vspC*. Black arrows on the edge of white boxes represent promoters except in a–c, where promoters are detailed; dotted lines indicate DNA inversion or recombination events; dotted arrows indicate reciprocal recombination events and arrows indicate the direction of antigenic variation events.

an unknown mechanism that prevents the expression of more than one *vlhA* gene in a given cell.

Phase variation promoted by frameshift mutations within coding sequences has also been described and predicted for several mycoplasma genes. Although in these cases the genes are still transcriptionally active, truncated proteins may be expressed that comprise little more than the membrane anchor, thereby mimicking an OFF expression state. An advantage of this pseudo-OFF variation is that the membrane retains its inner structural integrity (i.e. experiences no loss in membrane proteins per se), but may alter the composition of membrane components by exposing previously hidden epitopes or removing present epitopes.

The mechanical switch that dictates full length expression of the *vaa* (variable adherence-associated) gene that encodes the 'Vaa surface lipoprotein adhesin' of *M. hominis* is likewise based on insertion/deletion of adenines at a poly-A ($A_8$) tract. Mutations occur in its N-terminal encoding region located 166 nt downstream of the ATG start codon (Zhang and Wise, 1997). High frequency deletion or addition of a single adenine at this site results in an in-frame stop codon so that short, 64 or 79 amino acid proteins are expressed (Fig. 9.2d). A similar event was described for the surface lipoprotein P78 of *M. fermentans* (Theiss and Wise, 1997). The corresponding gene lies at the 3′ end of a four-gene operon encoding a putative ABC type sugar transport system. High frequency alterations in a poly-A ($A_7$) tract in the N-terminal region of the *p78* gene generates pseudo-ON/OFF variation of P78, yielding a truncated 10-kDa protein, but has however no effect on the expression of the first gene in the operon (Theiss and Wise, 1997).

Aside from these well described systems, strand slippage at short sequence repeats both within coding regions and potential promoters have been predicted for *M. pulmonis, M. genitalium, M. pneumoniae, Ureaplasma urealyticum* and *M. mycoides* subsp. *mycoides* SC (Rocha and Blanchard, 2002; Westberg *et al.*, 2004). A newly identified set of 19 dispersed lipoprotein encoding genes in the sequenced genome of *M. bovis* PG45, each with a rare homopolymeric tract of 7 to 12 G or C residues in the N-terminal coding region, represent contingency loci subject to frameshift mutations (Wise *et al.*, 2011). Fifteen of these genes (annotated as authentic frameshifts) are predicted to be in the OFF configuration.

A high frequency reversible point mutation responsible for generating alternating adhesion phenotypes was shown to occur in *gap*A gene of *M. gallisepticum*, the first gene of a three gene operon encoding the major adhesin of the organism (Winner *et al.*, 2003). The base substitution CAA→TAA, which occurs at amino acid residue 335, results in truncation of the otherwise 1122 amino acid long protein and a haemadsorption negative phenotype. Moreover, the mutation leads to early termination of transcription with consequential lack of *crm*A mRNA, which represents the next gene in the operon.

Other nucleotide substitutions, such as the nonsense mutation GAA→TAA were shown to generate antigenic variation in expression of the *pvpA* gene encoding the 'phase-variant protein A' of *M. gallisepticum* and of the U172 phase-variable element of *U. parvum* (Boguslavsky *et al.*, 2000; Zimmerman *et al.*, 2011). Interestingly, although the nonsense mutation in the *pvpA* gene seems to take place at a high frequency of about $10^{-3}$ to $10^{-4}$ per cell per generation, the event is either irreversible or reversion occurs at low frequency.

## Surface antigenic variation via DNA rearrangements (site-specific DNA inversions)

A number of *Mycoplasma* species have the potential to achieve phase variation by DNA rearrangement, more precisely by DNA inversion via conservative site-specific recombination. This cut-and-paste mechanism is generally processed by a site-specific recombinase that utilizes two specific and inverted DNA sequence to alternate silent genes behind a functional promoter. Three mechanisms have been described that slightly differ from one another; in two of them, a single promoter that lies in a gene cluster is alternated between gene members and in another all members of a cluster possess their own alternating promoter.

The processed clusters may consist of either a pair or a group of coding sequences (CDSs). Phase-variable elements composed of pairs have recently been described for the *mba* and UU172 phase-variable loci of *U. parvum* (Zimmerman *et al.*, 2009, 2011). Well-described systems comprising groups in which a single gene is placed behind a promoter while the others are silent, are the *vsa* locus of *M. pulmonis* (Bhugra *et al.*, 1995; Shen *et al.*, 2000), the *vsp* locus of *M. bovis* (Lysnyansky *et al.*, 1999, 2001a,b) and the *vpma* locus of *M. agalactiae* (Glew *et al.*, 2002; Flitman-Tene *et al.*, 2003). Although these three loci are very similar in appearance, they differ in their genetic setups and phase variation is believed to be driven either by site-specific DNA inversions which link the ORF of a silent but fully equipped (i.e. with signal peptide encoding sequence, membrane anchor and C-terminal sequence) gene to a unique active promoter, as observed in the *vsp* and the *vpma* genes (Fig. 9.2e), or by juxtaposition of a DNA sequence containing promoter, ribosome binding site and 5′ terminal region of an ORF in front of the 3′ region of a previously silent gene, as reported for the *vsa* genes (Fig. 9.2f). The latter mechanism has recently been suggested to occur within the *mba* locus and the UU172 phase-variable element of *U. parvum* serovar 3. The reason for the two different mechanistic events can be found in the location of the inversion sites, which are located in upstream sequence cassettes, 5′ to every *vsp* and *vpma* gene. All four organisms possess a site-specific recombinase that recognizes a specific short DNA sequence as a target for DNA rearrangements. In the genomes of *M. bovis, M. agalactiae* and *M pulmonis,* the recombinase gene lies adjacent to their phase-variable locus (Ron *et al.*, 2002).

In *M. agalactiae,* the DNA inversion is processed by the Xer1 recombinase. Activity of this recombinase was proven by generating a *xer*1 knockout and complementation of Vpma phase-locked mutants with an actively expressed *xer*1 gene (Chopra-Dewasthaly *et al.*, 2008). Two short 21-bp inverted repeats are sufficient for Xer1 mediated DNA inversion (Czurda *et al.*, 2010). The inversion sites of the *vpma* cluster are almost identical to those of the *vsp* cluster in the phylogenetically closely related species. The site-specific DNA recombinase HvsR of *M. pulmonis* has dual substrate specificity, catalysing independent DNA inversions at distinct recombination sites in non-homologous loci, one encoding the variable surface proteins and the other the restriction and modification system from the two loci *hsd*1 and *hsd*2 (Fig. 9.2g) (Sitaraman *et al.*, 2002). Mutants generated by transposon mutagenesis, which expressed only a truncated HvsR protein, were phase locked for VsaA expression and incapable of *hsd*1 and *hsd*2 inversion.

A third mechanism, discovered in *M. penetrans,* is the inversion of individual promoters in the *mpl* gene cluster, where the orientation of each gene remains unaltered. Both, phase-variable gene clusters and the hereupon acting recombinase have been identified in this organism (Horino *et al.*, 2009). A total of 38 potentially phase-variable *mpl* genes are distributed in three clusters (Sasaki *et al.*, 2002), each gene equipped with its own invertible promoter possessing two inversion sites (Fig. 9.2h) (Horino *et al.*, 2003). A possible reason why inversions are restricted to the individual promoter regions rather than occurring in between promoters of different genes might be due to sequence alterations in each of the inversion sites. On the contrary, site-specific inversions at these variable sequences suggest that the recombinase of *M. penetrans,* like HvsR of *M. pulmonis,* possesses multiple substrate specificity.

## Domain shuffling via homologous recombination

A third type of genetic mechanism for generating antigenic variation involves high frequency homologous recombination. At least three different mechanistic events have been described in mycoplasmas that lead to domain shuffling: non-reciprocal recombination (gene conversion), reciprocal recombination and intrachromosomal recombination. Non-reciprocal recombination is a unidirectional event, meaning that one sequence (the donor) will remain unchanged after recombination. This also implies that the recombination event itself is unidirectional and irreversible (Fig. 9.2i). On the contrary, reciprocal recombination involves sequence exchanges and can be reversible (Fig. 9.2j). Intrachromosomal homologous recombination can result in the generation of chimeric genes and/or deletions within gene clusters (Fig. 9.2k).

Gene conversion has been demonstrated for the *vlh*A gene of *M. synoviae* and the MgPa encoding gene from *M. genitalium* and was suggested for the RepMP elements of *M. pneumoniae*, while reciprocal recombination has only been demonstrated for the repetitive MgPar elements of *M. genitalium* (Noormohammadi *et al.*, 2000; Iverson-Cabral *et al.*, 2006, 2007; Ma *et al.*, 2007; Spuesens *et al.*, 2009, 2010).

The *vlh*A family of *M. synoviae* is a large cluster, comprising a single transcriptionally active expression unit and multiple highly homologous pseudogenes located 5′ to the *vlh*A gene that all lack the 5′ end coding sequence. Antigenic variation of the VlhA product is the result of unidirectional recombination events occurring between the donor pseudogenes and the functional *vlhA* copy, resulting in the duplication of the pseudogene sequence and loss of the corresponding region in the previously expressed gene (Noormohammadi *et al.*, 1998, 2000). Proximity and opposite orientation of the pseudogenes to the *vlhA* gene seem necessary for this high-frequency variation.

*M. genitalium* and *M. pneumoniae* also use homologous recombination to achieve antigenic variation in their major adhesion protein. Their genomes possess multiple repeat regions termed MgPar and RepMP respectively, which contain homologous sequences to distinct regions on the genes encoding the immunogenic adhesion proteins. Serving as a pool of donor sequences, the repetitive elements are exploited through both reciprocal and non-reciprocal recombination to generate antigenic variation (Iverson-Cabral *et al.*, 2006, 2007; Ma *et al.*, 2007; Spuesens *et al.*, 2009, 2010).

High frequency intrachromosomal homologous recombination within the *vsp* locus of *M. bovis* has been shown to generate a chimeric variable lipoprotein (Lysnyansky *et al.*, 2001a). Each potential *vsp* gene within the cluster is preceded by a 35-bp homologous sequence (*vis*), which, when appearing in inverse orientation, can serves as a region for site-specific DNA inversion. Homologous regions within genes and possibly also *vis* sites, that are oriented as direct repeats, may promote intrachromosomal recombination through which chimeric genes are generated and the chromosomal region between the recombination sites is lost. Such an event has been suggested to occur between homologous regions within the genes *vspO* and *vspA*, leading to the generation of the chimeric gene *vspC* (Fig. 9.2k). Supporting data for such an event comes from experimental results obtained in *M. agalactiae*, where Xer1 mediated excision between two direct repeats could be demonstrated using a *lacZ* reporter system (Czurda *et al.*, 2010).

## Epitope masking and unmasking

Epitope or domain masking and unmasking is an indirect consequence resulting from genetic events that contribute to antigenic variation and for which the affected gene must not necessarily be the target of genetic alterations. This is an interesting phenomenon, as it demonstrates how constitutively expressed surface proteins may contribute to the plasticity of the cell surface. Since the masking/unmasking is only a side effect, it often remains difficult to identify the partners that are directly involved in this type of phase variation and are the true targets of the genetic alteration.

Epitope masking has been described for the P56 protein of *M. hominis*. ON/OFF switching in the expression of the surface exposed P120 protein had been shown to precisely correlate with changes in the surface accessibility of P56 epitopes, despite continuous expression of the P56 product (Zhang and Wise, 2001). Masking of P56 by P120 appeared to be selective for P56 and the striking inverse correlation between P56 accessibility and P120 expression suggests that the two antigens may specifically interact as nearest neighbours.

The P29 cytadhesin of *M. fermentans* is another interesting protein affected by this phenomenon, although the masking components affecting the surface display of P29 epitopes have not been identified (Theiss *et al.*, 1996). High frequency variations in display of epitopes associated with both the C- and N-terminal region were shown to occur independently of each other and the altered exposure of these epitopes was not the result from changes in the primary or secondary structure of the protein. As the functional binding domain of P29 is localized in a central 78 amino acid region, it is likely that the ligand binding domain is differentially presented or subject to phase-variable presentation that would lead to a phase-variable phenotype, although this has not been proven (Leigh and Wise, 2002).

Domain masking has also been proposed for the Maa1 protein of *M. arthritidis*, and the phase-variable Maa2 protein has been suggested as one of its masking component. Both Maa1 and Maa2 are adhesins mediating cytadhesion to rat lung cells. Maa1 is produced constitutively and does not vary in size, while Maa2 is subject to both size and phase variation as a consequence of strand slippage of tandem 264-bp repeats in the coding sequence and/or of thymidines in a poly-T tract within the promoter region (Washburn *et al.*, 1998). The idea that Maa1 and Maa2 are cooperating partners in epitope and/or domain masking was based on the finding of differential cytadhesion strength in Maa2 mutants that constitutively expressed Maa1. Loss of Maa2 enhanced cytadhesion fivefold, while complementation of the mutants with the wild-type *maa2* reduced cytoadhesion to wild-type levels (Bird *et al.*, 2008).

## Genetic events mediating phase and antigenic variation in selected Mycoplasma species

In the following section, selected examples of loci will be presented that represent each of the three described general mechanisms governing phase variation: slipped-strand mispairing, site-specific DNA inversion and homologous recombination.

## Strand-slippage at poly-TA sequences

### The *vmm* gene of *Mycoplasma mycoides* subspecies *mycoides* SC type

Two genes have been reported to be involved in antigenic variation in *M. mycoides* subsp. *mycoides* SC (*Mmymy*SC) strain PG1; these are *vmm* (MSC_0390) and MSC_0364, both encoding phase-variable lipoprotein precursors (Persson *et al.*, 2002; Hamsten *et al.*, 2008). The Vmm prolipoprotein is 59 amino acids long and the mature protein migrates at 15 kDa. Expression of Vmm is regulated at the transcription level and is dictated by alternations in a poly-TA tract between the −10 and −35 box of its putative promoter (Persson *et al.*, 2002). The gene is transcriptionally active when the poly-TA region contains 10 TAs and inactive when the region contains 6, 7, 12 or 13 TAs. Other potential phase-variable genes underlying the same mechanism were identified in the course of genome sequencing (Westberg *et al.*, 2004). Among these, five genes (MSC_0117, MSC_0364, MSC_1005, MSC_1033, and MSC_1058) encoding prolipoproteins also have promoters with a poly-$TA_{(5-12)}$ tract, some of which showed variation in length among different clones (Westberg *et al.*, 2004; Hamsten *et al.*, 2008). Seven additional putative promoters containing poly-TA repeats seem to have been interrupted by IS*Mmy1* elements in the *Mmymy*SC str. PG1 genome. Three of these interrupted promoters are located upstream of genes encoding membrane-associated proteins, while four more lack a corresponding gene (Westberg *et al.*, 2004). In the putative promoters of nine surface protein genes (MSC_0809, MSC_810, MSC_812, MSC_813, MSC_815, MSC_816, MSC_817, MSC_818, and MSC_847), poly-$A_{(15-23)}$ tracts were detected, which may also be involved in transcriptional control (Westberg *et al.*, 2004).

Aside from the phase-variable loci, at least two surface protein encoding genes have been predicted to undergo size variation at poly-$T_{(10 \text{ and } 14)}$ stretches (Westberg *et al.*, 2004). Variable expression was also observed for the gene (MSC_0860) encoding the glucose-specific IIBC component of the PTS permease system (Gaurivaud *et al.*, 2004). A single base substitution (TGA to TAA) at amino acid residue 208, upstream of the immunodominant epitope of the protein, leads to premature termination of translation. Although this mutation does not happen at a poly-nucleotide tract, it has been demonstrated to undergo reversion at low frequency.

### The *vmc* clusters of *Mycoplasma capricolum*

*Vmm*-like genes have been demonstrated in the genomes of *M. capricolum* subsp. *capricolum*, *M. capricolum* subsp. *capripneumoniae*, *M. leachii* and *M. putrefaciens* by Southern blot analyses with a probe against *vmm* (Persson *et al.*, 2002).

In the *M. capricolum* subsp. *capricolum* chromosome, six ORFs representing a potential family of genes encoding variable lipoproteins (VmcA-F) were localized in two loci that are separated by ~40 kb of intervening sequence; four (*vmcA-D*) occur at one locus and two (*vmcE* and *vmcF*) at another. The 5′ flanking regions of *vmcA* and *vmm* from *Mmymy*SC PG1 share synteny and the two chromosomal regions first diverge within these two genes (Wise *et al.*, 2006). The 5′ flanking sequences of all six *vmc* genes are highly conserved and each contains a poly-TA tract between the −10 and −35 sequence motifs. All *vmc* genes encode prolipoproteins with very similar signal peptide encoding sequences to *vmm*. Tandemly repeating in-frame DNA sequences encoding size-variable tandem repeating domains occur in *vmc* coding regions (except in *vmcC*) of strain ATCC27343. Longer repeat units occur in VmcA [5 times 80 amino acids (aa)], VmcB (2 times 69 aa), and VmcD (3 times 68 aa), whereas products of the other *vmc* locus, VmcE (16 times 10 aa) and VmcF (11 times 11 aa), contain shorter repeat units. The first 11 N-terminal residues of the mature VmcE and VmcF polypeptides are identical to those of Vmm, which is interesting, as the *vmm* gene of *Mmymy*SC PG1 and the *vmc*E/F cluster lie in a different locus (Wise *et al.*, 2006). The predicted phase-variable expression of *vmc* genes is governed by a reversible promoter mutation mechanism analogous to that occurring in the *vmm* gene. Expression of VmcE and VmcF has been demonstrated to follow a non-coordinate pattern, indicating that combinations of all six lipoproteins are possible (Wise *et al.*, 2006).

## Site-specific DNA inversions

### The vpma clusters of *Mycoplasma agalactiae*

*Mycoplasma agalactiae* is a ruminant pathogen that causes contagious agalactia in sheep and goats and exhibits antigenic diversity by site-specific DNA rearrangements within a pathogenicity island-like gene locus, the *vpma* locus (Glew *et al.*, 2000b, 2002; Ron *et al.*, 2002; Flitman-Tene *et al.*, 2003). This gene family comprises six distinct but related genes in strain PG2 that encode the major immunodominant membrane lipoproteins Vpma U-Z 'variable proteins of *Mycoplasma agalactiae*' (Sirand-Pugnet *et al.*, 2007). The 5′ untranslated regions and those encoding the signal peptide are conserved within the *vpma* gene family and share a high identity to the *vsp* system of the phylogenetically related *M. bovis* (Flitman-Tene *et al.*, 2000). Both *vsp* and *vpma* genes contain repeated sequences, exhibit the same lipoprotein cleavage motif and encode similar potential short cytadherence epitopes (Glew *et al.*, 2002). Due to these similarities and the very close phylogenetic relationship of the two species, it has been suggested that the *vpma* and *vsp* loci might have derived from a common ancestor. However, beyond the highly conserved 5′ untranslated and N-terminal regions, the *vsp* and *vpma* genes share no significant homology with each other.

Vpmas vary in expression at high frequency, and only one *vpma* gene is transcribed at a time from a single promoter present in the locus, while the other genes are silent (Glew *et al.*, 2002; Flitman-Tene *et al.*, 2003). The gene encoding the Xer1 recombinase lies in the vicinity of the *vpma* locus and its product is responsible for mediating DNA inversions at the 21-bp consensus sequence 5′-TTGATATTTATTAATAGATTT-3′ that lies in the 5′ untranslated regions of each Vpma protein (Glew *et al.*, 2002; Czurda *et al.*, 2010). Targeted knockouts of the *xer1* gene, generated by homologous recombination, have been shown to prevent Vpma switching and produced Vpma phase-locked mutants steadily expressing a single *vpma* gene. Through complementation with the wild-type *xer1* gene in these mutants, Vpma phase variation could be restored (Chopra-Dewasthaly *et al.*, 2008).

The Xer1 recombinase, just like the putative tyrosine recombinases Mbr of *M. bovis*, HvsR of *M. pulmonis* and MYPE2900 of *M. penetrans*, belongs to the λ-integrase family of site-specific recombinases. Members of this family share four conserved residues located in the C-terminal half of the protein sequences in the order Arg, His–X–X–Arg and Tyr, with Tyr closest to the C-terminus (Esposito and Scocca, 1997; Ron *et al.*, 2002). These residues are directly involved in the recombination reaction and recombination occurs by formation and resolution of a Holliday junction intermediate involving a covalent linkage between the recombinase and the DNA via the tyrosine residue. Xer1 recognizes the recombination site, causing cleavage and strand exchange within the conserved 21-bp region. This was demonstrated in an *in vitro E. coli* model by placing two *vpma* derived recombination sites on a plasmid along with the *xer1* gene and monitoring inversion by PCR and restriction analysis (Czurda *et al.*, 2010). Interestingly, Xer1 also mediates excisions when the recombination sites are arranged as direct repeats. This could be demonstrated by placing the *lacZ* gene, flanked by two recombination sites that were oriented in the same direction, into the left IS element of the transposon *Tn4001*, and inserting this construct into the chromosome of *M. agalactiae* (Czurda *et al.*, 2010). Excision of the *lacZ* gene was observed through blue-white screening and PCR analysis and could be proven in Xer1 expressing clones but not in Xer1 deletion mutants.

The genome sequence of another *M. agalactiae* strain (strain 5632) revealed that the *vpma* cluster must not be limited to the six previously described genes. Strain 5632 possesses a total of 23 *vpma* genes distributed in two loci ($I_{5632}$ and $II_{5632}$) that are 250 kbp apart, of which only two genes (*vpmaW* and *vpmaX*) have significant similarity with those in strain PG2 (Nouvel *et al.*, 2009, 2010). Locus $I_{5632}$ is 19,453 bp long, comprises 16 *vpma* genes and is the counterpart of the PG2 *vpma* locus. Besides the difference in *vpma* gene numbers, locus $I_{5632}$ is flanked by an IS element and contains two additional non-*vpma* ORFs that are not present in the *vpma* locus of PG2. Regarding these differences, locus $I_{5632}$ resembles the *vsp* locus of *M. bovis* strain PG45. The size and the gene repertoire of locus $I_{5632}$ is especially interesting in respect to the size limitations of phase-variable loci that operate via site-specific DNA inversion. The size of locus $I_{5632}$ and the finding that DNA inversions of greater than 13-kbp can occur, as in the *vsp* locus of *M. bovis*, suggests that these loci can expand to unlimited dimensions.

Locus $II_{5632}$ is 8462 kb long, clustered between two IS*Mag1* copies and contains seven *vpma* genes which are identical to seven *vpma* genes of locus $I_{5632}$. Locus $II_{5632}$ is thought to have arisen by a duplication-insertion mechanism involving a circular intermediate carrying a set of *vpma* genes together with the IS*Mag1* element from locus I. Of the 16 potential Vpma proteins in strain 5632, 13 could be identified by nano-liquid chromatography coupled to ion-trap tandem mass spectrometry (Nouvel *et al.*, 2009). Although expansion and contraction of the variable gene repertoires, such as the *vpma* clusters observed in strain 5632, are expected to be rare events and responsible for intrastrain diversity rather than intraclonal variation, their occurrence might be promoted by mobile IS elements that inserted nearby phase-variable loci. Locus expansions may have a drastic impact on the possible antigenic mosaics displayed at the surface, increasing a species' virulence potential and hampering its elimination by the host.

## The *vsp* cluster of *Mycoplasma bovis*

*M. bovis* is the most important aetiological agent of various bovine diseases, such as mastitis in cows and pneumonia and arthritis in calves as well as genital disorders, which are altogether responsible for considerable economic losses in cattle and milk production. A chronic infection with *M. bovis* is linked to the organism's capability of undergoing high frequency antigenic variation in the *vsp* gene cluster. This cluster comprises 13 potential size and phase-variable *vsp* genes in strain PG45 (Lysnyansky *et al.*, 1999) that encode the 'variable surface proteins' (Behrens *et al.*, 1994). Site-specific DNA inversion, intrachromosomal recombination, strand slippage (Behrens *et al.*, 1994; Rosengarten *et al.*, 1994; Lysnyansky *et al.*, 1996) and gene locus expansion (Nussbaum *et al.*, 2002), all contribute to the variable expression of the Vsp proteins.

Each *vsp* gene is preceded by a highly conserved 5′ non-coding sequence that can be divided into two cassettes (Lysnyansky *et al.*, 1999, 2001b). Cassette 1 comprises 71 base pairs upstream of the ATG initiation codon and exhibits 99% homology among all *vsp* genes. This region contains a putative ribosome binding site and a 35-bp region from nucleotide −37 to −71, designated *vis* (*vsp* inversion site), which was identified as the potential sequence for site-specific DNA inversion. Cassette 2 is between 50 to 180 bp long and is more divergent among the *vsp* genes than cassette 1. The promoter responsible for *vsp* transcription lies in a cassette 2 designated $A_2$, as it was initially identified in a VspA-ON locus configuration. In this configuration, all other *vsp* genes were found to be transcriptionally silent. Juxtaposition of silent genes with the promoter or inversion of the promoter to silent genes occurs at inverted *vis* sites with a frequency of $10^{-2}$ to $10^{-3}$ per cell per generation (Fig. 9.2k). Interestingly, genomic inversion may concern short 545 bp regions, as between the *vis* sites of *vspE* and *vspC* or long 13.4 kbp regions, as between *vis* sites of *vspA* and *vspO*, which raises the question as to where is the size limitation of site-specific DNA inversion.

Approximately 80% of each Vsp protein is made up of tandem repeats that extend from the N-to the C-terminal end, the shortest repeat being 6 and the longest 87 amino acids in length (Lysnyansky *et al.*, 1999). Some of the 18 distinctive repetitive domains that have been identified among the *vsp* genes of strain PG45 were found to be present in only one Vsp family member, whereas other repeats occurred at variable locations in several Vsps (Lysnyansky *et al.*, 1999). This genetic composition not only allows size variation to happen via spontaneous slipped-strand mispairing within coding sequences, but also homologous recombination to take place between genes. Indeed, high frequency intrachromosomal homologous recombination between the two closely related members, *vspA* and *vspO*, explains the generation of a new chimeric and functional *vsp* gene, designated *vspC* (Lysnyansky *et al.*, 2001a). The authors suggested *vspC* to have evolved through a non-conservative recombination event between the *vis* sites of *vspO* and *vspA* and between two homologous sequences within the two genes when the genes are oriented in opposite direction. Because this mechanism, however, fails to explain the loss of the genetic fragment containing *vspM* and *vspN*, it is more likely that *vspC* evolved through an intrachromosomal recombination event that included only the two homologous sequences $R_O$ and $R_A$ when both genes have the same orientation (Fig. 9.3). The equal orientation of the homologous sequences would argue in favour of a looping out configuration by which a genomic fragment could be lost.

Comparison of the *vsp* locus from strain PG45 with that of another strain, designated 422, revealed that different strains may possess modified versions of the locus (Nussbaum *et al.*, 2002). A cluster of 11 Vsp-related open reading frames were identified in strain 422. High sequence homology of 99% exists among cassette 1 between both strains and among the N-terminal regions encoding the first 32 amino acids of each *vsp* gene. The distribution of the repeats found in both strains within individual Vsps, as well as their copy numbers, however varied considerably between the two *vsp* gene families. Altogether, up to 30 different repetitive units of different lengths (ranging from 4 to 127 amino acids) could be identified among the two strains; nine repeats were unique for strain 422 and 13 repeats were found only in the PG45 strain.

## The *vsa* cluster of *Mycoplasma pulmonis*

*M. pulmonis* is the etiologic agent of murine respiratory tract mycoplasmosis in rats and mice. The first documented examples of genes involved in phase variation based on site-specific DNA inversions in mycoplasma were the *vsa* system, encoding surface lipoproteins, and the *hsd* systems, encoding subunits of the type I restriction modification systems, of *M. pulmonis*.

To date, the *vsa* system is one of the most thoroughly investigated in respect to both mechanism and biological significance of the expressed proteins.

The *vsa* system differs from the *vpma* system of *M. agalactiae* and the *vsp* system of *M. bovis* in two major ways, the location of the inversion sites and the gene integrity. In the two latter systems, each CDS represents a fully equipped gene containing signal peptide and lipid anchor encoding sequences. In the *vsa* system, only one transcriptionally active gene possesses this sequence, which is alternated through DNA inversion with the transcriptionally silent CDS within the cluster. Hence, the proposed 31-bp long inversion sites (5′-CATCAAATAATGAACAAAGTGGAAATAATTC-3′) (*vrs* boxes: '*vsa* recombination sites') are located at the 5′ end of each silent CDS and within the N-terminal encoding sequence of the expressed gene. A 6-bp core sequence within the *vrs* box was identified as central to the recombination event (Shen *et*

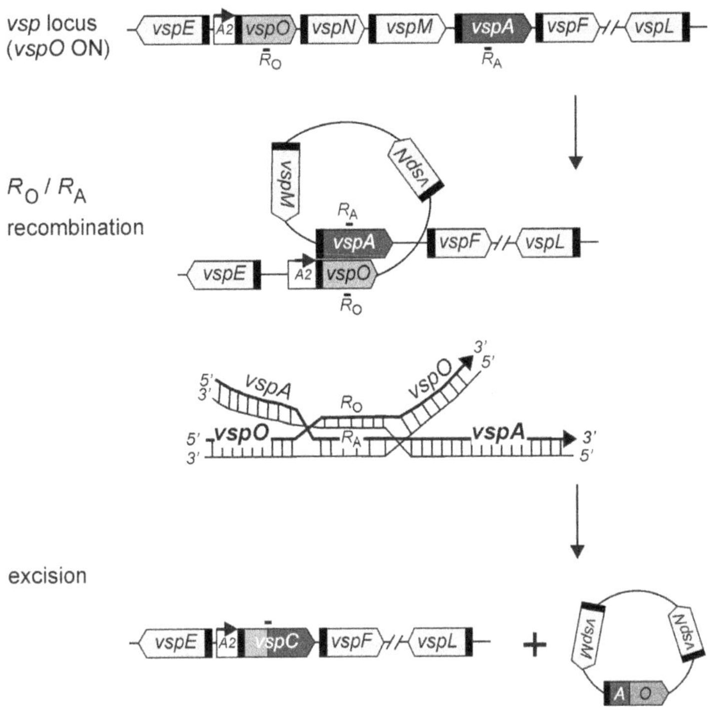


**Figure 9.3** A putative intrachromosomal recombination event within the *vsp* locus of *M. bovis*. Looping out of a region within the *vsp* locus, promoting intrachromosomal recombination between the homologous, repetitive domains $R_O$ and $R_A$ of the genes, *vspO* and the *vspA*. The event results in the generation of the chimeric, chromosomal *vspC* gene and a circular DNA containing the genes *vspM*, *vspN* and a potential chimeric *vspA/O*.

*al.*, 2000). Site-specific DNA inversions at the *vrs* boxes are *recA* independent and are mediated by the site-specific DNA recombinase HvsR '*hsd* and *vsa* site-specific recombinase', which is located adjacent to the *vsa* locus (Gumulak-Smith *et al.*, 2001; Ron *et al.*, 2002; Sitaraman *et al.*, 2002). Knockout mutants of the *hvsR* gene obtained by transposon mutagenesis were shown to be phase-locked for VsaA expression and unable to undergo both *vsa* and *hsd* switching.

The *vsa* locus of *M. pulmonis* strain KD735–15 contains at least 11 potential *vsa* genes (*vsa*A-F), some of which have homologous sequences and were therefore designated with numbers. Most of the genes (except *vsa*E1) encode proteins with variable close tandem repeats which are between 11 and 17 amino acids in length (Shen *et al.*, 2000). The genes *vsaA*, *-B*, and *-F* are distinct and encode proteins with unique tandem repeat domains. Gene duplication(s) must have occurred of the three *vsaC*, three *vsaE*, and two *vsaD* genes as a large block, as the intergenic regions between the three sets of *vsaC* and *-E* genes and the intergenic regions between the sets of *vsaE* and *-D* genes are highly similar. Moreover, the repeat units of the tandem repeat region of the three *vsaC* genes are nearly identical. Similarly, the two *vsaD* genes have identical repeat units and differ only in respect to the number of repeats.

The *vsa* locus from another strain (UAB CTIP) showed that neither the gene order nor the numbers of repeated units nor the sequences of these units in individual genes are conserved between this strain and KD735–15 (Shen *et al.*, 2000; Chambaud *et al.*, 2001). The *vsa* cluster of strain UAB CTIP contains only seven *vsa* genes (*vsa*A, -C, -E, -F, -G, -H and -I), which supports the hypothesis of gene duplication of *vsa* genes in strain KD735–15. The VsaH protein has the shortest C-terminal region within the family and VsaE lacks tandem repeats.

## The hsd clusters of *Mycoplasma pulmonis*

*M. pulmonis* strain UAB CTIP harbours three type I restriction modification (R-M) *hsd* loci, of which two were shown to be site-specific DNA inversion systems (Sitaraman and Dybvig, 1997; Dybvig *et al.*, 1998; Chambaud *et al.*, 2001). Type I R-M systems encode multi-functional protein complexes composed of subunits for sequence specificity (HsdS), methylation (HsdM) and restriction (HsdR). These subunits act as complexes with the organizations $M_2S$, composed of two HsdM units and one HsdS unit, or $R_2M_2S$, composed of two HsdR, two HsdM and one HsdS subunit [for review see: (Brocchi *et al.*, 2007; Srikhanta *et al.*, 2010)]. Both the S and M subunits are necessary and sufficient for the methyltransferase activity, which occurs at specific adenine residues. The nuclease reaction, which occurs at random sites thousands of base pairs away from the recognition site, requires a complex made up of all three types of subunits.

Both *hsd* loci in *M. pulmonis* are made up of two *hsdS* genes flanking *hsdR* and *hsdM* (Fig. 9.2g). Gene rearrangements are site-specific DNA inversions that occur at inversion sites located in the *hsdS* genes, designated *vip* 'vipareetus' and *hrs* '*hsd* recombination site' (Sitaraman and Dybvig, 1997; Dybvig *et al.*, 1998). Since the HsdS subunit determines the DNA recognition sequence specificity of the restriction enzyme, it is speculated that the generation of *hsdS* variants results in the phase-variable production of a family of restriction enzymes of differing specificities. Each *hsd* locus has four *vip* sites (two per *hsdS* gene) referred to as *vipα*, -β, -δ, and -γ and three *hrs* 1–3 sites, where *vipα* overlaps with *hrs1* and *vipγ* overlaps with *hrs3*. Among the *vip* sites, four types of DNA inversions were identified; these are: *vipα*/*hrs1* and *vipγ*/*hrs3* (α-γ inversion), *vipα* and *vipδ* (α-δ inversion), *vipβ* and *vipγ* (β-γ inversion), and *vipβ* and *vipδ* (β-δ inversion) (Sitaraman and Dybvig, 1997). Among the *hrs* sites, DNA inversion was identified only between *hrs1* and *hrs2* (Dybvig *et al.*, 1998). *In vitro* studies, where *E. coli* was co-transformed with plasmids containing pair–wise combinations of recombination sites together with a plasmid that expressed the HvsR recombinase, and where site-specific inversions were monitored by PCR, revealed that the minimal 12-bp homologous *vip* (5′-CAAAGTGCAATA-3′) site was not sufficient for site-specific DNA inversion (Sitaraman *et al.*, 2002). HvsR, however, catalysed inversions between two *vip*/*hrs* sequences or between a *vip*/*hrs* (5′-CAAAGTGCAATATAATTAAGATTATTGAACCT-3′) and a *hrs* (5′-TAATTAAGATTATTGAACCT-3′) sequence in *E. coli*.

## The *mba* and UU172 phase-variable loci of *Ureaplasma parvum*

*U. parvum* and *U. urealyticum* are potential pathogens of the human genital tract and are associated with non-gonococcal, non-chlamydial urethritis in men, chorioamnionitis in pregnant women as well as bronchopulmonary dysplasia in newborn infants. Both species express a distinct immunodominant phase- and size-variable surface protein, the multiple-banded antigen (MBA), whose gene is one member of a paralogous family of genes dispersed throughout the chromosome (Teng *et al.*, 1994; Zheng *et al.*, 1994, 1995; Glass *et al.*, 2000; Kong *et al.*, 2000; Monecke *et al.*, 2003). Molecular genotyping methods based on the *mba* gene and its 5′ end, as well as on other genomic regions, have been explored as a replacement for the antibody-based phenotyping methods (Xiao *et al.*, 2010).

The *mba* gene encodes a prolipoprotein with a signal peptide, an approximately 120 amino acids long non-variable part and a size-variable part comprising a number of uniform repeating units, which, depending on the serovar, are made up of 4 to 10 amino acids. In *U. parvum* serovar 3, the phase-variable *mba* locus comprises the two open reading frames (ORFs) UU375 (*mba*) and UU376, whereas in *U. urealyticum* serovar 10 the equivalent locus comprises up to six potentially phase-variable ORFs. UU376 has no homology to the *mba* gene and contains no repeated sequences. Phase variation between UU375 and UU376 is achieved by site-specific DNA rearrangement in a 24-bp-long inverted repeat (5′-ATTTGAATTATCAAACAGAAAAAG-3′) and occurs when the ORFs are oriented in opposite directions (Zimmerman *et al.*, 2009). The inverted region includes the sequence encoding the non-variable part and a 330-bp intergenic spacer region that contains the promoter.

A second, more conserved phase-variable locus among the two species is the 'UU172 phase-variable element' (Zimmerman *et al.*, 2011). Like the *mba* locus of *U. parvum* serovar 3, the UU172 phase-variable element comprises two CDSs (UU172 and UU171) which are oriented in opposite direction. UU172, an MBA N-terminal paralogue, shows high sequence similarity to UU375 in the N-terminal region (residues 1–160 are 47% similar to residues 1–160 of UU375) and UU171 shows high sequence similarity to UU376 (residues 1–187 are 60% similar to residues 45–224 of UU376). Moreover, two inverted repeats (5′-ATAATTTAAATTAT-CAAACAGTAACTTTTGAACAAGTTCCT-3′), which share partial homology (in bold) to the inverted repeats of the *mba* locus, are located in this region; one in the 5′ sequence of UU172 and another in the intergenic spacer region between UU172 and UU171. Phase-variable expression of the UU172 element is governed by site-specific DNA inversion analogous to that occurring in the *mba* locus. Phase variation of UU172 has also been predicted as a consequence of slipped-strand mispairing at a poly-AT tract located in the intergenic region between UU171 and UU172 (Rocha and Blanchard, 2002). Although this has not been demonstrated, it is a likely mechanism for obtaining ON/OFF switching in expression in *Ureaplasma*. Several poly-T and poly-TA tracts were located in the intergenic regions 5′ to genes encoding putative membrane lipoproteins; some of these regions show strong similarity to the *vmm* promoter of *M. mycoides* subsp. *mycoides* SC.

## The mpl clusters of *Mycoplasma penetrans*

*M. penetrans* was first isolated from the urine sample of a HIV (Human Immunodeficiency Virus)-infected patient (Lo *et al.*, 1992, 1993). The characteristics of this bacterium (e.g. invasion of eukaryotic cells, toxicity to chick embryo, and haemolytic and haemoxidative activity) (Lo *et al.*, 1993; Giron *et al.*, 1996; Kannan and Baseman, 2000) suggest its potential pathogenicity to humans. *M. penetrans* expresses an immunodominant protein, the P35 lipoprotein, which is surface-exposed and has been used as the major antigen for serological diagnosis of *M. penetrans* infection (Wang *et al.*, 1993; Ferris *et al.*, 1995; Grau *et al.*, 1995; Neyrolles *et al.*, 1999a,b). The *p35* gene (MYPE6810) is a member of the largest paralogous gene family in *M. penetrans* designated as the '*p35* gene family' or '*mpl* genes', which comprises a total of 44 CDSs in the sequenced *M. penetrans* HF-2 genome; all members are homologous to the *p35* gene with a degree of identity to the P35 lipoprotein ranging from 34% to 70% (Sasaki *et al.*, 2002). Of these 44 CDSs, 38 possess at their N-terminal a highly conserved amino acid sequence containing a cysteine residue which is responsible for membrane anchorage of the mature product once linked to a fatty acid chain. The remaining six CDSs lack this N-terminal sequence and were therefore designated 'signal-peptide-less *p35* gene homologues' (Sasaki *et al.*, 2002). The 44 CDSs are clustered at four different loci in the genome, with two loci each containing four CDSs, one containing the six signal-peptide-less homologues and the largest locus containing 30 CDSs including the *p35* gene and the previously described lipid-associated membrane proteins (LAMPs) P34A and P38 (Sasaki *et al.*, 2002). P35 and several LAMPs have been shown to undergo ON/OFF phase variation with a frequency ranging from $10^{-2}$ to $10^{-3}$ per cell per generation (Ferris *et al.*, 1995; Neyrolles *et al.*, 1999a; Roske *et al.*, 2001). Phase variation of the proteins is achieved by inversion of individual promoter regions in front of the genes. These promoters are located in ca. 135 bp long regions, are flanked by 12 bp inverted repeats (IR) and contain a unique sequence that form a terminator-like structure which has been proposed as preventing read-through transcription from preceding genes or antisense transcription from OFF configuration promoters (Horino *et al.*, 2003). The *p34* or the *p42* promoter inversion was monitored in two model organisms (*Escherichia coli* and *M. pneumoniae*) in the presence of either of two *M. pneumoniae* recombinases (MYPE2900 and MYPE8180) and shown to occur only in presence of the MYPE2900 protein, indicating that the promoter inversion is catalysed by a single factor. As the IRs of *p42* and *p35* are slightly different, it was suggested that the recombinase has flexibility in recognizing the recombination site sequences in all *mpl* promoters. An alignment of 30 IR sequences found in the *mpl* promoter regions showed that the nucleotides TAA and ATTA are conserved at both ends of 12 bp-IRs; however, the five nucleotides in the middle are variable with the consensus sequence reading TAAYNNNDATTA (Y = C or T; D = A, G or T) (Horino *et al.*, 2009). Although the recombinase seems to have this flexibility in sequence recognition, inversions between IRs of different promoters have not been reported. In contrast to the *vsa, vsp* and *vpma* systems where the recombinase gene lies in proximity to the corresponding loci, the MYPE2900 gene is located approximately 500 kb away from the major *mpl* gene cluster.

## DNA rearrangements by homologous recombination

### The vlhA cluster of *Mycoplasma synoviae*

*M. synoviae* is an economically important pathogen of poultry, causing synovitis, chronic respiratory tract disease, and retarded growth in chickens and turkeys. The genome of strain 53 has been sequenced and comparative analysis with that of the avian pathogen *M. gallisepticum* provided evidence for horizontal gene transfer between the two phylogenetically distant species (Vasconcelos *et al.*, 2005; Sirand-Pugnet *et al.*, 2007). Among the genes that could have arisen by horizontal gene transfer are those encoding haemagglutinins, which represent the most important surface proteins involved in colonization and virulence of avian mycoplasmas (Bencina, 2002). In *M. synoviae, M. gallisepticum* and *M. imitans* haemagglutinins are encoded by related sequences of multigene families referred to as *vlhA* genes (Markham *et al.*, 1994, 1999; Noormohammadi *et al.*, 1998; Bencina *et al.*, 1999). Unlike *M. gallisepticum,* in which the *vlhA* genes are located in five distinct loci around the chromosome and in which antigenic variation is generated by alternating transcription of over 40 translationally competent genes (Glew *et al.*, 1998; Markham *et al.*, 1999), *M. synoviae* has all of the *vlhA* sequences (one gene and several tandemly oriented pseudogenes) clustered together. In strain 53, the *vlhA* cluster is confined to a 69-kb region with only one copy being expressed. The *vlhA* promoter region of *M. synoviae* has no significant identity to any of the *vlhA* promoter regions of *M. gallisepticum* and contains no extended tri-nucleotide repeat motif (Noormohammadi *et al.*, 2000). The *vlh*A pseudogenes of *M. synoviae* contain variable regions but also homologous regions among each other. Antigenic variation of the *M. synoviae* VlhA products is the result of unidirectional recombination occurring between members of the multiple pseudogene copies and the functional *vlhA* copy (Noormohammadi *et al.*, 2000). Recombination events among the pseudogenes have not been reported suggesting that the opposite orientation of the pseudogenes to the active *vlhA* gene might be an essential factor for the gene conversion event. A model has been suggested in which part, or all, of a pseudogene sequence is first duplicated and this duplicate recombines with at least one of three specific 5′ sites (codon for amino acid 136, 356 or 442) and at one of two specific 3′ sites (codon for amino acid 612 or the carboxyl-terminal amino acid) in the expressed *vlhA* gene (Noormohammadi *et al.*, 2000). A possible model of the gene conversion is illustrated in Fig. 9.4.

The expressed VlhA is post-translationally cleaved after amino acid residue 344 resulting into an N-terminal lipoprotein (MSPB) and a C-terminal haemagglutinin protein (MSPA) (Bencina *et al.*, 1999). Both MSPA and MSPB are surface-exposed proteins and exhibit high frequency antigenic variation, but only MSPA is responsible for haemagglutination and mediates binding to erythrocytes (Noormohammadi *et al.*, 1997, 1998). Several expressed *vlh*A gene variants (*vlh*A1–5) have been characterized in *M. synoviae* strain WVU 1853 (Noormohammadi *et al.*, 2000); their genes are equally sized and they share a highly conserved 5′ region (nucleotides 1–700) and 3′ region (nucleotides 2244-stop codon) but differ from each other in a semi-variable region (nucleotides 720–1100) and a highly variable region (nucleotides 1100–2244). In a derivative of strain WVU 1853, another *vlhA* variant termed

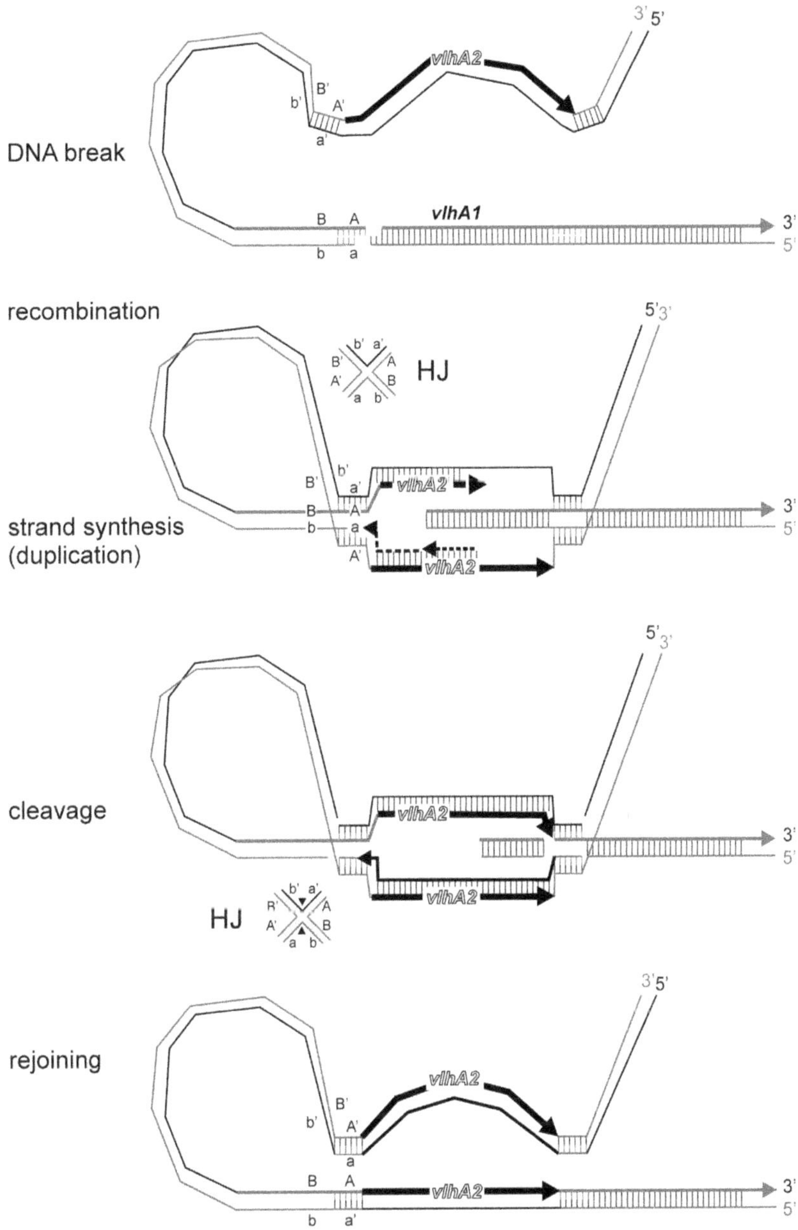


**Figure 9.4** A putative gene conversion event as in the *vlhA* locus of *M. synoviae*. Sequences within the conserved regions of the *vlhA1* gene might be broken, followed by 5′→3′ exonuclease digestion. The free 3′ ssDNA tail that is generated by this process scans the pseudogenes (e.g. *vlhA2*) for homologous sequences. After homologous recombination, the pseudogene region *vlhA2* is duplicated through DNA extension on both strands. During this process, the *vlhA1* sequence to be exchanged might be degraded by nucleases. Holliday junctions (HJ) are cleaved (▲) by an HJ resolvase and nicked ends are relegated. A colour version of this figure is available in the plate section at the back of the book.

MS2/28.1 was discovered whose haemagglutinin region has undergone considerable size reduction (Ben Abdelmoumen *et al.*, 1999). Its deduced 604-amino-acid sequence shows a perfect sequence identity to the previously reported *vlhA* expressed genes along the first 224 residues, then highly diverges with only 37.6% identity. The highly divergent and considerably shorter (60 amino acids) C-terminal haemagglutinin product was found to be expressed at the surface of the bacterium and was able to haemagglutinate chicken erythrocytes (Khiari *et al.*, 2010).

## The mgp genes of *Mycoplasma genitalium*

The protein MgPa (also referred to as P140) of *M. genitalium* corresponds to the protein P1 of *M. pneumoniae,* sharing identical functional characteristics, such as the important role in the attachment of the organism to host epithelial cells (Hu *et al.*, 1987; Morrison-Plummer *et al.*, 1987). Like the P1 protein, MgPa is one of the major surface proteins eliciting a strong host immune response (Svenstrup *et al.*, 2006). The gene encoding MgPa (*mgpB*) is located in an operon comprising three genes in the order of MG190 (*mgpA*), MG191 (*mgpB*) and MG192 (*mgpC*) (Inamine *et al.*, 1989; Fraser *et al.*, 1995). The *mgpC* gene encodes the P110 protein (also referred to as P114). Mutants that fail to express MgPa and/or P110 are non-adherent and lack the characteristic attachment organelle (Burgos *et al.*, 2006). The complete operon is present only once in the genome, but there are nine repetitive elements in the form of truncated (partial) copies of the *mgpB* and *mgpC* genes dispersed throughout the genome. These repetitive elements have been designated as 'MgPa repeats' or 'MgPar' sequences and have not been predicted to serve as functional expression sites due to multiple stop codons in every frame and to the lack of over half of the expression site (Fraser *et al.*, 1995; Peterson *et al.*, 1995; Iverson-Cabral *et al.*, 2006). Initially, the *mgpB* gene was subdivided into 10 regions (A-J) with sequence repetition throughout the genome occurring only of the regions B, EF and G (Dallo and Baseman, 1991; Iverson-Cabral *et al.*, 2006). Sequence repetition of *mgpC* on the genome is confined to its KLM region.

Recent studies have focused on the recombination that occurs between the nine MgPar regions and the discrete regions located in the genes *mgpB* and *mgpC* and demonstrated that both reciprocal and non-reciprocal recombination can occur between them *in vitro* and *in vivo* (Iverson-Cabral *et al.*, 2006, 2007; Ma *et al.*, 2007). Sequence analysis of subclonal populations of the type strain G37$^{T}$ showed that reciprocal recombination between regions B, G and KLM with partial sequences from MgPar 7, MgPar 3 and MgPar 1 and 8 respectively occurs *in vitro* while passaging the organism in broth (Iverson-Cabral *et al.*, 2007). Results also suggested that an intra-MgPar recombination event happened between MgPar 1 and 8 prior to recombination with the KLM region. Evaluation of heterogeneity and variability of the *mgpB* sequence from a woman persistently infected with a single *M. genitalium* strain identified 17 different *mgpB* region B variants that had developed over a time period of almost 2 years (Iverson-Cabral *et al.*, 2006). The newly identified region B variants contained partial sequences deriving from MgPar 1, 2, 4, 5, 7, and 9. A similar study on region KLM of *mgpC* identified variants that could contain partial sequences from all nine MgPar elements (Iverson-Cabral *et al.*, 2007). Novel sequences, non-homologous to any sequences in the sequenced G37$^{T}$ genome were also discovered in the regions B and KLM (Iverson-Cabral *et al.*, 2006, 2007).

## The P1 gene and RepMP elements of *Mycoplasma pneumoniae*

### Recombination of RepMP2/3 and RepMP4

*M. pneumoniae* strains are divided into the two groups M129 (subtype 1) and FH (subtype 2), based on sequence variations of the gene encoding the major adhesin P1 (Dallo *et al.*, 1990; Su *et al.*, 1990b). The P1 protein corresponds to the adhesin molecule MgPa of *M. genitalium.* Like MgPa, the P1 gene (MPN141, also known as ORF5) is located in an operon that includes MPN140 and MPN142 (also known as ORF6). The P1 gene contains two distinct repetitive DNA sequence (RepMP2/3-d and RepMP4-c) while MPN142 has only one (RepMP5-c) (Colman *et al.*, 1990; Ruland *et al.*, 1990, 1994). There are at least 10 copies of RepMP2/3 (Fig. 9.2a–j), eight copies of RepMP4 (Fig. 9.2a–h) and 8 copies of RepMP5 (Fig. 9.2a–h) distributed throughout the genome with predicted lengths between 1–2 kbp (Himmelreich *et al.*, 1996; Spuesens *et al.*, 2009, 2010). It has long been hypothesized that numerous copies of these RepMPs provide a pool of sequences for homologous recombination which could enable the organism to undergo antigenic variation (Wenzel and Herrmann, 1988; Su *et al.*, 1990a, 1993; Forsyth and Geary, 1996; Rocha and Blanchard, 2002). This hypothesis was supported by the finding of new sequence patterns of the P1 gene in different isolates (Kenri *et al.*, 1999; Dorigo-Zetsma *et al.*, 2001), which led to a temporary subdivision of the two subtypes into eight P1 types (1a–e and 2a–c) (Dorigo-Zetsma *et al.*, 2000). Among these, one isolate showed that the RepMP2/3-d element could acquire sequences deriving from the distant RepMP2/3-a (Kenri *et al.*, 1999).

In a recent study, a more thorough sequence analysis and comparison of the RepMP2/3 and RepMP4 regions was carried out. Results showed that all RepMP2/3 and RepMP4 elements carry either a subtype 1 or a subtype 2-specific signature sequence, but that the occurrence of intragenomic recombination of the RepMP2/3 elements might be a relatively common phenomenon (Spuesens *et al.*, 2009). Unlike the RepMP2/3 elements, the RepMP4 elements within 23 studied strains did not show any indication of an inter-element recombination event. However, results suggested that RepMP4 elements can be donated to RepMP4-c of the P1 gene. The highest level of variation among strains was observed in RepMP2/3-d and -c. In four of the investigated strains, including three subtype 1 strains and one subtype 2 strain, sequences from element RepMP2/3-h appeared to be transferred to the RepMP2/3-c element. As the original sequence of element RepMP2/3-c seemed to be missing in these strains and as the RepMP2/3-h element remained unaltered, it was inferred that the elements have recombined by gene conversion. Another possible unidirectional recombination event was the donation of sequences from RepMP2/3-e to the RepMP2/3-d element within the P1 gene.

Other than for the MgPar regions of *M. genitalium*, where homologous recombinations had been monitored in clonal lineages (Iverson-Cabral *et al.*, 2006, 2007), recombination events in the P1 gene have thus far not been directly demonstrated and the proposed unidirectional gene conversion among RepMP elements is solely based on sequence comparisons of different isolates. It therefore remains to be explained why certain P1 configurations from different strains (e.g. isolate Mp3896) cannot be classified as either subtyp1 or 2. Another interesting exception is the P1 sequence of strain Mp4817, which contains a hybrid sequence that possesses both subtype 1 and subtype 2-specific sequences (Dorigo-Zetsma *et al.*, 2001; Spuesens *et al.*, 2009).

Based on the finding of little sequence diversity among *M. pneumoniae* strains, this species is considered as being genetically highly stable among mycoplasmas, a property that has been attributed to its truncated recombination machinery and non-functional holiday junction resolving enzyme RecU (Sluijter *et al.*, 2010). In 68 analysed subtype 1 strains, a translation termination codon was found in the corresponding gene (MPN528a) that leads to the expression of a short 60 amino acid long polypeptide. Interestingly, while subtype 2 strains do have the capacity to express a full-length RecU, this protein was found to be inactive in *in vitro* DNA binding and cleavage assays, in contrast to its counterpart from *M. genitalium*. The difference in enzymatic activity of the RecU protein might explain, in part, why the MgPar elements in *M. genitalium* appear to rearrange with a considerably higher frequency than the RepMP elements in *M. pneumoniae* (Sluijter *et al.*, 2010).

**Recombination of RepMP5**

A recombination event between RepMP5-c and a distant donor sequence which eventually led to the MPN142 gene of *M. pneumoniae* strain FH (subtype 2) had been previously suggested (Ruland *et al.*, 1994). Spruesens *et al.* (2010) identified this potential donor sequence as RepMP5-h. So far, all subtype 2 strains analysed have a similar RepMP5-c sequence to that of strain FH, suggesting that the putative recombination event must have taken place early on in the evolution of *M. pneumoniae*. Other potential recombination events between RepMP5 elements were suggested to have occurred more recently (Spuesens *et al.*, 2010). Among such, sequence donations seem to have occurred from RepMP5-h (199 bp) to RepMP5-d, from either RepMP5-f or -h (127 bp) to RepMP5-g, from RepMP5-e (61 bp) to RepMP5-h and from RepMP5-a or -f (45 bp) to the RepMP5-c element of MPN142 (Spuesens *et al.*, 2010). As the 'donor' DNA elements appeared to have remained unaltered by the proposed recombination events, the results suggest that the rearrangement of RepMP5 sequences occur in a nonreciprocal, unidirectional fashion, similarly to that proposed for the RepMP2/3 and RepMP4 elements (Kenri *et al.*, 1999; Spuesens *et al.*, 2009).

### Recombination within RepMP1

A further, large repetitive element in the *M. pneumoniae* genome, which is however not found within the P1 operon, is RepMP1 (Wenzel and Herrmann, 1988). Analysis of the complete genomic sequence of the M129 reference strain revealed 14 copies of RepMP1 (Himmelreich *et al.*, 1996), several of which seem to encode parts of expressed proteins (Musatovova *et al.*, 2008). In addition to the 14 identified RepMP1 elements, two previously unrecognized elements were localized in the MPN137 and MPN138 genes (Musatovova *et al.*, 2008). Musatovova *et al.* proposed a recombination event in a subtype 1 strain that seemingly had occurred between RepMP1 elements located within MPN130, MPN137 and MPN138. These authors report on the replacement of 888 nucleotides in the MPN137 and MPN138 genes by 49 nucleotides derived from MPN130, an event that had led to the generation of a new fused reading frame (Musatovova *et al.*, 2008).

## Biological role and significance of mycoplasma surface antigenic variation

For most variable membrane components of host-adapted bacteria, it has been suggested that the main purpose of their phase variation is to evade the host immune response. Convincing support for this view comes from *in vivo* and *in vitro* experiments where phase variation was observed either as a result of an active adaptive immune system or as a consequence of antibody pressure directed against specific surface exposed proteins. In this sense, phase-variable proteins operate as a dynamic armour by which pathogens compulsorily camouflage or protect themselves when being attacked. Biological significance of antigenic variation should, however, not be confused with the biological function(s) of a variable protein, although these readily seem to overlap. In many cases proteins may serve diverse functions or play a secondary role in more complex biological systems. Moreover, the biological significance of the antigenic switch often manifests itself when a phase-variable protein is not expressed and the cell is devoid of its primary function.

Among mycoplasmas, the most extensive investigations regarding the biological significance of surface antigenic variation have been so far undertaken on the *vsa* system of *M. pulmonis.* This research has shown that phase and size variation of the Vsa proteins affects cell growth, biofilm formation, resistance to complement and haemadsorption aside from their immune evasive role *in vivo* (Dybvig *et al.*, 1989; Gumulak-Smith *et al.*, 2001; Simmons and Dybvig, 2003, 2007; Simmons *et al.*, 2004, 2007; Denison *et al.*, 2005). Moreover, it was observed that phase variation of the *vsa* system was concurrently taking place with phase variation of the host specific restriction-modification system (hsd) in the same organism, which alludes to the complexity of phase variation in regard to its biological significance.

The significance of antigenic variation systems for the survival of the mycoplasma pathogen in the immunocompetent host was recently illustrated in an *M. agalactiae* infection study. The selective pressure of the host immune response was such that it triggered alternative secondary mechanisms of Vpma switching, which were never before observed during numerous *in vitro* passages of these *xer1*-disrupted phase-locked mutants that were otherwise incapable of the usual Xer1-mediated site-specific recombinations (R. Chopra-Dewasthal, unpublished). Furthermore, this selective pressure was instrumental in inducing complex gene rearrangements, such as chimeras, duplications and deletions in the *vpma* locus of mycoplasma reisolates, a causal relationship that was so far never shown for *M. agalactiae.* These are indeed interesting new features of surface antigenic variation systems in pathogenic mycoplasmas.

In many cases, the role that phase-variable proteins play in regard to the biological significance remains speculative and phase variation should therefore be seen as a global strategy used by bacteria to survive environmental conditions or to colonize new habitats by creating heterogeneous populations.

## Immune evasion and protection

For many *Mycoplasma* species, appropriate model organisms for conducting *in vivo* experiments are not available and phase variation, therefore, has to be monitored in *in vitro* experiments. A classical way to induce antigenic variation is through application of host derived serum antibodies or antibodies directed against specific surface exposed proteins in broth or on agar. Such an approach has been undertaken to demonstrate phase variation of the *vlp* system of *M. hyorhinis* (Citti *et al.*, 1997), the *vlhA* system of *M. gallisepticum* (Markham *et al.*, 1998) and the *mba* system of *U. parvum* and *urealyticum* (Monecke *et al.*, 2003; Zimmerman *et al.*, 2009). In all cases, antibody pressure was circumvented by the expression of size variations or alternative proteins within the same protein families.

The importance of antigenic variation in providing mycoplasmas with a means to escape host antibody attack was first supported by experimental evidence acquired from serum antibody-treated *M. hyorhinis*, which showed that the growth inhibitory effect of serum antibodies collected from an infected swine was Vlp length-dependent. Clonal variants expressing a long Vlp were resistant, while those displaying a short version of the same Vlp were susceptible (Citti *et al.*, 1997). Further, propagation of susceptible variants in the presence of these host antibodies resulted in resistant populations that escaped the pressure by expressing either a longer variation of an already expressed Vlp or by expressing an alternative Vlp. When a clonal population with the Vlp profile A87-B30+C37- was propagated in 10% immune serum, half of the output population had completely switched OFF VlpB and expressed only the 87-kDa version of VlpA. Similarly, an input population with the profile A87-B66-C37+ switched to either $A_{87}^{-}B_{66}^{+}C_{37}^{+}$ or $A_{87}^{+}B_{66}^{-}C_{37}^{+}$ expression after antibody pressure (Citti *et al.*, 1997). Depleting the host derived serum of specific Vlp-directed antibodies failed to eliminate the growth-inhibiting activity against susceptible populations expressing the short Vlp version. This meant that the Vlps themselves were not the antigens targeted by the serum but rather that long Vlps serve a masking role to other critical, yet unknown, components.

Antigenic switching as an escape mechanism to selective antibody pressure could be demonstrated for the *vlhA* system of *M. gallisepticum* and the *mba* system of *U. parvum*. Antibody pressure with either, mono- or polyclonal antibodies directed against an expressed VlhA resulted in the expression of alternative VlhAs (Markham *et al.*, 1998). Likewise, antibody pressure with monospecific antibodies directed against either of the two proteins expressed from the *mba* locus of *U. parvum* resulted in antigenic switching and expression of the protein not exposed to pressure (Zimmerman *et al.*, 2009). These experiments, of course, have not demonstrated the significance of size variation of the MBA protein or a possible protective, masking role of long MBA variants.

The potential role that Vsa size variation might play in protecting *M. pulmonis* from the immune system is intriguing. *In vitro* experiments indicated that size variation of individual Vsa proteins correlates with differential adhesion strength, biofilm development and susceptibility to complement. *M. pulmonis* variants expressing a short Vsa with a few tandem repeats were shown to be susceptible to complement killing, while long Vsa variants were highly resistant (Simmons and Dybvig, 2003; Simmons *et al.*, 2004). This raises the question of why an organism would risk a decrease in protein size when thereby becoming more vulnerable to elimination. Size reduction is, therefore, only tolerable if, through the incidence, growth advantages or alternative means of protection emerge that will compensate for the loss of the previous protection. Indeed, this seems to be the case in Vsa size variation. Variants expressing short Vsa proteins were shown to have higher adhesion strength to erythrocytes and polystyrene than variants expressing a long version of the same protein (Simmons and Dybvig, 2003; Simmons *et al.*, 2004). In addition, variants expressing the short Vsa proteins were shown to have the potential to form biofilms on polystyrene and glass, which protected the cells from the lytic effect of complement (Simmons and Dybvig, 2007; Simmons *et al.*, 2007). Variants that produced a long Vsa protein formed microcolonies instead of biofilms in the same environment. These results suggest that the vulnerability obtained through Vsa size reduction to the host immune system is balanced by the ability to form protective biofilms. Whether *M. pulmonis* really experiences this protective benefit *in vivo* is not known.

*In vivo* experiments with *M. pulmonis* using wild-type mice versus mice deficient in B and T-cells ($RAG^{-}/^{-}$) indicated that the adaptive immune system exerts a selection pressure on the phase variation, but not on size variation, of the *vsa* system (Denison *et al.*, 2005). The Vsa profile of output populations from infected $RAG^{-}/^{-}$ mice varied little from that of the initial inoculum, while that from wild-type mice varied significantly. Selective pressure in the wild-type host was thought to be driven by specific antibodies, as the timing of the appearance of phase variants was consistent with the production of specific antibodies in the host (Gumulak-Smith *et al.*, 2001; Denison *et al.*, 2005). Phenotypic switching of VlhA antigens in *M. gallisepticum* could similarly be traced *in vivo*, but seems to be independent of the host immune response, as rapid VlhA switching could be observed even before serum antibodies were detectable (Glew *et al.*, 2000a). This finding indicates that the phenotypic switching is a stochastic event and only through selective pressure, escape variants will propagate. Variation in the antigenic repertoire is therefore a combination of the frequency of the genetic switch and the selective pressure of the environment for a particular surface protein.

## Variation without loss of function

Many of the phase-variable membrane proteins identified in bacterial pathogens have proven to be highly immunogenic in their respective hosts and therefore offer easy targets to the host immune system. For the host-adapted microorganisms this means that such components must be either temporarily dispensable but recoverable or capable of variation without loss of their principal function. High-frequency ON/OFF switching via DNA inversion and strand slippage at hot-spots are secure mechanisms to achieve this state. A more risky way to acquire variation without loss of function is through homologous recombination events, as these can introduce disadvantageous alterations in essential domains.

High-frequency sequence variations in surface proteins in pathogenic *Mycoplasma* species without loss of function have been demonstrated for the haemagglutinin of *M. synoviae* and the major adhesins of *M. genitalium* and *M. pneumoniae*. The *vlhA* gene of *M. synoviae* encodes a protein that is cleaved into an N-terminal (MSPB) product and a C-terminal (MSPA) haemagglutinin. Several expressed *vlhA* variants with similar lengths but variable sequences have been characterized, all encoding an MSPA region of approximately 45–50 kDa (Noormohammadi *et al.*, 2000). Khiari *et al.* identified a *vlhA* variant that encoded a considerably shorter (~60 aa) MSPA protein and demonstrated that this protein may vary both in size and in sequence composition without compromising the haemagglutination activity (Khiari *et al.*, 2010).

Similarly, *M. pneumoniae* can vary its RepMP5 region of the MPN142 gene without losing the adhesive properties of the P1 protein (Catrein *et al.*, 2004). MPN142 is a member of the P1 operon that encodes one of the most prominent adherence proteins in the attachment organelle of *M. pneumoniae*. Two cleavage products (P40 and P90) derive from MPN142 that serve as accessory proteins in the P1 complex and which are required for the correct localization of P1 at the attachment organelle (Layh-Schmitt and Harkenthal, 1999; Catrein *et al.*, 2005). The MPN142 gene contains one (RepMP5-c) of eight RepMP5 repetitive elements that are distributed throughout the chromosome. In a complementation experiment, a cytadherence-negative mutant was converted to a cytadherence-positive strain by exchanging its RepMP5-c element with the RepMP5-b, such as to mimic a putative recombination event (Catrein *et al.*, 2004). Also, the recombinant strain carrying this modified MPN142 gene was found to be virulent in an animal model and expressed both the P40 and P90 protein (Catrein *et al.*, 2004).

## The 'phasevarion'

Phase variation is not limited to membrane proteins. In several host-adapted pathogens, phase variation has been found to occur in genes that encode methyltransferases associated with restriction–modification (R-M) systems. These methyltransferases, encoded by members of the *mod* gene family, have the potential to regulate the expression of multiple genes, a genetic system that has received the term 'phasevarion' or 'phase-variable regulon'. Since phase-variable *mod* family genes are widely distributed, the phasevarion may be a common strategy used by host-adapted bacterial pathogens to randomly switch between distinct cell types (Srikhanta *et al.*, 2010). The function of the phase varion might be diverse, reaching from cellular defence against foreign DNA (such as phage DNA, other but similar R-M systems or horizontally transferred DNA) to epigenetic gene regulation (Brocchi *et al.*, 2007). R-M systems are present in all *Mycoplasma* species sequenced so far, but only in *M. pulmonis* has it been shown to undergo phase variation.

Gumuklak-Smith *et al.* (2001) demonstrated that there is a correlation between the phase switching of the *vsa* locus and the phase variation of the *hsd* R-M system in *M. pulmonis* during *in vivo* infection, which suggests a possible relationship between these mechanisms and virulence (Gumulak-Smith *et al.*, 2001). In this study, rats were infected intranasally with a VsaA expressing population, isolates were taken from nose, trachea and lung tissues at 7 and 14 days post infection and both, the R-M activity and the Vsa variation, were analysed. Isolates collected from the nose lacked R-M activity and still expressed the VsaA protein, while isolates from the lower respiratory tract were more heterogeneous, showing activity of the R-M system and expression of alternative Vsa proteins (VsaB, VsaC and VsaE). It was speculated that the altered specificity of the Hsd holoenzyme might result in new methylation patterns in chromosomal DNA, providing an epigenetic mechanism for regulation of expression of other genes. Phase variation of the R-M system in *M. pulmonis* has been described as a consequence of DNA inversion. Strand slippage at short sequence repeats (poly-AG) was proposed as a mechanism for generating phase variation of the *hsd*S gene in *U. parvum*, of several genes involved in an R-M type III system of *M. pulmonis* and of the *mod* genes of *M. hyopneumoniae* (Rocha and Blanchard, 2002; Brocchi *et al.*, 2007; Srikhanta *et al.*, 2010).

## Growth competition and mobility

High-frequency phase variation reported *in vivo* or *in vitro* during selective antibody pressure conditions, is often not observed when cells are grown in broth in standard conditions. For instance *M. pulmonis* populations frequently switch to VsaA expression when passaged in broth and arrest in this phase (Gumulak-Smith *et al.*, 2001). Similarly, *M. gallisepticum* had been shown to revert to the expression of a certain VlhA protein after removal of selective antibody pressure (Markham *et al.*, 1998). Also, the expression of certain Vsa proteins (Vsa A, -G and -I) were shown to be more stable than others (VsaC and -H) throughout prolonged laboratory passaging (Denison *et al.*, 2005). The biological significance of this switch and arrest is not clear, but it was speculated that cells producing a certain protein might have a growth advantage in culture (Dybvig *et al.*, 1989). In other words, with the relief from selective pressure the organism might experience an involuntary arrest in that phase which is metabolically most favourable for growth in a defined medium. In this circumstance, both phase and size variations observed in broth might have a completely different biological significance than they have *in vivo*.

An often, disregarded stress factor during proliferation is the growth competition that a pathogen has with its own subclonal population. Fast growing organisms deplete their habitat of nutrients and a previously balanced environment can quickly become a niche unfavourable for growth. To sustain an exponential growth phase, the development of a strategy by which part of the subpopulation can be sacrificed is of advantage for a persisting infection. ON/OFF-switching of major cytadhesins could provide this ability and indeed several *Mycoplasma* species have developed systems by which an alternating adhesion-strength can be achieved through phase variation. *M. hominis*, for instance, reduces adhesion-strength by expressing a truncated form of its variable adhesion protein Vaa (see 'Surface antigen ON/OFF switching via DNA slippage and point mutations' above), which may provide part of the population with the flexibility to transfer among niches with different requirements for colonization, aside from the often discussed evasion of host immune response. Phase variation of the major adhesion component of the avian pathogen *M. gallisepticum* is likewise governed by a point mutation. The adhesin is homologous to the P1 protein of *M. pneumoniae* and is also expressed from an operon. The three genes composing this operon 5′ to 3′ are *mgc2*, *gap*A and *crm*A. A reversible nonsense mutation (CAA→TAA) in the 5′ region of *gap*A results in GapA-negative and CrmA-negative variants, which are also haemadsorption-negative *in vitro* (Winner *et al.*, 2003). Further examples of phase-variable adhesion proteins have already been introduced earlier (see 'Domain shuffling via homologous recombination' above).

## Outlook and future perspectives

Despite their small genomes and the lack of a protecting cell wall many *Mycoplasma* species are able to establish successful, chronic infections in diverse hosts. Their success in colonizing and persisting in the in-host environment, i.e. in the presence of specific immune response, is attributed in part to the clever use of their limited genomic coding capacity, namely to rapidly alter the antigenic make-up on their membrane surface via a number of sophisticated genetic mechanisms. These complex mechanisms are stochastic events that can affect the expression and/or structure of surface lipoproteins that are usually highly immunogenic. The use of *M. agalactiae* Vpma phase-locked mutants in animal studies clearly demonstrates the important role of the lipoprotein phase variation systems during the infection process. Although the Xer1 recombinase is the sole factor responsible for Vpma switching *in vitro*, other alternative molecular switches operate in its absence *in vivo* under the selective pressure of the host immune response that is instrumental in inducing complex DNA rearrangements leading to novel chimeric *vpma* genes (R. Chopra-Dewasthaly and S. Czurda, unpublished). These results attest not only of the *in vivo* significance of the Vpma surface antigenic variation system of *M. agalactiae* and new features of its regulation under immune pressure, but will also have direct impact on understanding the *in vivo* role of surface antigenic variation systems and their regulation in other pathogenic *Mycoplasma* species.

Modern bioinformatics tools frequently used in current mycoplasma research facilitate both, identification and prediction of the underlying systems. For instance, short sequence repeats and close repeats are highly indicative of strand slippage, inverted repeats of DNA rearrangements and large repeats of homologous recombinations. When such repeats are located within or around potential membrane protein encoding genes, the probability that these are involved in surface antigenic variation is high. Sequence comparison between the genomes of *M. genitalium, M. pneumoniae, M. pulmonis,* and *U. parvum* revealed that the largest number of repeated elements can be found in the lipoprotein encoding genes of *M. pulmonis* (Rocha and Blanchard, 2002). The high potential of antigenic variability in this organism is probably linked to its ability of colonizing different body sites in at least two rodents (rat and mouse). Since high-frequency surface antigenic variation and adaptation to in-host environments are strongly associated, it would be surprising if any of the host-adapted *Mycoplasma* species were not subject to this type of variation. Phase and antigenic variation of surface membrane proteins is fascinating, and many questions still remain to be answered about the diverse genetic mechanisms and the biological significance of these sophisticated systems. But of similar importance is the potential phase variation of non- membrane encoding genes. Intriguing examples are the short sequence repeat $CAAC_4$ found in the 5′ region of the *rpoA* gene of *M. genitalium* or the $GCT_4$ repeat in the *rpl7* gene of *U. parvum,* as alterations in these sequences would result in frameshift mutations (Rocha and Blanchard, 2002). Also, not all short sequence repeats found within a mycoplasma genome are prone to size variation, and a possible reversible switching in the described examples would have to be tested experimentally.

## References

Behrens, A., Heller, M., Kirchhoff, H., Yogev, D., and Rosengarten, R. (1994). A family of phase- and size-variant membrane surface lipoprotein antigens (Vsps) of *Mycoplasma bovis*. Infect. Immun. *62*, 5075–5084.

Ben Abdelmoumen, B., Roy, R.S., and Brousseau, R. (1999). Cloning of *Mycoplasma synoviae* genes encoding specific antigens and their use as species-specific DNA probes. J. Vet. Diagn. Invest. *11*, 162–169.

Bencina, D. (2002). Haemagglutinins of pathogenic avian mycoplasmas. Avian Pathol. *31*, 535–547.

Bencina, D., Narat, M., Dovc, P., Drobnic-Valic, M., Habe, F., and Kleven, S.H. (1999). The characterization of *Mycoplasma synoviae* EF-Tu protein and proteins involved in hemadherence and their N-terminal amino acid sequences. FEMS Microbiol. Lett. *173*, 85–94.

Bhugra, B., Voelker, L.L., Zou, N., Yu, H., and Dybvig, K. (1995). Mechanism of antigenic variation in *Mycoplasma pulmonis*: interwoven, site-specific DNA inversions. Mol. Microbiol. *18*, 703–714.

Bird, D.W., Graber, K., Knutson, A., and Washburn, L.R. (2008). Mutation of two *Mycoplasma arthritidis* surface lipoproteins with divergent functions in cytadherence. Infect. Immun. *76*, 5768–5776.

Boguslavsky, S., Menaker, D., Lysnyansky, I., Liu, T., Levisohn, S., Rosengarten, R., Garcia, M., and Yogev, D. (2000). Molecular characterization of the *Mycoplasma gallisepticum pvpA* gene which encodes a putative variable cytadhesin protein. Infect. Immun. *68*, 3956–3964.

Brocchi, M., de Vasconcelos, A.T.R., and Zaha, A. (2007). Restriction-modification systems in *Mycoplasma* spp. Genet. Mol. Biol. *30*, pp. 236–244.

Burgos, R., Pich, O.Q., Ferrer-Navarro, M., Baseman, J.B., Querol, E., and Pinol, J. (2006). *Mycoplasma genitalium* P140 and P110 cytadhesins are reciprocally stabilized and required for cell adhesion and terminal organelle development. J. Bacteriol. *188*, 8627–8637.

Calcutt, M.J., Kim, M.F., Karpas, A.B., Muhlradt, P.F., and Wise, K.S. (1999). Differential posttranslational processing confers intraspecies variation of a major surface lipoprotein and a macrophage-activating lipopeptide of *Mycoplasma fermentans*. Infect. Immun. *67*, 760–771.

Catrein, I., Dumke, R., Weiner, J. 3rd., Jacobs, E., and Herrmann, R. (2004). Cross-complementation between the products of the genes P1 and ORF6 of *Mycoplasma pneumoniae* subtypes 1 and 2. Microbiology *150*, 3989–4000.

Catrein, I., Herrmann, R., Bosserhoff, A., and Ruppert, T. (2005). Experimental proof for a signal peptidase I like activity in *Mycoplasma pneumoniae*, but absence of a gene encoding a conserved bacterial type I SPase. FEBS J. *272*, 2892–2900.

Chambaud, I., Heilig, R., Ferris, S., Barbe, V., Samson, D., Galisson, F., Moszer, I., Dybvig, K., Wroblewski, H., Viari, A., *et al.* (2001). The complete genome sequence of the murine respiratory pathogen *Mycoplasma pulmonis*. Nucleic Acids Res. *29*, 2145–2153.

Chopra-Dewasthaly, R., Citti, C., Glew, M.D., Zimmermann, M., Rosengarten, R., and Jechlinger, W. (2008). Phase-locked mutants of *Mycoplasma agalactiae*: defining the molecular switch of high-frequency Vpma antigenic variation. Mol. Microbiol. *67*, 1196–1210.

Citti, C., and Wise, K.S. (1995). *Mycoplasma hyorhinis vlp* gene transcription: critical role in phase variation and expression of surface lipoproteins. Mol. Microbiol. *18*, 649–660.

Citti, C., Kim, M.F., and Wise, K.S. (1997). Elongated versions of Vlp surface lipoproteins protect *Mycoplasma hyorhinis* escape variants from growth-inhibiting host antibodies. Infect. Immun. *65*, 1773–1785.

Citti, C., Watson-McKown, R., Droesse, M., and Wise, K.S. (2000). Gene families encoding phase- and size-variable surface lipoproteins of *Mycoplasma hyorhinis*. J. Bacteriol. *182*, 1356–1363.

Citti, C., Nouvel, L.X., and Baranowski, E. (2010). Phase and antigenic variation in mycoplasmas. Future Microbiol. *5*, 1073–1085.

Colman, S.D., Hu, P.C., and Bott, K.F. (1990). Prevalence of novel repeat sequences in and around the P1 operon in the genome of *Mycoplasma pneumoniae*. Gene *87*, 91–96.

Czurda, S., Jechlinger, W., Rosengarten, R., and Chopra-Dewasthaly, R. (2010). Xer1-mediated site-specific DNA inversions and excisions in *Mycoplasma agalactiae*. J. Bacteriol. *192*, 4462–4473.

Dallo, S.F., and Baseman, J.B. (1991). Adhesin gene of *Mycoplasma genitalium* exists as multiple copies. Microb. Pathog. *10*, 475–480.

Dallo, S.F., Horton, J.R., Su, C.J., and Baseman, J.B. (1990). Restriction fragment length polymorphism in the cytadhesin P1 gene of human clinical isolates of *Mycoplasma pneumoniae*. Infect. Immun. *58*, 2017–2020.

Davis, K.L., and Wise, K.S. (2002). Site-specific proteolysis of the MALP-404 lipoprotein determines the release of a soluble selective lipoprotein-associated motif-containing fragment and alteration of the surface phenotype of *Mycoplasma fermentans*. Infect. Immun. *70*, 1129–1135.

de Castro, L.A., Rodrigues Pedroso, T., Kuchiishi, S.S., Ramenzoni, M., Kich, J.D., Zaha, A., Henning Vainstein, M., and Bunselmeyer Ferreira, H. (2006). Variable number of tandem aminoacid repeats in adhesion-related CDS products in *Mycoplasma hyopneumoniae* strains. Vet. Microbiol. *116*, 258–269.

Deitsch, K.W., Lukehart, S.A., and Stringer, J.R. (2009). Common strategies for antigenic variation by bacterial, fungal and protozoan pathogens. Nat. Rev. Microbiol. *7*, 493–503.

Denison, A.M., Clapper, B., and Dybvig, K. (2005). Avoidance of the host immune system through phase variation in *Mycoplasma pulmonis*. Infect. Immun. *73*, 2033–2039.

Dorigo-Zetsma, J.W., Dankert, J., and Zaat, S.A. (2000). Genotyping of *Mycoplasma pneumoniae* clinical isolates reveals eight P1 subtypes within two genomic groups. J. Clin. Microbiol. *38*, 965–970.

Dorigo-Zetsma, J.W., Wilbrink, B., Dankert, J., and Zaat, S.A. (2001). *Mycoplasma pneumoniae* P1 type

1- and type 2-specific sequences within the P1 cytadhesin gene of individual strains. Infect. Immun. *69*, 5612–5618.

Dufour-Gesbert, F., Kempf, I., De Simone, F., and Kobisch, M. (2001). Antigen heterogeneity and epitope variable expression in *Mycoplasma meleagridis* isolates. Vet. Microbiol. *78*, 261–273.

Dybvig, K., Simecka, J.W., Watson, H.L., and Cassell, G.H. (1989). High-frequency variation in *Mycoplasma pulmonis* colony size. J. Bacteriol. *171*, 5165–5168.

Dybvig, K., Sitaraman, R., and French, C.T. (1998). A family of phase-variable restriction enzymes with differing specificities generated by high-frequency gene rearrangements. Proc. Natl. Acad. Sci. U.S.A. *95*, 13923–13928.

Esposito, D., and Scocca, J.J. (1997). The integrase family of tyrosine recombinases: evolution of a conserved active site domain. Nucleic Acids Res. *25*, 3605–3614.

Ferris, S., Watson, H.L., Neyrolles, O., Montagnier, L., and Blanchard, A. (1995). Characterization of a major *Mycoplasma penetrans* lipoprotein and of its gene. FEMS Microbiol. Lett. *130*, 313–319.

Flitman-Tene, R., Levisohn, S., Lysnyansky, I., Rapoport, E., and Yogev, D. (2000). A chromosomal region of *Mycoplasma agalactiae* containing *vsp*-related genes undergoes in vivo rearrangement in naturally infected animals. FEMS Microbiol. Lett. *191*, 205–212.

Flitman-Tene, R., Mudahi-Orenstein, S., Levisohn, S., and Yogev, D. (2003). Variable lipoprotein genes of *Mycoplasma agalactiae* are activated in vivo by promoter addition via site-specific DNA inversions. Infect. Immun. *71*, 3821–3830.

Forsyth, M.H., and Geary, S.J. (1996). The repetitive element Rep MP 1 of *Mycoplasma pneumoniae* exists as a core element within a larger, variable repetitive mosaic. J. Bacteriol. *178*, 917–921.

Fraser, C.M., Gocayne, J.D., White, O., Adams, M.D., Clayton, R.A., Fleischmann, R.D., Bult, C.J., Kerlavage, A.R., Sutton, G., Kelley, J.M., *et al.* (1995). The minimal gene complement of *Mycoplasma genitalium*. Science *270*, 397–403.

Gaurivaud, P., Persson, A., Grand, D.L., Westberg, J., Solsona, M., Johansson, K.E., and Poumarat, F. (2004). Variability of a glucose phosphotransferase system permease in *Mycoplasma mycoides* subsp. *mycoides* Small Colony. Microbiology *150*, 4009–4022.

Giron, J.A., Lange, M., and Baseman, J.B. (1996). Adherence, fibronectin binding, and induction of cytoskeleton reorganization in cultured human cells by *Mycoplasma penetrans*. Infect. Immun. *64*, 197–208.

Glass, J.I., Lefkowitz, E.J., Glass, J.S., Heiner, C.R., Chen, E.Y., and Cassell, G.H. (2000). The complete sequence of the mucosal pathogen *Ureaplasma urealyticum*. Nature *407*, 757–762.

Glew, M.D., Markham, P.F., Browning, G.F., and Walker, I.D. (1995). Expression studies on four members of the pMGA multigene family in *Mycoplasma gallisepticum* S6. Microbiology *141*, 3005–3014.

Glew, M.D., Baseggio, N., Markham, P.F., Browning, G.F., and Walker, I.D. (1998). Expression of the pMGA genes of *Mycoplasma gallisepticum* is controlled by variation in the GAA trinucleotide repeat lengths within the 5′ noncoding regions. Infect. Immun. *66*, 5833–5841.

Glew, M.D., Browning, G.F., Markham, P.F., and Walker, I.D. (2000a). pMGA phenotypic variation in *Mycoplasma gallisepticum* occurs in vivo and is mediated by trinucleotide repeat length variation. Infect. Immun. *68*, 6027–6033.

Glew, M.D., Papazisi, L., Poumarat, F., Bergonier, D., Rosengarten, R., and Citti, C. (2000b). Characterization of a multigene family undergoing high-frequency DNA rearrangements and coding for abundant variable surface proteins in *Mycoplasma agalactiae*. Infect. Immun. *68*, 4539–4548.

Glew, M.D., Marenda, M., Rosengarten, R., and Citti, C. (2002). Surface diversity in *Mycoplasma agalactiae* is driven by site-specific DNA inversions within the *vpma* multigene locus. J. Bacteriol. *184*, 5987–5998.

Grau, O., Slizewicz, B., Tuppin, P., Launay, V., Bourgeois, E., Sagot, N., Moynier, M., Lafeuillade, A., Bachelez, H., Clauvel, J.P., *et al.* (1995). Association of *Mycoplasma penetrans* with human immunodeficiency virus infection. J. Infect. Dis. *172*, 672–681.

Gumulak-Smith, J., Teachman, A., Tu, A.H., Simecka, J.W., Lindsey, J.R., and Dybvig, K. (2001). Variations in the surface proteins and restriction enzyme systems of *Mycoplasma pulmonis* in the respiratory tract of infected rats. Mol. Microbiol. *40*, 1037–1044.

Hamsten, C., Westberg, J., Bolske, G., Ayling, R., Uhlen, M., and Persson, A. (2008). Expression and immunogenicity of six putative variable surface proteins in *Mycoplasma mycoides* subsp. *mycoides* SC. Microbiology *154*, 539–549.

Himmelreich, R., Hilbert, H., Plagens, H., Pirkl, E., Li, B.C., and Herrmann, R. (1996). Complete sequence analysis of the genome of the bacterium *Mycoplasma pneumoniae*. Nucleic Acids Res. *24*, 4420–4449.

Horino, A., Sasaki, Y., Sasaki, T., and Kenri, T. (2003). Multiple promoter inversions generate surface antigenic variation in *Mycoplasma penetrans*. J. Bacteriol. *185*, 231–242.

Horino, A., Kenri, T., Sasaki, Y., Okamura, N., and Sasaki, T. (2009). Identification of a site-specific tyrosine recombinase that mediates promoter inversions of phase-variable *mpl* lipoprotein genes in *Mycoplasma penetrans*. Microbiology *155*, 1241–1249.

Hu, P.C., Schaper, U., Collier, A.M., Clyde, W.A., Jr., Horikawa, M., Huang, Y.S., and Barile, M.F. (1987). A *Mycoplasma genitalium* protein resembling the *Mycoplasma pneumoniae* attachment protein. Infect. Immun. *55*, 1126–1131.

Inamine, J.M., Loechel, S., Collier, A.M., Barile, M.F., and Hu, P.C. (1989). Nucleotide sequence of the MgPa (*mgp*) operon of *Mycoplasma genitalium* and comparison to the P1 (mpp) operon of *Mycoplasma pneumoniae*. Gene *82*, 259–267.

Iverson-Cabral, S.L., Astete, S.G., Cohen, C.R., Rocha, E.P., and Totten, P.A. (2006). Intrastrain heterogeneity of the *mgpB* gene in *Mycoplasma genitalium* is extensive in vitro and in vivo and suggests that variation is generated via recombination with repetitive chromosomal sequences. Infect. Immun. *74*, 3715–3726.

Iverson-Cabral, S.L., Astete, S.G., Cohen, C.R., and Totten, P.A. (2007). *mgpB* and *mgpC* sequence diversity in *Mycoplasma genitalium* is generated by segmental reciprocal recombination with repetitive chromosomal sequences. Mol. Microbiol. *66*, 55–73.

Kannan, T.R., and Baseman, J.B. (2000). Hemolytic and hemoxidative activities in *Mycoplasma penetrans*. Infect. Immun. *68*, 6419–6422.

Kenri, T., Taniguchi, R., Sasaki, Y., Okazaki, N., Narita, M., Izumikawa, K., Umetsu, M., and Sasaki, T. (1999). Identification of a new variable sequence in the P1 cytadhesin gene of *Mycoplasma pneumoniae*: evidence for the generation of antigenic variation by DNA recombination between repetitive sequences. Infect. Immun. *67*, 4557–4562.

Khiari, A.B., Gueriri, I., Mohammed, R.B., and Mardassi, B.B. (2010). Characterization of a variant *vlhA* gene of *Mycoplasma synoviae*, strain WVU 1853, with a highly divergent haemagglutinin region. BMC Microbiol. *10*, 6.

Kong, F., Ma, Z., James, G., Gordon, S., and Gilbert, G.L. (2000). Molecular genotyping of human *Ureaplasma* species based on multiple-banded antigen (MBA) gene sequences. Int. J. Syst. Evol. Microbiol. *50*, 1921–1929.

Layh-Schmitt, G., and Harkenthal, M. (1999). The 40- and 90-kDa membrane proteins (ORF6 gene product) of *Mycoplasma pneumoniae* are responsible for the tip structure formation and P1 (adhesin) association with the Triton shell. FEMS Microbiol. Lett. *174*, 143–149.

Leigh, S.A., and Wise, K.S. (2002). Identification and functional mapping of the *Mycoplasma fermentans* P29 adhesin. Infect. Immun. *70*, 4925–4935.

Liu, L., Dybvig, K., Panangala, V.S., van Santen, V.L., and French, C.T. (2000). GAA trinucleotide repeat region regulates M9/pMGA gene expression in *Mycoplasma gallisepticum*. Infect. Immun. *68*, 871–876.

Liu, L., Panangala, V.S., and Dybvig, K. (2002). Trinucleotide GAA repeats dictate pMGA gene expression in *Mycoplasma gallisepticum* by affecting spacing between flanking regions. J. Bacteriol. *184*, 1335–1339.

Liu, W., Fang, L., Li, S., Li, Q., Zhou, Z., Feng, Z., Luo, R., Shao, G., Wang, L., Chen, H., *et al.* (2010). Complete genome sequence of *Mycoplasma hyorhinis* strain HUB-1. J. Bacteriol. *192*, 5844–5845.

Lo, S.C., Hayes, M.M., Tully, J.G., Wang, R.Y., Kotani, H., Pierce, P.F., Rose, D.L., and Shih, J.W. (1992). *Mycoplasma penetrans* sp. nov., from the urogenital tract of patients with AIDS. Int. J. Syst. Bacteriol. *42*, 357–364.

Lo, S.C., Hayes, M.M., Kotani, H., Pierce, P.F., Wear, D.J., Newton, P.B., 3rd., Tully, J.G., and Shih, J.W. (1993). Adhesion onto and invasion into mammalian cells by *Mycoplasma penetrans*: a newly isolated mycoplasma from patients with AIDS. Mod. Pathol. *6*, 276–280.

Lysnyansky, I., Rosengarten, R., and Yogev, D. (1996). Phenotypic switching of variable surface lipoproteins in *Mycoplasma bovis* involves high-frequency chromosomal rearrangements. J. Bacteriol. *178*, 5395–5401.

Lysnyansky, I., Sachse, K., Rosenbusch, R., Levisohn, S., and Yogev, D. (1999). The *vsp* locus of *Mycoplasma bovis*: gene organization and structural features. J. Bacteriol. *181*, 5734–5741.

Lysnyansky, I., Ron, Y., Sachse, K., and Yogev, D. (2001a). Intrachromosomal recombination within the *vsp* locus of *Mycoplasma bovis* generates a chimeric variable surface lipoprotein antigen. Infect. Immun. *69*, 3703–3712.

Lysnyansky, I., Ron, Y., and Yogev, D. (2001b). Juxtaposition of an active promoter to *vsp* genes via site-specific DNA inversions generates antigenic variation in *Mycoplasma bovis*. J. Bacteriol. *183*, 5698–5708.

Ma, L., Jensen, J.S., Myers, L., Burnett, J., Welch, M., Jia, Q., and Martin, D.H. (2007). *Mycoplasma genitalium*: an efficient strategy to generate genetic variation from a minimal genome. Mol. Microbiol. *66*, 220–236.

Markham, P.F., Glew, M.D., Sykes, J.E., Bowden, T.R., Pollocks, T.D., Browning, G.F., Whithear, K.G., and Walker, I.D. (1994). The organisation of the multigene family which encodes the major cell surface protein, pMGA, of *Mycoplasma gallisepticum*. FEBS Lett. *352*, 347–352.

Markham, P.F., Glew, M.D., Browning, G.F., Whithear, K.G., and Walker, I.D. (1998). Expression of two members of the pMGA gene family of *Mycoplasma gallisepticum* oscillates and is influenced by pMGA-specific antibodies. Infect. Immun. *66*, 2845–2853.

Markham, P.F., Duffy, M.F., Glew, M.D., and Browning, G.F. (1999). A gene family in *Mycoplasma imitans* closely related to the pMGA family of *Mycoplasma gallisepticum*. Microbiology *145*, 2095–2103.

Monecke, S., Helbig, J.H., and Jacobs, E. (2003). Phase variation of the multiple banded protein in *Ureaplasma urealyticum* and *Ureaplasma parvum*. Int. J. Med. Microbiol. *293*, 203–211.

Morrison-Plummer, J., Lazzell, A., and Baseman, J.B. (1987). Shared epitopes between *Mycoplasma pneumoniae* major adhesin protein P1 and a 140-kilodalton protein of *Mycoplasma genitalium*. Infect. Immun. *55*, 49–56.

Musatovova, O., Kannan, T.R., and Baseman, J.B. (2008). Genomic analysis reveals *Mycoplasma pneumoniae* repetitive element 1-mediated recombination in a clinical isolate. Infect. Immun. *76*, 1639–1648.

Neyrolles, O., Chambaud, I., Ferris, S., Prevost, M.C., Sasaki, T., Montagnier, L., and Blanchard, A. (1999a). Phase variations of the *Mycoplasma penetrans* main surface lipoprotein increase antigenic diversity. Infect. Immun. *67*, 1569–1578.

Neyrolles, O., Eliane, J.P., Ferris, S., da Cunha, R.A., Prevost, M.C., Bahraoui, E., and Blanchard, A. (1999b). Antigenic characterization and cytolocalization of P35, the major *Mycoplasma penetrans* antigen. Microbiology *145*, 343–355.

Noormohammadi, A.H., Markham, P.F., Whithear, K.G., Walker, I.D., Gurevich, V.A., Ley, D.H., and Browning, G.F. (1997). *Mycoplasma synoviae* has two distinct phase-variable major membrane antigens, one of which is a putative hemagglutinin. Infect. Immun. *65*, 2542–2547.

Noormohammadi, A.H., Markham, P.F., Duffy, M.F., Whithear, K.G., and Browning, G.F. (1998). Multigene families encoding the major hemagglutinins in phylogenetically distinct mycoplasmas. Infect. Immun. *66*, 3470–3475.

Noormohammadi, A.H., Markham, P.F., Kanci, A., Whithear, K.G., and Browning, G.F. (2000). A novel mechanism for control of antigenic variation in the haemagglutinin gene family of *Mycoplasma synoviae*. Mol. Microbiol. *35*, 911–923.

Nouvel, L.X., Marenda, M., Sirand-Pugnet, P., Sagne, E., Glew, M., Mangenot, S., Barbe, V., Barre, A., Claverol, S., and Citti, C. (2009). Occurrence, plasticity, and evolution of the vpma gene family, a genetic system devoted to high-frequency surface variation in *Mycoplasma agalactiae*. J. Bacteriol. *191*, 4111–4121.

Nouvel, L.X., Sirand-Pugnet, P., Marenda, M.S., Sagne, E., Barbe, V., Mangenot, S., Schenowitz, C., Jacob, D., Barre, A., Claverol, S., *et al.* (2010). Comparative genomic and proteomic analyses of two *Mycoplasma agalactiae* strains: clues to the macro- and micro-events that are shaping mycoplasma diversity. BMC Genomics *11*, 86.

Nussbaum, S., Lysnyansky, I., Sachse, K., Levisohn, S., and Yogev, D. (2002). Extended repertoire of genes encoding variable surface lipoproteins in *Mycoplasma bovis* strains. Infect. Immun. *70*, 2220–2225.

Papazisi, L., Gorton, T.S., Kutish, G., Markham, P.F., Browning, G.F., Nguyen, D.K., Swartzell, S., Madan, A., Mahairas, G., and Geary, S.J. (2003). The complete genome sequence of the avian pathogen *Mycoplasma gallisepticum* strain R(low). Microbiology *149*, 2307–2316.

Persson, A., Jacobsson, K., Frykberg, L., Johansson, K.E., and Poumarat, F. (2002). Variable surface protein Vmm of *Mycoplasma mycoides* subsp. *mycoides* small colony type. J. Bacteriol. *184*, 3712–3722.

Peterson, S.N., Bailey, C.C., Jensen, J.S., Borre, M.B., King, E.S., Bott, K.F., and Hutchison, C.A., 3rd. (1995). Characterization of repetitive DNA in the *Mycoplasma genitalium* genome: possible role in the generation of antigenic variation. Proc. Natl. Acad. Sci. U.S.A. *92*, 11829–11833.

Rocha, E.P., and Blanchard, A. (2002). Genomic repeats, genome plasticity and the dynamics of *Mycoplasma* evolution. Nucleic Acids Res. *30*, 2031–2042.

Ron, Y., Flitman-Tene, R., Dybvig, K., and Yogev, D. (2002). Identification and characterization of a site-specific tyrosine recombinase within the variable loci of *Mycoplasma bovis*, *Mycoplasma pulmonis* and *Mycoplasma agalactiae*. Gene 292, 205–211.

Rosengarten, R., and Wise, K.S. (1990). Phenotypic switching in mycoplasmas: phase variation of diverse surface lipoproteins. Science *247*, 315–318.

Rosengarten, R., and Wise, K.S. (1991). The Vlp system of *Mycoplasma hyorhinis*: combinatorial expression of distinct size variant lipoproteins generating high-frequency surface antigenic variation. J. Bacteriol. *173*, 4782–4793.

Rosengarten, R., Behrens, A., Stetefeld, A., Heller, M., Ahrens, M., Sachse, K., Yogev, D., and Kirchhoff, H. (1994). Antigen heterogeneity among isolates of *Mycoplasma bovis* is generated by high-frequency variation of diverse membrane surface proteins. Infect. Immun. *62*, 5066–5074.

Roske, K., Blanchard, A., Chambaud, I., Citti, C., Helbig, J.H., Prevost, M.C., Rosengarten, R., and Jacobs, E. (2001). Phase variation among major surface antigens of *Mycoplasma penetrans*. Infect. Immun. *69*, 7642–7651.

Ruland, K., Wenzel, R., and Herrmann, R. (1990). Analysis of three different repeated DNA elements present in the P1 operon of *Mycoplasma pneumoniae*: size, number and distribution on the genome. Nucleic Acids Res. *18*, 6311–6317.

Ruland, K., Himmelreich, R., and Herrmann, R. (1994). Sequence divergence in the ORF6 gene of *Mycoplasma pneumonia*. J. Bacteriol. *176*, 5202–5209.

Sasaki, Y., Ishikawa, J., Yamashita, A., Oshima, K., Kenri, T., Furuya, K., Yoshino, C., Horino, A., Shiba, T., Sasaki, T., *et al.* (2002). The complete genomic sequence of *Mycoplasma penetrans*, an intracellular bacterial pathogen in humans. Nucleic Acids Res. *30*, 5293–5300.

Shen, X., Gumulak, J., Yu, H., French, C.T., Zou, N., and Dybvig, K. (2000). Gene rearrangements in the vsa locus of *Mycoplasma pulmonis*. J. Bacteriol. *182*, 2900–2908.

Simmons, W.L., and Dybvig, K. (2003). The Vsa proteins modulate susceptibility of *Mycoplasma pulmonis* to complement killing, hemadsorption, and adherence to polystyrene. Infect. Immun. *71*, 5733–5738.

Simmons, W.L., and Dybvig, K. (2007). Biofilms protect *Mycoplasma pulmonis* cells from lytic effects of complement and gramicidin. Infect. Immun. *75*, 3696–3699.

Simmons, W.L., Zuhua, C., Glass, J.I., Simecka, J.W., Cassell, G.H., and Watson, H.L. (1996). Sequence analysis of the chromosomal region around and within the V-1-encoding gene of *Mycoplasma pulmonis*: evidence for DNA inversion as a mechanism for V-1 variation. Infect. Immun. *64*, 472–479.

Simmons, W.L., Denison, A.M., and Dybvig, K. (2004). Resistance of *Mycoplasma pulmonis* to complement lysis is dependent on the number of Vsa tandem repeats: shield hypothesis. Infect. Immun. *72*, 6846–6851.

Simmons, W.L., Bolland, J.R., Daubenspeck, J.M., and Dybvig, K. (2007). A stochastic mechanism for biofilm formation by *Mycoplasma pulmonis*. J. Bacteriol. *189*, 1905–1913.

Sirand-Pugnet, P., Lartigue, C., Marenda, M., Jacob, D., Barre, A., Barbe, V., Schenowitz, C., Mangenot, S., Couloux, A., Segurens, B., *et al.* (2007). Being pathogenic, plastic, and sexual while living with a nearly minimal bacterial genome. PLoS Genet. *3*, e75.

Sitaraman, R., and Dybvig, K. (1997). The *hsd* loci of *Mycoplasma pulmonis*: organization, rearrangements and expression of genes. Mol. Microbiol. *26*, 109–120.

Sitaraman, R., Denison, A.M., and Dybvig, K. (2002). A unique, bifunctional site-specific DNA recombinase from *Mycoplasma pulmonis*. Mol. Microbiol. *46*, 1033–1040.

Sluijter, M., Kaptein, E., Spuesens, E.B., Hoogenboezem, T., Hartwig, N.G., Van Rossum, A.M., and Vink, C. (2010). The *Mycoplasma genitalium* MG352-encoded protein is a Holliday junction resolvase that has a non-functional orthologue in *Mycoplasma pneumoniae*. Mol. Microbiol. *77*, 1261–1277.

Spuesens, E.B., Oduber, M., Hoogenboezem, T., Sluijter, M., Hartwig, N.G., van Rossum, A.M., and Vink, C. (2009). Sequence variations in RepMP2/3 and RepMP4 elements reveal intragenomic homologous DNA recombination events in *Mycoplasma pneumoniae*. Microbiology *155*, 2182–2196.

Spuesens, E.B., van de, Kreeke, N., Estevao, S., Hoogenboezem, T., Sluijter, M., Hartwig, N., van Rossum, A.M., and Vink, C. (2010). Variation in a surface-exposed region of the *Mycoplasma pneumoniae* P40 protein as a consequence of homologous DNA recombination between RepMP5 elements. Microbiology *157*, 473–483.

Srikhanta, Y.N., Fox, K.L., and Jennings, M.P. (2010). The phasevarion: phase variation of type III DNA methyltransferases controls coordinated switching in multiple genes. Nat. Rev. Microbiol. *8*, 196–206.

Su, C.J., Chavoya, A., Dallo, S.F., and Baseman, J.B. (1990a). Sequence divergency of the cytadhesin gene of *Mycoplasma pneumoniae*. Infect. Immun. *58*, 2669–2674.

Su, C.J., Dallo, S.F., and Baseman, J.B. (1990b). Molecular distinctions among clinical isolates of *Mycoplasma pneumoniae*. J. Clin. Microbiol. *28*, 1538–1540.

Su, C.J., Dallo, S.F., Chavoya, A., and Baseman, J.B. (1993). Possible origin of sequence divergence in the P1 cytadhesin gene of *Mycoplasma pneumoniae*. Infect. Immun. *61*, 816–822.

Svenstrup, H.F., Jensen, J.S., Gevaert, K., Birkelund, S., and Christiansen, G. (2006). Identification and characterization of immunogenic proteins of *Mycoplasma genitalium*. Clin. Vaccine Immunol. *13*, 913–922.

Teng, L.J., Zheng, X., Glass, J.I., Watson, H.L., Tsai, J., and Cassell, G.H. (1994). *Ureaplasma urealyticum* biovar specificity and diversity are encoded in multiple-banded antigen gene. J. Clin. Microbiol. *32*, 1464–1469.

Theiss, P., and Wise, K.S. (1997). Localized frameshift mutation generates selective, high-frequency phase variation of a surface lipoprotein encoded by a mycoplasma ABC transporter operon. J. Bacteriol. *179*, 4013–4022.

Theiss, P., Karpas, A., and Wise, K.S. (1996). Antigenic topology of the P29 surface lipoprotein of *Mycoplasma fermentans*: differential display of epitopes results in high-frequency phase variation. Infect. Immun. *64*, 1800–1809.

Tortschanoff, M., Aurich, C., Rosengarten, R., and Spergser, J. (2005). Phase and size variable surface-exposed proteins in equine genital mycoplasmas. Vet. Microbiol. *110*, 301–306.

van der Woude, M.W., and Bäumler, A.J. (2004). Phase and antigenic variation in bacteria. Clin. Microbiol. Rev. *17*, 581–611, table of conten.

Vasconcelos, A.T., Ferreira, H.B., Bizarro, C.V., Bonatto, S.L., Carvalho, M.O., Pinto, P.M., Almeida, D.F., Almeida, L.G., Almeida, R., Alves-Filho, L., *et al.* (2005). Swine and poultry pathogens: the complete genome sequences of two strains of *Mycoplasma hyopneumoniae* and a strain of *Mycoplasma synoviae*. J. Bacteriol. *187*, 5568–5577.

Wang, R.Y., Shih, J.W., Weiss, S.H., Grandinetti, T., Pierce, P.F., Lange, M., Alter, H.J., Wear, D.J., Davies, C.L., Mayur, R.K., and *et al.* (1993). *Mycoplasma penetrans* infection in male homosexuals with AIDS: high seroprevalence and association with Kaposi's sarcoma. Clin. Infect. Dis. *17*, 724–729.

Washburn, L.R., Weaver, K.E., Weaver, E.J., Donelan, W., and Al-Sheboul, S. (1998). Molecular characterization of *Mycoplasma arthritidis* variable surface protein MAA2. Infect. Immun. *66*, 2576–2586.

Wenzel, R., and Herrmann, R. (1988). Repetitive DNA sequences in *Mycoplasma pneumoniae*. Nucleic Acids Res. *16*, 8337–8350.

Westberg, J., Persson, A., Holmberg, A., Goesmann, A., Lundeberg, J., Johansson, K.E., Pettersson, B., and Uhlen, M. (2004). The genome sequence of *Mycoplasma mycoides* subsp. *mycoides* SC type strain PG1T, the causative agent of contagious bovine pleuropneumonia (CBPP). Genome Res. *14*, 221–227.

Wilton, J.L., Scarman, A.L., Walker, M.J., and Djordjevic, S.P. (1998). Reiterated repeat region variability in the ciliary adhesin gene of *Mycoplasma hyopneumoniae*. Microbiology *144*, 1931–1943.

Winner, F., Markova, I., Much, P., Lugmair, A., Siebert-Gulle, K., Vogl, G., Rosengarten, R., and Citti, C. (2003). Phenotypic switching in *Mycoplasma gallisepticum* hemadsorption is governed by a high-frequency, reversible point mutation. Infect. Immun. *71*, 1265–1273.

Wise, K.S., Foecking, M.F., Roeske, K., Lee, Y.J., Lee, Y.M., Madan, A., and Calcutt, M.J. (2006). Distinctive repertoire of contingency genes conferring mutation-based phase variation and combinatorial expression of surface lipoproteins in *Mycoplasma capricolum* subsp. *capricolum* of the *Mycoplasma mycoides* phylogenetic cluster. J. Bacteriol. *188*, 4926–4941.

Wise, K.S., Calcutt, M.J., Foecking, M.F., Roeske, K., Madupu, R., and Methe, B.A. (2011). Complete Genome Sequence of *Mycoplasma bovis* Type Strain PG45 (ATCC 25523). Infect. Immun. *79*, 982–983.

Wisniewski-Dye, F., and Vial, L. (2008). Phase and antigenic variation mediated by genome modifications. Antonie Van Leeuwenhoek *94*, 493–515.

Xiao, L., Glass, J.I., Paralanov, V., Yooseph, S., Cassell, G.H., Duffy, L.B., and Waites, K.B. (2010). Detection and characterization of human *Ureaplasma* species and serovars by real-time PCR. J. Clin. Microbiol. *48*, 2715–2723.

Yogev, D., Rosengarten, R., Watson-McKown, R., and Wise, K.S. (1991). Molecular basis of *Mycoplasma* surface antigenic variation: a novel set of divergent genes undergo spontaneous mutation of periodic coding regions and 5′ regulatory sequences. EMBO J. *10*, 4069–4079.

Yogev, D., Rosengarten, R., and Wise, K.S. (1993). Variation and genetic control of surface antigen expression in mycoplasmas: the Vlp system of *Mycoplasma hyorhinis*. Zentralbl. Bakteriol. *278*, 275–286.

Yogev, D., Watson-McKown, R., Rosengarten, R., Im, J., and Wise, K.S. (1995). Increased structural and combinatorial diversity in an extended family of genes encoding Vlp surface proteins of *Mycoplasma hyorhinis*. J. Bacteriol. *177*, 5636–5643.

Zhang, Q., and Wise, K.S. (1996). Molecular basis of size and antigenic variation of a *Mycoplasma hominis* adhesin encoded by divergent *vaa* genes. Infect. Immun. *64*, 2737–2744.

Zhang, Q., and Wise, K.S. (1997). Localized reversible frameshift mutation in an adhesin gene confers a phase-variable adherence phenotype in *Mycoplasma*. Mol. Microbiol. *25*, 859–869.

Zhang, Q., and Wise, K.S. (2001). Coupled phase-variable expression and epitope masking of selective surface lipoproteins increase surface phenotypic diversity in *Mycoplasma hominis*. Infect. Immun. *69*, 5177–5181.

Zheng, X., Teng, L.J., Glass, J.I., Blanchard, A., Cao, Z., Kempf, M.C., Watson, H.L., and Cassell, G.H. (1994). Size variation of a major serotype-specific antigen of *Ureaplasma urealyticum*. Ann. N. Y. Acad. Sci. *730*, 299–301.

Zheng, X., Teng, L.J., Watson, H.L., Glass, J.I., Blanchard, A., and Cassell, G.H. (1995). Small repeating units within the *Ureaplasma urealyticum* MB antigen gene encode serovar specificity and are associated with antigen size variation. Infect. Immun. *63*, 891–898.

Zimmerman, C.U., Stiedl, T., Rosengarten, R., and Spergser, J. (2009). Alternate phase variation in expression of two major surface membrane proteins (MBA and UU376) of *Ureaplasma parvum* serovar 3. FEMS Microbiol. Lett. *292*, 187–193.

Zimmerman, C.U., Rosengarten, R., and Spergser, J. (2011). *Ureaplasma* antigenic variation beyond MBA phase variation: DNA inversions generating chimeric structures and switching in expression of the MBA N-terminal paralogue UU172. Mol. Microbiol. *79*, 663–676.

# Spiroplasma Transmission from Insects to Plants: *Spiroplasma citri* Proteins Involved in Transmission by Leafhopper Vectors

10

Laure Béven, Saskia Hogenhout, Fabien Labroussaa, Nathalie Arricau-Bouvery and Colette Saillard


## Abstract

Members of the genus *Spiroplasma* are motile, helical, wall-free eubacteria that are associated primarily with insects. Three spiroplasma species associated with plant disease are phloem restricted and transmitted in a persistent manner by sap-feeding leafhopper vectors. The spiroplasmas are acquired by the leafhoppers, traverse various insect tissues and then are transmitted to plants after having reached the leafhopper salivary glands. Thus, the spiroplasmas have to successfully passage different physical barriers of the gut and salivary glands. This involves specific protein–protein interactions between the spiroplasmas and insect tissues. Indeed, transmission electron microscopy studies indicated that spiroplasma invasion of gut epithelial cells occurs primarily by endocytosis at the brush border membranes and that upon invasion oval or flask-shaped spiroplasmas are inside membrane-bound vesicles. Confocal microscopy analyses of *Spiroplasma citri*-infected *C. haematoceps* salivary glands and cultured cell line (Ciha-1) experimentally infected by *S. citri* provided evidence that spiroplasmas associate with cell actin microfilaments. Adhesin-like proteins namely SARP1, ScARPs and SkARP are encoded on spiroplasma plasmids pBJS-O, pSci1–6 and pSKU146. Several other proteins including spiralin, ABC transporter solute binding protein and a phosphoglycerate kinase required for efficient transmission of *S. citri* by leafhoppers have been identified. The involvement of all these proteins in an interaction between the spiroplasma and the leafhopper cells is discussed in this chapter.


## Introduction

The insect-transmitted plant pathogenic spiroplasmas and phytoplasmas have a remarkable life cycle that involves invasion and (intracellular) replication in plants and insects. Both groups of bacteria are located in, and restricted to, the phloem sieve tubes, whereas they can invade multiple organs and tissue types within the insect host. They are obligate colonizers of their plant hosts and insect vectors and require both for successful dispersal in nature (reviewed in Hogenhout *et al.*, 2008). Nonetheless, spiroplasmas and phytoplasmas can be directly transmitted from plant to plant via grafting, parasitic plants, such as dodder (*Cuscuta* spp.) and tissue culturing but this frequently requires human intervention, and the bacteria often lose their ability of insect transmission over a period of time.

In the Class Mollicutes, *Spiroplasma* is one of the largest genera and comprises 37 described species (Williamson *et al.*, 2011). Although most of them are found in arthropods, especially in insects, only three spiroplasmas, *Spiroplasma citri* (Saglio *et al.*, 1973), *Spiroplasma kunkelii* (Whitcomb *et al.*, 1986), and *Spiroplasma phoeniceum* (Saillard *et al.*, 1987), are plant pathogenic bacteria. They are transmitted from plant to plant by phloem-feeding leafhoppers in a circulative, propagative, persistent manner (Purcell, 1983).

*S. citri* was the first plant pathogenic Mollicute to be cultured and characterized (Saglio *et al.*, 1971, 1973). It is an important pathogen, causing citrus stubborn disease in the Mediterranean area and California (Calavan and Bové, 1989) as well as horseradish brittle root disease in the United States (Fletcher *et al.*, 1981). It also infects many other plants, including carrots (*Daucus carota*) (Mello *et al.*, 2009) and periwinkles (*Catharanthus roseus*), in which it induces symptoms such as stunting, leaf yellowing, and wilting, leading to plant death. It is naturally transmitted from plant to plant by two *Circulifer* spp. (Oldfield *et al.*, 1976; Fos *et al.*, 1986) and three *Scaphytopius* spp. (Oldfield *et al.*, 1977; Kaloostian *et al.*, 1979).

## Infection and replication sites of spiroplasmas in leafhopper vectors

### Spiroplasma cell morphology

Electron microscopy observations revealed that spiroplasmas are highly pleiomorphic with oval, flask-shaped and helical/spiral cell morphologies, while phytoplasmas cells are oval and flask-shaped and do not form helices (Ammar *et al.*, 2004; Hogenhout *et al.*, 2008). Length of the spiroplasma helices can vary depending on the species with 3.33 μm and 4.88 μm for *S. citri* (strain ASP 1) and *S. kunkelii* (CSS-M), respectively, while their width does not differ averaging 115–160 nm (Ammar *et al.*, 2004). Flask-shaped spiroplasma and phytoplasma cells have globular structures of 350–500 nm widths with short or long tubular extensions of 115–160 nm widths, and oval shapes of these bacteria are approximately 500 nm in diameter (Ozbek *et al.*, 2003; Ammar *et al.*, 2004; Ammar and Hogenhout, 2006). One end of the helical cells and the tubular extensions of flask-shaped cells often possess a tapered end referred to as the tip structure (Ammar *et al.*, 2004). This tip structure is constricted and has a different internal structure from the rest of the cell, and is 225–260 long and 89–100 nm wide. *S. kunkelii* also produces thin pili-like structures on their surface when in culture and during leafhopper invasion (Ozbek *et al.*, 2003; Ammar *et al.*, 2004).

### Colonization of the insect vector

Insects acquire the plant Mollicutes whilst feeding upon an infected plant. Particularly, a naive (or uninfected) leafhopper may acquire the Mollicutes while ingesting the phloem sap of a diseased plant, an event referred to as acquisition feeding. Upon ingestion by the insect, the bacteria must traverse the gut epithelial cell layer to access the circulatory system (haemolymph) of the host. In the midgut lumen, the tip structures of helical spiroplasmas are aligned between microvilli close to the apical plasma membrane in the brush border of midgut epithelium cells (Ozbek *et al.*, 2003; Ammar *et al.*, 2004). It was observed that *S. citri* and *S. kunkelii*, as well as commensal/symbiotic spiroplasma-like organisms, align with microvilli in this way in leafhoppers (Ozbek *et al.*, 2003; Ammar *et al.*, 2004, 2011). However, *S. citri* and *S. kunkelii* invade the cytoplasm of gut cells while the commensal spiroplasmas remain confined to the gut lumen. The progression of *S. citri* and *S. kunkelii* through the gut appears to require endocytosis by epithelial cells. It was observed that *S. citri* and *S. kunkelii, as well as* commensal/symbiotic spiroplasma-like organisms, align with microvilli in this way in leafhoppers midgut, and the subsequent relocation of the bacteria from membrane-bound compartments to the basal side of the midgut (Fletcher *et al.*, 1998; Kwon *et al.*, 1999; Ozbek *et al.*, 2003). An investigation of *S. kunkelii* distribution in the leafhopper vector *Dalbulus maidis* revealed that the membrane-bound compartments at the apical side of the epithelial cells contain single oval or flask-shaped spiroplasmas while those at the basal (haemolymph) side more often contain multiple spiroplasma cells (Ozbek *et al.*, 2003; Ammar *et al.*, 2004). The latter spiroplasmas often have flask-shaped structures with long tubular extensions that reach out of the membrane-bound compartments into the intercellular area between two epithelial cells or between the epithelial cell plasma membrane and the basal lamina, which aligns the epithelial cells at the haemolymph side (Ozbek *et al.*, 2003). The spiroplasmas are located between infoldings of the epithelial cell plasma membrane and between the basal lamina and the external lamina, which aligns the muscle cells that lie adjacently to the epithelial cells at the haemolymph side of the gut tissue (Ozbek *et al.*, 2003). The muscle cells appear to be heavily infected by the spiroplasmas; they were found to predominantly accumulate between the external lamina and the muscle cell membrane and complete muscle cells seemed to have disappeared and replaced by the spiroplasmas (Ozbek *et al.*, 2003; Ammar and Hogenhout, 2006). These spiroplasmas have helical, flask-shaped and oval morphologies and appear to use their tip structures to disrupt the basal/external lamina or to traverse intact lamina for reaching the haemolymph (Ozbek *et al.*, 2003; Ammar and Hogenhout, 2006). *S. kunkelii* also produces hair-like pili that adhere to the basal/external lamina of insect cells and larger structures that resemble conjugation pili (Ozbek *et al.*, 2003; Ammar *et al.*, 2004).

To be transmitted to plants, spiroplasmas must reach the salivary glands and ultimately be incorporated into salivary secretions before being inoculated by its insect vector into the host plant. *S. citri* is able to invade all eight cell types of the salivary gland cells of the leafhopper vector, *C. tenellus*, thereby causing cytopathological effects such as the loss of plasma membrane and basal membrane integrity as well as the disorganization of the endoplasmic reticulum (Wayadande *et al.*, 1997; Kwon *et al.*, 1999). The time period extending between insect vector acquisition of *S. citri* and the ability of the vector to transmit the infection is called the latency period, and this is a temperature-dependent process that will also vary according to the species involved (host and pathogen). The latency period may comprise only a few days or require many (up to 12) weeks (Murral *et al.*, 1996; Ammar and Hogenhout, 2006; Berho *et al.*, 2006a). Phytoplasmas follow a route of infection in their insect vectors similar to that observed for spiroplasmas though they seem to accumulate at greater numbers in the cytoplasm of midgut epithelial cells that appear to survive despite the phytoplasma infestation (Ammar and Hogenhout, 2006).

Spiroplasmas infect other organs that are accessible from the haemolymph, including the filter chamber, other gut areas of the intestinal tract, the lobulated and tubular parts of the Malpighian tubules, the haemocytes and the fat tissues, but the most extensively infected tissues in *D. maidis* seemed to be the muscle and tracheal cells (Ammar and Hogenhout, 2005). *S. kunkelii* was not detected in nerve cells of the brain or other nerve ganglia although adjacent muscle, fat and connective tissues were infected (Ammar and Hogenhout, 2005, 2008; Todd *et al.*, 2010).

Experimentally all *Dalbulus* and *Baldulus* species appear to be capable of transmitting *S. kunkelii* to maize plants; however, the maize leafhopper *D. maidis* is the natural and most efficient vector of *S. kunkelii* (> 80% transmission rates) compared to the other *Dalbulus* spp. (9–38% transmission rates) for several reasons. First, *D. maidis* benefits from the *S. kunkelii* infection as its survival and longevity is significantly improved when deprived of maize plants and/or at lower sub-optimal temperatures in laboratory or field conditions in comparison to non-infected insects in similar conditions (Ebbert and Nault, 1994, 2001). Secondly, *S. kunkelii* reduces the survival and longevity of all other *Dalbulus* and *Baldulus* spp. tested, including *D. gelbus*; these leafhoppers die at the end of the latency period (Madden and Nault, 1983; Madden *et al.*, 1984). It was observed that *S. kunkelii* accumulates in haemolymph and in cells of the midgut, salivary glands, muscles, trachea and neuromuscular junctions of *D. gelbus*. However, fewer *S. kunkelii* were found in the salivary gland cells and a much higher number of this Mollicute was present in muscle tissues of *D. gelbus* compared to *D. maidis* (Ammar and Hogenhout, 2006). This may partly explain why *S. kunkelii* reduces the survival and longevity of *D. gelbus* but not that of *D. maidis*.

## *Spiroplasma citri* transmissible and non transmissible strains

Once infected the leafhopper vectors remain infectious for their entire life. Some *S. citri* strains multiply to high titers in insects ($10^6$–$10^7$ spiroplasma cells/insect) but fail to be inoculated into plants. Some *S. citri* strains are well known to have lost their transmissibility due to serial passages in artificial media or maintenance for a long time in the same plant without reinoculation by the vector into healthy plants. The *S. citri* BR3-3X strain, first isolated from horseradish plants (*Armoracia rusticana*) affected with brittle root disease (Fletcher, 1983), was maintained until 1991 by leafhopper transmission to turnip plants and was designed BR3-T. The original BR3 strain, introduced by leafhoppers into periwinkles and maintained by grafting in periwinkles for 10 years (designed strain BR3-G), lost insect transmissibility (Wayadande and Fletcher, 1995).

Similarly the Iranian strain '44' kept on a citrus tree for more than 20 years without undergoing cycles through the insect cannot be transmitted by *C. haematoceps* to a plant. The successful transmission of *S. citri* by its leafhopper relies on the ability of the spiroplasmas to cross two physical barriers, the gut epithelial and the salivary gland cell layers. These non transmissible strains seem to have lost the ability to cross these two barriers. A model has been proposed in which spiroplasmas adhere to receptors on the gut epithelium, and on the plasmalemma outer surface of the salivary glands (Fletcher *et al.*, 1998).

## Development and use of leafhopper cell culture to characterize spiroplasmas–vector interactions

Early events that occur during the spiroplasmas infection process are directly linked to the ability of these bacteria to adhere to and invade insect cells. One of the best ways to study interactions between bacteria and their host at the cellular and molecular levels is the use of cells in culture. Cellular models are currently used to study the mechanisms of cell infection by animal pathogens, including mycoplasmas (Giron *et al.*, 1996; Winner *et al.*, 2000; Fleury *et al.*, 2002; Svenstrup *et al.*, 2002; Burnett *et al.*, 2006; Drasbek *et al.*, 2007). However, established leafhopper cell lines are still very few (Omura and Kimura, 1994). Here, we deal with the more recent cell lines that have been developed to describe interaction between *S. citri* strains with the cells of their insect vector. These cells were used to describe and quantify adhesion to and invasion of insect cells by transmissible and non-transmissible strains.

CT 1 and Ciha-1 cell lines were developed from *C. tenellus* (Wayadande and Fletcher, 1998) and *C. haematoceps* (Duret *et al.*, 2010), respectively, the natural vectors of *S. citri* in the USA and in Mediterranean countries. The establishment of these leafhopper cell lines was difficult and time consuming, but successful and useful to characterize the interaction between *S. citri* and insect cells in culture. The first cell line, CT 1, consists of a heterogeneous cell line made up of two types of epithelial cells, one type of fibroblast-like cells, and rounded and deeply pigmented cells. The second cell line, Ciha-1, is composed of a cell population consisting of 90% of cells with epithelial-type morphology. Interaction of insect cells in culture with spiroplasmas was studied by electron microscopy. Small invaginations of the CT 1 cell plasmalemma in close contact with both strains of *S. citri* BR3 isolated from infected horseradish plants (Fletcher, 1983), the strains BR3-T (insect-transmissible,) and BR3-G (graft-transmissible, non-insect-transmissible), strongly suggested an adhesion of both *S. citri* strains to insect cells (Wayadande and Fletcher, 1998). Such adhesion was also observed when the Ciha-1 cells were infected with the transmissible strain *S. citri* GII3, originally isolated from *C. haematoceps* captured in Morocco (Vignault *et al.*, 1980; Duret *et al.*, 2010). The apparent invagination that occurs during the adhesion of *S. citri* BR3 and GII3 to the plasmalemmae of the insect cells in culture assay is consistent with the hypothesis that a receptor-mediated endocytosic event is involved in the entry of *S. citri* into host cells. In addition, quantification of *S. citri* ability to adhere to insect cells was performed by counting colony-forming units (CFUs). This binding assay was first carried out at 4°C, a temperature that prevents spiroplasmal internalization. The *S. citri* adhesion of the non-insect-transmissible strain 44, isolated from a stubborn-disease sweet orange tree in Iran (Hosseini Pour, 2000) was significantly lower (0.4% of cells with adherent spiroplasmas at a multiplicity of infection (m.o.i.) of 30; 2.7% at a m.o.i. of 200, at 4°C) than that of *S. citri* GII3 (3.2% of cells with adherent spiroplasmas at a m.o.i. of 40; 20% at a m.o.i. of 300, at 4°C) (Duret *et al.*, 2010). This adhesion rate increased with temperature (5–10 times higher at 32°C) but was still significantly lower for the non-insect-transmissible strain 44 than for the insect-transmissible *S. citri* GII3. Differences between the adhesion rate of strains

GII3 and 44 to insect cells in culture assay suggest that the non-insect-transmissible strain 44 lacks factors that permit the interaction of the bacteria with the insect cell receptor.

BR3-G bacteria were also observed in membrane-bound vesicles within the cytoplasm of CT 1 cells, showing the ability of these spiroplasmas to enter insect cells in culture. Double immunofluorescence assay was performed to visualize intra-cellular *S. citri* GII3 (Duret *et al.*, 2010). Through the double staining method using fluorescent antibodies and observations using confocal laser scanning microscope, intracellular spiroplasmas could clearly be distinguished from those remaining outside the cells. The presence of *S. citri* GII3 within the cells was detected as early as 1 hour after inoculation and the morphology of the spiroplasmas changed during this process. While in cell culture medium spiroplasmas displayed a helical morphology as in the insect haemolymph, they lost their helical morphology and appeared round or pleomorphic once inside the insect cell (Duret *et al.*, 2010). These changes in morphology resemble those observed during the passage through the intestinal cells (Ammar *et al.*, 2004). Thus this cell line offers a complementary approach to decipher the cascade of events that governs the spiroplasmal morphological changes during the adhesion and entry process into insect cells.

As adhesion is a prerequisite to cell- invasion, the differences in adhesion between the insect-transmissible strain *S. citri* GII3 and the non-insect-transmissible *S. citri* 44 implied that the *S. citri* 44 should be affected in its entry capability into insect cells. Gentamicin protection assays were used to quantify and compare the ability of both *S. citri* GII3 and 44 to enter the leafhopper cells in culture. Gentamycin is commonly used to only kill extracellular bacteria and gentamycin protection assays were successfully applied to insect cells infected with spiroplasmas. *S. citri* GII3 was able to invade the Ciha-1 cells and was significantly more invasive than the strain 44 (Duret *et al.*, 2010). *S. citri* GII3 survived within the cell for at least 2 days. The finding that the non-insect-transmissible *S. citri* 44 was less adherent and invasive than GII3 suggests a good correlation between the ability to invade insect cells in culture assay and leafhopper vector cells *in vivo*. Thus, use of the leafhopper cell lines proved to be of great interest to learn more about the precise mechanisms by which spiroplasmas pass through insect cells to be transmitted to the plant.

## Spiroplasma plasmids play a key role in the leafhopper transmission

*S. citri* strain BR3–3X harbours a plasmid designated pBJS-O (Joshi *et al.*, 2005) that codes the spiroplasma adhesion-related protein, SARP1, involved in *S. citri* adhesion to cell culture of its leafhopper vector *C. tenellus*. The pSKU146 plasmid from *S. kunkelii* strain CR2–3X encodes a homologue of SARP1, SkARP1 (Davis *et al.*, 2005). Characteristics of SARP1 and SkARP1 are detailed in the paragraph related to the candidate proteins involved in the *S. citri* transmission. The involvement of these plasmids in spiroplasma transmission is not yet demonstrated.

*S. citri* strain GII3 has seven plasmids, designated pSciA of 79 kbp and pSci1 to pSci6 which sizes range from 12.9 to 35.3 kbp (Saillard *et al.*, 2008). All plasmids were detected as multiple copies in strain GII3. Plasmids pSci1 to pSci6 were estimated at 10 to 14 copies per spiroplasma cell and therefore represent 1.6 Mb of extrachromosomal DNA. Plasmid DNA should approximately correspond to 47% of total DNA of *S. citri* GII3. Sequence analysis revealed a mosaic gene organization of the six plasmids pSci1–6 constituted by blocks of sequences with sequence similarities of 80%. The smallest plasmid pSciA for which the number of copies was estimated at 2 or 3 displayed no nucleotide similarity with the other pScis.

Genes encoding proteins of the TrsE-TraE, Mob, TraD-TraG, and Soj-ParA protein families were predicted in most of the pSci sequences in addition to 14 protein families of unknown function. Plasmids pSci1 to pSci5 encode eight different *S. citri* adhesion-related proteins, designated ScARPs homologous to the previously described *S. citri* BR3 adhesion related membrane protein P89/SARP1 (Berg *et al.*, 2001) and *S. kunkelii* SkARP1 protein (Davis *et al.*, 2005). Two additional heavily truncated ScARP CDS were detected on pSci6 that carries the cytosolic protein P32 gene tentatively designed as a marker of insect transmissibility (Killiny *et al.*, 2006). Plasmid pSci6 also contains 16 truncated parts of various CDS, including disrupted copies of *soj* and *mob*. Most of the plasmid CDS have no orthologue in other bacterial genera with the exceptions of *soj*, *Trs*E, *Tra*G and *mob*, the designation of which indicates that they encode proteins closely related to proteins involved in partitioning (Soj/ParA family COG1192) and transfer (TrsE, TraG and mob) of DNA molecules. Each plasmid has its own partition system involving the soj protein (Breton *et al.*, 2008). None of the plasmid CDS was homologous to Rep proteins involved in the replication. It is noteworthy that pSci plasmids do not share sequence similarities with phytoplasma plasmids (Liefting *et al.*, 2006) that encode putative replication proteins (Rep) having the conserved motives associated with the rolling circle mechanism of replication (Oshima *et al.*, 2001). Plasmids pSciA and pSci1-6 possess the *soj* and *pE* genes, which are also conserved in plasmids pBJS-O from *S. citri* BR3 (Joshi *et al.*, 2005) and pSKU146 from *S. kunkelii* (Davis *et al.*, 2005). Interestingly, *pE* and *soj* are the only CDS that are conserved, uninterrupted, in all seven plasmids, pSciA and pSci1-6. Preliminary data strongly suggested that the replication region of plasmid pSci2 was located within an 8 kbp restriction fragment (Berho *et al.*, 2006b). This fragment encodes seven putative polypeptides, including the pE hypothetical protein. Characterization of this replication region has revealed that pE is essential and is the only plasmid-encoded protein required for plasmid replication (Breton *et al.*, 2008).

The role of plasmid encoded determinants in insect transmission of *S. citri* was first suggested from the observation that, in contrast to *S. citri* GII3, the non-insect-transmissible strains did not express ScARPs and P32 (Killiny *et al.*, 2006) and did not possess the pSci1-6 plasmids encoding these proteins (Berho *et al.*, 2006a). Such a correlation between plasmid occurrence and the ability of the spiroplasma strains to be transmitted by the leafhopper vector suggested that plasmids pSci1-6 encoded genetic determinants might be essential for insect-transmission. To further investigate this hypothesis, plasmids from insect-transmissible strain GII3 were introduced by electroporation into the plasmid-free, non-transmissible strain 44. Successful transmission of *S. citri* 44 transformants to the host plant via injection into the leafhopper vector indicated that genetic determinants required for insect transmission of *S. citri* 44 are encoded by plasmid pSci6 (Berho *et al.*, 2006b).

To delineate the pSci6 region containing the genetic determinants required for insect transmission, a large collection of *S. citri* GII3 mutants differing in their plasmid contents (Fig. 10.1) were constructed by developing a plasmid curing/replacement strategy based on the fact that plasmids having identical replication regions are incompatible. This plasmid incompatibility was also used to replace the wild type pSci6 plasmid with its mutated/deleted derivatives (Breton *et al.*, 2010).

Experimental transmission of these mutants through injection to or ingestion by the leafhopper vector revealed that pSci6 and more precisely CDS pSci6_06 annotated *traG* (Carle *et al.*, 2010) was required for insect transmission (Breton *et al.*, 2010). This CDS encodes a 287-amino-acid protein sharing limited homology with the C-terminal part of ATPases of the TraG/VirD4 family involved in type IV secretion systems (Rogowsky *et al.*, 1990; Schroder *et al.*, 2002). However, unlike most TraG proteins, including the two (of 600 and 721 amino acids) encoded by the *S. citri* GII3 chromosome, the pSci6_06-encoded protein lacks a signal peptide, the N-terminal transmembrane segments, and the Walker A motif (i.e. one of two functional domains of the TraG proteins). Hence it is unlikely that this pSci6_06 protein is part of a type IV secretion system.

In summary, the pSci6_06 coding sequence encodes a protein of unknown function that was essential for acquisition and transmission. ScARPs and P32 proteins were not absolutely necessary for transmission although *S. citri* mutants lacking pSci1 to 5 (encoding ScARPs) were acquired and transmitted at lower efficiencies than the wild-type strain.

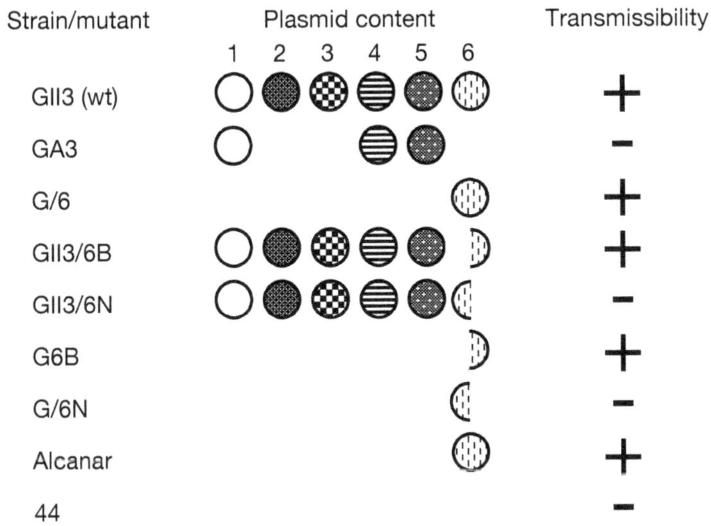


**Figure 10.1** Insect-transmissibility of plasmid mutants. A large set of spiroplasmal mutants differing in their plasmid contents (Breton *et al.*, 2010) could be tested for their transmissibility by *C. haematoceps*. Circles depict native plasmids pSci1–6 and half-circles represent pSci6 truncated derivatives (unpublished figure courtesy of Dr Marc Breton University Bordeaux Segalen, France). Mutant GA3 contains native pSci1, pSci4 and pSci5 and mutant G/6 carries only pSci6. Mutants GII3/6B and GII3/6N contain native plasmids pSci1–5, the difference between them concerns the plasmid pSci6. Mutant GII3/6B contains a pSci6 deletion derivative plasmid with a region containing *traG* sequence whereas the mutant GII3/N carry a pSci6 deletion derivative plasmid lacking the *traG* region. Mutants G6B and G/6N do not carry plasmids pSci1–5 but carry a pSci6 deletion derivative plasmid with or without the region containing *traG* sequence. *S. citri* strains carrying either native pSci6 or pSci6B were efficiently transmitted by the leafhopper, while the strains lacking the *traG*-containing fragment of pSci6 were not.

## *Spiroplasma citri* candidate proteins involved in insect transmission

### Spiralin

Spiralin has first been identified as the most abundant protein of the *S. citri* membrane, since it accounts for more than 20% of the membrane total protein mass (Wróblewski *et al.*, 1977). Later, spiralin was shown to be shared by several transmissible and non-transmissible strains of *S. citri* (Foissac *et al.*, 1996), as well as by other spiroplasma species. Spiralin is specific to spiroplasmas and the set of available nucleotide sequences deposited in Genbank includes the spiralin-encoding genes isolated from the three phytopathogenic species, i.e. *S. citri* (Chevalier *et al.*, 1990; Foissac *et al.*, 1996), *S. kunkelii* and *S. phoeniceum* (Foissac *et al.*, 1997). In these three species, the protein products comprise 237–244 amino acids. *S. citri* spiralin (strain R8A2) shows 65% and 60% identity with spiralins of *S. phoeniceum* (strain P40) and *S. kunkelii* (strain E275), respectively. Spiralin encoding genes have also been identified and sequenced in *S. melliferum*, which is a non phytopathogenic spiroplasma but belongs to the same serogroup (Group I) (Williamson *et al.*, 2011) as *S. citri* (Foissac *et al.*, 1997). Spiralin from *S. citri* R8A2 shows 76% identity with the spiralin from *S. melliferum* BC3. However, spiralin does not seem to be specific to spiroplasmas from Group I, as evidences for the presence of sequences encoding putative spiralin-like proteins in spiroplasmas from distinct serovars have been obtained (Bové *et al.*, 1993).

Spiralin can be quantitatively purified under non-denaturing conditions (Wróblewski *et al.*, 1977, 1989). Consequently, this protein has been extensively studied for its topology and arrangement in membranes, its antigenicity, its composition and its secondary structure. Spiralin is a highly amphiphilic, antigenic protein (Wróblewski, 1979, 1984), and its polypeptide chain is exposed on the spiroplasma membrane surface (Castano et al., 2002). It is produced as a prolipoprotein containing a conserved N-terminal signal peptide with a typical cleavage site for the signal peptidase II (SPaseII) (Le Henaff et al., 1991; Béven et al., 1996; Foissac et al., 1996). The protein is covalently lipid-modified (Wróblewski, 1979; Wróblewski et al., 1989; Foissac et al., 1996) with a diacylglyceryl cysteine as the N-terminal residue. Acyl moiety favours the anchorage of the hydrophilic polypeptide in the outer lipid leaflet of the membrane. The presence of an additional amide-linked fatty acid in the mature form has been proposed (Le Henaff and Fontenelle, 2000), but the gene encoding the N-acyltransferase responsible for the catalysis of this reaction has not yet been identified. Also, spiralin was shown to self-assemble into multimers in the membrane (Wróblewski, 1981) and a 'carpet-model' for the organization of the lipoprotein was proposed. In this model, the hydrophilic, surface-exposed moiety of spiralin is made of two co-linear domains and covers most lipids of the membrane outer leaflet (Castano et al., 2002). The amphiphilic character as well as the presence of a lipoprotein signal peptide, are common features to spiralins encoded in the different spiroplasma species.

It has long been suggested that spiralin was essential for shape maintenance and motility of spiroplasmas. However, these hypotheses had to be rejected since the strain *S. citri* GII3–9a2, which lacks a functional spiralin gene, is helical and motile (Duret et al., 2003). Furthermore, the non-helical strain *S. citri* ASP1 still produces the lipoprotein (Townsend et al., 1977). The spiralin-less mutant GII3–9a2 proved to be a valuable tool for assessing the role of spiralin in transmission of *S. citri* by its vector. Indeed, experimental assays, in which the wild-type strain *S. citri* GII3 and the spiralin-disrupted strain GII3-9a2 were transmitted (two-weeks transmission period) to periwinkle plants through injection into the leafhopper vector C. haematoceps were carried out. The control and mutant strains did not differ in their ability to multiply in the insects. While each plant developed severe symptoms within two weeks after transmission with the control strain, only one or two plants out of five infected with *S. citri* GII3–9a2 showed symptoms. In addition, the development of symptoms was significantly delayed with the spiralin deficient mutant, as compared to the wild-type strain. These highly reproducible results indicated that spiralin is not essential for pathogenicity of *S. citri* but is required for efficient transmission of the spiroplasma by its insect vector (Duret et al., 2003). More recently, spiroplasma overlay assays of protein blots (far Western assays) were performed to screen *C. haematoceps* proteins as putative *S. citri*-binding molecules (Killiny et al., 2005). Spiralin could bind *in vitro* to *C. haematoceps* glycoproteins with apparent masses of 50 kDa and 60 kDa. These *N*-glycoproteins displayed a high content of mannose with fucose-linked α(1,6) to *N*-acetylglucosamine. Given that spiralin showed a significant affinity for these glycoproteins, it was hypothesized that this lipoprotein could act as a lectin allowing the attachment of spiroplasmas to cell surface-exposed insect glycoproteins (Killiny et al., 2005). Such a role for spiralin could explain that *S. citri* GII3–9A2 was less efficiently transmitted than the wild type in the experimental transmission assay (Duret et al., 2003). Thus a prominent role for this lipoprotein during the transmission process and, more specifically, in crossing gut and salivary gland epithelial barriers is suspected (Killiny et al., 2005). Nevertheless, owing to its abundance, a structural and/or mechanical role for spiralin in the control of membrane integrity and organization, in addition to its lectin function can not yet be ruled out.

## *Spiroplasma citri* Sc76 protein

A gentamicin resistant mutant G76 (Boutareaud et al., 2004) was obtained by random insertion of transposon Tn*4001* into the spiroplasmal genome of the wild-type (wt) *S. citri* GII3.

Growth of mutant G76 in gentamicin-SP4 medium was similar to that of wt GII3 in gentamicin-free SP4 medium ($5 \times 10^8$ CFU/ml). This mutant was screened for its ability to be transmitted by the leafhopper vector *C. haematoceps* and to induce symptoms in periwinkle plants. At day 13 after microinjection into the insects, the final titre obtained for mutant G76 was found to be half that of wt GII3 ($6 \times 10^5$ CFU/insect for G76 and $1.3 \times 10^6$ for wt GII3).

One hundred per cent of the plants infected by

the wt GII3 always developed symptoms during the second week following insect transmission, while the first symptoms on G76-infected periwinkles appeared approximately at week 7 or 8 and became severe from week 9 after the transmission period.

Furthermore with mutant G76 only 50–70% of the periwinkle plants showed symptoms, the symptomless plants contained no spiroplasma even after the 16 weeks of monitoring. No spiroplasma revertant was detected by PCR in crude extract of G76-infected plants at 16 weeks after the transmission period. Spiroplasma titres of mutant G76 in symptomatic leaves were about 20–30 times lower than those of the wt GII3.

The delay observed in symptom development with mutant G76-infected plants could be due to a lower multiplication rate of G76 in the plants than that of wt GII3 or to a less efficient transmission of spiroplasmas. When mutant G76 and wt GII3 were transmitted to periwinkle plants by graft inoculation, symptoms appeared at the same time on both plants even though the titre of G76 was lower than that of GII3. This suggests that the multiplication of mutant G76 in the plant is not responsible for the delay in symptom development and, thus, that the leafhoppers are less efficient in injecting mutant G76 into plants than wt GII3.

In the insect, spiroplasmas in the haemolymph have to cross the salivary glands cell membrane, multiply in these glands before they can be released in the salivary duct and be transmitted to the plant during a phloem-sap feeding of the insect. In the insect, mutant G76 and wt GII3 multiplied at the same titre ($1.2 \times 10^6$ CFU for G76 and $2.5 \times 10^6$ for GII3), but the number of mutant G76 within the salivary glands was 10 times lower than that of wt GII3 ($10^4$ CFU for G76 and $1.5 \times 10^5$ for GII3). In addition, the ratios of spiroplasmas present in leafhopper heads were similar for the mutant G76 and wt GII3. The low G76 spiroplasma titre in leafhoppers salivary glands could be explained by the two following hypotheses: (i) mutant G76 is defective in its ability to move from the haemolymph into the salivary gland or (ii) mutant G76 has a reduced multiplication level within the salivary gland cells compared to strain GII3.

The gene inactivated by insertion of Tn*4001* into the genome of mutant G76 was identified and named *sc76* (Boutareaud *et al.*, 2004). Functional complementation of the mutant G76 with the *sc76* gene restores completely the wild phenotype indicating that SC76 protein is involved in transmission of *S. citri* by the leafhopper vector *C. haematoceps*.

As determined by ScanProsite analyses, the SC76 protein (51.8 kDa) had a typical signal sequence and a signal peptidase II cleavage site followed by a cysteine residue. This putative signal peptide comprised two basic amino acids at positions 2 and 3, followed by 20 mainly hydrophobic amino acids and a cysteine in position 24 representing a potential acylation site (+1 after the peptide site cleavage). The composition of this signal peptide is in agreement with the consensus sequence described for bacterial lipoproteins (Hayashi and Wu, 1990). Using the TMpred program, the protein was predicted to contain three transmembrane helices.

In silico analyses indicated that the SC76 protein could be a Solute Binding Protein (SBP) of a sugar ABC transporter. The superfamily of ABC transporters played an important role for the export of proteins and polysaccharides and for the import of sugars, inorganic ions, and oligopeptides (Quentin *et al.*, 1999). The SBP proteins function in conjunction with ABC transporters and are anchored on the outer surface of the cell membrane. If the SC76 protein is a SBP of a sugar ABC transporter, it may have affinity for sugars present in the basal lamina of salivary glands epithelium. It may be required for sugar uptake in addition to or instead of adherence. Growth comparison between mutant G76, mutant G76 complemented with *sc76* and *S. citri* GII3 in different media supplemented with fructose, trehalose or glucose suggested that Sc76 might be involved in glucose transport. Glucose and trehalose being the main sugars in insects (Becker *et al.*, 1996), the lack of SC76 could explain the low multiplication rate of G76 mutant in cells of the salivary gland. Another possibility, not exclusive of the previous one, is that SC76 functions as an adhesin interacting with glucose on the salivary gland surface. Following this hypothesis, SC76 protein implicated in transmission of *S. citri* to plants by leafhoppers would have a bifunctional role: it could act as a glucose SBP component of an ABC transporter and as an adhesin facilitating penetration of the spiroplasmas into salivary glands. Evidence for a single gene affording dual functions has been already described in *S. citri* (Le Dantec *et al.*, 1998). It is interesting to note that in *Mycoplasma hominis*, OppA the substrate-binding subunit of the oligopeptide permease which is a member of the ABC transporter family (Henrich *et al.*, 1999), is a multifunctional lipoprotein involved in cytadherence, but also functions as an ecto-ATPase inducing apoptose of cells (Hopfe and Henrich, 2004) thus suggesting its role as a virulence factor in *M. hominis*.

Recently, *S. citri* GII3 chromosome sequence (Carle *et al.*, 2010) has revealed two additional CDS predicted to encode lipoproteins with similar sequence to Sc76. In the G76 mutant reduced insect transmissibility was associated with the deletion of Sc76 protein alone. Thus, the 2 similar lipoproteins to Sc76 were not sufficient to conserve the transmissibility of the spiroplasma by leafhoppers. These data suggest a specific role for the protein SC76 in transmission, and that this function is not shared by the two Sc76-like proteins.

## The dual role of the spiroplasma phosphoglycerate kinase

Salivary gland invasion is the ultimate step of the phytopathogenic Mollicute cycle in its leafhopper vector before transmission to plant. Several lines of evidence suggest that host–pathogen interactions such as attachment to host cells could be a prerequisite for invasion and successful colonization of salivary glands. In most Mollicutes, this event is mediated by surface proteins and among which adhesins play an important role.

Interactions between *S. citri* proteins and proteins of the leafhopper salivary glands were investigated using protein overlay assays (far Western assays). Spiroplasma proteins were found to bind to a set of insect host proteins having an apparent molecular masses of 42, 35, 30, 27, 25 kDa (Killiny *et al.*, 2005; Labroussaa *et al.*, 2010). Among these, a product of 42 kDa was identified by LC-MS/MS analysis as the actin protein, a major component of the insect cell cytoskeleton. Confocal observations of *C. haematoceps* leafhopper salivary glands and *C. haematoceps* 'Ciha-1' cell line, both experimentally infected with *S. citri*, revealed that spiroplasmas were located along the insect actin microfilaments (Fig. 10.2). These *in vitro* and *in vivo* results are in good agreement with those obtained with phytoplasmas. Previous studies have shown an interaction between the immunodominant membrane protein (Amp) from *Candidatus* Phytoplasma asteris (OY strain) and three leafhopper proteins involved in a microfilament complex (Suzuki *et al.*, 2006). According to the authors, these interactions determined the insect-vector specificity for the phytoplasmas and played a major role in the transmission process.

The location of spiroplasmas along actin filaments, in particular with those close to the host cell surface, suggested that entry into the host cells may involve cytoskeleton rearrangements at the point of contact between spiroplasma and the host cell membrane.

Protein overlay assays revealed that the *S. citri* phosphoglycerate kinase (PGK) of 44 kDa was involved in the interaction with the actin cytoskeleton (Labroussaa *et al.*, 2010). PGK is a key enzyme of the glycolysis by catalysing the transfer of a phosphate group from 1,3-biphosphoglycerate to ADP, to form 3-phosphoglycerate and ATP. Only one copy of the corresponding gene is present in the *S. citri* genome (Carle *et al.*, 2010) but the deduced amino acid sequence showed no N-terminal signal peptide and no C-terminal anchor sequence or typical transmembrane domain required for the protein to be surface or membrane located. Thus PGK joins the group of enzymes including enolase (Yavlovich *et al.*, 2007), glyceraldehyde-3-phosphate dehydrogenase (Bergmann *et al.*, 2004), and pyruvate dehydrogenase E1 β-subunit (Dallo *et al.*, 2002) that besides their apparent metabolic functions may have some other roles in bacteria. The finding that *S. citri* PGK, a glycolytic enzyme, interacts with eukaryotic actin is supported by earlier studies: several other glycolytic enzymes from rabbit muscle and PGK from bacteria such as *Streptococcus agalactiae* act as actin-binding proteins (Arnold and Pette, 1968; Burnham *et al.*, 2005).

In an attempt to identify PGK regions that interact with actin, several overlapping PGK truncated

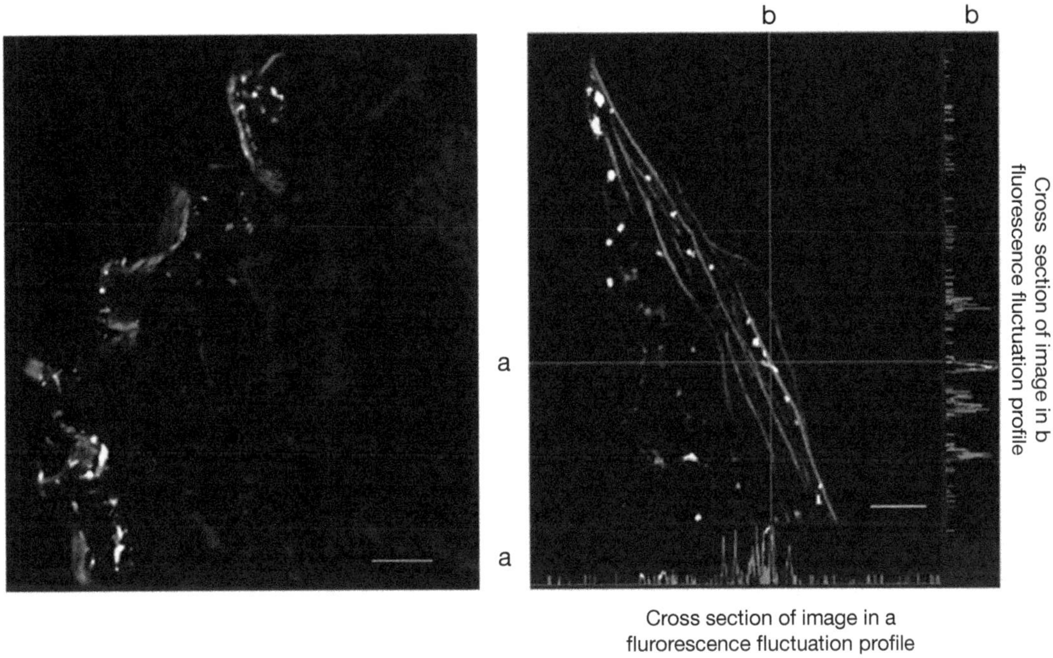


**Figure 10.2** Spiroplasmas (green) and actin filaments (red) in *C. haematoceps* salivary gland (A) and *S. citri*-infected Ciha-1 (B) after infection with *S. citri* GII3, as visualized by immuno-fluorescence confocal laser microscopy. Infected cells nuclei were stained with DAPI (blue). Scale bar, 8µm. Spiroplasmas are localized along the actin cytoskeleton in insect infected cells. A colour version of this figure is available in the plate section at the back of the book.

peptides were generated. Binding assays using the truncated peptides as probes on insect protein blots revealed that the PGK actin-binding region was located on one truncated peptide designated PGK-FL5 (amino acids 49–154) (Labroussaa *et al.*, 2011).

### *Ex vivo* effect of PGK and its truncated versions in attachment to and internalization in Ciha 1 cells

To investigate the role of PGK–actin interaction, competitive spiroplasma attachment and internalization assays were performed in which recombinant PGK and its truncated versions were individually added to Ciha-1 cells prior to infection with *S. citri*. Results showed that PGK and all individual truncated versions (Labroussaa *et al.*, 2011) had no effect on spiroplasma attachment to the leafhopper cell surface. In contrast, PGK and truncated peptide PGK-FL5 containing aa 49 to 154, reduced *S. citri* internalization into Ciha-1 cells in a dose dependent manner.

### *In vivo* effect of the PGK actin binding region in *Spiroplasma citri* transmission process

The *in vivo* effect of PGK and PGK-FL5 was also confirmed by transmission studies. The ability of *S. citri* to be inoculated in SP4 medium by infected leafhoppers injected with phosphate buffer versus infected leafhoppers injected with PGK or PGK truncations were compared. Injection with tagged PGK protein or PGK-FL5 peptide in the *S. citri* infected leafhoppers resulted in a significant reduction of inoculation in SP4, compared with controls (52% and 59%, respectively).

## Adhesion-related proteins and *Spiroplasma citri* p32 protein

As mentioned above, insights of the putative *S. citri* genetic determinants involved in transmission were obtained by comparing the protein profile of the non-insect-transmissible strain *S. citri* 44 with that of the wild type *S. citri* GII3. ScARPs and P32 were not expressed in *S. citri* non-insect-transmissible strains while they were detected in the proteome of the transmissible strain GII3 (Killiny *et al.*, 2006). The lack of expression of these proteins is due to the absence of pSci1–5 encoding ScARPs and of pSci6 carrying the P32-encoding gene. These plasmids are also absent in other non-insect-transmissible *S. citri* strains (Berho *et al.*, 2006a).

P32 is a 238 aa, hydrophilic, cytoplasmic protein with a theoretical molecular mass of about 28 kDa and eight putative sites of phosphorylation. This protein is probably specific to *S. citri*, as no homologue to the p32 gene could be identified in any other organism. The transmissibility of *S. citri* 44 is not restored by functional complementation with the p32 gene. Nevertheless, the *S. citri* strains GII3, 44 and the complemented strain 44-P32 significantly differ in their distribution in salivary glands of *C. haematoceps*. While the wild-type spiroplasmas were mainly found at the periphery of the salivary cells in cytoplasmic vesicles, strains 44 and 44-P32 were located between the basal lamina and the plasmalemma with none being detected inside the cell cytoplasm. Nonetheless, few spiroplasmas were in close contact with the plasmalemma with the complemented strain 44-P32, but not with strain 44. Thus, despite the fact that no precise function could be assigned to P32, this protein most likely plays a role in the transmission process and could mediate the attachment of *S. citri* to the salivary gland cells of its insect host (Killiny *et al.*, 2006).

Since functional complementation of the P32-encoding gene in the non-insect transmissible strain 44 did not restore transmissibility, the involvement in *S. citri* transmissibility of other plasmid-encoded proteins had to be considered. A spiroplasma adhesion-related protein, P89/SARP1, was first identified in *S. citri* BR3-T (Yu *et al.*, 2000), a strain which is naturally transmitted by *C. tenellus*. To better understand the involvement of *S. citri* surface proteins in the interaction of the pathogen with *C. tenellus* cells, Yu *et al.* investigated spiroplasma binding to cultured *C. tenellus* cells (Yu *et al.*, 2000). The pre-treatment of the bacteria with proteinase K significantly reduced the cell adherence of *S. citri*. Such a treatment triggered the proteolysis of SARP1 and led to the production of a SARP1 breakdown product of 46 kDa. It was hypothesized that SARP1 was a surface protein, which may act as an adhesin to mediate the attachment of spiroplasmas to insect cells. The characterization of *S. citri* BR3-T *arp1* gene (Berg *et al.*, 2001) revealed that SARP1 contained a domain made of six repeats of 39–42 amino acids (sarpin repeats) each predicted to form a propeller-like structure, and a putative transmembrane alpha-helix close to its C-terminal end. SARP1 is expected to contain a large surface-exposed hydrophilic domain and a short cytoplasmic C-terminal tail. In *S. citri* BR3–3X, the plasmid pBJS-O carrying the SARP1-encoding gene has been isolated and characterized (Joshi *et al.*, 2005). The authors also reported the presence on the BR3–3X chromosome of the *arp2* gene encoding a putative protein SARP2, which differs from SARP1 by its N-terminal part (Joshi *et al.*, 2005). An adhesion-related protein showing high homology with SARP1, SkARP1, was also found in the second phytopathogenic, plasmid-harbouring spiroplasma species, namely *S. kunkelii* CR2–3X. SkARP1 contains seven sarpin repeats and is encoded by a gene located on plasmid pSKU146 (Davis *et al.*, 2005). Two other sequences encoding putative ARP proteins are located on the chromosome of *S. kunkelii* (Bai *et al.*, 2004). In *S. citri* GII3, the presence of proteins (ScARPs) highly homologous to SARP1 was first evidenced in 2006 (Berho *et al.*, 2006a). ScARPs are major antigenic proteins of *S. citri* GII3 membrane. Eight ScARPs encoding genes, which may result from gene duplication and recombination events, are distributed on plasmids pSci1-5 in *S. citri* GII3. Based on their similarities to SARP1, ScARPs have been classified in four families named ScARP2 to ScARP5. In addition, two C-terminal truncated *scarp* CDS were identified on pSci6 (Saillard *et al.*, 2008). Since *scarp* genes were detected in *S. citri* GII3 but not in three different non insect-transmissible strains (Berho *et al.*, 2006a), ScARPs have been considered as excellent candidates for having a determinant role in *S. citri* GII3 insect transmission. ScARPs share 40–77% and 43–56% sequence identity with SARP1 and SkARP1, respectively. Similarly to SARP1, ScARPs show typical features of surface-exposed proteins. In their N-terminal part, except for ScARP4a, 39- to 42-amino-acid-long repeated sequences have been identified and their consensus sequence has been reported (Saillard *et al.*, 2008). Owing to the lack of repeated motives in its sequence, ScARP4a (683 amino acids) is the shortest ScARP while ScARP5a (683 residues) with eight repeats is the longest. According to analyses using SignalP (Bendtsen *et al.*, 2004), the ScARPs share a well-conserved signal peptide, which is expected to be cleaved after alanine 23 and to be responsible for the surface-exposure of a large, hydrophilic domain comprising *n* (6–8) repeats, a central conserved region (CR) of about 340 amino acids and a 110 amino acids containing variable region (VR). VR sequences are of two types; VR1 being shared by ScARP2 and ScARP3 families and VR2 found in ScARP4a and ScARP5a. Downstream of the VR sequences, a transmembrane helix is predicted to be responsible for the membrane-anchorage of ScARPs. The short cytoplasmic C-terminal moiety comprises one or two domains enriched in polar amino acids. The first one is common to all eight ScARPs and is rich in the basic amino acids lysine and arginine, while the second one, found in 6 ScARPs, contains a high proportion of lysyl-aspartyl-glutamyl residues (Saillard *et al.*, 2008) (Fig. 10.3).

As mentioned above, a role for adhesion-related proteins in spiroplasma interaction with insect cells is largely considered. More precisely, these proteins are suspected to function as adhesins in the phytopathogenic spiroplasmas *S. citri* and *S. kunkelii*. Indeed, the repetitive domain of ScARP3d was recently shown to trigger entry of *S. citri* into cultured cells of the vector *Circulifer haematoceps* (Béven *et al.*, 2012). Nevertheless, the nature and identity of their potential insect receptors are still unknown. The presence of ScARPs does appear to be essential for spiroplasma insect-transmissibility, as *S. citri* GII3 mutant strains lacking the ScARPs encoding plasmids pSci1–5 were less efficiently transmitted than the wild-type strain (Breton *et al.*, 2010). Nevertheless, ScARPs are not sufficient to confer insect-transmissibility, given that *scarp* genes have been identified in non-transmissible strains such as *S. citri* BR3-G (Joshi *et al.*, 2005). As well, an intriguing point resides in the presence of eight *scarp* genes in *S. citri* GII3. Because ScARPs are major membrane antigens, their expression likely contributes to shaping the spiroplasma surface. Thus, the co-occurrence of highly identical sequences raises the question of whether all eight

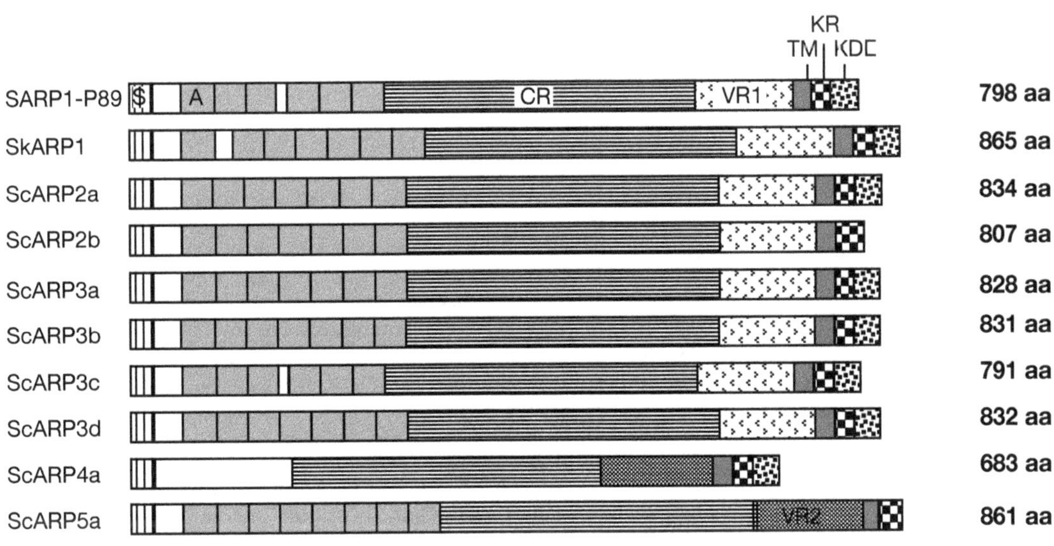


**Figure 10.3** Modular organization of spiroplasma adhesion-related proteins (ARPs). SARP1 and SkARP1 are found in *S. citri* BR3 and *S. kunkelii* CR2–3X, respectively. ScARPs 2a-5a are plasmid-encoded products in *S. citri* GII3. S, putative signal peptides; A, sarpin repeat; TM, hydrophobic transmembrane domains; CR, conserved region; VR1 and VR2, variable regions of type 1 or 2; KR and KDE: lysine-arginine and lysine-aspartate-glutamate rich domains, respectively.

gene products are simultaneously expressed in the different organs of the insect vector and whether they share the same function or not. Although spiroplasma ARPs share a high sequence identity, they also doubtlessly bear specific features that may correspond to different functional motives. Indeed, searches for conserved domains against the Conserved Domains Database (Marchler-Bauer *et al.*, 2011) indicated that SARP1 and five different ScARPs (2b, 3b, 3c, 3d, 5a) contain a region (repeated sequence) corresponding to a predicted ligand-binding sensor domain (COG3292), while a region containing a conserved domain common to AvrE (pathogenicity factor secreted by Gram-negative bacteria such as *Pseudomonas syringae* and *Erwinia amylovora*) was identified in ScARP2a sequence. Thus, the different ARPs may be responsible for different biological activities in *S. citri*. Finally, owing to the modular organization of spiroplasma ARPs, this protein family may also correspond to multifunctional proteins.

## Conclusion

This chapter reinforces the idea that *S. citri* transmission is a complex multi-step process, which involves proteins with additional roles to their known functions. PGK with its well-known function in glycolysis and its novel role in spiroplasma internalization is such a protein. Several outer-surface proteins involved in transmission such as spiralin, Sc76 and adhesins (ScARPs, P89) likely have diverse possibly unrelated functions. This chapter also described a crucial role played by several plasmid-encoded proteins (ScARPs, P32, CDS pSci6_06) in the transmission process. In conclusion, plasmid recruitment and rearrangements of genomic regions harbouring genes encoding proteins with diverse roles in the invasion processes are likely subjected to evolutionary events allowing spiroplasmas adaptation to specific plants and insect species.

## References


Ammar, E.D., and Hogenhout, S.A. (2005). Use of immunofluorescence confocal laser scanning microscopy to study distribution of the bacterium corn stunt spiroplasma in vector leafhoppers (*Hemiptera: Cicadellidae*) and in host plants. Ann. Entomol. Soc. Am. *98*, 820–826.

Ammar, E.D., and Hogenhout, S.A. (2006). Mollicutes associated with arthropods and plants. In Insect Symbiosis Vol 2, Bourtzis, K., and Miller, T.A., eds. (CRC Press, Florida), pp. 97–118.

Ammar, E.D., and Hogenhout, S.A. (2008). A neurotropic route for *Maize mosaic virus* (*Rhabdoviridae*) in its planthopper vector *Peregrinus maidis*. Virus Res. *131*, 77–85.

Ammar, E.D., Fulton, D., Bai, X., Meulia, T., and Hogenhout, S.A. (2004). An attachment tip and pili-like structures in insect- and plant-pathogenic spiroplasmas of the class Mollicutes. Arch. Microbiol. *181*, 97–105.

Ammar, E.D., Gasparich, G.E., Hall, D.G., and Hogenhout, S.A. (2011). Spiroplasma-like organisms closely associated with the gut in five leafhopper species (*Hemiptera: Cicadellidae*). Arch. Microbiol. *193*, 35–44.

Arnold, H., and Pette, D. (1968). Binding of glycolytic enzymes to structure proteins of the muscle. Eur. J. Biochem. *6*, 163–171.

Bai, X., Fazzolari, T., and Hogenhout, S.A. (2004). Identification and characterization of traE genes of *Spiroplasma kunkelii*. Gene *336*, 81–91.

Becker, A., Schloder, P., Steele, J.E., and Wegener, G. (1996). The regulation of trehalose metabolism in insects. Experientia *52*, 433–439.

Bendtsen, J.D., Nielsen, H., von Heijne, G., and Brunak, S. (2004). Improved prediction of signal peptides: SignalP 3.0. J. Mol. Biol. *340*, 783–795.

Berg, M., Melcher, U., and Fletcher, J. (2001). Characterization of *Spiroplasma citri* adhesion related protein SARP1, which contains a domain of a novel family designated sarpin. Gene *275*, 57–64.

Bergmann, S., Rohde, M., and Hammerschmidt, S. (2004). Glyceraldehyde-3-phosphate dehydrogenase of *Streptococcus pneumoniae* is a surface-displayed plasminogen-binding protein. Infect. Immun. *72*, 2416–2419.

Berho, N., Duret, S., and Renaudin, J. (2006a). Absence of plasmids encoding adhesion-related proteins in non-insect-transmissible strains of *Spiroplasma citri*. Microbiology *152*, 873–886.

Berho, N., Duret, S., Danet, J.L., and Renaudin, J. (2006b). Plasmid pSci6 from *Spiroplasma citri* GII-3 confers insect transmissibility to the non-transmissible strain S. citri 44. Microbiology *152*, 2703–2716.

Béven, L., Le Henaff, M., Fontenelle, C., and Wróblewski, H. (1996). Inhibition of spiralin processing by the lipopeptide antibiotic globomycin. Curr. Microbiol. *33*, 317–322.

Béven, L., Duret, S., Batailler, B., Dubrana, M.-P., Saillard, C., Renaudin, J., and Arricau-Bouvery, N. (2012). The repetitive domain of ScARP3d triggers entry of *Spiroplasma citri* into cultured cells of the vector *Circulifer haematoceps*. Plos One 7, e48606.

Boutareaud, A., Danet, J.L., Garnier, M., and Saillard, C. (2004). Disruption of a gene predicted to encode a solute binding protein of an ABC transporter reduces transmission of *Spiroplasma citri* by the leafhopper *Circulifer haematoceps*. Appl. Environ. Microbiol. *70*, 3960–3967.

Bové, J.M., Foissac, X., and Saillard, C. (1993). Spiralins. Subcell. Biochem. *20*, 203–223.

Breton, M., Duret, S., Arricau-Bouvery, N., Béven, L., and Renaudin, J. (2008). Characterizing the replication and stability regions of *Spiroplasma citri* plasmids identifies a novel replication protein and expands the genetic toolbox for plant-pathogenic spiroplasmas. Microbiology *154*, 3232–3244.

Breton, M., Duret, S., Danet, J.L., Dubrana, M.P., and Renaudin, J. (2010). Sequences essential for transmission of *Spiroplasma citri* by its leafhopper vector, *Circulifer haematoceps*, revealed by plasmid curing and replacement based on incompatibility. Appl. Environ. Microbiol. *76*, 3198–3205.

Burnett, T.A., Dinkla, K., Rohde, M., Chhatwal, G.S., Uphoff, C., Srivastava, M., Cordwell, S.J., Geary, S., Liao, X., Minion, F.C., *et al.* (2006). P159 is a proteolytically processed, surface adhesin of *Mycoplasma hyopneumoniae*: defined domains of P159 bind heparin and promote adherence to eukaryote cells. Mol. Microbiol. *60*, 669–686.

Burnham, C.A., Shokoples, S.E., and Tyrrell, G.J. (2005). Phosphoglycerate kinase inhibits epithelial cell invasion by group B streptococci. Microb. Pathog. *38*, 189–200.

Calavan, E.C., and Bové, J.M. (1989). Ecology of *Spiroplasma citri*. In The Mycoplasmas, Withcomb, R.F., and Tully, J.G., eds. (Academic Press, New York), pp. 425–485.

Carle, P., Saillard, C., Carrere, N., Carrere, S., Duret, S., Eveillard, S., Gaurivaud, P., Gourgues, G., Gouzy, J., Salar, P., *et al.* (2010). Partial chromosome sequence of *Spiroplasma citri* reveals extensive viral invasion and important gene decay. Appl. Environ. Microbiol. *76*, 3420–3426.

Castano, S., Blaudez, D., Desbat, B., Dufourcq, J., and Wróblewski, H. (2002). Secondary structure of spiralin in solution, at the air/water interface, and in interaction with lipid monolayers. Biochim. Biophys. Acta *1562*, 45–56.

Chevalier, C., Saillard, C., and Bové, J.M. (1990). Spiralins of *Spiroplasma citri* and *Spiroplasma melliferum*. Amino acid sequences and putative organization in the cell membrane. J. Bacteriol. *172*, 6090–6097.

Dallo, S.F., Kannan, T.R., Blaylock, M.W., and Baseman, J.B. (2002). Elongation factor Tu and E1 beta subunit of pyruvate dehydrogenase complex act as fibronectin binding proteins in *Mycoplasma pneumoniae*. Mol. Microbiol. *46*, 1041–1051.

Davis, R.E., Dally, E.L., Jomantiene, R., Zhao, Y., Roe, B., Lin, S., and Shao, J. (2005). Cryptic plasmid pSKU146 from the wall-less plant pathogen *Spiroplasma kunkelii* encodes an adhesin and components of a type IV translocation-related conjugation system. Plasmid *53*, 179–190.

Drasbek, M., Christiansen, G., Drasbek, K.R., Holm, A., and Birkelund, S. (2007). Interaction between the P1 protein of *Mycoplasma pneumoniae* and receptors on HEp-2 cells. Microbiology *153*, 3791–3799.

Duret, S., Berho, N., Danet, J.L., Garnier, M., and Renaudin, J. (2003). Spiralin is not essential for helicity, motility, or pathogenicity but is required for efficient transmission of *Spiroplasma citri* by its leafhopper vector *Circulifer haematoceps*. Appl. Environ. Microbiol. *69*, 6225–6234.

Duret, S., Batailler, B., Danet, J.L., Béven, L., Renaudin, J., and Arricau-Bouvery, N. (2010). Infection of the *Circulifer haematoceps* cell line Ciha-1 by *Spiroplasma citri*: the non insect-transmissible strain 44 is impaired in invasion. Microbiology *156*, 1097–1107.

Ebbert, M.A., and Nault, L.R. (1994). Improved overwintering ability in *Dalbulus maidis* (*Homoptera, Cicadellidae*) vectors infected with *Spiroplasma kunkelii* (*Mycoplasmatales, Spiroplasmataceae*). Environ. Entomol. 23, 634–644.

Ebbert, M.A., and Nault, L.R. (2001). Survival in *Dalbulus* leafhopper vectors improves after exposure to maize stunting pathogens. Entomol. Exp. Appl. *100*, 311–324.

Fletcher, J. (1983). Brittle root of horseradish in Illinois and the distribution of *Spiroplasma citri* in the United States. Phytopathology 73, 354–357.

Fletcher, J., Schultz, G.A., Davis, R.E., Eastman, C.E., and Goodman, R.M. (1981). Brittle root disease of horseradish. Evidence for an etiological role of *Spiroplasma citri*. Phytopathology *71*, 1073–1080.

Fletcher, J., Wayadande, A., Melcher, U., and Ye, F. (1998). The phytopathogenic Mollicute–insect vector interface: a closer look. Phytopathology *88*, 1351–1358.

Fleury, B., Bergonier, D., Berthelot, X., Peterhans, E., Frey, J., and Vilei, E.M. (2002). Characterization of P40, a cytadhesin of *Mycoplasma agalactiae*. Infect. Immun. *70*, 5612–5621.

Foissac, X., Saillard, C., Gandar, J., Zreik, L., and Bové, J.M. (1996). Spiralin polymorphism in strains of *Spiroplasma citri* is not due to differences in posttranslational palmitoylation. J. Bacteriol. *178*, 2934–2940.

Foissac, X., Bové, J.M., and Saillard, C. (1997). Sequence analysis of *Spiroplasma phoeniceum* and *Spiroplasma kunkelii* spiralin genes and comparison with other spiralin genes. Curr. Microbiol. 35, 240–243.

Fos, A., Bové, J.M., Lallemand, J., Saillard, C., Vignault, J.C., Ali, Y., Brun, P., and Vogel, R. (1986). The leafhopper *Neoaliturus hematoceps* is a vector of *Spiroplasma citri* in the mediterranean area. Ann. Inst. Pasteur. Microbiol. *137A*, 97–107.

Giron, J.A., Lange, M., and Baseman, J.B. (1996). Adherence, fibronectin binding, and induction of cytoskeleton reorganization in cultured human cells by *Mycoplasma penetrans*. Infect. Immun. *64*, 197–208.

Hayashi, S., and Wu, H.C. (1990). Lipoproteins in bacteria. J. Bioenerg. Biomembr. *22*, 451–471.

Henrich, B., Hopfe, M., Kitzerow, A., and Hadding, U. (1999). The adherence-associated lipoprotein P100, encoded by an opp operon structure, functions as the oligopeptide-binding domain OppA of a putative oligopeptide transport system in *Mycoplasma hominis*. J. Bacteriol. *181*, 4873–4878.

Hogenhout, S.A., Oshima, K., Ammar el, D., Kakizawa, S., Kingdom, H.N., and Namba, S. (2008). Phytoplasmas: bacteria that manipulate plants and insects. Mol. Plant Pathol. *9*, 403–423.

Hopfe, M., and Henrich, B. (2004). OppA, the substrate-binding subunit of the oligopeptide permease, is the major Ecto-ATPase of *Mycoplasma hominis*. J. Bacteriol. *186*, 1021–1928.

Hosseini Pour, A. (2000). Determination of some molecular and cellular characteristics of *Spiroplasma citri*, the causal agent of citrus stubborn disease in Kerman, Fars and Mazadaran provinces. (Tarbiat Modares University, Tehran).

Joshi, B.D., Berg, M., Rogers, J., Fletcher, J., and Melcher, U. (2005). Sequence comparisons of plasmids pBJS-O of *Spiroplasma citri* and pSKU146 of *S. kunkelii*: implications for plasmid evolution. BMC Genomics *6*, 175.

Kaloostian, G.H., Oldfield, G.N., Pierce, H.D., and Calavan, E.C. (1979). *Spiroplasma citri* and its transmission to citrus and other plants by leafhoppers. In Leafhopper Vectors and Plant Disease Agents, Maramorosch, K., and Harris, K.F., eds. (Academic Press, New York), pp. 447–450.

Killiny, N., Castroviejo, M., and Saillard, C. (2005). *Spiroplasma citri* spiralin acts in vitro as a lectin binding to glycoproteins from its insect vector *Circulifer haematoceps*. Phytopathology *95*, 541–548.

Killiny, N., Batailler, B., Foissac, X., and Saillard, C. (2006). Identification of a *Spiroplasma citri* hydrophilic protein associated with insect transmissibility. Microbiology *152*, 1221–1230.

Kwon, M.O., Wayadande, A.C., and Fletcher, J. (1999). *Spiroplasma citri* movement into the intestines and salivary glands of its leafhopper vector, *Circulifer tenellus*. Phytopathology *89*, 1144–1151.

Labroussaa, F., Arricau-Bouvery, N., Dubrana, M.P., and Saillard, C. (2010). Entry of *Spiroplasma citri* into *Circulifer haematoceps* cells involves interaction between spiroplasma phosphoglycerate kinase and leafhopper actin. Appl. Environ. Microbiol. *76*, 1879–1886.

Labroussaa, F., Dubrana, M.P., Arricau-Bouvery, N., Béven, L., and Saillard, C. (2011). Involvement of a minimal actin-binding region of *Spiroplasma citri* phosphoglycerate kinase in spiroplasma transmission by its leafhopper vector. PLoS One *6*, e17357.

Le Dantec, L., Castroviejo, M., Bové, J.M., and Saillard, C. (1998). Purification, cloning, and preliminary characterization of a *Spiroplasma citri* ribosomal protein with DNA binding capacity. J. Biol. Chem. *273*, 24379–24386.

Le Henaff, M., and Fontenelle, C. (2000). Chemical analysis of processing of spiralin, the major lipoprotein of *Spiroplasma melliferum*. Arch. Microbiol. *173*, 339–345.

Le Henaff, M., Brenner, C., Fontenelle, C., Delamarche, C., and Wróblewski, H. (1991). Does spiralin has a cleavable N-terminal signal sequence? C. R. Acad. Sci. III *312*, 189–195.

Liefting, L.W., Andersen, M.T., Lough, T.J., and Beever, R.E. (2006). Comparative analysis of the plasmids from two isolates of '*Candidatus Phytoplasma australiense*'. Plasmid *56*, 138–144.

Madden, L.V., and Nault, L.R. (1983). Differential pathogenicity of corn stunting Mollicutes to leafhopper vectors in *Dalbulus* and *Baldulus* species. Phytopathology *73*, 1608–1614.

Madden, L.V., Nault, L.R., Heady, S.E., and Styer, W.E. (1984). Effect of maize stunting Mollicutes on survival and fecundity of *Dalbulus* leafhopper vectors. Ann. Appl. Biol. *105*, 431–441.

Marchler-Bauer, A., Lu, S., Anderson, J.B., Chitsaz, F., Derbyshire, M.K., DeWeese-Scott, C., Fong, J.H., Geer, L.Y., Geer, R.C., Gonzales, N.R., *et al.* (2011). CDD: a conserved domain database for the functional annotation of proteins. Nucleic Acids Res. *39*, 225–229.

Mello, A.F., Wayadande, A.C., Yokomi, R.K., and Fletcher, J. (2009). Transmission of different isolates of *Spiroplasma citri* to carrot and citrus by *Circulifer tenellus* (*Hemiptera: Cicadellidae*). J. Econ. Entomol. *102*, 1417–1422.

Murral, D.J., Nault, L.R., Hoy, C.W., Madden, L.V., and Miller, S.A. (1996). Effects of temperature and vector age on transmission of two Ohio strains of aster yellows phytoplasma by the aster leafhopper (*Homoptera: Cicadellidae*). J. Econ. Entomol. *89*, 1223–1232.

Oldfield, G.N., Kaloostian, G.H., Pierce, H.D., Calavan, E.C., Granett, A.L., and Blue, R.L. (1976). Beet leafhopper transmits citrus stubborn disease. California Agricult.*30*, 15–15.

Oldfield, G.N., Kaloostian, G.H., Pierce, H.D., Calavan, E.C., Granett, A.L., Blue, R.L., Rana, G.L., and Gumpf, D.J. (1977). Transmission of *Spiroplasma citri* from citrus to citrus by *Scaphytopius nitridus*. Phytopathology *67*, 763–765.

Omura, T., and Kimura, I. (1994). Leafhopper cell culture for virus research. In Arthropod Cell Culture Systems, Maramorosch, K., and McIntosh, A.H., eds. (CRC Press, Philadelphia, PA), pp. 91–107.

Oshima, K., Shiomi, T., Kuboyama, T., Sawayanagi, T., Nishigawa, H., Kakizawa, S., Miyata, S., Ugaki, M., and Namba, S. (2001). Isolation and characterization of derivative lines of the '*Onion Yellows Phytoplasma*' that do not cause stunting or phloem hyperplasia. Phytopathology *91*, 1024–1029.

Ozbek, E., Miller, S.A., Meulia, T., and Hogenhout, S.A. (2003). Infection and replication sites of *Spiroplasma kunkelii* (Class: Mollicutes) in midgut and Malpighian tubules of the leafhopper *Dalbulus maidis*. J. Invertebr. Pathol. *82*, 167–175.

Purcell, A.H. (1983). Insect vector relationships with procaryotic plant-pathogens. Annu. Rev. Phytopathol. *20*, 397–417.

Quentin, Y., Fichant, G., and Denizot, F. (1999). Inventory, assembly and analysis of *Bacillus subtilis* ABC transport systems. J. Mol. Biol. *287*, 467–484.

Rogowsky, P.M., Powell, B.S., Shirasu, K., Lin, T.S., Morel, P., Zyprian, E.M., Steck, T.R., and Kado, C.I. (1990). Molecular characterization of the *vir* regulon of *Agrobacterium tumefaciens*: complete nucleotide sequence and gene organization of the 28.63-kbp regulon cloned as a single unit. Plasmid *23*, 85–106.

Saglio, P., Lhospital, M., Laflèche, M., Dupont, G., Bové, J.M., Tully, J.G., and Freundt, E.A. (1971). Culture in vitro des mycoplasmes associés au stubborn des agrumes et leur observation au microscope electronique. C. R. Acad. Sci. Paris 272, 1387–1390.

Saglio, P., Lhospital, M., Laflèche, D., Dupont, G., Bové, J.M., Tully, J.G., and Freundt, E.A. (1973). *Spiroplasma citri* gen., and sp. nov.: a mycoplasma-like organism associated with stubborn disease of citrus. Int. J. Syst. Bacteriol. *23*, 191–204.

Saillard, C., Vignault, J.C., Bové, J.M., Raie, A., Tully, J.G., Williamson, D.L., Fos, A., Garnier, M., Gadeau, A., Carle, P., *et al.* (1987). *Spiroplasma phoeniceum* sp-nov, a new plant-pathogenic species from Syria. Int. J. Syst. Bacteriol. *37*, 106–115.

Saillard, C., Carle, P., Duret-Nurbel, S., Henri, R., Killiny, N., Carrere, S., Gouzy, J., Bové, J.M., Renaudin, J., and Foissac, X. (2008). The abundant extrachromosomal DNA content of the *Spiroplasma citri* GII3-3X genome. BMC Genomics 9, 195.

Schroder, G., Krause, S., Zechner, E.L., Traxler, B., Yeo, H.J., Lurz, R., Waksman, G., and Lanka, E. (2002). TraG-like proteins of DNA transfer systems and of the *Helicobacter pylori* type IV secretion system: Inner membrane gate for exported substrates? J. Bacteriol. *184*, 2767–2779.

Suzuki, S., Oshima, K., Kakizawa, S., Arashida, R., Jung, H.Y., Yamaji, Y., Nishigawa, H., Ugaki, M., and Namba, S. (2006). Interaction between the membrane protein of a pathogen and insect microfilament complex determines insect-vector specificity. Proc. Natl. Acad. Sci. U.S.A. *103*, 4252–4257.

Svenstrup, H.F., Nielsen, P.K., Drasbek, M., Birkelund, S., and Christiansen, G. (2002). Adhesion and inhibition assay of *Mycoplasma genitalium* and *M. pneumoniae* by immunofluorescence microscopy. J. Med. Microbiol. *51*, 361–373.

Todd, J.C., Ammar, E.D., Redinbaugh, M.G., Hoy, C., and Hogenhout, S.A. (2010). Plant host range and leafhopper transmission of *Maize fine streak virus*. Phytopathology *100*, 1138–1145.

Townsend, R., Markham, P.G., Plaskitt, K.A., and Daniels, M.J. (1977). Isolation and characterization of a non-helical strain of *Spiroplasma citri*. J. Gen. Microbiol. *100*, 15–21.

Vignault, J.C., Bové, J.M., Saillard, C., Vogel, R., Farro, A., Venegas, L., Stemmer, W., Aoki, S., McCoy, R.E., Al-beldawi, A.S., *et al.* (1980). Mise en culture de spiroplasmes à partir de matériel végétal et d'insectes provenant de pays circum méditerranéens et du Proche Orient. C. R. Acad. Sci. Paris *290*, 775–780.

Wayadande, A.C., and Fletcher, J. (1995). Transmission of *Spiroplasma citri* lines and their ability to cross gut and salivary glands barriers within the leafhopper vector *Circulifer tenellus*. Phytopathology *85*, 1256–1259.

Wayadande, A.C., Baker, G.R., and Fletcher, J. (1997). Comparative ultrastructure of the salivary glands of two phytopathogen vectors, the beet leafhopper, *Circulifer tenellus* (Baker), and the corn leafhopper, *Dalbulus maidis* DeLong and Wolcott (*Homoptera: Cicadellidae*). J. Insect Morphol. Embryol. *26*, 113–120.

Wayadande, A.C., and Fletcher, J. (1998). Development and use of an established cell line of the leafhopper *Circulifer tenellus* to characterize *Spiroplasma citri*–vector interactions. J. Invertebr. Pathol. 72, 126–131.

Whitcomb, R.F., Chen, T.A., Williamson, D.L., Liao, C., Tully, J.G., Bové, J.M., Mouches, C., Rose, D.L., Coan, M.E., and Clark, T.B. (1986). *Spiroplasma kunkelii* sp-nov. Characterization of the etiologic agent of corn stunt disease. Int. J. Syst. Bacteriol. *36*, 170–178.

Williamson, D.L., Gasparich, G.E., Regassa, L.B., Saillard, C., Renaudin, J., Bové, J.M., and Whitcomb, R.F. (2011). Family II. *Spiroplasmataceae*. In Bergey's Manual of Systematic Bacteriology Vol 4, Krieg, N.R., Ludwig, W., Whitman, W.B., Hedlund, B., Paster, B.J., Staley, J.T., Ward, N., Brown, D.R., and Parte, A., eds. (Springer, New York), pp. 654–686.

Winner, F., Rosengarten, R., and Citti, C. (2000). In vitro cell invasion of *Mycoplasma gallisepticum*. Infect. Immun. *68*, 4238–4244.

Wróblewski, H. (1979). Amphilphilic nature of spiralin, the major protein of the *Spiroplasma citri* cell membrane. J. Bacteriol. *140*, 738–741.

Wróblewski, H. (1981). Electrophoretic analysis of the arrangement of spiralin and other major proteins in isolated *Spiroplasma citri* cell membranes. J. Bacteriol. *145*, 61–67.

Wróblewski, H., Johansson, K.E., and Hjerten, S. (1977). Purification and characterization of spiralin, the main protein of the *Spiroplasma citri* membrane. Biochim. Biophys. Acta *465*, 275–289.

Wróblewski, H., Robic, D., Thomas, D., and Blanchard, A. (1984). Comparison of the amino acid compositions and antigenic properties of spiralins purified from the plasma membranes of different spiroplasmas. Ann. Microbiol. (Paris) *135A*, 73–82.

Wróblewski, H., Nystrom, S., Blanchard, A., and Wieslander, A. (1989). Topology and acylation of spiralin. J. Bacteriol. *171*, 5039–5047.

Yavlovich, A., Rechnitzer, H., and Rottem, S. (2007). Alpha-enolase resides on the cell surface of *Mycoplasma fermentans* and binds plasminogen. Infect. Immun. *75*, 5716–5719.

Yu, J., Wayadande, A.C., and Fletcher, J. (2000). *Spiroplasma citri* surface protein P89 implicated in adhesion to cells of the vector *Circulifer tenellus*. Phytopathology *90*, 716–722.

# Organization of the Cytoskeletons of Diverse Mollicutes

11

Mitchell F. Balish

## Abstract

Like other organisms, mycoplasmas have cytoskeletons, which are proteinaceous, detergent-insoluble polymers that constitute important structural elements. As in other bacteria, these cytoskeletons play important roles in cell division and probably in chromosome segregation. Additionally, some groups of mycoplasmas have independently developed additional, novel cytoskeletal elements that control cell shape, polarity and movement, often with concomitant alterations to the conventional set of cytoskeletal proteins found within the bacterial domain. In this chapter the diversity, composition, organization, and assembly of the cytoskeletons of *Mycoplasma pneumoniae, Mycoplasma genitalium, Mycoplasma mobile, Mycoplasma insons, Mycoplasma penetrans, Mycoplasma iowae,* and *Spiroplasma melliferum* are discussed. Special attention is paid to the best-characterized cytoskeletal structure of mycoplasmas, the electron-dense core of the *M. pneumoniae* attachment organelle.

## Introduction

Proteins serve two principal, frequently overlapping, roles in cells: they are catalysts, facilitating the biochemical reactions and interactions that propel life; and they are structural elements, organizing the physical framework in which these reactions and interactions occur. Along with proteins that regulate the three-dimensional structure of nucleic acids and proteinaceous scaffolds that organize enzymes, the cytoskeleton is a major, complex structural feature, although many cytoskeletal proteins are also catalytic. The cytoskeleton consists of a series of proteinaceous, polymeric filaments and the proteins that associate with those filaments to effect numerous cellular and subcellular processes, including movement of cells, positioning of organelles, intracellular trafficking of membrane-bound compartments, chromosome segregation and cell division. Both eukaryotes and prokaryotes have cytoskeletons, but the mycoplasma cytoskeleton is unusually diverse in both function and form, probably because of the unusual demands of a bacterial cell that lacks a cell wall.

The cytoskeleton has hallmark features that facilitate its identification. Treatment of eukaryotic cells with non-ionic detergents, with Triton X-100 the conventional choice for cell biologists, causes dissolution of membranes, membrane-embedded proteins, proteins that are soluble in the cytoplasm, and, variably, nucleic acid molecules. The remaining insoluble material consists principally of cytoskeletal filaments, the large size and extensive intermolecular contacts of which preclude solubilization. Other large protein complexes with surface areas reduced by intermolecular contacts may also be variably preserved. Thus, detergent insolubility is one defining feature. However, there have been few reported efforts to observe cytoskeletal filaments of bacteria other than mycoplasmas in this manner; advances in cell imaging technology, discussed below, have been more enlightening. The second defining feature of the cytoskeleton is functional, necessitating a discussion of the cellular roles the cytoskeleton plays.

## Eukaryotic cytoskeletal structure and function

The eukaryotic cytoskeleton features three principal types of filaments: microtubules, microfilaments and intermediate filaments. Microtubules are polymers of a dimer of the proteins α-tubulin and β-tubulin (Dyer, 2009). They are critical for chromosome segregation during meiosis and mitosis, trafficking of intracellular vesicles, flagellar movement, and, in plant cells, organization of the cell wall synthetic apparatus. Most of these functions rely on molecular motors that associate with the microtubules. Though fairly rigid, microtubules are generally quite dynamic, polymerizing and depolymerizing in response to the GTPase activity of the constituent α-tubulin/β-tubulin dimers and the differences in physical properties between GTP-bound and GDP-bound filaments. This activity principally occurs at one end of these polarized structures, as the other end, anchored in microtubule-organizing centres, is relatively inert. A second type of filament, the microfilament, is a somewhat more flexible polymer composed of the protein actin (Pollard and Cooper, 2009). Microfilaments are involved in cell movement, including muscle movement, as well as membrane events like extension of projections, cell division in non-plant cells, endocytosis and exocytosis. They are also quite dynamic, with individual actin monomers functioning as ATPases. Like microtubules, the stability of a microfilament depends largely upon whether its subunits are bound to ATP or ADP. Motor proteins also interact with microfilaments, although motor-independent activities are more frequent for microfilaments than for microtubules. Unlike microtubules, microfilaments do not emanate from organizing centres; rather, their inherent polarity causes them to polymerize and depolymerize at opposite ends, resulting in treadmilling. Finally, many eukaryotic cells have a variety of intermediate filaments that are not dynamic, do not bind nucleotides, and have no association with motor proteins, but rather function strictly structurally (Herrmann *et al.*, 2009). Their monomers interact via extensive α-helical coiled-coil regions.

## Prokaryotic cytoskeletal structure and function

Prokaryotes were long thought to lack cytoskeletons, but in fact they have been revealed to possess a great diversity of cytoskeletal filaments, the functions of many of which parallel those of the eukaryotic cytoskeleton (Adams and Errington, 2009; Margolin, 2009). Furthermore, many prokaryotic cytoskeletal filaments are structurally similar to those of eukaryotes, suggesting an evolutionary link. Nonetheless, not all bacterial cytoskeletal polymers are composed of proteins resembling those of eukaryotes; bacteria have both eukaryotic-like cytoskeletal filaments and structures that have no homologues in eukaryotic organisms (Graumann, 2009). A significant difference between eukaryotic and prokaryotic cytoskeletal filaments is that the latter make no use of motor proteins, the general role of which is to provide cellular trafficking in cells that are too large for effective localization of subcellular components by diffusion alone.

The bacterial cytoskeleton functions in peptidoglycan synthesis, spatial organization of organelles, development and maintenance of cell polarity, membrane constriction and chromosome segregation; in contrast to the eukaryotic cytoskeleton, there is no evidence for involvement in motility. In the Mollicutes, many of these processes are poorly understood or, in the case peptidoglycan, absent entirely, and membrane-bound compartments have not been described in these organisms. Nonetheless, mycoplasmas face some of the challenges addressed in other bacteria by the cytoskeleton. Cell polarity is evident in those mycoplasmas that exhibit unidirectional motility along surfaces and/or have specialized polar structures involved in adherence and motility (Balish, 2006). Constriction of cell membranes has been observed during cytokinesis of some mycoplasma species (Maniloff, 1969), whereas others appear to use motility at least in part to separate dividing cells (Hasselbring *et al.*, 2006a; Lluch-Senar *et al.*, 2010). Although chromosome segregation in mycoplasmas has received some attention (Seto and Miyata, 1999), the underlying molecular mechanism has not been investigated. Thus, possible roles for the cytoskeleton in mycoplasmas include generation of cell polarity, membrane constriction and chromosome segregation. Furthermore, because mycoplasma motility appears to be driven by mechanisms that are distinct from those employed by other bacteria (Jarrell and McBride, 2008), it is possible that the mycoplasma cytoskeleton also plays a role in this process.

There has been a considerable focus on the cytoskeleton in mycoplasma research over a significantly longer period of time than for other bacteria. This work began independently in *Spiroplasma* species (Williamson, 1974) and in *Mycoplasma pneumoniae* (Neimark, 1977), but in recent years has expanded to include a wider variety of species. In *M. pneumoniae*, cytoskeletal research has focused principally on the polar attachment organelle, the interior of which is dominated by a Triton X-100-insoluble structure that is essential for construction of this appendage (Balish and Krause, 2006). More recently, mycoplasmologists have also begun studying the mycoplasma orthologues of the components of the conventional bacterial cytoskeleton.

## FtsZ

The dynamic GTPase FtsZ is related to tubulin and is essential for cell division in most bacteria (Adams and Errington, 2009; Dyer, 2009; Margolin, 2009). Not only are FtsZ filaments capable of constricting liposomes, suggesting that they are directly responsible for constricting the cell membrane, but they also serve as a scaffold for the synthesis of a molecular machine, the divisome, which synthesizes the peptidoglycan septum at and near the division site, in temporal and spatial coordination with membrane constriction (Osawa *et al.*, 2008; Adams and Errington, 2009; Margolin, 2009; Erickson *et al.*, 2010). The activity of FtsZ is largely controlled by proteins that interfere with its assembly, including MinC, MipZ, SlmA and Noc, the subcellular locations of which are controlled by unique dynamic processes. As a result, polymerized FtsZ is present predominantly in a single structure, the Z ring, located on the cytoplasmic surface of the membrane at the division plane (Adams and Errington, 2009; Löwe and Amos, 2009). However, FtsZ is also found throughout the cell between division cycles in a helix that constricts to form the Z ring at the appropriate time (Thanedar and Margolin, 2004; Peters *et al.*, 2007). FtsZ dynamics in living cells are poorly understood, but several lines of evidence point to the organization of FtsZ as a discontinuous series of relatively short filaments sometimes laterally associated with other filaments, all of which have short half-lives (Erickson *et al.*, 2010). Presumably they function to promote rapid local constriction and assembly of peptidoglycan synthesis sites, with each individual filament contributing to septum synthesis and membrane invagination.

In many bacteria, *ftsZ* lies in an operon with other genes associated with cell division, including divisome genes (Mingorance *et al.*, 2004) and genes of uncertain function, including *mraZ*, the orthologue of which in *M. pneumoniae* has been analysed structurally by X-ray crystallography (Chen *et al.*, 2004), and *mraW*, an S-adenosylmethionine-dependent methyltransferase with unidentified substrates (Carrión *et al.*, 1999). All sequenced Mollicutes genomes contain orthologues of *mraW*, and all but the *Acholeplasmatales* and *Mycoplasma penetrans* also contain *mraZ*. None have peptidoglycan biosynthetic genes. Analysis of the currently sequenced Mollicutes genomes reveals that *ftsZ* is absent from *M. mobile*, the ureaplasmas and the phytoplasmas, but present in all the others (unpublished observations). Our examination of these genomes has also revealed that an extremely poorly conserved gene, with some, low sequence similarity to *ftsA*, which in other bacteria encodes a protein that interacts with FtsZ to promote its association with the plasma membrane and with other components of the divisome (Graumann, 2007), is present in all of the genomes except those of the *Acholeplasmatales* and the *Mycoplasma neurolyticum* cluster, including *Mycoplasma conjunctivae*, *Mycoplasma hyopneumoniae* and *Mycoplasma hyorhinis* (Zhao *et al.*, 2004; unpublished observations). Interestingly, this *ftsA*-like gene is present in the ureaplasmas and *M. mobile*, suggesting a function independent of FtsZ. In most cases, all these genes lie in an apparent operon and analysis of their expression in *M. pneumoniae* and the closely related *Mycoplasma genitalium* has revealed that they are cotranscribed (Benders *et al.*, 2005).

In *Mycoplasma hominis*, FtsZ is detected at the

division site, consistent with a role in division (Vishniakov *et al.*, 2009). Likewise, a fusion of *Mycoplasma pulmonis* FtsZ to the membrane-association region of *E. coli* FtsA in *Escherichia coli* supports a role in cell division (Osawa and Erickson, 2006). However, in other mycoplasmas, including the entire pneumoniae group and the *M. neurolyticum* cluster, the FtsZ amino acid sequence is poorly conserved (unpublished observations), and a fusion of *M. pneumoniae* FtsZ to FtsA, as described for *M. pulmonis,* does not support cell division (Osawa and Erickson, 2006). This might indicate a reduction in FtsZ function in these species, perhaps preceding loss. Indeed, despite being cotranscribed in *M. pneumoniae,* the concentrations of *ftsA* and *ftsZ* mRNAs are substantially lower than those of *mraZ* and *mraW* mRNAs; indeed, less than one copy of *ftsZ* mRNA is present per cell (Benders *et al.*, 2005), and the protein is virtually undetectable by immunoblotting in *M. pneumoniae* (M.F. Balish and D.C. Krause, unpublished data).

Nevertheless, when *ftsZ* is disrupted, *M. genitalium,* though viable, exhibits a distinct division-associated phenotype, with the population containing a large number of dividing cells connected by cytoplasmic filaments (Lluch-Senar *et al.*, 2010). This is typical for *M. pneumoniae,* but not *M. genitalium* (Hatchel and Balish, 2008), except in gliding-impaired mutant strains (Pich *et al.*, 2006). Additionally, no spontaneous haemadsorption-negative mutants can be isolated against the *ftsZ*-null background (Lluch-Senar *et al.*, 2010). Non-adhering mutants of *M. pneumoniae* are often branched, also suggesting impairment of cytokinesis (Balish and Krause, 2006). These data suggest the synergistic contribution of FtsZ-mediated membrane constriction and adherence to cell division in motile mycoplasmas. Thus, the absence of *ftsZ* in *M. mobile* may reflect its loss as the organism evolved to become entirely dependent upon adherence for normal cell division. The absence of *ftsZ* in the non-motile ureaplasmas, however, remains unexplained.

## MreB

MreB and its homologues, found principally in non-coccoid bacteria, have a similar role to actin, localizing either adjacent to the Z ring or in cables that extend helically throughout the cell along the inner surface of the plasma membrane (Graumann, 2009; Margolin, 2009). MreB is essential for the shape of most rod-shaped bacteria, spatially coordinating synthesis of peptidoglycan subunits with incorporation of new peptidoglycan into the cell wall during cell elongation (White *et al.*, 2010). It interacts both with peptidoglycan synthetic enzymes in the cytoplasm and, through the transmembrane proteins RodZ and MreD (van den Ent *et al.*, 2010; White *et al.*, 2010), with MreC, which forms a cytoskeleton-like polymer on the outer surface of the cell membrane that interacts with cell wall transpeptidases and transglycosylases (Dye *et al.*, 2005; Leaver and Errington, 2005). MreB is a dynamic ATPase/GTPase (Esue *et al.*, 2006; Kim *et al.*, 2006) that exerts force on the plasma membrane (Defeu Soufo and Graumann, 2010).

Although most mycoplasmas lack *mreB, Spiroplasma citri* has five *mreB* genes (Bové *et al.*, 2003), suggesting a role for this cytoskeletal protein in the elongated spirillar morphology of species of the genus *Spiroplasma.* Recently, *Spiroplasma melliferum* has served as a model for understanding the spiroplasma cytoskeleton, which appears as a flattened, plasma membrane-associated ribbon that stretches along the shortest helical path within the cytoplasm (Trachtenberg and Gilad, 2001; Fig. 11.1A). This structure includes approximately half a dozen fibrils composed of paired polymers of the novel structural protein Fib, the functional unit of which could be a tetramer that undergoes contraction and expansion, propelling the cell through liquid media (Trachtenberg *et al.*, 2003). However, a second, thinner type of fibril, which is postulated to be composed of MreB, is also present, possibly on the interior of the cytoskeletal structure (Kürner *et al.*, 2005). Indeed, along with elongation factor Tu, itself suggested to double, in conjunction with MreB, as a cytoskeletal protein in several bacteria (Mayer, 2003; Defeu Soufo *et al.*, 2010), Fib and MreB are both highly enriched in the Triton X-100-insoluble fraction of *S. melliferum* (Trachtenberg *et al.*, 2008). However, direct evidence for MreB polymers in the spiroplasma cytoskeleton is still lacking. Given the absence of a cell wall in the Mollicutes, the role of MreB in spiroplasmas

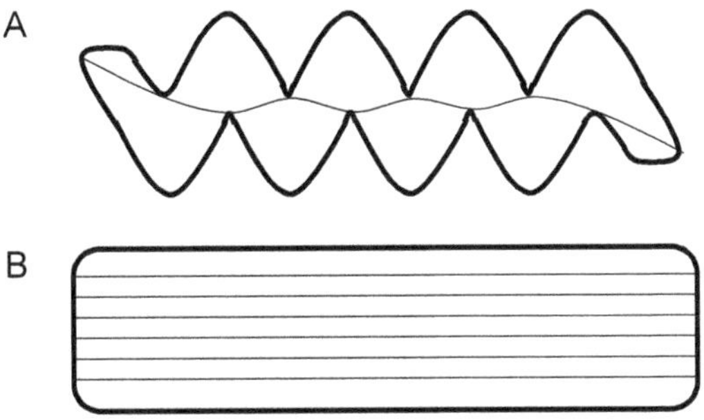


**Figure 11.1** Mycoplasma cytoskeletal features that span the length of the cell. (A) The *Spiroplasma* cytoskeleton (thin line) is a cytoplasmic ribbon, associated with the plasma membrane inner surface, that extends from one pole to the other along the shortest helical path. (B) The *M. insons* cytoskeleton (thin lines) is a series of filaments extending from one pole to the other. Although they have not been imaged in thin sections of intact cells, the filaments are probably parallel and associated with the inner surface of the plasma membrane.

is bound to be distinct from that which it plays in other bacteria.

*Mycoplasma insons* exhibits a distinct rod shape (May *et al.*, 2007), but despite the presence of numerous unidentified Triton X-100-insoluble filaments stretching between the two cell poles (Fig. 11.1B), the MreB antagonist A22 failed to affect cell shape (Relich *et al.*, 2009). These data indicate that *M. insons* either lacks MreB or that, if the cytoskeletal filaments are composed of MreB, their polymerization state is insensitive to this compound. In detergent-extracted samples there appear to be short, irregular filaments connecting the long ones at oblique angles. However, it is clear from images of partially extracted cells that the filaments are present in a three-dimensional array, possibly directly associated with the cytoplasmic surface of the plasma membrane. Therefore, it is possible that when the membrane is removed, the filaments on the top rupture as they fall onto the filaments on the bottom, resulting in images of long, unbroken filaments superimposed with fragments of short filaments. Therefore, we model the filaments of *M. insons* as parallel, cell-length, and membrane-associated (Fig. 11.1B). The unidirectional movement of this organism in the absence of a differentiated tip structure (Relich *et al.*, 2009) suggests a polarized distribution of molecules along the long axis of the cell and these cytoskeletal elements may play roles in establishing that gradient.

## Intermediate filament-like proteins

As in eukaryotes, intermediate filament proteins in bacteria are diverse and apparently scattered in distribution. Crescentin forms a static rod in *Caulobacter crescentus,* deforming one side of the cell through interactions with MreB, in turn locally altering peptidoglycan cross-linking during growth such that the characteristic curved shape of these cells is also conferred by the cell wall (Cabeen *et al.*, 2009; Charbon *et al.*, 2009). Some aspects of cell shape of diverse bacteria, including *Helicobacter pylori, Bdellovibrio bacteriovorus* and *Corynebacterium glutamicum* (Waidner *et al.*, 2009; Fenton *et al.*, 2010; Fiuza *et al.*, 2010), are also controlled by proteins with the properties of intermediate filament components, and the intermediate filament protein CfpA of *Treponema denticolum* may play a role in cell division (Izard, 2006). It is unclear whether these proteins are homologues or analogues of each other or of eukaryotic intermediate filament monomers, but given how commonly the α-helical coiled coil motif occurs in proteins, more than one evolutionary origin is conceivable.

It is difficult to identify intermediate filament proteins solely on the basis of amino acid sequence, but no proteins similar to intermediate filament proteins, such as crescentin, have been identified in the Mollicutes. However, some structural proteins of mycoplasma attachment organelle cytoskeletons, described below, are predicted to share the α-helical coiled coil structure with intermediate filament proteins. If, like intermediate filaments, these structures are not dynamic, which is suggested by their stability but has not been tested, then the proteins that constitute these unique mycoplasmal structures could be argued to represent one or more classes of bacterial intermediate filament.

## DNA segregation proteins

Segregation of some low-copy number plasmids in bacteria is mediated by proteins that are structurally similar to FtsZ/tubulins and MreB/actin, but that are, nonetheless, quite distinct from them. These plasmid-encoded proteins, TubZ and ParM, interact with proteins located at the plasmid centromere, driving the plasmids towards the poles as the filaments polymerize (Löwe and Amos, 2009; Gerdes *et al.*, 2010). There are also other actin-related proteins that appear to function similarly (Derman *et al.*, 2009). Other plasmids make use of a distinct polymerizing ATPase, ParA (Gerdes *et al.*, 2010). Like TubZ and ParM, ParA interacts with a centromeric protein, but the interaction promotes disassembly of ParA higher order structures. ParA causes plasmids to spread out approximately evenly across the chromosome to facilitate distribution during cytokinesis (Gerdes *et al.*, 2010). ParA is also used in bacteria to segregate chromosomes to opposite poles (Ptacin *et al.*, 2010; Schofield *et al.*, 2010; Shebelut *et al.*, 2010). At the onset of DNA segregation, the centrosome of one daughter chromosome is anchored to one cell pole. ParB, a centromere-binding protein on the unanchored chromosome, interacts with ParA, but that interaction causes depolymerization of ParA. As the ParA structure depolymerizes, shrinking towards the opposite cell pole, the chromosome continues to interact with ParA, resulting in movement of the chromosome towards that pole. Eventually the centromere is captured at the pole. In this manner, chromosomes are segregated to daughter cells.

Many mycoplasma genomes encode a homologue of ParA, usually designated Soj, in parallel with *Bacillus subtilis*. As in other bacteria, mycoplasmal *parA* is located near the putative origin of replication (unpublished observations). However, mycoplasmas lack a *parB* homologue and also lack sequences that resemble centromeric ParB binding sites. ParA might therefore be involved in mycoplasma chromosome segregation, but in a novel, ParB-independent manner, possibly using a non-orthologous replacement for ParB. Alternatively, as *B. subtilis* Soj also interacts with and modulates the activity of the DNA replication initiation protein DnaA (Murray and Errington, 2008), it is possible that the role of mycoplasma ParA is restricted to regulation of DNA replication.

## The polar cytoskeleton of Mycoplasma attachment organelles

Although the majority of mycoplasma species exist as pseudococcoidal cells that exhibit no evidence of polarity, several species of mycoplasmas, spanning different phylogenetic clusters, have a distinctly polar cellular organization (Fig. 11.2). In most cases this polarity is manifested as a differentiated tip structure, known as a terminal organelle, terminal bleb, head-like structure or attachment organelle (Balish, 2006). Despite the differences in terminology, which result from independent discovery and description in different mycoplasma species, all of these structures appear to be associated with adherence to host cells *in vivo* and inert surfaces *in vitro*. This is the case whether that adherence is mediated by proteins at the distal end of the structure, like P30 in *M. pneumoniae* (Baseman *et al.*, 1987; Seto and Miyata, 2003), along the sides, like Gli349 in *Mycoplasma mobile* (Uenoyama *et al.*, 2004), or both, like P1 in *M. pneumoniae* (Collier *et al.*, 1983). There are exceptional cases, such as *Mycoplasma alvi*, in which no adherence to cells or surfaces is observed, despite the presence of a differentiated tip structure (Gourlay *et al.*, 1977; Hatchel and Balish, 2008), but the absence of adherence could conceivably result from secondary loss or phase variation. The presence of a differentiated polar structure is usually associated with gliding motility; only in the absence of adherence to surfaces, such as after mutation of an adhesin gene, do mycoplasmas with differentiated tip structures fail to glide on surfaces, with the possible exception of *Mycoplasma sualvi* (K.A. Sibbing and M.F. Balish, unpublished data). Additionally, in rod-shaped *Mycoplasma insons*, unipolar adherence to surfaces and unidirectional motility are both observed (Relich *et al.*, 2009), suggesting polarization of function. However, this adherence-associated pole does not appear morphologically distinct from the other pole, resulting in the absence of a differentiated tip structure. In all cases, it is likely that the primary function of the distinct cell pole is attachment to surfaces, even if that function has been lost in some instances through mutation or phase variation. Thus, all of these structures can be

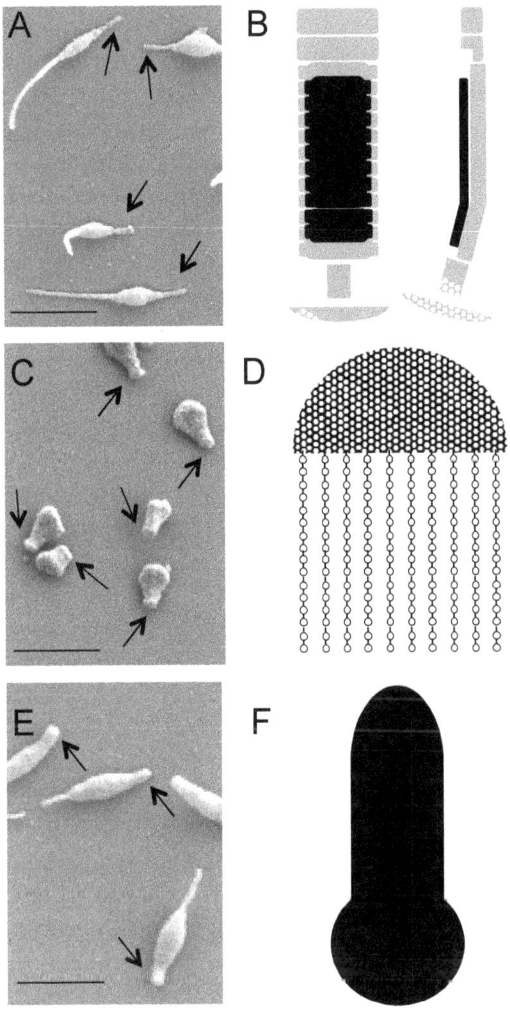


**Figure 11.2** Three polarized mycoplasmas, *M. pneumoniae* (A and B), *M. mobile* (C and D), and *M. penetrans*. A, C, and E scanning electron micrographs of *M. pneumoniae* (Panel A, courtesy of R.F. Relich), *M. mobile* (C, courtesy of D.A. Jurkovic), and *M. penetrans* (courtesy of D.A. Jurkovic). Panels B, D, and F, schematics of cytoskeletal structures present in attachment organelles, based on published reconstructions (Henderson and Jensen, 2006; Nakane and Miyata, 2007; Seybert *et al.*, 2007) and other images (Jurkovic *et al.*, 2012). The *M. pneumoniae* structure (Panel B) is shown from two angles, rotated 90° around the long axis relative to each other; differential colouring is used to illustrate the two parallel rod structures. The two large segments at the top constitute the terminal button. The elements at the bottom constitute the base, including the bowl-like structure at the very bottom. It is not clear which substructure the DNA is associated with. The *M. mobile* structure (Panel D) consists of the bell, with its hexagonal array, and the tentacles. The *M. penetrans* structure (F) has not been well-defined. Arrows, attachment organelles. Scale bars, 1 µm.

described as attachment organelles, whether they are differentiated or cryptic, whether they mediate adherence using their distal or lateral surfaces, and whether they always function in adherence or only sometimes. We therefore will make no effort to distinguish attachment organelles from terminal blebs or head-like structures in our discussion of these cellular features.

Cell polarity, whether in eukaryotes or prokaryotes, is generally the product of a cytoskeleton. Accordingly, cytoskeletal elements are major features of mycoplasma attachment organelles, with significant roles in attachment organelle biogenesis, maintenance and function. Most of what is known about mycoplasma cytoskeletons relates to the study of attachment organelles.

## *Mycoplasma pneumoniae* attachment organelle

The prominent feature of *M. pneumoniae* cells is a polar attachment organelle (Fig. 11.2A). The insolubility in Triton X-100 of a sizable insoluble structure within it (Meng and Pfister, 1980; Göbel *et al.*, 1981) suggests that the cytoskeleton of *M. pneumoniae* is a significant feature of this attachment organelle. This insoluble structure is called the electron-dense core (Biberfeld and Biberfeld, 1970; Fig. 11.2B). The Triton X-100-insoluble fraction of *M. pneumoniae* contains proteins known from other work to be structural elements of the attachment organelle, as well as other proteins without an apparent role in the attachment organelle (Regula *et al.*, 2001). In *M. pneumoniae*, the attachment organelle, which is approximately 290 nm long and 80 nm wide (Hatchel and Balish, 2008), functions in both adherence and motility (Balish and Krause, 2006). Two transmembrane adhesins, P1 and P30, are concentrated there (Krause and Balish, 2004). P30 is found exclusively at the distal terminus (Seto and Miyata, 2003), while P1, presumably in a complex with proteins B and C (Layh-Schmitt *et al.*, 2000; Nakane *et al.*, 2010), also called P90 and P40 (Waldo *et al.*, 2005), occurs all over the attachment organelle, and at lower concentrations over the entire cell surface (Seto *et al.*, 2001; Seto and Miyata, 2003). Proteins B and C are derived from a precursor that is proteolytically cleaved (Layh-Schmitt, 1993) but the *M. genitalium* orthologue, P110, is not (Dhandayuthapani *et al.*, 1999). Adherence requires normal localization of both adhesins, suggesting synergistic activity (Krause and Balish, 2004). Antisera against either protein interfere with adherence to host cells (Krause and Baseman, 1983; Morrison-Plummer *et al.*, 1986). *M. pneumoniae* cells interact with host cells via their attachment organelles (Collier and Clyde, 1971), although the presence of fibronectin-binding proteins over the entire cell surface (Dallo *et al.*, 2002) suggests that the principal adherence role of the attachment organelle may be in the early stages of interaction with the host cell.

The attachment organelle is the leading end of the *M. pneumoniae* cell during gliding motility on surfaces *in vitro* (Bredt, 1968), which appears not to be chemotactic (Harwick *et al.*, 1977), consistent with the absence of homologues of chemotaxis genes (Himmelreich *et al.*, 1996). The presence of the motor activity in the attachment organelle is evidenced by a study of a mutant strain lacking a protein, P41, which is located near the junction between the attachment organelle and the cell body (Kenri *et al.*, 2004). When P41 is absent, a gliding cell that encounters an obstacle fails to move over it, and instead stretches until the attachment organelle becomes detached (Hasselbring and Krause, 2007a). While the cell body snaps back to its original shape, the free attachment organelle continues to glide for a time. This observation suggests strongly that the attachment organelle contains everything necessary for motility, and that P41 is important for the mechanical integrity of the cell during movement. A different *M. pneumoniae* mutant, lacking the Triton X-100-insoluble protein P200, exhibits reduced gliding speed as its principal distinguishing phenotype *in vitro* (Jordan *et al.*, 2007). This strain has a normal capacity to adhere to cells, but a greatly reduced capacity to colonize differentiated bronchial epithelium, suggesting a role for motility in infection that is likely to be independent of a role in adherence.

The adherence and motility functions of the attachment organelle both appear to play significant roles in cell division in *M. pneumoniae* and its relatives. In coordination with initiation of DNA replication, *M. pneumoniae* constructs a new attachment organelle adjacent to the pre-existing one (Seto *et al.*, 2001). Evidence from both *M. pneumoniae* (Seybert *et al.*, 2006) and *Mycoplasma gallisepticum* (Nakane and Miyata, 2009) suggests that the old core serves as a template for the growth of the new core, although the successful complementation of mutant strains that lack cores by molecular manipulation indicates that *de novo* construction of cores is also possible. Observation of yellow fluorescent protein-tagged P30 in living *M. pneumoniae* cells (Hasselbring *et al.*, 2006a) suggests that the newly synthesized attachment organelle is initially incapable of motility, remaining in place as the old attachment organelle pulls the predivisional cell past it. This process results in a dividing cell with an attachment organelle at either pole, one motile and the other fixed. Division itself appears to involve the stretching apart of the two cell bodies through the force of the gliding attachment pole, causing a cytoplasmic filament to develop between them. This filament ultimately ruptures, resulting in cytokinesis. The cell with the original attachment organelle glides away, while the cell with the new one rests for several hours, after which it becomes capable of gliding (Fig. 11.3). Taken together with the requirement in *M. genitalium* for either FtsZ or attachment for growth (Lluch-Senar *et al.*, 2010), these data suggest that the role of adherence in cell division of these organisms is to provide a substrate against which the force for pulling dividing cells apart can be generated, manifested *in vitro* as gliding motility. Thus, this form of cell division, which has not been described in other

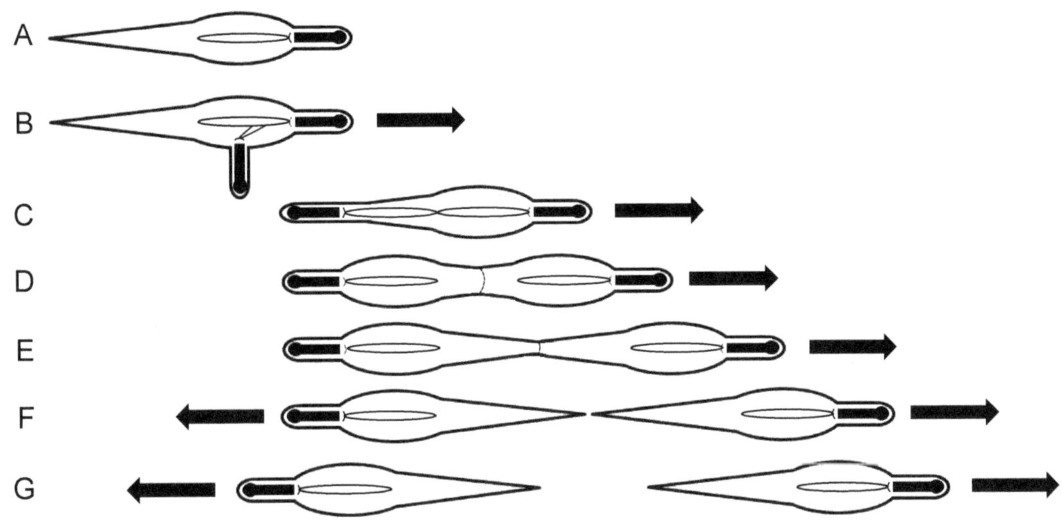


**Figure 11.3** Model of *M. pneumoniae* cell division in seven stages. See the text for evidence of each event. (A) A single cell has an electron-dense core (solid black) in a single attachment organelle, with the nucleoid (oval) attached to the attachment organelle base (arc), which might be the bowl structure observed in images (Henderson and Jensen, 2006). (B) The core has duplicated, resulting in a new attachment organelle with newly replicated DNA attached. (C) Forward motion of the old attachment organelle has driven the new attachment organelle to the rear of the cell. (D) Growth of the dividing cell results in two approximately even halves that will be separated by continued forward motion of the old attachment organelle together with constriction of FtsZ (dashed arc) at midcell. (E) A filament connecting the nascent daughter cells develops as forward motion and constriction continue. (F) Rupture of the connecting filament results in cytokinesis; the cell with the old attachment organelle continues forward motion. (G) The new attachment organelle becomes activated for motility, and both daughter cells are now motile. This model is an extension of that of Hasselbring *et al.* (2006a).

prokaryotes, enrols motility mediated through the adherence function of the attachment organelle.

The timing of events during cell division is not always as linear as represented in our model (Fig. 11.3). It is quite common to observe *M. pneumoniae* cells with three or more attachment organelles. Observation of these cells suggests that, in a dividing cell, duplication of one or more attachment organelles may occur prior to separation of daughter cells (Hasselbring *et al.*, 2006a). Cells bearing multiple attachment organelles are observed with considerable frequency in some species, including *M. pneumoniae* and *Mycoplasma amphoriforme* (Hatchel *et al.*, 2006). In *M. genitalium*, on the other hand, it is unusual to encounter such cells other than in a mutant strain in which concentrations of P140 and P110, the orthologues of *M. pneumoniae* P1 and the B/C precursor, are markedly reduced (Pich *et al.*, 2009), suggesting a role for these proteins in regulating attachment organelle duplication.

## *Mycoplasma pneumoniae* electron-dense core

The interior of the *M. pneumoniae* attachment organelle, as observed by transmission electron microscopy of thin sections, is highly differentiated from the cytoplasm of the main cell body and contains an electron-dense core surrounded by an electron-lucent space and lacks large particles like ribosomes (Biberfeld and Biberfeld, 1970). The dimensions of this electron-dense core match those of the Triton X-100-insoluble structure. Cryoelectron microscopic imaging, which circumvents the dehydration-associated artefacts associated with conventional electron microscopy, has revealed the detailed structure of the *M. pneumoniae* core to consist of several discrete elements (Henderson and Jensen, 2006; Seybert *et al.*, 2006; Fig. 11.2B). The most prominent element is a pair of rectangular plates separated by 7 nm, with about twelve distinct segments, taking the form of evenly spaced perpendicular striations. One plate is thicker than the other, with a 30° bend approximately one third of the way from the proximal end. The thinner plate is on the inner curvature. The distal end consists of a knob-like structure, the terminal button, which consists of two connected masses. At the proximal end is a discrete bowl-like structure, its opening facing distally. Although no material is visible in the surrounding space, several other lines of evidence suggest the presence of fine structures connecting the core to the membrane. Extraction of *M. pneumoniae* and six of its close relatives with Triton X-100 reveals the presence of similar structures in each, with varying dimensions (Hatchel and Balish, 2008), and images of an eighth species, *Mycoplasma alvi*, indicate the presence of a similar structure (Gourlay *et al.*, 1977), suggesting that the electron-dense core was inherited from the common ancestor of the *M. pneumoniae* cluster. The species-specific variations include size of the terminal button, the rod width, and especially the dimensions of the base of the rod, which in these preparations is probably the most proximal region of the rod, rather than the bowl-like structure, which is not preserved (Hatchel and Balish, 2008). These differences in the relative dimension of the substructures are confirmed by transmission electron microscopic imaging of *M. gallisepticum* (Nakane and Miyata, 2009). Among these various species, the overall dimensions of both the entire attachment organelle and the internal core are most similar in the most closely related species, further suggesting common inheritance of the attachment organelle and its core (Hatchel and Balish, 2008). In many cases the length of the core exceeds the length of the attachment organelle itself, suggesting that, in these species, which include *M. gallisepticum*, *Mycoplasma imitans*, *M. amphoriforme* and *Mycoplasma testudinis*, the cores actually project into the cell body.

## Composition and assembly

Identification of the components of the *M. pneumoniae* attachment organelle core has proceeded principally through studies of mutants with defects in attachment organelle function, and these studies have been supplemented by lines of experimentation with genetic and biochemical components. These findings have been expanded through studies of relatives of *M. pneumoniae*, especially *M. genitalium* and *M. gallisepticum*.

Early studies aimed at determining attachment organelle components resulted in the discovery of adhesins, as well as identification of cytadherence accessory proteins, which are required for attachment organelle function but not directly involved in adherence as adhesins. Several cytadherence accessory proteins, themselves Triton X-100-insoluble because they comprise the electron-dense core, were initially identified by screening for spontaneous mutants that formed colonies incapable of haemadsorption (Krause *et al.*, 1982). These proteins include HMW1, HMW2, HMW3 and P65. Subsequent studies based on transposon-mediated mutagenesis and sequence analysis, and further analyses of the original set of mutants, identified P200, TopJ, P41, and P24 as significant components of the *M. pneumoniae* attachment organelle, as described below. Roles for each of these proteins in attachment organelle biology have been confirmed through studies of mutants, including phenotypic analysis and immunocytochemistry. Numerous other proteins have been identified as components of either the *M. pneumoniae* attachment organelle or the Triton X-100-insoluble fraction, but functional analyses have not been carried out for these (Regula *et al.*, 2001; Hasselbring *et al.*, 2006b). With the exception of P24, all these proteins have orthologues in *M. genitalium* and *M. gallisepticum* (unpublished observations), suggesting that they constitute integral components of the mechanisms for attachment organelle assembly and function in the *M. pneumoniae* phylogenetic cluster.

### Core components

HMW2 is a large, alkaline protein predicted to contain extensive α-helical coiled coil structure (Krause *et al.*, 1997). *M. pneumoniae* haemadsorption mutant I-2, in which HMW2 is absent and other proteins, including HMW1, HMW3, P65 and adhesin P30 are at reduced steady-state levels (Krause *et al.*, 1982; Jordan *et al.*, 2001), lacks an electron-dense core (Seto and Miyata, 2003) and tends to be filamentous with no clearly defined tip structure (Baseman *et al.*, 1982; Balish *et al.*, 2003a). This phenotype is also seen in mutants with a transposon inserted in the HMW2 gene

(Krause *et al.*, 1997). Loss of the orthologous *M. genitalium* gene, *mg218*, similarly results in loss of haemadsorption and instability of other attachment organelle proteins (Dhandayuthapani *et al.*, 1999), including P140 and P110, in this species. In the absence of HMW2, HMW1 and HMW3 are synthesized in normal amounts, but are unstable (Popham *et al.*, 1997). A version of HMW2 lacking 81% from the central region of the protein restores the stability of HMW3 and P65, but not HMW1 (Balish *et al.*, 2003a), indicating that the determinant of HMW1 stability, but not stability of the others, is located in the missing region, whereas stabilization of the other proteins is conferred by a region near one terminus or the other. Thus, HMW2 exhibits modular organization, and the lack of a core in its absence suggests that it is either a major structural element or a master organizer of the structural elements of the core. Structures that resemble partially assembled cores are found in an HMW2 carboxyl-terminal truncation mutant (Bose *et al.*, 2009). Both a green fluorescent protein fusion and immunocytochemical experiments confirm that HMW2 is a core component, specifically associated with the rod region, although competing models describing its orientation within that structure have been proposed, including those in which domains of HMW2 are also present in the electron-lucent space (Balish *et al.*, 2003b; Bose *et al.*, 2009). Unlike in *M. pneumoniae*, amino-terminal fragments of the *M. genitalium* HMW2 orthologue are stable to proteolysis and allow formation of short cores. Consequently they retain gliding motility, albeit with significantly reduced speed (Pich *et al.*, 2008).

An internal, alternative translation start within the HMW2-encoding gene results in production of a protein, P28 or HMW2-s, the sequence of which is identical to the 198 carboxyl-terminal residues of HMW2 (Krause *et al.*, 1997; Boonmee *et al.*, 2009). Its loss does not contribute appreciably to attachment organelle function, despite its involvement in attachment organelle biology, as evidenced by its requirement for attachment organelle protein P41 for stability (Hasselbring and Krause, 2007b). P41 is localized at the base of the attachment organelle (Kenri *et al.*, 2004), suggesting that P28 is also located there. Interestingly, loss of P28 is accompanied by slight shortening of the core (Bose *et al.*, 2009), suggesting that it is a structural element.

The stabilizing interaction between HMW1 and HMW2 is mutual, in that HMW1 is also required for HMW2 stabilization (Willby *et al.*, 2004), as determined by study of the non-haemadsorbing M6 mutant, in which HMW1 is not produced and P30 includes an internal deletion (Layh-Schmitt *et al.*, 1995). Expression of full-length P30 in the M6 mutant allows the study of cells lacking only HMW1. These cells, like those without HMW2, lack cores, normal attachment organelles, normal localization of attachment organelle component proteins and normal morphology, and other attachment organelle proteins are unstable (Hahn *et al.*, 1998; Seto and Miyata, 2003; Willby *et al.*, 2004). The same is true for *M. genitalium* cells lacking the HMW1 orthologue MG312 (Burgos *et al.*, 2007). Like HMW2, HMW1 is localized in the attachment organelle (Stevens and Krause, 1991), but at steady state a significant minority of HMW1 is Triton X-100-soluble, whereas all HMW2 is insoluble (Stevens and Krause, 1990). The relationship between the soluble and insoluble forms of HMW1 is complex. About half of newly synthesized HMW1 is detergent-insoluble, but over time it shifts from the soluble to the insoluble fraction, whereas in the absence of HMW2 further insolubilization of HMW1 does not occur, and HMW1 is instead degraded, accounting for its reduced steady state levels (Balish *et al.*, 2001). Thus, HMW1 becomes incorporated into the cytoskeleton in two phases, the second one requiring HMW2. Although analysis of the process of stabilization of HMW2 by HMW1 has not been documented, these data suggest that core formation depends on an interaction between HMW1 and HMW2. HMW1 is clearly modular in its organization: starting from the amino-terminus it has an amino-terminal domain; an EAGR (enriched in aromatic and glycine residues) box, which is an approximately thirty-amino acid motif of unknown function present only in attachment organelle proteins; an APR (acidic, proline-rich) domain, which is very large but the sequence of which is not conserved across species; and a carboxyl-terminal domain that includes predicted α-helical coiled coils (Dirksen *et al.*, 1994;

Proft *et al.*, 1996; Balish *et al.*, 2001). Multiple EAGR boxes are present in the *M. gallisepticum* orthologue of HMW1. Other than similar composition to other attachment organelle proteins, the sequence reveals little to suggest specific function. Analysis of deletion mutants lacking various regions of MG312 in *M. genitalium* reveals specific roles for the amino-terminal domain and the EAGR box in gliding motility, as speed, but not adherence, was reduced in the absence of either region. A Walker A-like sequence, suggesting ATP binding and hydrolysis, is present in MG312 (Pich *et al.*, 2008), but not conserved across species. The failure of P1 to cluster at the attachment organelle when HMW1 is not stably insolubilized (Balish *et al.*, 2003a) suggests a specific role for HMW1 in localization of this adhesin.

An *M. pneumoniae* transposon insertion mutant that fails to produce HMW3 forms cores, but they are irregular, frequently bifurcating from the proximal end (Willby and Krause, 2002). Absence of the HMW3 orthologue in *M. genitalium*, MG317, likewise permits formation of cores, but rather than tending to split, the shorter cores noticeably lack mass in the terminal button region (Pich *et al.*, 2008). Both of these phenotypes are consistent with localization of *M. pneumoniae* HMW3 by immunogold labelling to the terminal button (Stevens and Krause, 1992), as well as localization by immunofluorescence microscopy to the distal portion of the attachment organelle (Seto *et al.*, 2003). Taken together these data suggest a role for HMW3 in forming some portion of the terminal button, which has a stabilizing effect on the core. Interactions between HMW3 and HMW2 are suggested by the instability of HMW3 in the absence of HMW2 (Popham *et al.*, 1997), restorable by production of various portions of HMW2 in both *M. pneumoniae* and *M. genitalium* (Balish *et al.*, 2003a; Pich *et al.*, 2008). In both species, loss of HMW3 is accompanied by reduction, but not abrogation, of haemadsorption (Balish *et al.*, 2003a; Pich *et al.*, 2008). The *M. genitalium* mutant lacking HMW3 also exhibits slower gliding (Pich *et al.*, 2008), a phenotype not reported for *M. pneumoniae*. HMW3 contains an amino-terminal predicted amphipathic α-helix, an APR domain that, at least in *M. pneumoniae*, consists of a region enriched in aspartic acid and another region enriched in glutamic acid, and an extensive predicted α-helical coiled coil region (Ogle *et al.*, 1992; unpublished observations). In *M. genitalium* alone, the APR domain also contains significant proportions of basic amino acids, rendering the pI of this protein neutral (Reddy *et al.*, 1995). *M. pneumoniae* HMW3 is cotranscribed with P30, a transmembrane adhesin that is located exclusively at the tip of the attachment organelle, physically close to P30 (Waldo *et al.*, 1999; Seto and Miyata, 2003).

Although there are no data suggesting stabilizing interactions between HMW3 and P30, both are required for stabilizing a third protein, P65 (Jordan *et al.*, 2001; Willby and Krause, 2002), which is also located in the vicinity of the terminal button (Jordan *et al.*, 2001; Seto and Miyata, 2003). This requirement suggests interactions between the three proteins. P65 consists of an APR domain and a predicted α-helical coiled coil region (Proft *et al.*, 1995). Because P65 is encoded upstream of HMW2 in *M. pneumoniae, M. genitalium* and *M. gallisepticum*, it might interact with HMW2 to form a bridge between HMW2 and HMW3 that is affected by the interaction between HMW3 and P30, thereby constituting a critical link between the adhesins and the cytoskeleton. Studies in *M. genitalium* implicate involvement of P65 in processes related to the shape of the attachment organelle, presumably a function of the cytoskeleton. Loss of the *M. genitalium* P65 orthologue MG217 results in a tendency of the attachment organelle, which is normally curved (Hatchel *et al.*, 2008), to straighten, changing the trajectories of gliding cells from generally circular paths to random paths similar to *M. pneumoniae* (Burgos *et al.*, 2008), the attachment organelle of which is normally straight (Hatchel *et al.*, 2008). However, the gliding speed is unaffected (Burgos *et al.* 2008). Control of directionality during mycoplasma gliding is not understood, but P65 might regulate this attachment organelle function through interactions with the cytoskeletal core. However, the attachment organelles of most close relatives of *M. genitalium* are not appreciably curved (Hatchel and Balish, 2008), so it is unclear how P65 might contribute to attachment organelle function in these organisms if this is its only role. The *M. gallisepticum* orthologue of P65,

PlpA, is implicated in fibronectin-binding via sequences in its APR domain, implying an extracellular location (May *et al.*, 2006). At least some P65 is surface-exposed in *M. pneumoniae* (Proft *et al.*, 1995), but no extracellular P65 is detectable in *M. genitalium* (Burgos *et al.*, 2008). Thus, P65 might play multiple roles in mycoplasma adherence and motility.

In all species in the *M. pneumoniae* cluster that have been investigated, DNA is associated with the end of the core that presumably faces the cytoplasm. For *M. pneumoniae* this association is either weaker or less frequent (Hatchel *et al.*, 2006), but for six other species, every core extracted with Triton X-100 is physically associated with DNase-sensitive material, presumably the chromosome (Hatchel and Balish, 2008). How this association is mediated or modulated during cell growth is unknown. Nevertheless, given the division mechanism, in which attachment organelles are move to opposite cell poles before cytokinesis (Hasselbring *et al.*, 2006a), physical association of chromosomes with the core could provide a mechanism for chromosomal segregation during cell division.

Several proteins have been localized to the base region of the electron-dense core in *M. pneumoniae* by fluorescence microscopy using yellow fluorescent protein fusion (Kenri *et al.*, 2004). One is P41, encoded immediately downstream of HMW2 (Krause *et al.*, 1997). Like HMW2, P41 is composed of extensive predicted α-helical coiled coil structure (unpublished observation). Aside from a tendency to grow in chains of connected cells and a reduced gliding speed, the most prominent feature of cells in which transposon mutagenesis prevents production of P41 is the 'terminal organelle detachment' phenotype (Hasselbring and Krause, 2007a). Gliding cells lacking P41 that encounter obstacles stretch and ultimately break, with the detached attachment organelle continuing to glide and the cell body remaining immotile. Eventually a new attachment organelle will form *de novo*. Thus, P41 serves a critical mechanical role, maintaining the integrity of the cell during gliding, presumably preventing inappropriate cytokinesis. Its dependence upon HMW2 for stability (Krause *et al.*, 1997), and the dependence of P28 on P41 for stability (Hasselbring and Krause, 2007b), suggests physical interactions, direct or indirect, between the three.

P24 is encoded immediately downstream of the gene for P41 (Krause *et al.*, 1997) and the two are translationally coupled (Hasselbring and Krause, 2007b). P24 is present at reduced levels in the absence of HMW2 (Krause *et al.*, 1997). Although in *M. genitalium* and *M. gallisepticum* a gene occupies this position, P24 has no detectable sequence similarity with the products of these genes (Krause *et al.*, 1997; unpublished observation). Thus, either the function of P24 is not conserved, the function of P24 is carried out by non-orthologous replacements, or the proteins have similar folds but extensive sequence divergence. A transposon insertion mutant of *M. pneumoniae*, in which P24 is not expressed, exhibits wild-type gliding speed, but attachment organelle formation is impaired, occurring markedly less frequently, resulting in a reduced proportion of gliding cells within the population (Hasselbring and Krause, 2007b). Unusually for an attachment organelle protein, P24 is only partly localized to the proximal region of the attachment organelle, with a significant fraction also appearing in an adjacent focus that is devoid of the attachment organelle protein P30. This localization is dependent upon P41. In its absence, P24 appears in multiple foci along the length of the cell. Furthermore, new attachment organelles frequently form at sites not adjacent to the old attachment organelle in the absence of P41. Populations lacking P24 also have a lower proportion of gliding cells. These data suggest that P24 is recruited by P41 to the attachment organelle base and plays a significant early role in the formation and placement of new attachment organelles, as well as their activation for gliding motility.

Stability of P24 depends upon the presence of another cytoskeletal protein at the proximal end of the attachment organelle, TopJ (Cloward and Krause, 2009). Unlike P24, TopJ appears to act late in the assembly process, as fully formed cores are frequently present in cells lacking TopJ, but their placement within the cells is erratic, resulting in non-adherent cells without attachment organelles (Cloward and Krause, 2010). These mutants are otherwise similar in phenotype to other cytadherence accessory protein mutants, with loss of haemadsorption, poor adherence, no motility and irregular morphology (Cloward and Krause, 2009). Loss of the *M. genitalium* homologue, MG200, results in similar phenotypes, but adherence and motility are reduced, not abolished (Pich *et al.*, 2006). Although TopJ is structurally similar to other cytadherence accessory proteins, with an EAGR box and an APR domain, it also contains an amino-terminal J domain with substantial homology to that of DnaJ (Balish *et al.*, 2001). These domains are associated with co-chaperone activity, suggesting a role for TopJ in correct folding of attachment organelle components (Cloward and Krause, 2010). However, mutational analysis reveals that P24 stability is independent of the J domain, instead depending upon localization of TopJ to the *M. pneumoniae* attachment organelle, which is conferred by the other regions of TopJ. Nonetheless, the loss of adherence and other normal functions of the attachment organelle is associated with the J domain. In the absence of TopJ, cells are distinctly larger, less dense, and leakier than wild-type cells (Cloward and Krause, 2010), but the significance of these observations is not clear. It is likely that TopJ is an attachment organelle-specific co-chaperone involved in correct organization of the attachment organelle once the core has assembled.

*M. pneumoniae* P200 is a Triton X-100-insoluble protein with attachment organelle-associated motifs, including an APR domain and several EAGR boxes (Proft *et al.*, 1996; Balish *et al.*, 2001). Accordingly, it localizes to the attachment organelle, although in mutants with disrupted attachment organelle function, especially one lacking P1, P200 is more dispersed (Jordan *et al.*, 2007). Its specific location within the attachment organelle has not been demonstrated. In a mutant lacking P200, gliding motility is slower and less frequent than in wild-type cells, but haemadsorption and adherence to A549 epithelial cells are unaffected (Jordan *et al.*, 2007), suggesting a specific role for P200 in motility. The normal haemadsorption and reduced motility phenotypes are also seen in *M. genitalium* in the absence of the P200 orthologue MG386 (Pich *et al.*, 2008). Significantly, colonization of normal human bronchial epithelial (NHBE) cells by *M. pneumoniae* is markedly impaired in the absence of P200 (Jordan *et al.*, 2007), suggesting that normal motility is important for productive infection.

In agreement with observations that some *M. pneumoniae* Triton X-100-insoluble proteins are phosphorylated (Dirksen *et al.*, 1994), several proteins of the *M. pneumoniae* attachment organelle, including cytadherence accessory proteins, require the protein kinase PrkC for phosphorylation (Schmidl *et al.*, 2010). The absence of PrkC simulates loss of HMW2, with post-transcriptional loss of HMW2 and a reduction in steady-state levels of HMW1, HMW3 and P65. and these and other attachment organelle proteins are hypophosphorylated (Schmidl *et al.*, 2010). Accordingly, *M. pneumoniae* cells lacking PrkC cause markedly reduced cytotoxicity. Loss of the corresponding protein phosphatase, PrpC, allows function, but HMW1, HMW3 and the adhesin P1 are hyperphosphorylated (Schmidl *et al.*, 2010). These data suggest a structural role for the level of phosphorylation of cytoskeletal and other attachment organelle components by the PrkC–PrpC pair, and do not preclude an additional regulatory role.

## Overall organization of the *Mycoplasma pneumoniae* attachment organelle cytoskeleton

Establishing the interactions between *M. pneumoniae* attachment organelle cytoskeletal proteins and the timeline of their assembly is both inherently interesting as a structural biology problem and of practical value as it may reveal targets for novel therapeutic agents against *M. pneumoniae* and its other pathogenic relatives. However, at present there are substantial obstacles precluding direct analysis of these interactions. Firstly, exogenous production of these proteins in other organisms is complicated by the large size of many of the proteins, their insolubility, their dependence upon each another for stability, and the alternative genetic code of mycoplasmas, which requires another layer of effort for expression. Their unusual primary structure suggests the potential for novel folds that might require the presence of cytoplasmic agents that are not available in distantly related organisms. Secondly, the

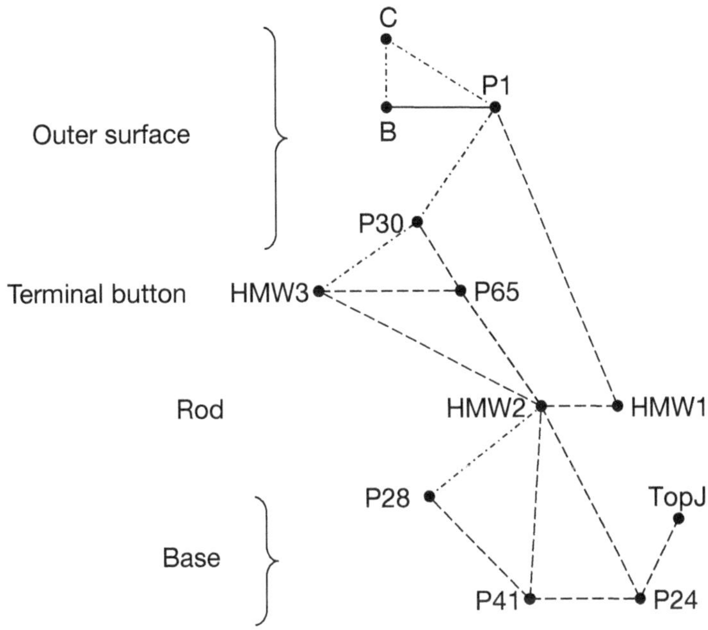


**Figure 11.4** Model of interactions among *M. pneumoniae* attachment organelle proteins. See text for evidence of each interaction. Solid line, direct interaction demonstrated biochemically. Dashed line, interaction, possibly direct, based on instability and/or abnormal localization of one partner in the absence of the other partner. Dashed/dotted line, speculated interaction based on cotranscription of genes and similar location. TopJ might have additional interactions with other components. Information about interactions of P200 with other proteins is lacking.

low density ceiling for growth of *M. pneumoniae*, combined with the expense of the components required for growing this organism prevent accumulation of enough cellular material to develop protocols for purifying sufficient protein for direct analysis, given typical resources, although it has been possible to purify a complex of P1 and B. Therefore, biochemical evaluations of interactions between these proteins have never been performed.

Nevertheless, as described above, alternative data have provided a framework for proposing likely interactions between cytoskeletal components of the *M. pneumoniae* electron-dense core. These data include immunocytochemical localization of proteins within the organelle, operon structure, and analysis of cytoskeletal protein stability and localization in attachment organelle mutants. These interactions are summarized in Fig. 11.4, excluding a specific location for P200. How these proteins, either individually or as an overall structure, promote localization of transmembrane adhesins remains unknown.

It is noteworthy that, in addition to association with the internal structure of the attachment organelle, both HMW1 and P65 have been identified as surface components, although this varies across species (Stevens and Krause, 1991; Proft *et al.*, 1996; Balish *et al.*, 2001; May *et al.*, 2006; Burgos *et al.*, 2007, 2008). Although no general mechanism to explain dual localization has been proposed, secondary surface localization of cytosolic proteins, in which these proteins carry out functions related to adherence, is found among other mycoplasma proteins, as well as proteins of other bacteria. The surface-exposed population of the *M. gallisepticum* orthologue of P65, PlpA, has been assigned a role in binding fibronectin (May *et al.*, 2006), similar to the role of moonlighting housekeeping proteins in the binding of fibronectin and mucin by *M. pneumoniae* and/or *M. genitalium* (Dallo et al., 2002; Alvarez et al., 2003). The use of the surface-exposed CARDS toxin of *M. pneumoniae* (Kannan *et al.*, 2010) to bind to surfactant protein A (Kannan *et al.*, 2005) might also be in this category.

## *Mycoplasma mobile* attachment organelle

The *M. mobile* attachment organelle structure is quite distinct from that of *M. pneumoniae* and its relatives. Although, as in most species of the *M. pneumoniae* cluster, its distal tip is wider than its shaft, rather than being spheroidal, this tip has a distinct, shallow cone shape, with the vertex of the cone the most distal part of the structure (Kirchhoff and Rosengarten, 1984; Fig. 11.2C). Additionally, it is much larger than any attachment organelle described for the *M. pneumoniae* cluster, albeit highly variable in size, with lengths frequently exceeding 1 μm.

The protein associated directly with *M. mobile* adherence to both glass and cells, Gli349, is not at the distal region of the attachment organelle, instead being distributed along its lateral surfaces and at the junction between the tip structure and the cell body, designated the neck region (Uenoyama *et al.*, 2004). This adhesin, which is also directly involved in gliding motility, is present in the Triton X-100-insoluble fraction of *M. mobile* cells, possibly because of its presence in a complex with other proteins, including Gli521, Gli123 and the ATPase P42, which are involved in localization of Gli349, as well as its ability to transmit force (Seto *et al.*, 2005; Uenoyama and Miyata, 2005; Ohtani and Miyata, 2007; Uenoyama *et al.*, 2009). Gli349 can be visualized as spikes, 50 nm in length, protruding from the cell membrane, but these structures are absent from Triton X-100-insolubilized material (Miyata and Petersen, 2004; Adan-Kubo *et al.*, 2006). At higher concentrations of detergent, Gli349, Gli521, and Gli123 are no longer present in the insoluble fraction (Nakane and Miyata, 2007), suggesting that these proteins are not part of the cytoskeleton of *M. mobile.*

Instead, treatment of *M. mobile* cells with Triton X-100 reveals a novel structure, designated the jellyfish structure because of its resemblance to that organism (Nakane and Miyata, 2007). Successive treatments with increasing concentrations of detergent clearly indicate the orientation of this asymmetrical cytoskeletal structure within the cell (Fig. 11.2D). It consists of two principal substructures. The larger mass, designated the bell by analogy with jellyfish, is an ovoid structure 235 nm wide and 155 nm high. The bell occupies the distal region of the attachment organelle, possibly corresponding to the conical tip. It appears to be constructed of a 12-nm hexagonal lattice. The other substructure is a series of tentacle-like filaments emanating from the proximal face of the bell and extending deep into the cell body cytoplasm. The surfaces of these filaments, which number in the dozens, contain pairs of densities connected by strands that form elliptical rings. These ring structures are approximately 20 nm wide and are repeated at 30-nm intervals. In the absence of Gli521, the tentacles appear shorter and the ring structures are discontinuous (Nakane and Miyata, 2007), suggesting interactions between the cytoskeleton and the surface adhesin complex.

*M. mobile* lacks homologues of the *M. pneumoniae* cytoskeletal proteins (Jaffe *et al.*, 2004). Mass spectrometric analysis of 10 proteins that are enriched in the Triton X-100-insoluble fraction of *M. mobile* identifies them as a mixture of proteins encoded at five genetic loci with both known and unknown functions (Nakane and Miyata, 2007). The best-characterized is phosphoglycerate kinase, which generates ATP during glycolysis and could potentially concentrate ATP at the motor complex. A set of proteins with considerable sequence similarity to components of the membrane ATP synthase is also among the detergent-insoluble proteins. These proteins are found in a large number of mycoplasmas (unpublished observations), suggesting that they serve some function, perhaps involving transport, that *M. mobile* has co-opted for use in the cytoskeleton. Another protein has significant similarity to the extracytoplasmic substrate-binding component of an ABC transporter for xylose. Whether its function in the cytoskeleton is related to this process or whether it has an alternative function is unknown. The remaining proteins are novel. Immunocytochemical localization of some of these proteins reveals distinct distribution throughout the cell, but localization of individual proteins to cytoskeletal substructures was not achieved (Nakane and Miyata, 2007). Interestingly, with the exception of phosphoglycerate kinase, none of these proteins have orthologues in *M. pneumoniae* (Nakane and Miyata, 2007), suggesting independent, convergent evolution of attachment organelles and cytoskeletons in two different mycoplasma lineages.

## *Mycoplasma penetrans*/ *Mycoplasma iowae* attachment organelle

Yet another case of a distinct polar structure mediating both adherence and motility is found in both *Mycoplasma penetrans* and *Mycoplasma iowae*, members of the *Mycoplasma muris* cluster within the pneumoniae group. Both species have been observed attaching to host cells through these structures (Mirsalimi *et al.*, 1989; Lo *et al.*, 1991), and in both species, these poles are the leading end during gliding motility (Jurkovic *et al.*, 2012), but much less is known about the structures in these species than in *M. pneumoniae* and *M. mobile*. The attachment organelle of both *M. penetrans* and *M. iowae* is more pleomorphic than in other polarized mycoplasma species (Fig. 11.2E).

As in *M. pneumoniae* and *M. mobile*, both *M. penetrans* and *M. iowae* have structures that occupy the attachment organelle (Jurkovic *et al.*, 2012). These structures have been observed in thin sections of both species as a fine-grained material distinct from the cytoplasm of the cell body (Mirsalimi *et al.*, 1989; Lo *et al.*, 1991; Neyrolles *et al.*, 1998), but the nature of this material has been unclear. However, Triton X-100 extraction of these cells revealed objects with dimensions similar to those of the attachment organelle (Jurkovic *et al.*, 2012; Fig. 11.2F), albeit narrower. The narrowness could possibly be explained by a model in which the insoluble material is comprised of parallel filaments that contract from dehydration during processing for electron microscopy, or more trivially that there is an outer component not retained during processing. Regardless, this structure is quite distinct from that found in other mycoplasmas, consistent with an absence of *M. pneumoniae* and *M. mobile* cytoskeletal protein homologues in the *M. penetrans* genome. Most of the objects in the material observed by scanning electron microscopy are 440–450 nm long and have a wider and narrow substructure (Jurkovic *et al.*, 2012). The widths differ between *M. penetrans* and *M. iowae*, reminiscent of the differences in electron-dense core dimensions among the species of the *M. pneumoniae* cluster (Hatchel and Balish, 2008). The specific composition of this material has not been reported.

## Evolution of the Mycoplasma cytoskeleton

Most mycoplasmas lack specialized polar structures. Whether these species have underlying Triton X-100-insoluble filaments has not been not reported. However, we were unable to detect such structures in one species that exhibits pseudococcoidal morphology, *Mycoplasma gallinarum* (J.P. Norton and M.F. Balish, unpublished). If *M. gallinarum* is representative of non-polarized mycoplasma species, it suggests that the common ancestor of mycoplasmas lacked a specialized cytoskeleton, harbouring only FtsZ and perhaps MreB. The appearance of distinct types of cytoskeletal structures and components in different mycoplasma lineages, including the *M. pneumoniae* cluster, the *M. muris* cluster, M. *mobile* and *M. insons*, suggests that these organisms have independently evolved specialized cytoskeletal apparatus multiple times. If so, then one of the principal defining features of mycoplasmas, the absence of any kind of cell wall, may have allowed this group of organisms to outpace other bacteria in terms of innovation in the area of cell structure and organization. Mycoplasmas therefore will continue to be a fascinating and informative subject of study by the scientific community in efforts towards understanding the nature and evolution of the cytoskeleton.

## References


Adams, D.W., and Errington, J. (2009). Bacterial cell division: assembly, maintenance and disassembly of the Z ring. Nat. Rev. Microbiol. *7*, 642–653.

Adan-Kubo, J., Uenoyama, A., Arata, T., and Miyata, M. (2006). Morphology of isolated Gli349, a leg protein responsible for *Mycoplasma mobile* gliding via glass binding, revealed by rotary shadowing electron microscopy. J. Bacteriol. *188*, 2821–2828.

Alvarez, R.A., Blaylock, M.W., and Baseman, J.B. (2003). Surface localized glyceraldehyde-3-phosphate dehydrogenase of *Mycoplasma genitalium* binds mucin. Mol. Microbiol. *48*, 1417–1425.

Balish, M.F. (2006). Subcellular structures of mycoplasmas. Front. Biosci. *11*, 2017–2027.

Balish, M.F., and Krause, D.C. (2006). Mycoplasmas: a distinct cytoskeleton for wall-less bacteria. J. Mol. Microbiol. Biotechnol. *11*, 244–255.

Balish, M.F., Hahn, T.W., Popham, P.L., and Krause, D.C. (2001). Stability of *Mycoplasma pneumoniae* cytadherence-accessory protein HMW1 correlates with its association with the triton shell. J. Bacteriol. *183*, 3680–3688.

Balish, M.F., Ross, S.M., Fisseha, M., and Krause, D.C. (2003a) Deletion analysis identifies key functional domains of the cytadherence-associated protein HMW2 of *Mycoplasma pneumoniae*. Mol. Microbiol. *50*, 1507–1516.

Balish, M.F., Santurri, R.T., Ricci, A., Lee, K.K., and Krause, D.C. (2003b) Localization of *Mycoplasma pneumoniae* cytadherence-associated protein HMW2 by fusion with green fluorescent protein: implications for attachment organelle structure. Mol. Microbiol. *47*, 49–60.

Baseman, J.B., Morrison-Plummer, J., Drouillard, D., Puleo-Scheppke, B., Tryon, W., and Holt, S.C. (1987). Identification of a 32-kilodalton protein of *Mycoplasma pneumoniae* associated with hemadsorption. Isr. J. Med. Sci. *23*, 474–479.

Benders, G.A., Powell, B.C., and Hutchison, C.A. 3rd (2005). Transcriptional analysis of the conserved *ftsZ* gene cluster in *Mycoplasma genitalium* and *Mycoplasma pneumoniae*. Proc. Natl. Acad. Sci. U.S.A. *187*, 4542–4551.

Biberfeld, G., and Biberfeld, P. (1970). Ultrastructural features of *Mycoplasma pneumoniae*. J. Bacteriol. *102*, 855–861.

Boonmee, A., Ruppert, T., and Herrmann, R. (2009). The gene *mpn310* (*hmw2*) from *Mycoplasma pneumoniae* encodes two proteins, HMW2 and HMW2-s, which differ in size but use the same reading frame. FEMS Microbiol. Lett. *290*, 174–181.

Bose, S.R., Balish, M.F., and Krause, D.C. (2009). *Mycoplasma pneumoniae* cytoskeletal protein HMW2 and the architecture of the terminal organelle. J. Bacteriol. *191*, 6741–6748.

Bové, J.M., Renaudin, J., Saillard, C., Foissac, X., and Garnier, M. (2003). *Spiroplasma citri*, a plant pathogenic Mollicute: relationships with its two hosts, the plant and the leafhopper vector. Annu. Rev. Phytopathol. *41*, 483–500.

Bredt, W. (1968). Motility and multiplication of *Mycoplasma pneumoniae*. A phase contrast study. Pathol. Microbiol. *32*, 321–326.

Burgos, R., Pich, O.Q., Querol, E., and Piñol, J. (2007). Functional analysis of the *Mycoplasma genitalium* MG312 protein reveals a specific requirement of the MG312 N-terminal domain for gliding motility. J. Bacteriol. *189*, 7014–7023.

Burgos, R., Pich, O.Q., Querol, E., and Piñol, J. (2008). Deletion of the *Mycoplasma genitalium* MG_217 gene modifies cell gliding behaviour by altering terminal organelle curvature. Mol. Microbiol. *69*, 1029–1040.

Cabeen, M.T., Charbon, G., Vollmer, W., Born, P., Ausmees, N., Weibel, D.B., and Jacobs-Wagner, C. (2009). Bacterial cell curvature through mechanical control of cell growth. EMBO J. *28*, 1208–1219.

Carrión, M., Gómez, M.J., Merchante-Schubert, R., Dongarrá, S., and Ayala, J.A. (1999). *mraW*, an essential gene at the *dcw* cluster of *Escherichia coli* codes for a cytoplasmic protein with methyltransferase activity. Biochimie *81*, 879–888.

Charbon, G., Cabeen, M.T., and Jacobs-Wagner, C. (2009). Bacterial intermediate filaments: in vivo assembly, organization, and dynamics of crescentin. Genes Dev. *23*, 1131–1144.

Chen, S., Jancrick, J., Yokota, H., Kim, R., and Kim, S.H. (2004). Crystal structure of a protein associated with cell division from *Mycoplasma pneumoniae* (GI: 13508053): a novel fold with a conserved sequence motif. Proteins *55*, 785–791.

Cloward, J.M., and Krause, D.C. (2009). *Mycoplasma pneumoniae* J-domain protein required for terminal organelle function. Mol. Microbiol. *71*, 1296–1307.

Cloward, J.M., and Krause, D.C. (2010). Functional domain analysis of the *Mycoplasma pneumoniae* co-chaperone TopJ. Mol. Microbiol. *77*, 158–169.

Collier, A.M., and Clyde, W.A., Jr. (1971). Relationships between *Mycoplasma pneumoniae* and human respiratory epithelium. Infect. Immun. *3*, 694–701.

Collier, A.M., Hu, P.C., and Clyde, W.A., Jr. (1983). Location of attachment moiety on *Mycoplasma pneumoniae*. Yale J. Biol. Med. *56*, 671–677.

Dallo, S.F., Kannan, T.R., Blaylock, M.W., and Baseman, J.B. (2002). Elongation factor Tu and E1 beta subunit of pyruvate dehydrogenase complex act as fibronectin binding proteins in *Mycoplasma pneumoniae*. Mol. Microbiol. *46*, 1041–1051.

Defeu Soufo, H.J., and Graumann, P.L. (2010). *Bacillus subtilis* MreB paralogues have different filament architectures and lead to shape remodelling of a heterologous cell system. Mol. Microbiol. *78*, 1145–1158.

Defeu Soufo, H.J., Reimold, C., Linne, U., Knust, T., Gescher, J., and Graumann, P.L. (2010). Bacterial translation elongation factor EF-Tu interacts and colocalizes with actin-like MreB protein. Proc. Natl. Acad. Sci. U.S.A. *107*, 3163–3168.

Derman, A.I., Becker, E.C., Truong, B.D., Fujioka, A., Tucey, T.M., Erb, M.L., Patterson, P.C., and Pogliano, J. (2009). Phylogenetic analysis identifies many uncharacterized actin-like proteins (Alps) in bacteria: regulated polymerization, dynamic instability and treadmilling in Alp7A. Mol. Microbiol. 73, 534–552.

Dhandayuthapani, S., Rasmussen, W.G., and Baseman, J.B. (1999). Disruption of gene *mg218* of *Mycoplasma genitalium* through homologous recombination leads to an adherence-deficient phenotype. Proc. Natl. Acad. Sci. U.S.A. *96*, 5227–5232.

Dirksen, L.B., Krebes, K.A., and Krause, D.C. (1994). Phosphorylation of cytadherence-accessory proteins in *Mycoplasma pneumoniae*. J. Bacteriol. *176*, 7499–7505.

Dirksen, L.B., Proft, T., Hilbert, H., Plagens, H., Herrmann, R., and Krause, D.C. (1996). Sequence analysis and characterization of the hmw gene cluster of *Mycoplasma pneumoniae*. Gene *171*, 19–25.

Dye, N.A., Pincus, Z., Theriot, J.A., Shapiro, L., and Gitai, Z. (2005). Two independent spiral structures control cell shape in *Caulobacter*. Proc. Natl. Acad. Sci. U.S.A. *102*, 18608–18613.

Dyer, N. (2009). Tubulin and its prokaryotic homologue FtsZ: a structural and functional comparison. Sci. Prog. *92*, 113–137.

Erickson, H.P., Anderson, D.E., and Ozawa, M. (2010). FtsZ in bacterial cytokinesis: cytoskeleton and force generator all in one. Microbiol. Mol. Biol. Rev. *74*, 504–528.

Esue, O., Wirtz, D., and Tseng, Y. (2006). GTPase activity, structure, and mechanical properties of filaments assembled from bacterial cytoskeleton protein MreB. J. Bacteriol. *188*, 968–976.

Fenton, A.K., Hobley, L., Butan, C., Subramaniam, S., and Sockett, R.E. (2010). A coiled-coil-repeat protein 'Ccrp' in *Bdellovibrio bacteriovorus* prevents cellular indentation, but is not essential for vibroid cell morphology. FEMS Microbiol. Lett. *313*, 89–95.

Fiuza, M., Letek, M., Leiba, J., Villadangos, A.F., Vaquera, J., Zanella-Cléon, I., Mateos, L.M., Molle, V., and Gil, J.A. (2010). Phosphorylation of a novel cytoskeletal protein (RsmP) regulates rod-shaped morphology in *Corynebacterium glutamicum*. J. Biol. Chem. *285*, 29387–29397.

Gerdes, K., Howard, M., and Szardenings, F. (2010). Pushing and pulling in prokaryotic DNA segregation. Cell *141*, 927–942.

Göbel, U., Speth, V., and Bredt, W. (1981). Filamentous structures in adherent *Mycoplasma pneumoniae* cells treated with nonionic detergents. J. Cell. Biol. *91*, 537–543.

Gourlay, R.N., Wyld, S.G., and Leach, R.H. (1977). *Mycoplasma alvi*, a new species from bovine intestinal and urogenital tracts. Int. J. Syst. Bacteriol. *27*, 86–96.

Graumann, P.L. (2009). Dynamics of bacterial cytoskeletal elements. Cell. Motil. Cytoskel. *66*, 909–914.

Hahn, T.W., Willby, M.J., and Krause, D.C. (1998). HMW1 is required for cytadhesin P1 trafficking to the attachment organelle in *Mycoplasma pneumoniae*. J. Bacteriol. *180*, 1270–1276.

Harwick, H.J., Kalmanson, G.M., and Guze, L.B. (1977). Chemotactic activity of L-forms and mycoplasma. Aust. J. Exp. Biol. Med. Sci. *55*, 431–434.

Hasselbring, B.M., Jordan, J.L., Krause, R.W., and Krause, D.C. (2006a) Terminal organelle development in the cell wall-less bacterium *Mycoplasma pneumoniae*. Proc. Natl. Acad. Sci. U.S.A. *103*, 16478–16483.

Hasselbring, B.M., Page, C.A., Sheppard, E.S., and Krause, D.C. (2006b) Transposon mutagenesis identifies genes associated with *Mycoplasma pneumoniae* gliding motility. J. Bacteriol. *188*, 6335–6345.

Hasselbring, B.M., and Krause, D.C. (2007a) Cytoskeletal protein P41 is required to anchor the terminal organelle of the wall-less prokaryote *Mycoplasma pneumoniae*. Mol. Microbiol. *63*, 44–53.

Hasselbring, B.M., and Krause, D.C. (2007b) Proteins P24 and P41 function in the regulation of terminal-organelle development and gliding motility in *Mycoplasma pneumoniae*. J. Bacteriol. *189*, 7442–7449.

Hatchel, J.M., and Balish, M.F. (2008). Attachment organelle ultrastructure correlates with phylogeny, not gliding motility properties, in *Mycoplasma pneumoniae* relatives. Microbiology *154*, 286–295.

Hatchel, J.M., Balish, R.S., Duley, M.A., and Balish, M.F. (2006). Ultrastructure and gliding motility of *Mycoplasma amphoriforme*, a possible human respiratory pathogen. Microbiology *152*, 2181–2189.

Henderson, G.P., and Jensen, G.J. (2006). Three-dimensional structure of *Mycoplasma pneumoniae*'s attachment organelle and a model for its role in gliding motility. Mol. Microbiol. *60*, 376–385.

Herrmann, H., Strelkov, S.V., Burkhard, P., and Aebi, U. (2009). Intermediate filaments: primary determinants of cell architecture and plasticity. J. Clin. Invest. *119*, 1772–1783.

Himmelreich, R., Hilbert, H., Plagens, H., Pirkl, E., Li, B.C., and Herrmann, R. (1996). Complete sequence analysis of the genome of the bacterium *Mycoplasma pneumoniae*. Nucleic Acids Res. *15*, 4420–4449.

Izard, J. (2006). Cytoskeletal cytoplasmic filament ribbon of *Treponema*: a member of an intermediate-like filament protein family. J. Mol. Microbiol. Biotechnol. *11*, 159–166.

Jaffe, J.D., Stange-Thomann, N., Smith, C., DeCaprio, D., Fisher, S., Butler, J., Calvo, S., Elkins, T., FitzGerald, M.G., Hafez, N., *et al.* (2004). the complete genome and proteome of *Mycoplasma mobile*. Genome Res. *14*, 1447–1461.

Jarrell, K.F., and McBride, M.J. (2008). The surprisingly diverse ways that prokaryotes move. Nat. Rev. Microbiol. *6*, 466–476.

Jordan, J.L., Berry, K.M., Balish, M.F., and Krause, D.C. (2001). Stability and subcellular localization of cytadherence-associated protein P65 in *Mycoplasma pneumoniae*. J. Bacteriol. *183*, 7387–7391.

Jordan, J.L., Chang, H.Y., Balish, M.F., Holt, L.S., Bose, S.R., Hasselbring, B.M., Waldo, R.H. 3rd, Krunkosky, T.M., and Krause, D.C. (2007). Protein P200 is dispensable for *Mycoplasma pneumoniae* hemadsorption but not gliding motility or colonization of differentiated bronchial epithelium. Infect. Immun. 75, 518–522.

Jurkovic, D.A., Newman, J.T., and Balish, M.F. (2012). Conserved terminal organelle morphology and function in *Mycoplasma penetrans* and *Mycoplasma iowae*. J. Bacteriol. *194*, 2877–2883.

Kannan, T.R., Provenzano, D., Wright, J.R., and Baseman, J.B. (2005). Identification and characterization of human surfactant protein A binding protein of *Mycoplasma pneumoniae*. Infect. Immun. 73, 2828–2834.

Kannan, T.R., Musatovova, O., Balasubramanian, S., Cagle, M., Jordan, J.L., Krunkosky, T.M., Davis, A., Hardy, R.D., and Baseman, J.B. (2010). *Mycoplasma pneumoniae* Community Acquired Respiratory Distress Syndrome toxin expression reveals growth phase and infection-dependent regulation. Mol. Microbiol. *76*, 1127–1141.

Kenri, T., Seto, S., Horino, A., Sasaki, Y., Sasaki, T., and Miyata, M. (2004). Use of fluorescent-protein tagging to determine the subcellular localization of *Mycoplasma pneumoniae* proteins encoded by the cytadherence regulatory locus. J. Bacteriol. *186*, 6944–6955.

Kim, S.Y., Gitai, Z., Kinkhabwala, A., Shapiro, L., and Moerner, W.E. (2006). Single molecules of the bacterial actin MreB undergo directed treadmilling motion in *Caulobacter crescentus*. Proc. Natl. Acad. Sci. U.S.A. *103*, 10929–10934.

Kirchhoff, H., and Rosengarten, R. (1984). Isolation of a motile mycoplasma from a fish. J. Gen. Microbiol. *130*, 2439–2445.

Krause, D.C., and Balish, M.F. (2004). Cellular engineering in a minimal microbe: structure and assembly of the terminal organelle of *Mycoplasma pneumoniae*. Mol. Microbiol. *51*, 917–924.

Krause, D.C., and Baseman, J.B. (1983). Inhibition of *Mycoplasma pneumoniae* hemadsorption and adherence to respiratory epithelium by antibodies to a membrane protein. Infect. Immun. *39*, 1180–1186.

Krause, D.C., Leith, D.K., Wilson, R.M., and Baseman, J.B. (1982). Identification of *Mycoplasma pneumoniae* proteins associated with hemadsorption and virulence. Infect. Immun. *35*, 809–817.

Krause, D.C., Proft, T., Hedreyda, C.T., Hilbert, H., Plagens, H., and Herrmann, R. (1997). Transposon mutagenesis reinforces the correlation between *Mycoplasma pneumoniae* cytoskeletal protein HMW2 and cytadherence. J. Bacteriol. *179*, 2668–2677.

Kürner, J., Frangakis, A.S., and Baumeister, W. (2005). Cryo-electron tomography reveals the cytoskeletal structure of *Spiroplasma melliferum*. Science *307*, 436–438.

Layh-Schmitt, G. (1993). The ORF6 gene product of the P1 operon of *Mycoplasma pneumoniae*. Zentralbl. Bakteriol. *278*, 287–295.

Layh-Schmitt, G., Hilbert, H., and Pirkl, E. (1995). A spontaneous hemadsorption-negative mutant of *Mycoplasma pneumoniae* exhibits a truncated adhesin-related 30-kilodalton protein and lacks the cytadherence-accessory protein HMW1. J. Bacteriol. *177*, 843–846.

Layh-Schmitt, G., Podtelejnikov, A., and Mann, M. (2000). Proteins complexed to the P1 adhesin of *Mycoplasma pneumoniae*. Microbiology *146*, 741–747.

Leaver, M., and Errington, J. (2005). Roles for MreC and MreD proteins in helical growth of the cylindrical cell wall of *Bacillus subtilis*. Mol. Microbiol. *57*, 1196–1209.

Lluch-Senar, M., Querol, E., and Piñol, J. (2010). Cell division in a minimal bacterium in the absence of FtsZ. Mol. Microbiol. *78*, 278–289.

Lo, S.C., Hayes, M.M., Wang, R.Y., Pierce, P.F., Kotani, H., and Shih, J.W. (1991). Newly discovered mycoplasma isolated from patients infected with HIV. Lancet *338*, 1415–1418.

Löwe, J., and Amos, L.A. (2009). Evolution of cytomotive filaments: the cytoskeleton from prokaryotes to eukaryotes. Int. J. Biochem. Cell. Biol. *41*, 423–429.

Maniloff, J. (1969). Electron microscopy of small cells: *Mycoplasma hominis*. J. Bacteriol. *100*, 1402–1408.

Margolin, W. (2009). Sculpting the bacterial cell. Curr. Biol. *19*, R812-R822.

May, M., Papazisi, L., Gorton, T.S., and Geary, S.J. (2006). Identification of fibronectin-binding proteins in *Mycoplasma gallisepticum* strain R. Infect. Immun. *74*, 1777–1785.

May, M., Ortiz, G.J., Wendland, L.D., Rotstein, D.S., Relich, R.F., Balish, M.F., and Brown, D.R. (2007). *Mycoplasma insons* sp. nov., a twisted mycoplasma from green iguanas (*Iguana iguana*). FEMS Microbiol. Lett. *274*, 298–303.

Mayer, F. (2003). Cytoskeletons in prokaryotes. Cell. Biol. Int. *27*, 429–438.

Meng, K.E., and Pfister, R.M. (1980). Intracellular structures of *Mycoplasma pneumoniae* revealed after membrane removal. J. Bacteriol. *144*, 390–399.

Mingorance, J., Tamames, J., and Vicente, M. (2004). Genomic channeling in bacterial cell division. J. Mol. Recognit. *17*, 481–487.

Mirsalimi, S.M., Rosendal, S., and Julian, R.J. (1989). Colonization of the intestine of turkey embryos exposed to *Mycoplasma iowae*. Avian Dis. *33*, 310–315.

Miyata, M., and Petersen, J.D. (2004). Spike structure at the interface between gliding *Mycoplasma mobile* cells and glass surfaces visualized by rapid-freeze-and-fracture electron microscopy. J. Bacteriol. *186*, 4382–4386.

Morrison-Plummer, J., Leith, D.K., and Baseman, J.B. (1986). Biological effects of anti-lipid and anti-protein monoclonal antibodies on *Mycoplasma pneumoniae*. Infect. Immun. *53*, 398–403.

Murray, H., and Errington, J. (2008). Dynamic control of the DNA replication initiation protein DnaA by Soj/ParA. Cell *135*, 74–84.

Nakane, D., and Miyata, M. (2007). Cytoskeletal 'jellyfish' structure of *Mycoplasma mobile*. Proc. Natl. Acad. Sci. U.S.A. *104*, 19518–19523.

Nakane, D., and Miyata, M. (2009). Cytoskeletal asymmetrical dumbbell structure of a gliding mycoplasma, *Mycoplasma gallisepticum*, revealed by negative-staining electron microscopy. J. Bacteriol. *191*, 3256–3264.

Nakane, D., Adan-Jubo, J., Kenri, T., and Miyata, M. (2010). Isolation and characterization of P1 adhesin, leg protein of gliding bacterium, *Mycoplasma pneumoniae*. J. Bacteriol. *193*, 715–722.

Neimark, H.C. (1977). Extraction of an actin-like protein from the prokaryote *Mycoplasma pneumoniae*. Proc. Natl. Acad. Sci. U.S.A. *74*, 4041–4045.

Neyrolles, O., Brenner, C., Prevost, M., Fontaine, T., Montaginer, L., and Blanchard, A. (1998). Identification of two glycosylated components of *Mycoplasma penetrans*: a surface-exposed capsular polysaccharide and a glycolipid fraction. Microbiology *144*, 1247–1255.

Ogle, K.F., Lee, K.K., and Krause, D.C. (1992). Nucleotide sequence analysis reveals novel features of the phase-variable cytadherence accessory protein HMW3 of *Mycoplasma pneumoniae*. Infect. Immun. *60*, 1633–1641.

Ohtani, N., and Miyata, M. (2007). Identification of a novel nucleoside triphosphatase from *Mycoplasma mobile*: a prime candidate motor for gliding motility. Biochem. J. *403*, 71–77.

Osawa, M., and Erickson, H.P. (2006). FtsZ from divergent foreign bacteria can function for cell division in *Escherichia coli*. J. Bacteriol. *188*, 7132–7140.

Osawa, M., Anderson, D.E., and Erickson, H.P. (2008). Reconstitution of contractile FtsZ rings in liposomes. Science *320*, 792–794.

Peters, P.C., Migocki, M.D., Thoni, C., and Harry, E.J. (2007). A new assembly pathway for the cytokinetic Z ring from a dynamic helical structure in vegetatively growing cells *of Bacillus subtilis*. Mol. Microbiol. *64*, 487–499.

Pich, O.Q., Burgos, R., Ferrer-Navarro, M., Querol, E., and Piñol, J. (2006). *Mycoplasma genitalium mg200* and *mg386* genes are involved in gliding motility but not in cytadherence. Mol. Microbiol. *60*, 1509–1519.

Pich, O.Q., Burgos, R., Ferrer-Navarro, M., Querol, E., and Piñol, J. (2008). Role of *Mycoplasma genitalium* MG218 and MG317 cytoskeletal proteins in terminal organelle organization, gliding motility and cytadherence. Microbiology *154*, 3188–3198.

Pich, O.Q., Burgos, R., Querol, E., and Piñol, J. (2009). P110 and P140 cytadherence-related proteins are negative effectors of terminal organelle duplication in *Mycoplasma genitalium*. PLoS One *4*, e7452.

Pollard, T.D., and Cooper, J.A. (2009). Actin, a central player in cell shape and movement. Science *326*, 1208–1212.

Popham, P.L., Hahn, T.W., Krebes, K.A., and Krause, D.C. (1997). Loss of HMW1 and HMW3 in noncytadhering mutants of *Mycoplasma pneumoniae* occurs post-translationally. Proc. Natl. Acad. Sci. U.S.A. *94*, 13979–13984.

Proft, T., Hilbert, H., Layh-Schmitt, G., and Herrmann, R. (1995). The proline-rich P65 protein of *Mycoplasma pneumoniae* is a component of the Triton X-100-insoluble fraction and exhibits size polymorphism in the strains M129 and FH. J. Bacteriol. *177*, 3370–3378.

Proft, T., Hilbert, H., Plagens, H., and Herrmann, R. (1996). The P200 protein of *Mycoplasma pneumoniae* shows common features with the cytadherence-associated proteins HMW1 and HMW3. Gene *171*, 79–82.

Ptacin, J.L., Lee, S.F., Garner, E.C., Toro, E., Eckart, M., Comolli, L.R., Moerner, W.E., and Shapiro, L. (2010). A spindle-like apparatus guides bacterial chromosome segregation. Nat. Cell. Biol. *12*, 791–798.

Reddy, S.P., Rasmussen, W.G., and Baseman, J.B. (1995). Molecular cloning and characterization of an adherence-related operon of *Mycoplasma genitalium*. J. Bacteriol. *177*, 5943–5951.

Regula, J.T., Boguth, G., Görg, A., Hegermann, J., Mayer, F., Frank, R., and Herrmann, R. (2001). Defining the mycoplasma 'cytoskeleton': the protein composition of the Triton X-100 insoluble fraction of the bacterium *Mycoplasma pneumoniae* determined by 2-D gel electrophoresis and mass spectrometry. Microbiology *147*, 1045–1057.

Relich, R.F., Friedberg, A.J., and Balish, M.F. (2009). Novel cellular organization in a gliding mycoplasma, *Mycoplasma insons*. J. Bacteriol. *191*, 5312–5314.

Schmidl, S.R., Gronau, K., Hames, C., Busse, J., Becher, D., Hecker, M., and Stülke, J. (2010). The stability of cytadherence proteins in *Mycoplasma pneumoniae* requires activity of the protein kinase PrkC. Infect. Immun. *78*, 184–192.

Schofield, W.B., Lim, H.C., and Jacobs-Wagner, C. (2010). Cell cycle coordination and regulation of bacterial chromosome segregation dynamics by polarly localized proteins. EMBO J. *29*, 3068–3081.

Seto, S., and Miyata, M. (1999). Partitioning, movement, and positioning of nucleoids in *Mycoplasma capricolum*. J. Bacteriol. *181*, 6073–6080.

Seto, S., and Miyata, M. (2003). Attachment organelle formation represented by localization of cytadherence proteins and formation of the electron-dense core in wild-type and mutant strains of *Mycoplasma pneumoniae*. J. Bacteriol. *185*, 1082–1091.

Seto, S., Layh-Schmitt, G., Kenri, T., and Miyata, M. (2001). Visualization of the attachment organelle and cytadherence proteins of *Mycoplasma pneumoniae* by immunofluorescence microscopy. J. Bacteriol. *183*, 1621–1630.

Seto, S., Uenoyama, A., and Miyata, M. (2005). Identification of a 521-kilodalton protein (Gli521) involved in force generation or force transmission for *Mycoplasma mobile* gliding. J. Bacteriol. *187*, 3502–3510.

Seybert, A., Herrmann, R., and Frangakis, A.S. (2006). Structural analysis of *Mycoplasma pneumoniae* by cryo-electron tomography. J. Struct. Biol. *156*, 342–354.

Shebelut, C.W., Guberman, J.M., van Teeffelen, S., Yakhnina, A.A., and Gitai, Z. (2010). *Caulobacter* chromosome segregation is an ordered multistep process. Proc. Natl. Acad. Sci. U.S.A. *107*, 14194–14198.

Stevens, M.K., and Krause, D.C. (1990). Disulfide-linked protein associated with *Mycoplasma pneumoniae* cytadherence phase variation. Infect. Immun. *58*, 3430–3433.

Stevens, M.K., and Krause, D.C. (1991). Localization of the *Mycoplasma pneumoniae* cytadherence-accessory proteins HMW1 and HMW4 in the cytoskeletonlike Triton shell. J. Bacteriol. *173*, 1041–1050.

Stevens, M.K., and Krause, D.C. (1992). *Mycoplasma pneumoniae* cytadherence phase-variable protein HMW3 is a component of the attachment organelle. J. Bacteriol. *174*, 4265–4274.

Thanedar, S., and Margolin, W. (2004). FtsZ exhibits rapid movement and oscillation waves in helix-like patterns in *Escherichia coli*. Curr. Biol. *14*, 1167–1173.

Trachtenberg, S., and Gilad, R. (2001). A bacterial linear motor: cellular and molecular organization of the contractile cytoskeleton of the helical bacterium *Spiroplasma melliferum* BC3. Mol. Microbiol. *41*, 827–848.

Trachtenberg, S., Gilad, R., and Geffen, N. (2003). The bacterial linear motor of *Spiroplasma melliferum* BC3: from single molecules to swimming cells. Mol. Microbiol. 47, 671–697.

Trachtenberg, S., Dorward, L.M., Speransky, V.V., Jaffe, H., Andrews, S.B., and Leapman, R.D. (2008). Structure of the cytoskeleton of *Spiroplasma melliferum* BC3 and its interactions with the cell membrane. J. Mol. Biol. *378*, 778–789.

Uenoyama, A., and Miyata, M. (2005). Identification of a 123-kilodalton protein (Gli123) involved in machinery for gliding motility of *Mycoplasma mobile*. J. Bacteriol. *187*, 5578–5584.

Uenoyama, A., Kusumoto, A., and Miyata, M. (2004). Identification of a 349-kilodalton protein (Gli349) responsible for cytadherence and glass binding during gliding of *Mycoplasma mobile*. J. Bacteriol. *186*, 1537–1545.

Uenoyama, A., Seto, S., Nakane, D., and Miyata, M. (2009). Regions on Gli349 and Gli521 protein molecules directly involved in movements of *Mycoplasma mobile* gliding machinery, suggested by use of inhibitory antibodies and mutants. J. Bacteriol. *191*, 1982–1985.

van den Ent, F., Johnson, C.M., Persons, L., de Boer, P., and Löwe, J. (2010). Bacterial actin MreB assembles in complex with cell shape protein RodZ. EMBO J. *29*, 1081–1090.

Vishniakov, I.E., Borkhsenius, S.N., Basovski, I., Levitski, S.A., Lazarev, V.N., Snigirevskaia, E.S., and Komissarchik, I. (2009). Localization of the division protein FtsZ in mycoplasma cells *Mycoplasma hominis*. Tsitologiia *51*, 247–256.

Waidner, B., Specht, M., Dempwolff, F., Haeberer, K., Schaetzle, S., Speth, V., Kist, M., and Graumann, P.L. (2009). A novel system of cytoskeletal elements in the human pathogen *Helicobacter pylori*. PLoS Pathog. *5*, e1000669.

Waldo, R.H. 3rd, Popham, P.L., Romero-Arroyo, C.E., Mothershed, E.A., Lee, K.K., and Krause, D.C. (1999). Transcriptional analysis of the *hmw* gene cluster of *Mycoplasma pneumoniae*. J. Bacteriol. *181*, 4978–4985.

Waldo, R.H. 3rd, Jordan, J.L., and Krause, D.C. (2005). Identification and complementation of a mutation associated with loss of *Mycoplasma pneumoniae* virulence-specific proteins B and C. J. Bacteriol. *187*, 747–751.

White, C.L., Kitich, A., and Gober, J.W. (2010). Positioning cell wall synthetic complexes by the bacterial morphogenetic proteins MreB and MreD. Mol. Microbiol. *76*, 616–633.

Willby, M.J., and Krause, D.C. (2002). Characterization of a *Mycoplasma pneumoniae* HMW3 mutant: implication for attachment organelle assembly. J. Bacteriol. *184*, 3061–3068.

Willby, M.J., Balish, M.F., Ross, S.M., Lee, K.K., Jordan, J.L., and Krause, D.C. (2004). HMW1 is required for stability and localization of HMW2 to the attachment organelle of *Mycoplasma pneumoniae*. J. Bacteriol. *186*, 8221–8228.

Williamson, D.L. (1974). Unusual fibrils from the spirochete-like sex ratio organism. J. Bacteriol. *117*, 904–906.

Zhao, Y., Hammond, R.W., Lee, I.M., Roe, B.A., Lin, S., and Davis, R.E. (2004). Cell division gene cluster in *Spiroplasma kunkelii*: functional characterization of *ftsZ* and the first report of *ftsA* in Mollicutes. DNA Cell Biol 23, 127–134.

# Gliding Mechanism of the *Mycoplasma pneumoniae* Subgroup – Implications from Studies on *Mycoplasma mobile*

12

Makoto Miyata and Daisuke Nakane


## Abstract

More than ten Mollicute species classified into the *hominis* and *pneumoniae* subgroups form a membrane protrusion at a cellular pole, and glide in this direction using mechanisms not seen in other bacterial genera. Our studies are unveiling the gliding mechanism of *Mycoplasma mobile*, the most rapidly motile species which belongs to the *hominis* subgroup. However, the gliding mechanism used by members of the *pneumoniae* subgroup are still unclear. Here, we review the gliding of members of the *pneumoniae* subgroup, in reference to that of *M. mobile*, with respect to features of their motility, their surface and cytoskeletal structures, their component proteins and their targets for binding to enable gliding. While the major features appear to be common between these two subgroups, no similarities have been found in the amino acid sequences of the component proteins involved in gliding motility.


## Introduction

Gliding motility is a striking feature of certain Mollicute species (Kirchhoff, 1992; Miyata, 2005, 2007, 2008, 2010). These species form a membrane protrusion at a cell pole and exhibit gliding motility in the direction of the protrusion. They can glide on various solid surfaces, including glass, plastic and animal cells, and the motility is thought to be involved in pathogenicity. *M. mobile*, a fish pathogen, glides smoothly and continuously on glass at an average speed of 2.0 to 4.5 μm per second, or 3 to 7 times the length of the cell per second, exerting a force of up to 27 pN (Fig. 12.1) (Rosengarten and Kirchhoff, 1987; Miyata *et al.*, 2002; Hiratsuka *et al.*, 2005, 2006) (many video files have been deposited on YouTube at http://www.youtube.com/user/MycoplasmaGliding).

The more than 200 Mollicute species described thus far are classified phylogenetically into four subgroups, *hominis*, *pneumoniae*, *spiroplasma* and *phytoplasma*, on the basis of their 16s rRNA and genome sequences (Weisburg *et al.*, 1989; Barre *et al.*, 2004) (Molligen, http://cbi.labri.fr/outils/molligen/). To date, over ten Mollicute species are known to glide (Hatchel and Balish, 2008; Relich *et al.*, 2009). These species belong to two phylogenetic subgroups, the *hominis* subgroup, represented by *M. mobile*, and the *pneumoniae* subgroup, both of which are phylogenetically distant from each other.

## Purpose and pathogenicity

Generally, motile bacteria move to access nutrients and escape from wastes and predators using 'two-component systems' (Jarrell and McBride, 2008), but mycoplasma gliding does not exhibit any obvious chemotaxis. This raises the question of the purpose of mycoplasma gliding. We cannot rule out the possibility of chemotaxis in mycoplasmas, because the failure to detect chemotaxis may be a result of the experimental conditions used in these studies. However, if they are chemotactic, the mechanism used is novel and distinct from that used by other bacteria, because no homologues of two-component system genes can be identified in mycoplasma genomes (Dandekar *et al.*, 2000; Barre *et al.*, 2004; Jaffe *et al.*, 2004b). Mycoplasmas

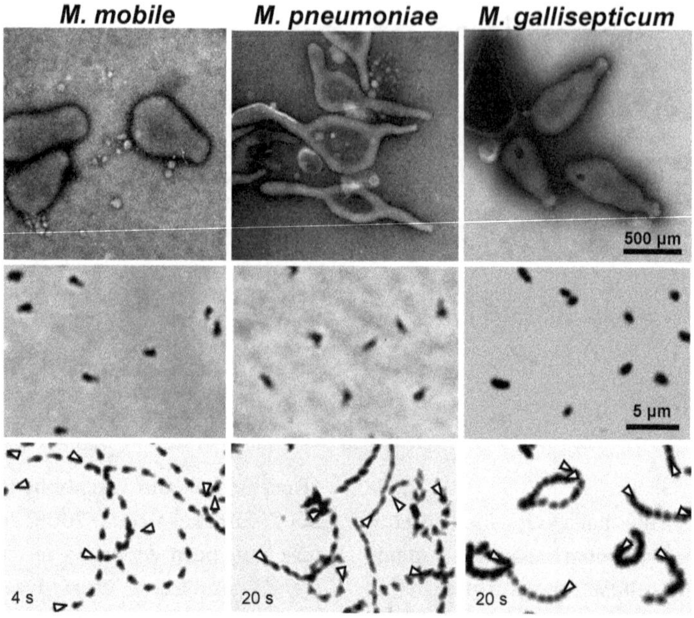


**Figure 12.1** Three gliding species. Upper panels: Electron microscopy. Middle panels: Phase contrast microscopy. Lower panels: Integrated video image. Movement over time is indicated with ten integrated video frames. The first images of a cell are marked with a white triangle.

repeatedly bind sialylated oligosaccharides on host cell surfaces to enable adhesion and gliding, as discussed below (Nagai and Miyata, 2006). The binding affinity and gliding properties depend on structures that vary significantly in different tissues and in different animals. Mycoplasmas may detect the differences in the structure and reach the tissue covered with the sialylated oligosaccharide for which they have the highest affinity. Even if they move randomly, such movements will enable them to reach conditions that are better for propagation, because gliding mycoplasmas bind to animal tissues tightly and cluster on them. If they stay in the same position without moving, they will consume the nutrients and pollute their environment with their waste.

*M. mobile* exhibits rheotaxis, moving upstream in a liquid flow, an environment typically present on the surfaces of fish (Rosengarten *et al.*, 1988). *M. mobile* is a pathogen isolated from the gills of a freshwater fish (Kirchhoff and Rosengarten, 1984; Stadtlander and Kirchhoff, 1995; Stadtlander *et al.*, 1995), so this response may help to prevent the organism from being swept out of the gills. The influence of the direction of flow of the liquid environment on the gliding direction of this organism can be explained by physical mechanisms, specifically the biased subcellular position of the adhesion protein (Miyata and Uenoyama, 2002; Kusumoto *et al.*, 2004; Uenoyama *et al.*, 2004) and the fact that gliding always occurs in the direction of the head of the mycoplasma cell. The head of the mycoplasma may be turned around by the force of the liquid flow, resulting in movement upstream (Miyata *et al.*, 2002).

Cytadherence of mycoplasmas, which is linked to gliding motility, has a well-established role in parasitism and pathogenicity (Razin *et al.*, 1998). When a mycoplasma strain loses its capacity for cytadherence, it is easily removed from tissue by the host. Many mycoplasmas that have been confirmed to be capable of gliding motility are pathogenic, and *M. pneumoniae* cells have been shown to glide when they move to the cell surface after having attached to the tip of the cilia of tracheal cells (Krunkosky *et al.*, 2007). However, the contribution of gliding to pathogenicity is less clear than that of cytadherence, because both functions are closely related.

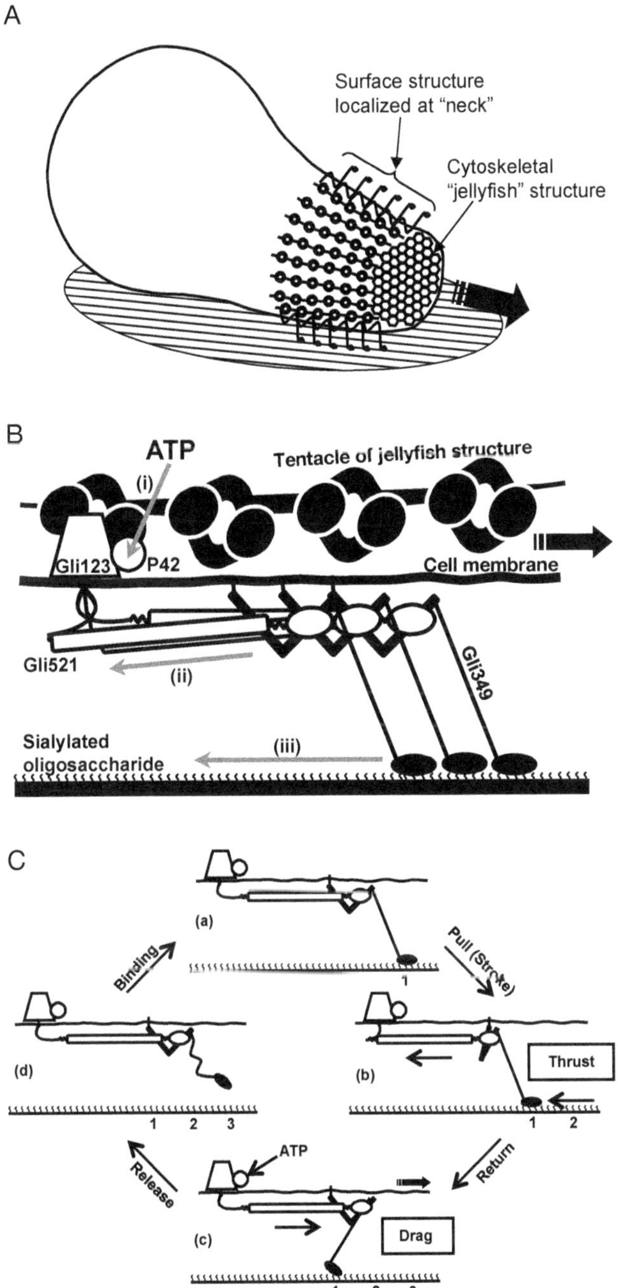


**Figure 12.2** Gliding mechanism of *M. mobile*. (A) Cellular architecture and gliding direction. The gliding machinery located at the membrane protrusion is composed of surface and jellyfish structures. The surface structure is located at the 'neck' adjacent to the subcellular position of jellyfish tentacles. (B) Structure of small numbers of the machinery unit. The cell is gliding to the right. The Gli123, Gli521 and Gli349 proteins, encoded by MMOB1020, MMOB1030 and MMOB1040, respectively, from the surface structure, supported by the tentacle part of the jellyfish structure. The putative motor, P42, encoded by MMOB1050, may form a complex with these three proteins. The tip of Gli349, the foot, catches and releases the sialylated oligosaccharide fixed on the solid surface. The possible sequence of movements for gliding are represented by the numbered arrows (i-iii). (C) Working 'centipede' model to explain gliding. Each gliding unit operates in a mechanical cycle composed of a series of states (a–d) and transition steps consisting of pull (stroke), return, release and binding. The leg (Gli349) binds tightly to sialylated oligosaccharide on the solid surface (a). The force applied to the leg from the front triggers a conformational change in the upper short arms of the leg, causing the pull (stroke), using energy from ATP hydrolysis. Cell movement generated by the other legs pull the units forward and return them to the initial conformation with short arms (c). A new molecule of ATP is required for release. The continuous pull in the forward direction removes the foot from the sialylated oligosaccharide (d). The foot can be removed preferentially in the forward direction, and this directional binding causes directional movement with a directional stroke. The detached foot then rebinds to sialylated oligosaccharide at the new position. For more detail, see our previous articles (Miyata, 2008; Chen *et al*., 2009; Uenoyama *et al*., 2009; Miyata, 2010).

## Mechanism used for gliding

Mycoplasmas do not have flagella, pili, or any proteins homologous with those used in known mechanisms involved in bacterial motility. In addition, they have no known homologues of the conventional motor proteins that are usually involved in eukaryotic motility (Dandekar *et al.*, 2000; Barre *et al.*, 2004; Jaffe *et al.*, 2004b). Therefore, the mechanism used by mycoplasmas to achieve gliding has, until recently, remained a mystery. In the last decade, we have studied the gliding mechanism of *M. mobile* and have proposed a working model based on our experimental data (Miyata, 2007, 2008, 2010; Chen *et al.*, 2009). The gliding motility of members of the *hominis* subgroup could be expected to be based on a mechanism similar to that of *M. mobile,* because most genes associated with gliding in *M. mobile* can be found in the genome of *M. pulmonis.* However, this mechanism cannot be readily correlated with that used for gliding in members of the *pneumoniae* subgroup, because no homologues of the genes involved in gliding in *M. mobile* can be found in the genomes of members of the *pneumoniae* subgroup.

In this review, we will briefly explain our current working model to explain the gliding mechanism of *M. mobile* (Miyata, 2010), and then discuss the information currently available on the gliding of members of the *pneumoniae* subgroup.

## Gliding mechanism of *Mycoplasma mobile*

The gliding machinery is located at the base of the protrusion, which has been designated the neck (Fig. 12.2A). String-like leg structures approximately 50 nm in length protrude from the membrane in the neck area, and their distal ends bind to the surface of the host or the glass (Miyata and Petersen, 2004). The machinery is formed by approximately 450 units, composed of similar ratios of four proteins, namely Gli349 (the leg), Gli521 (the crank), Gli123 (the mount) and P42 (the motor) (Uenoyama *et al.*, 2004; Metsugi *et al.*, 2005; Seto *et al.*, 2005b; Uenoyama and Miyata, 2005b; Adan-Kubo *et al.*, 2006; Ohtani and Miyata, 2007; Lesoil *et al.*, 2009; Uenoyama *et al.*, 2009; Nonaka *et al.*, 2010). The cytoskeletal structure, a 'jellyfish' composed of a characteristic 'bell' and 'tentacles', supports the machinery from inside the cell (Nakane and Miyata, 2007). The movements are thought to progress as illustrated by the numbered arrows in Fig. 12.2B (Miyata, 2010; Nonaka *et al.*, 2010). A working model has been proposed in which cells are propelled by legs composed of Gli349 proteins which repeatedly bind to and release from the solid surface (Miyata, 2008, 2010; Chen *et al.*, 2009; Uenoyama *et al.*, 2009). This model has been called the centipede model (or the power stroke model), because the many legs that protrude from the neck of the cell repeat the power strokes and keep binding to sialylated oligosaccharide, as shown in Fig. 12.2C. Recently, the legs have been suggested to stroke independently, based on the detailed analyses of cell pivoting during gliding (Nakane and Miyata, 2012).

## Gliding mechanism of *Mycoplasma pneumoniae*

More than 10 proteins are thought to be involved in gliding and adherence in *M. pneumoniae,* based on their subcellular localization and the effects of mutations on gliding and adherence. However, none of them has any amino acid sequence similarity to the proteins involved in gliding in *M. mobile* (Dandekar *et al.*, 2000; Barre *et al.*, 2004; Jaffe *et al.*, 2004b). As a result it has been suggested that these two subgroups of mycoplasmas glide using different mechanisms (Kusumoto *et al.*, 2004; Burgos *et al.*, 2008). From the perspective of the genes and proteins involved, the gliding mechanisms of those two subgroups can be defined as 'different'. However, they share many features, as discussed below, particularly when compared with the surface motility of other bacteria, including *Myxococcus, Pseudomonas* and *Flavobacterium* species (Jarrell and McBride, 2008).

### Gliding properties

The human pathogens *M. pneumoniae* and *M. genitalium,* and *M. gallisepticum,* an avian pathogen, are classified into the *pneumoniae* subgroup and appear to glide in a similar manner to that of *M. mobile.* They form a membrane protrusion at one pole, attach to solid surfaces at the protrusion and glide in the direction of the protrusion. Detailed analyses of gliding in *M. pneumoniae* and *M. gallisepticum* has shown that the cells glide continuously and pivot around the protrusion with varying frequencies, as observed in *M. mobile* (Fig. 12.3) (Nakane and Miyata, 2009, 2012). The gliding speeds of *M. pneumoniae, M. genitalium* and *M. gallisepticum* are 0.64, 0.15 and 0.41 μm/s, respectively (Pich *et al.*, 2008; Nakane and Miyata, 2009), much slower than that of *M. mobile,* 3.4 μm/s (Rosengarten and Kirchhoff, 1987; Miyata *et al.*, 2002). However, the ranges in speed are not subgroup specific, as *M. testudinis,* which was isolated from turtles and which belongs to the *pneumoniae* subgroup, glides at 3.0 μm/s, and *M. pulmonis* glides slowly, at 0.4–0.7 μm/s (Bredt and

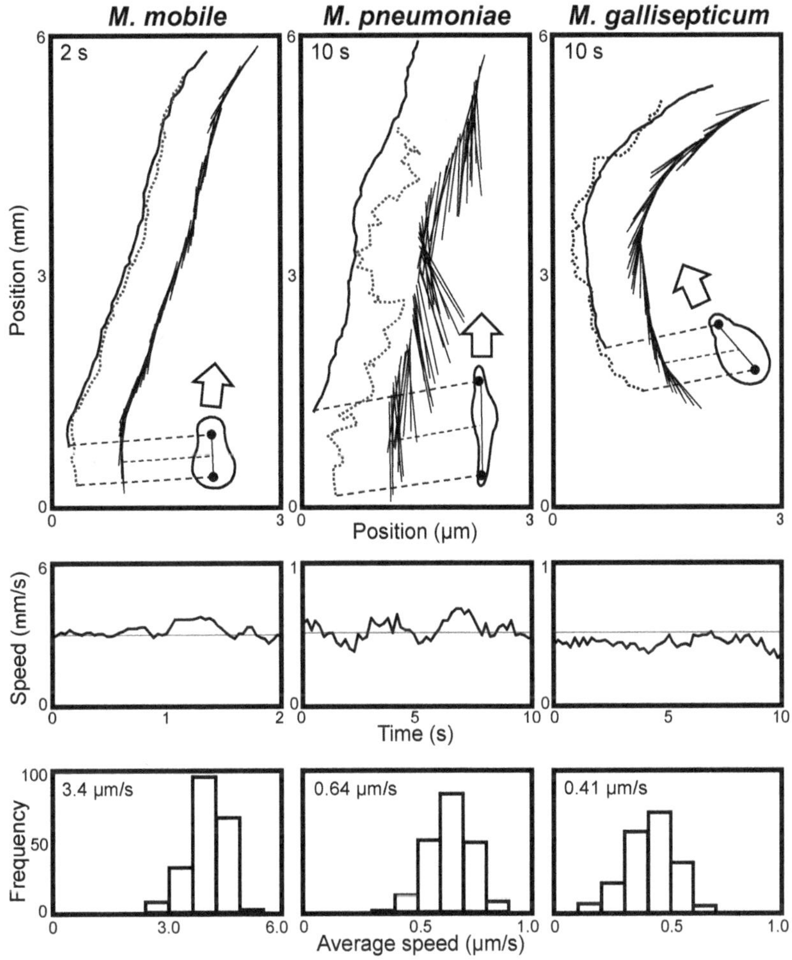


**Figure 12.3** Detailed analyses of gliding. Upper panels: A trace of cell over time as indicated in the upper left of each panel showing the relative position of the cell every 0.03 and 0.17 seconds, respectively, for *M. mobile* and two other species. The solid and dashed traces on the left in each panel present movements of the front and back positions of a cell, respectively. The series of bars in the right of each panel indicate the corresponding movements, by connecting the front and back positions of the cell. Middle panels: Changes in gliding speed for the movements corresponding to the upper panels. The average speed for 0.17 and 0.83 seconds is plotted for every 0.03 and 0.17 seconds, respectively, for *M. mobile* and the other two species. Lower panels: Distributions of gliding speed. The average speeds are shown in the upper left for each species. The cell movements have been analysed as described previously (Nakane and Miyata, 2009, 2012).

Radestock, 1977). These observations suggest that the species in the *pneumoniae* subgroup glide in a similar manner to *M. mobile.*

## Surface structure

As mycoplasmas bind to solid surfaces at their protrusion, surface structures responsible for binding can be expected to be located on this organelle. On *M. mobile,* complicated filamentous structures can be seen on the surface at the neck of the cell in negatively stained cells using electron microscopy (EM) (Fig. 12.4, upper left). These structures cannot be observed clearly, probably because of the presence of multiple layers of filaments. However, rapid-freeze-and-freeze-fracture rotary-shadow electron microscopy has demonstrated a plausible leg structure, with the moving legs frozen at one point, and revealed by fracturing (Fig. 12.4, upper right) (Miyata and Petersen, 2004). The putative legs are 50 nm long string-like

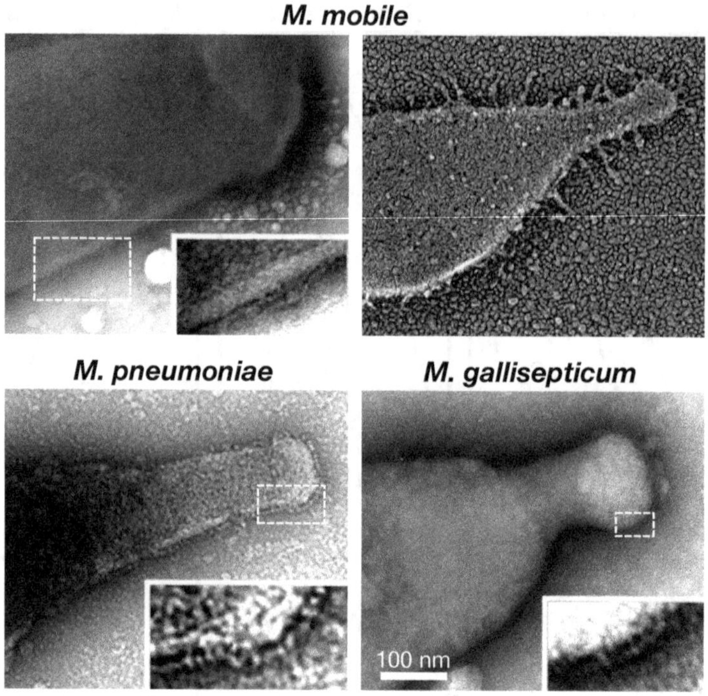


**Figure 12.4** Surface structures of gliding machinery observed by electron microscopy. *M. mobile* cells (upper panels) were examined by negative staining (left) and rapid-freeze-and-freeze-fracture rotary-shadow electron microscopy (right). Two other species are shown in the lower panels. Insets show magnified images of the boxed area. For more detail, see our previous articles (Miyata and Petersen, 2004; Nakane and Miyata, 2007, 2009; Nakane *et al.*, 2011).

structures, protruding from the membrane and attaching to the solid surface at their distal end. We infer that this structure is responsible for binding and movement in gliding.

In *M. pneumoniae*, nap structures, reminiscent of the raised surface of a cloth, can be observed at the surface of the membrane protrusion, the 'attachment organelle', by negatively stained EM (Hu *et al.*, 1982; Kirchhoff *et al.*, 1984). Similar structures can be seen in other species in the *pneumoniae* subgroup. The nap structures cannot be clearly seen, because of the presence of the multiple layers. However, the nap images of *M. pneumoniae* could be reconstructed using cryo-electron microscopy, in which the images of frozen specimens are captured at a series of angles relative to the electron beam, and the three dimensional structure is calculated. They were shown to be a 'knob' shape, with a length of 4 to 8 nm and a diameter of 8 nm (Seybert *et al.*, 2006).

## Binding target for gliding

To identify the binding target used in the gliding of *M. mobile*, we examined the factors affecting the binding of cells to solid surfaces and concluded that N-acetylneuraminyllactose (or sialyllactose, a sialylated oligosaccharide) can mediate binding to glass, based on four lines of evidence: (i) glass binding is inhibited by N-acetylneuraminidase, which degrades sialylated oligosaccharides; (ii) glass binding is inhibited by the addition of N-acetylneuraminyllactose; (iii) binding occurs on glass coated with N-acetylneuraminyllactose; and (iv) gliding speed depends on the concentration of N-acetylneuraminyllactose on the glass (Nagai and Miyata, 2006; Kasai *et al.*, 2013). *M. mobile* cells appear to bind various surfaces, including glass, mica and plastics, but a representative sialoprotein, fetuin, found in serum used in culture medium, is easily adsorbed onto these solid surfaces and forms the scaffold for gliding.

As the gliding activity of *M. pneumoniae* is less than that of *M. mobile,* detailed analyses cannot be as readily performed. However, we have found that the addition of free N-acetylneuraminyllactose removes gliding *M. pneumoniae* cells from the surface of the glass (Kasai *et al.*, 2013). It is probable that the gliding legs bind to the free N-acetylneuraminyllactose when they are displaced from the binding target on the glass, as seen in *M. mobile.* These observations are consistent with observations showing that *M. pneumoniae* adheres to animal cells by attaching to sialylated oligosaccharides (Baseman *et al.*, 1982a; Roberts *et al.*, 1989).

## Leg protein

In *M. mobile* gliding, Gli349 is believed to have the role of 'leg and foot' based on the following observations: (i) Gli349 is essential for binding and gliding (Uenoyama *et al.*, 2004); (ii) mutations in the gene encoding Gli349 modify the gliding speed and binding to solid surfaces during gliding (Miyata *et al.*, 2000; Uenoyama *et al.*, 2004, 2009); (iii) monoclonal antibodies against Gli349 inhibit movement and binding for gliding (Kusumoto *et al.*, 2004; Uenoyama *et al.*, 2004, 2009); (iv) Gli349 is localized at the position of the gliding machinery (Kusumoto *et al.*, 2004; Uenoyama *et al.*, 2004); (v) the music note-like molecular shape of isolated Gli349 is similar to that of the leg structure observed on surface of the neck of the cell by EM (Fig. 12.5) (Miyata and Petersen, 2004; Adan-Kubo *et al.*, 2006; Nakane and Miyata, 2007); and (vi) atomic force microscopy (AFM) has demonstrated the binding of Gli349 to sialylated oligosaccharides,

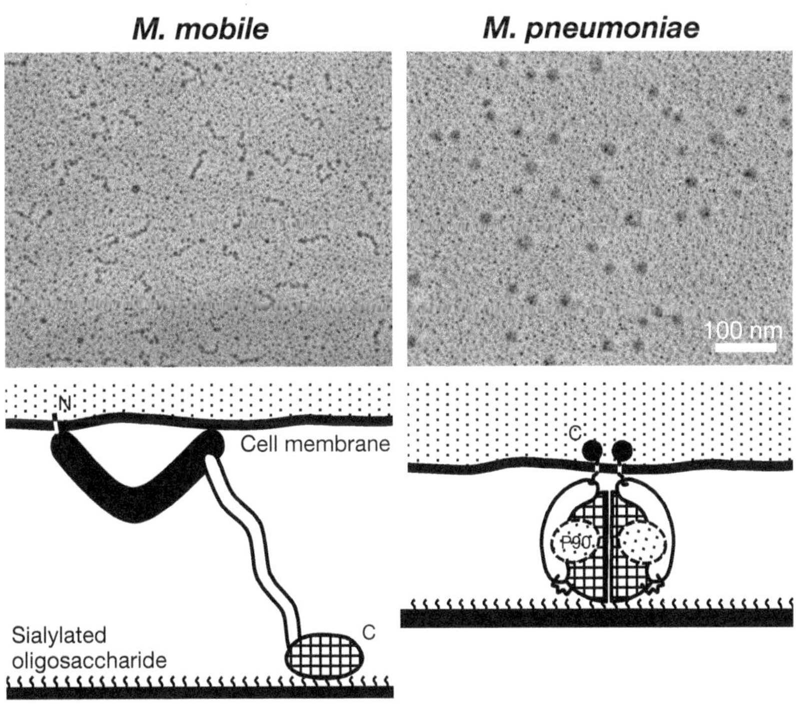


**Figure 12.5** Structures of leg proteins. Upper panels: EM images of isolated leg proteins, Gli349 of *M. mobile* (left) and the P1 adhesin complex of *M. pneumoniae* (right). Lower panels: Schematic diagrams of leg proteins. The Gli349 molecule (left) is divided into three parts, an amino-terminal rigid section, the middle flexible region, and the carboxyl-terminal globular section. A transmembrane segment is predicted at the amino-terminus. The P1 adhesin complex (right) is composed of two molecules each of the P1 adhesin and P90. The P1 adhesin molecule is divided into three domains, the amino-terminal conserved region, the middle variable section, which has a transmembrane segment near the C-terminal end, and the carboxyl-terminal highly conserved region. For more detail, see our previous articles (Metsugi *et al.*, 2005; Adan-Kubo *et al.*, 2006; Nakane *et al.*, 2011).

the binding target for gliding (Lesoil *et al.*, 2009). The molecular shape of Gli349 is similar to a protein involved in surface variation, MvspI, featured with rod parts composed of repeat sequences and a terminal globule (Adan-Kubo *et al.*, 2012; Wu *et al.*, 2012; Wu and Miyata, 2012). The common features in the structures of these proteins may suggest that they evolved from a common ancestor. Alternatively, these features are essential for the surface-exposed proteins of *M. mobile* and became common among unrelated proteins as a result of convergent evolution.

P1 adhesin (MPN141) of *M. pneumoniae* has been identified as the adhesin involved in static binding (Baseman *et al.*, 1982b; Feldner *et al.*, 1982; Hu *et al.*, 1982). This protein is also act as the leg in the gliding of *M. pneumoniae,* based on the following observations: (i) the P1 adhesin is essential for binding and gliding; (ii) the P1 adhesin is localized on the surface of the attachment organelle; (iii) monoclonal antibodies against P1 inhibit movement and eventually remove gliding cells from the surface of glass (Seto *et al.*, 2005a); and (iv) the nap structure localized on the attachment organelle is composed of P1 (unpublished data).

We isolated the P1 adhesin complex from cultures (Fig. 12.5) (Nakane *et al.*, 2011), and found that the complex contains two molecules each of P1 and P90 (protein B, MPN142) and has a molecular mass of about 480 kDa. The complex is a sphere 20 nm in diameter under rotary-shadowing electron microscopy. P90 is encoded in tandem with P1, and cleaved from another protein, P40 (protein C), after translation (Layh-Schmitt and Herrmann, 1992; Catrein *et al.*, 2005). This cleavage is specific to *M. pneumoniae,* and is not seen in the orthologue of this protein in *M. genitalium,* a species closely related to *M. pneumoniae,* or in the orthologue in *M. gallisepticum.* Chemical crosslinking studies and fluorescence microscopy suggest that the P1 complex also contains P40 (Layh-Schmitt *et al.*, 2000; Seto *et al.*, 2001). The P1 adhesin molecule can be divided into three domains. Analyses of amino acid sequences have shown that domains I and III are well conserved between different species, and that there is a transmembrane region within domain III. The binding site for sialylated oligosaccharide may lay in domain I or III, because the amino acid sequences of the binding site would not be expected to evolve rapidly. Although the P1 adhesin complex on a cell would be expected to bind sialylated oligosaccharide, we have not been able to detect such binding activity with the isolated complex. Other factors, such as the cell membrane or other proteins may be essential for binding activity. A deletion in the carboxyl-terminal region of P110, the orthologue of P90 in *M. genitalium,* increases the number of attachment organelles, suggesting physical interactions between the P1 adhesin complex and the cytoskeletal structures supporting the attachment organelle (Pich *et al.*, 2009). It is possible that the binding activity may require tension, like a receptor of sialylated oligosaccharide on animal cells, selectin (Springer, 2009). Recently, many receptors have been shown to demonstrate tension sensitive binding, (catch binding), a mechanism distinct from that seen in conventional binding (slip binding) (Sokurenko *et al.*, 2008). Analyses of P1 and Gli349 have not detected any similarities with the known receptors for sialylated oligosaccharides, suggesting that these proteins are novel receptors (Nakane *et al.*, 2011).

The sequence of the P1 adhesin varies between clinical strains, resulting in structural changes in immunodominant epitopes of the adhesin, enabling evasion of the host immune system (Nakane *et al.*, 2011). The variation in the P1 adhesin is thought to be generated by intragenomic recombination for the following reasons: (i) all P1 genotypes of *M. pneumoniae* found in clinical isolates can be explained by recombination between the P1 adhesin gene and one of the paralogous DNA sequences on the genome (Kenri *et al.*, 1999; Spuesens *et al.*, 2009); (ii) in *M. genitalium,* adherence-deficient mutants have been generated by recombination between the MgPa gene, the orthologue of the P1 adhesin, and a paralogous sequence encoded in a neighbouring region downstream of the gene (Peterson *et al.*, 1995; Burgos *et al.*, 2006; Iverson-Cabral *et al.*, 2006).

P30 (MPN453), which is comprised of 274 amino acid residues, has a transmembrane segment beginning from amino acid residue 72, with the carboxyl terminus oriented towards the outside of the cell. The subcellular localization of this protein is limited to the end of the attachment organelle, distinct from that of P1. In-frame deletions of 11 amino acid residues of this protein mostly disrupt binding and gliding activities, although the stability of the protein and its localization are not significantly affected (Hasselbring *et al.*, 2005; Chang *et al.*, 2011). Chemical cross-linking studies suggest close proximity between P30 and P1 (Layh-Schmitt *et al.*, 2000). These molecules may interact physically when they localize at the same subcellular position on the cell surface.

## Energy source and force generation

Information about the energy source and force generation is indispensable in clarifying the mechanisms involved in any kinds of motor. In *M. mobile*, the energy is provided by ATP (Jaffe *et al.*, 2004a; Uenoyama and Miyata, 2005a) and a maximum force of 27 pN is generated per cell when they are stalled (Miyata *et al.*, 2002). *M. mobile* has been proven to be driven by the energy of ATP through observations that motility of cellular 'ghosts' with their membrane damaged by exposure to the detergent Triton X-100 could be reactivated by the addition of ATP (Uenoyama and Miyata, 2005a).

We have tried similar experiments with *M. pneumoniae*, but were unable to reactivate the motility of 'ghosts' derived from this species. However, this does not suggest that *M. pneumoniae* is driven by a different energy source, because of the complexities involved in the generation of 'gliding ghost' cells. All the essential parts of the gliding machinery are required after the damage is inflicted on the cell envelope by the detergent, and prolonged treatment of *M. mobile* cells with Triton X-100 significantly reduces the proportion of reactivatable ghosts, so the failure of these experiments may reflect greater sensitivity of the gliding mechanism of *M. pneumoniae* to the detergent. Similarly, the stall force of gliding cannot be measured for *M. pneumoniae*, because its gliding is much less active than that of *M. mobile*. Examination of more actively gliding species in the *pneumoniae* subgroup, such as *M. testudinis*, may allow such detailed measurements to be taken.

## Cytoskeletal structures

As mycoplasmas lack a rigid cell wall, other structures are likely to be required to support the cell shape and the force for gliding motility. Removal of the cell membrane and cytosol of *M. mobile* using Triton X-100 reveals a striking cytoskeletal structure composed of a bell and tentacles, reminiscent of a jellyfish (Figs. 12.2A and B, 12.6, left, and 12.7A) (Nakane and Miyata, 2007). The bell, a solid structure, is located at the cell head and dozens of tentacles are bound to it. The jellyfish structure supports the cell membrane and the surface structure, including the leg and crank, which are directly associated with the gliding mechanism (Nakane and Miyata, 2007; Miyata, 2008, 2010; Sato *et al.*, 2012).

The cytoskeletal structure of *M. pneumoniae* was identified in the 1980s, much earlier than that of *M. mobile* (Meng and Pfister, 1980; Göbel *et al.*, 1981), and its detail was examined in the 2000s by cryoelectron tomography (CET) (Henderson and Jensen, 2006; Seybert *et al.*, 2006). The structure, a 'rod', can be readily seen by EM in the centre of the attachment organelle of the species in the *pneumoniae* subgroup after the treatment of the cells with Triton X-100 (Fig. 12.6) (Hatchel and Balish, 2008; Nakane and Miyata, 2009; Relich *et al.*, 2009). As this structure has a high electron density compared with other parts of the cell, it is also referred as the 'electron dense core' in sectioning images (Wilson and Collier, 1976; Seto and Miyata, 2003). The rod can be presumed to support the attachment organelle, because the mutants lacking the rod cannot form the attachment organelle (Krause and Balish, 2004; Miyata, 2005, 2007, 2008; Balish and Krause, 2006; Miyata and Ogaki, 2006). The structure can be divided into three parts, the 'terminal button', the 'paired plate' and the 'wheel (bowl)' (Fig. 12.7B). These features are more evident in the cytoskeletal structure of *M. gallisepticum*, which was recently described as an 'asymmetrical dumbbell' (Nakane and Miyata, 2009) (Fig. 12.6, right).

### Terminal button

A small piece of the membrane of Triton X-100-treated cells sometimes appears to be attached to the terminal button, suggesting a complex structure that includes the polar cell membrane.

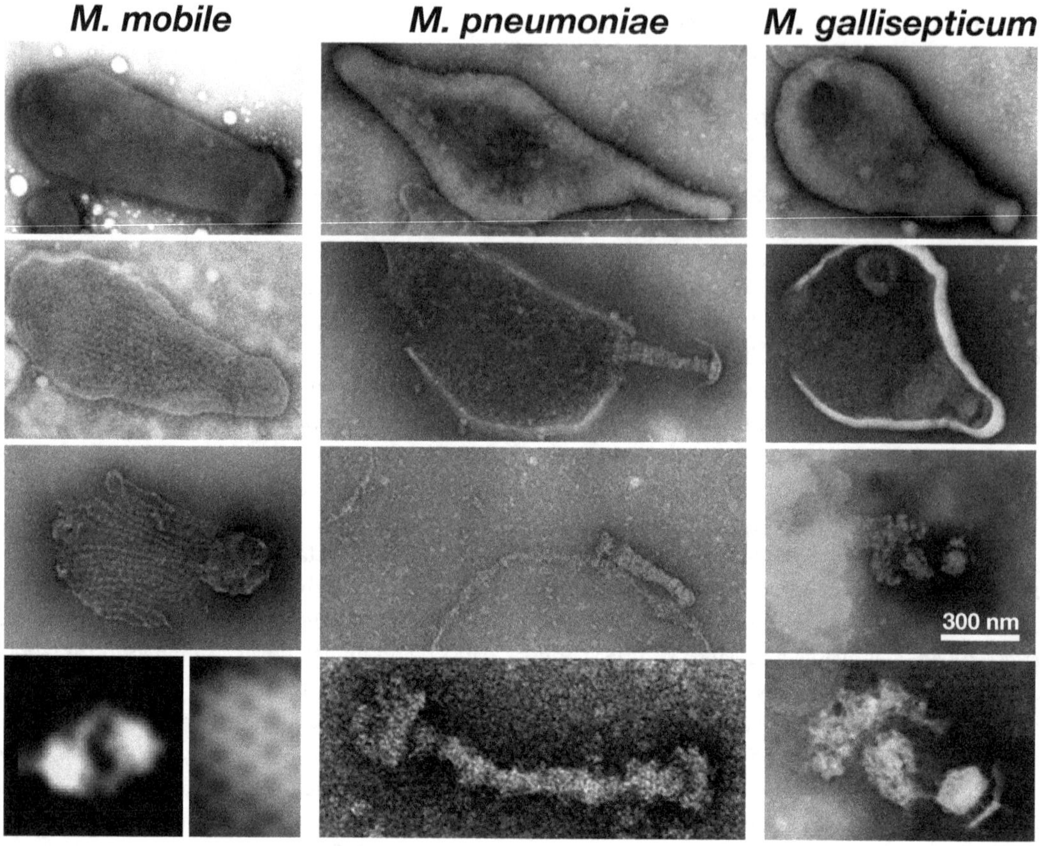


**Figure 12.6** Cytoskeletal structures of three mycoplasma species observed by EM. Upper panels: intact cells, with the direction of gliding oriented towards the lower right corner. Second row: cells with slightly damaged membranes; the contrast of the image of the cell is improved over that of the intact cells. Third row: the cytoskeletal structures remain after the cell membrane has been removed by the detergent. Bottom row: Magnified images of cytoskeletal structures. For *M. mobile*, an averaged image of particles of the 'jellyfish tentacles' and an image of the 'jellyfish bell' reconstructed using 'Fourier transformation' are presented on the left and the right, respectively. For more detail, see our previous articles (Nakane and Miyata, 2007, 2009; Nakane *et al.*, 2011).

The terminal button can be divided into three major parts, the most distal of which is attached to the inner layer of peripheral membrane proteins. The localization of the protein that makes up the terminal button, P65 (MPN309), has been demonstrated using fluorescence microscopy, while HMW3 (MPN452) is found around the boundary to the striated paired plate, as summarized in Table 12.1 (Seto *et al.*, 2001; Seto and Miyata, 2003; Kenri *et al.*, 2004; Nakane and Miyata, 2009). P65 may determine the gliding direction by modifying the angle of the organelle relative to the cell axis, because in *M. genitalium* the deletion of MG217, the orthologue of P65, does not affect the gliding speed, but the mutant cells glide in straighter lines than the wild type strain (Burgos *et al.*, 2008).

## Paired plate

The middle part of the rod, the paired plates, are composed of paired striated plates separated by a gap of about 7 nm (Henderson and Jensen, 2006; Seybert *et al.*, 2006). The paired plates appear flexible and bend approximately 150 degrees just proximal to their middle. This bend suggests that *M. pneumoniae* cells have three axes, front-back, upper-lower and left–right. The alignment of these axes relative to the solid surface is unknown. Using

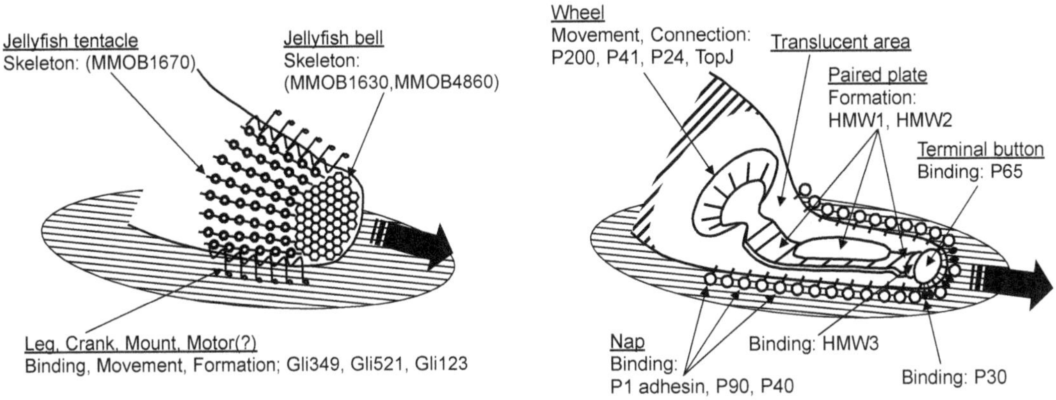


**Figure 12.7** Comparative illustration of cellular architectures of *M. mobile* (A) and *M. pneumoniae* (B). Each part is marked with an arrow and the roles and component proteins are indicated. The gliding direction is indicated by a black arrow. (A) The cytoskeletal 'jellyfish structure' can be divided into the bell and tentacles. The bell is composed of a hexagonal lattice with a periodicity of 12nm, and the tentacles are covered with particles 20nm in diameter at intervals of approximately 30nm. The genes encoding the component proteins are indicated in brackets. Seven proteins, encoded by MMOB1620, MMOB1640, MMOB1650, MMOB1660, MMOB0150, MMOB4530 and MMOB5430, have been identified as components of the jellyfish structure, while their localization on the jellyfish is still unknown. The proteins more directly related to the gliding mechanism, namely Gli123, Gli349 and Gli521, encoded by MMOB1020, MMOB1030 and MMOB1040, form a complex at the cell surface. Note that the gliding mechanism complex cannot be seen on the surface of the 'bell'. (B) The cytoskeletal 'rod' in the attachment organelle can be divided into three parts, the terminal button, the paired plate and the wheel. See the text and Table 12.1 for a list of the component proteins.

**Table 12.1** Proteins associated with gliding and their orthologues in three species of mycoplasmas

| | *M. pneumoniae* | | *M. genitalium* | | *M. gallisepticum* | |
|---|---|---|---|---|---|---|
| Subcellular position | Name | MPN Code | Name | Mg Code | Name | MGA code |
| Surface on whole organelle | P1 adhesin | 141 | P140, MgPa, MgpB | 191 | GapA | 0928 |
| | P90, P40 | 142 | P110, MgpC | 192 | CrmA | 0939 |
| Surface at tip end | P30 | 453 | | 318 | MGC2 | 0932 |
| Terminal button | P65 | 309 | | 217 | PlpA | 1199 |
| Terminal button–paired plate interface | HMW3 | 452 | | 317 | Hlp3 | 0928 |
| Paired plate | HMW1 | 447 | | 312 | | 0306 |
| | HMW2 | 310 | | 218 | | 1203 |
| Wheel | TopJ | 119 | | 200 | | 1228 |
| | P200 | 567 | | 386 | | 0205 |
| | P41 | 311 | | 218–1 | | NF* |
| | P24 | 312 | | NF | | NF |

NF, not found. The genomes of *M. pneumoniae*, *M. genitalium* and *M. gallisepticum* have sizes of 816,394, 580,074 and 996,422 base pairs, respectively (Fraser *et al.*, 1995; Dandekar *et al.*, 2000; Papazisi *et al.*, 2003).

fluorescence microscopy, HMW1 (MPN447) and HMW2 (MPN310), required for the early stage of organelle formation (Popham *et al.*, 1997; Hahn *et al.*, 1998; Seto *et al.*, 2001; Burgos *et al.*, 2007, 2008), can be localized to this region (Seto *et al.*, 2001; Balish *et al.*, 2003; Seto and Miyata, 2003; Kenri *et al.*, 2004; Bose *et al.*, 2009). The amino acid sequence of HMW2 is predicted to result in a high content of coiled-coils (as high as 69%) (Krause and Balish, 2004), suggesting that the paired plate is mainly assembled from this protein.

### Wheel complex

Hegermann *et al.* suggested a striking model based on examination of cryosections by EM (Hegermann *et al.*, 2002; Mayer, 2006). In their model, the striated paired plate is attached at the proximal end to a 'wheel complex' with fibrils, which connect the complex to the periphery of the cell. A structure, called the 'bowl', has been found at the position of the striated paired plates by CET, although the fibrils were not seen (Henderson and Jensen, 2006; Seybert *et al.*, 2006). Fibrils could also be observed in negatively stained EM images of the 'asymmetrical dumbbell' isolated from *M. gallisepticum* cells, suggesting that fibrils are likely to be present in the rod in *M. pneumoniae* (Nakane and Miyata, 2009). While the original structure is well conserved in studies using CET, the reduced image contrast renders identification of thin fibres in a cell more difficult.

P41(MPN311), P24 (MPN312) P200 (MPN567), and TopJ (MPN119) are localized to the proximal end of the attachment organelle, corresponding to the wheel position (Kenri *et al.*, 2004; Jordan *et al.*, 2007; Cloward and Krause, 2009), suggesting that these are the components of the wheel. The wheel is likely to have a role in connecting the attachment organelle to the other parts of the cell, because in mutant strains lacking P41, which is encoded in tandem with P24, the attachment organelle occasionally detaches from the cell body and glides independently (Hasselbring and Krause, 2007a,b). The wheel may also be involved in the generation or transmission of force, because mutation of either P200 or its orthologue in *M. genitalium*, mg386, results in mutants with an adhesive, but non-gliding, phenotype (Pich *et al.*, 2006; Jordan *et al.*, 2007).

### Translucent area

In images of sections, the electron-dense core corresponding to the 'rod' in negatively stained images is surrounded by an electron lucent area, from which the dense complexes are excluded (Wilson and Collier, 1976; Shimizu and Miyata, 2002; Seto and Miyata, 2003). Hegermann *et al.* examined the structure of this area by treating fixed cells with Triton X-100, and suggested that spoke structures connect the electron-dense core and the periphery of the cell (Hegermann *et al.*, 2002). Using CET, Seybert *et al.* found the proximal end of the electron-dense core to be attached to the membrane, suggesting the formation of a special space (Seybert *et al.*, 2006). However, Henderson and Jensen proposed that the translucent area was caused by the exclusion of macromolecules by the repeated movement of the electron-dense core (Henderson and Jensen, 2006). Recently, we examined cells fixed under various conditions and concluded that the translucent area is much smaller than that observed in the images of sections (unpublished).

## Conclusions about the cellular structure

The 'attachment and gliding organelle' of *M. pneumoniae* shares many of the features of that of *M. mobile*. Firstly, the organelle is formed by the cytoskeletal and surface structures and excludes the cellular components unrelated to binding and gliding (Miyata and Ogaki, 2006; Seto *et al.*, 2001). Secondly, the organelle is supported by a cytoskeleton featuring a solid end structure connected to a longitudinal rod or fibrils. If the tentacles of *M. mobile* were bundled, the cytoskeletal jellyfish structure would appear similar to the rod of *M. pneumoniae*, although it would lack the wheel. As the cytoskeletal structures of the species in the *pneumoniae* subgroup vary, and some species have a cytoskeletal structure with greater similarity to that of *M. mobile* (Hatchel and Balish, 2008; Nakane and Miyata, 2009).

## Prospects

We have been unable to detect any similarities between the amino acid sequences of proteins associated with the gliding mechanisms of *M. mobile* and those associated with gliding in species in the *pneumoniae* subgroup (Dandekar *et al.*, 2000; Barre *et al.*, 2004; Jaffe *et al.*, 2004b; Miyata, 2010). However, this does not rule out the possibility that the organelles of gliding in mycoplasma species have evolved from the same origins. If the whole structure of each part of the machinery, rather than the spatial positioning of amino acid side chains is more critical for its function, it is possible that the amino acid sequences of homologous proteins may have differentiated during the extended evolution of these systems. Moreover, rapidly growing organisms like bacteria evolve faster than organisms with more extended intergenerational times, sometimes resulting in difficulty tracing the evolution of genes through the predicted amino acid sequences of their products. The adhesion systems of Mollicutes might be expected to evolve very rapidly, because adhesion is a crucial determinant of survival during their interaction with their hosts. It is possible that the three dimensional structures of regions required for specific functions, such as ATP hydrolysis and binding to sialylated oligosaccharides, may be conserved. Clarification of the structures of the different components of the gliding machinery may answer these questions, as has been seen previously in studies of cytoskeletal proteins and conventional motor proteins (Kull *et al.*, 1996; Michie and Lowe, 2006).

The information essential for further elucidation of the gliding mechanism of members of the *pneumoniae* subgroup can be classified into two categories. Firstly, how do the legs perform the displacements along the sialylated oligosaccharides fixed on solid surfaces? As the P1 adhesin complex is responsible for this displacement, elucidation of the binding site for the sialylated oligosaccharide on the P1 adhesin complex will be critical to obtaining a concrete image of this process. The role of P30 in gliding and the mechanism used to determine the direction of movement may also be critical. Secondly, how is the force generated by the 'motor' transmitted through the other structures to the leg and the cell body? To answer this, we need information about the energy source, the motor protein, and the roles of the rod in this process. Genetic studies can often provide such information efficiently in bacteria, because the role of the proteins can be inferred from the phenotypes of the mutants. However, it is necessary to distinguish direct effects from those derived from indirect effects, such as errors in assembly resulting from aberrant structure. Integrating information from different fields of studies, including genetics, structural studies, biochemical measurements and microscopy is likely to yield the correct answers.

## Acknowledgements

We are grateful to our collaborators who shared in this exciting research and to colleagues who provided valuable comments and encouragement. We would also like to thank Dr Tsuyoshi Kenri, Ms Lisa Matsuo and Mr Taishi Kasai for valuable discussion. Our studies have been supported by grants from the Ministry of Education, Science, Sports, Culture, and Technology of Japan and by a grant from the Institute for Fermentation, Osaka.

## References

Adan-Kubo, J., Uenoyama, A., Arata, T., and Miyata, M. (2006). Morphology of isolated Gli349, a leg protein responsible for glass binding of *Mycoplasma mobile* gliding revealed by rotary-shadowing electron microscopy. J. Bacteriol. *188*, 2821–2828.

Adan-Kubo, J., Yoshii, S.H., Kono, H., and Miyata, M. (2012). Molecular Structure of Isolated MvspI, a Variable Surface Protein of the Fish Pathogen *Mycoplasma mobile*. J. Bacteriol. *194*, 3050–3057.

Balish, M.F., and Krause, D.C. (2006). Mycoplasmas: a distinct cytoskeleton for wall-less bacteria. J. Mol. Microbiol. Biotechnol. *11*, 244–255.

Balish, M.F., Santurri, R.T., Ricci, A.M., Lee, K.K., and Krause, D.C. (2003). Localization of *Mycoplasma pneumoniae* cytadherence-associated protein HMW2 by fusion with green fluorescent protein: implications for attachment organelle structure. Mol. Microbiol. 47, 49–60.

Barre, A., de Daruvar, A., and Blanchard, A. (2004). MolliGen, a database dedicated to the comparative genomics of Mollicutes. Nucleic Acids Res. 32, D307–310.

Baseman, J.B., Banai, M., and Kahane, I. (1982a). Sialic acid residues mediate *Mycoplasma pneumoniae* attachment to human and sheep erythrocytes. Infect. Immun. 38, 389–391.

Baseman, J.B., Cole, R.M., Krause, D.C., and Leith, D.K. (1982b). Molecular basis for cytadsorption of *Mycoplasma pneumoniae*. J. Bacteriol. *151*, 1514–1522.

Bose, S.R., Balish, M.F., and Krause, D.C. (2009). *Mycoplasma pneumoniae* cytoskeletal protein HMW2 and the architecture of the terminal organelle. J. Bacteriol. *191*, 6741–6748.

Bredt, W., and Radestock, U. (1977). Gliding motility of *Mycoplasma pulmonis*. J. Bacteriol. *130*, 937–938.

Burgos, R., Pich, O.Q., Ferrer-Navarro, M., Baseman, J.B., Querol, E., and Pinol, J. (2006). *Mycoplasma genitalium* P140 and P110 cytadhesins are reciprocally stabilized and required for cell adhesion and terminal-organelle development. J. Bacteriol. *188*, 8627–8637.

Burgos, R., Pich, O.Q., Querol, E., and Pinol, J. (2007). Functional analysis of the *Mycoplasma genitalium* MG312 protein reveals a specific requirement of the MG312 N-terminal domain for gliding motility. J. Bacteriol. *189*, 7014–7023.

Burgos, R., Pich, O.Q., Querol, E., and Pinol, J. (2008). Deletion of the *Mycoplasma genitalium* MG_217 gene modifies cell gliding behaviour by altering terminal organelle curvature. Mol. Microbiol. *69*, 1029–1040.

Catrein, I., Herrmann, R., Bosserhoff, A., and Ruppert, T. (2005). Experimental proof for a signal peptidase I like activity in *Mycoplasma pneumoniae*, but absence of a gene encoding a conserved bacterial type I SPase. FEBS J. 272, 2892–2900.

Chang, H.Y., Jordan, J.L., and Krause, D.C. (2011). Domain analysis of protein P30 in *Mycoplasma pneumoniae* cytadherence and gliding motility. J. Bacteriol. *193*, 1726–1733.

Chen, J., Neu, J., Miyata, M., and Oster, G. (2009). Motor–substrate interactions in *Mycoplasma* motility explains non-arrhenius temperature dependence. Biophys. J. *97*, 2930–2938.

Cloward, J.M., and Krause, D.C. (2009). *Mycoplasma pneumoniae* J-domain protein required for terminal organelle function. Mol. Microbiol. *71*, 1296–1307.

Dandekar, T., Huynen, M., Regula, J.T., Ueberle, B., Zimmermann, C.U., Andrade, M.A., Doerks, T., Sanchez-Pulido, L., Snel, B., Suyama, M., *et al.* (2000). Re-annotating the *Mycoplasma pneumoniae* genome sequence: adding value, function and reading frames. Nucleic Acids Res. *28*, 3278–3288.

Feldner, J., Gobel, U., and Bredt, W. (1982). *Mycoplasma pneumoniae* adhesin localized to tip structure by monoclonal antibody. Nature *298*, 765–767.

Fraser, C.M., Gocayne, J.D., White, O., Adams, M.D., Clayton, R.A., Fleischmann, R.D., Bult, C.J., Kerlavage, A.R., Sutton, G., Kelley, J.M., *et al.* (1995). The minimal gene complement of *Mycoplasma genitalium.* Science *270*, 397–403.

Göbel, U., Speth, V., and Bredt, W. (1981). Filamentous structures in adherent *Mycoplasma pneumoniae* cells treated with nonionic detergents. J. Cell. Biol. *91*, 537–543.

Hahn, T.W., Willby, M.J., and Krause, D.C. (1998). HMW1 is required for cytadhesin P1 trafficking to the attachment organelle in *Mycoplasma pneumoniae.* J. Bacteriol. *180*, 1270–1276.

Hasselbring, B.M., Jordan, J.L., and Krause, D.C. (2005). Mutant analysis reveals a specific requirement for protein P30 in *Mycoplasma pneumoniae* gliding motility. J. Bacteriol. *187*, 6281–6289.

Hasselbring, B.M., and Krause, D.C. (2007a). Cytoskeletal protein P41 is required to anchor the terminal organelle of the wall-less prokaryote *Mycoplasma pneumoniae.* Mol. Microbiol. *63*, 44–53.

Hasselbring, B.M., and Krause, D.C. (2007b). Proteins P24 and P41 function in the regulation of terminal-organelle development and gliding motility in *Mycoplasma pneumoniae.* J. Bacteriol. *189*, 7442–7449.

Hatchel, J.M., and Balish, M.F. (2008). Attachment organelle ultrastructure correlates with phylogeny, not gliding motility properties, in *Mycoplasma pneumoniae* relatives. Microbiology *154*, 286–295.

Hegermann, J., Herrmann, R., and Mayer, F. (2002). Cytoskeletal elements in the bacterium *Mycoplasma pneumoniae.* Naturwissenschaften *89*, 453–458.

Henderson, G.P., and Jensen, G.J. (2006). Three-dimensional structure of *Mycoplasma pneumoniae's* attachment organelle and a model for its role in gliding motility. Mol. Microbiol. *60*, 376–385.

Hiratsuka, Y., Miyata, M., and Uyeda, T.Q.P. (2005). Living microtransporter by uni-directional gliding of *Mycoplasma* along microtracks. Biochem. Biophys. Res. Commun. *331*, 318–324.

Hiratsuka, Y., Miyata, M., Tada, T., and Uyeda, T.Q.P. (2006). A microrotary motor powered by bacteria. Proc. Natl. Acad. Sci. U.S.A. *103*, 13618–13623.

Hu, P.C., Cole, R.M., Huang, Y.S., Graham, J.A., Gardner, D.E., Collier, A.M., and Clyde, W.A., Jr. (1982). *Mycoplasma pneumoniae* infection: role of a surface protein in the attachment organelle. Science *216*, 313–315.

Iverson-Cabral, S.L., Astete, S.G., Cohen, C.R., Rocha, E.P., and Totten, P.A. (2006). Intrastrain heterogeneity of the mgpB gene in *Mycoplasma genitalium* is extensive in vitro and in vivo and suggests that variation is generated via recombination with repetitive chromosomal sequences. Infect. Immun. *74*, 3715–3726.

Jaffe, J.D., Miyata, M., and Berg, H.C. (2004a). Energetics of gliding motility in *Mycoplasma mobile.* J. Bacteriol. *186*, 4254–4261.

Jaffe, J.D., Stange-Thomann, N., Smith, C., DeCaprio, D., Fisher, S., Butler, J., Calvo, S., Elkins, T., FitzGerald, M.G., Hafez, N., *et al.* (2004b). The complete genome and proteome of *Mycoplasma mobile.* Genome Res. *14*, 1447–1461.

Jarrell, K.F., and McBride, M.J. (2008). The surprisingly diverse ways that prokaryotes move. Nat. Rev. Microbiol. *6*, 466–476.

Jordan, J.L., Chang, H.Y., Balish, M.F., Holt, L.S., Bose, S.R., Hasselbring, B.M., Waldo, R.H., 3rd, Krunkosky, T.M., and Krause, D.C. (2007). Protein P200 is dispensable for *Mycoplasma pneumoniae* hemadsorption but not gliding motility or colonization of differentiated bronchial epithelium. Infect. Immun. *75*, 518–522.

Kasai, T., Nakane, D., Ishida, H., Ando, H., Kiso, M., and Miyata, M. (2013). Role of binding in *Mycoplasma mobile* and *Mycoplasma pneumoniae* gliding analyzed through inhibition by synthesized sialyated compounds. J. Bacteriol. *195(3)*, 429–435.

Kenri, T., Taniguchi, R., Sasaki, Y., Okazaki, N., Narita, M., Izumikawa, K., Umetsu, M., and Sasaki, T. (1999). Identification of a new variable sequence in the P1 cytadhesin gene of *Mycoplasma pneumoniae*: evidence for the generation of antigenic variation by DNA recombination between repetitive sequences. Infect. Immun. *67*, 4557–4562.

Kenri, T., Seto, S., Horino, A., Sasaki, Y., Sasaki, T., and Miyata, M. (2004). Use of fluorescent-protein tagging to determine the subcellular localization of *Mycoplasma pneumoniae* proteins encoded by the cytadherence regulatory locus. J. Bacteriol. *186*, 6944–6955.

Kirchhoff, H. (1992). Motility. In Mycoplasmas – Molecular Biology and Pathogenesis, Maniloff, J. McElhaney, R.N. Finch, L.R., and Baseman, J.B. eds. (American Society for Microbiology, Washington, D.C.), pp. 289–306.

Kirchhoff, H., and Rosengarten, R. (1984). Isolation of a motile mycoplasma from fish. J. Gen. Microbiol. *130*, 2439–2445.

Kirchhoff, H., Rosengarten, R., Lotz, W., Fischer, M., and Lopatta, D. (1984). Flask-shaped mycoplasmas: properties and pathogenicity for man and animals. Isr. J. Med. Sci. *20*, 848–853.

Krause, D.C., and Balish, M.F. (2004). Cellular engineering in a minimal microbe: structure and assembly of the terminal organelle of *Mycoplasma pneumoniae*. Mol. Microbiol. *51*, 917–924.

Krunkosky, T.M., Jordan, J.L., Chambers, E., and Krause, D.C. (2007). *Mycoplasma pneumoniae* host–pathogen studies in an air-liquid culture of differentiated human airway epithelial cells. Microb. Pathog. *42*, 98–103.

Kull, F.J., Sablin, E.P., Lau, R., Fletterick, R.J., and Vale, R.D. (1996). Crystal structure of the kinesin motor domain reveals a structural similarity to myosin. Nature *380*, 550–555.

Kusumoto, A., Seto, S., Jaffe, J.D., and Miyata, M. (2004). Cell surface differentiation of *Mycoplasma mobile* visualized by surface protein localization. Microbiology *150*, 4001–4008.

Layh-Schmitt, G., and Herrmann, R. (1992). Localization and biochemical characterization of the ORF6 gene product of the *Mycoplasma pneumoniae* P1 operon. Infect. Immun. *60*, 2906–2913.

Layh-Schmitt, G., Podtelejnikov, A., and Mann, M. (2000). Proteins complexed to the P1 adhesin of *Mycoplasma pneumoniae*. Microbiology *146*, 741–747.

Lesoil, C., Nonaka, T., Sekiguchi, H., Osada, T., Miyata, M., Afrin, R., and Ikai, A. (2009). Molecular shape and binding force of *Mycoplasma mobile's* leg protein Gli349 revealed by an AFM study. Biochem. Biophys. Res. Commun. *391*, 2930–2938.

Mayer, F. (2006). Cytoskeletal elements in bacteria *Mycoplasma pneumoniae, Thermoanaerobacterium* sp., and *Escherichia coli* as revealed by electron microscopy. J. Mol. Microbiol. Biotechnol. *11*, 228–243.

Meng, K.E., and Pfister, R.M. (1980). Intracellular structures of *Mycoplasma pneumoniae* revealed after membrane removal. J. Bacteriol. *144*, 390–399.

Metsugi, S., Uenoyama, A., Adan-Kubo, J., Miyata, M., Yura, K., Kono, H., and Go, N. (2005). Sequence analysis of the gliding protein Gli349 in *Mycoplasma mobile*. Biophysics *1*, 33–43.

Michie, K.A., and Lowe, J. (2006). Dynamic filaments of the bacterial cytoskeleton. Annu. Rev. Biochem. *75*, 467–492.

Miyata, M. (2005). Gliding motility of mycoplasmas – the mechanism cannot be explained by current biology. In Mycoplasmas: Pathogenesis, Molecular Biology, and Emerging Strategies for Control, Blanchard, A., and Browning, G., eds. (Horizon Biosciene, Norfolk, U.K.), pp. 137–163.

Miyata, M. (2007). Molecular mechanism of mycoplasma gliding – a novel cell motility system. In Cell Motility, Lenz, P., ed. (Springer, New York), pp. 137–175.

Miyata, M. (2008). Centipede and inchworm models to explain *Mycoplasma* gliding. Trends Microbiol. *16*, 6–12.

Miyata, M. (2010). Unique centipede mechanism of *Mycoplasma* gliding. Annu. Rev. Microbiol. *64*, 519–537.

Miyata, M., and Ogaki, H. (2006). Cytoskeleton of Mollicutes. J. Mol. Microbiol. Biotechnol. *11*, 256–264.

Miyata, M., and Petersen, J. (2004). Spike structure at interface between gliding *Mycoplasma mobile* cell and glass surface visualized by rapid-freeze and fracture electron microscopy. J. Bacteriol. *186*, 4382–4386.

Miyata, M., and Uenoyama, A. (2002). Movement on the cell surface of gliding bacterium, *Mycoplasma mobile*, is limited to its head-like structure. FEMS Microbiol. Lett. *215*, 285–289.

Miyata, M., Yamamoto, H., Shimizu, T., Uenoyama, A., Citti, C., and Rosengarten, R. (2000). Gliding mutants of *Mycoplasma mobile*: relationships between motility and cell morphology, cell adhesion and microcolony formation. Microbiology *146*, 1311–1320.

Miyata, M., Ryu, W.S., and Berg, H.C. (2002). Force and velocity of *Mycoplasma mobile* gliding. J. Bacteriol. *184*, 1827–1831.

Nagai, R., and Miyata, M. (2006). Gliding motility of *Mycoplasma mobile* can occur by repeated binding to N-acetylneuraminyllactose (sialyllactose) fixed on solid surfaces. J. Bacteriol. *188*, 6469–6475.

Nakane, D., and Miyata, M. (2007). Cytoskeletal 'Jellyfish' structure of *Mycoplasma mobile*. Proc. Natl. Acad. Sci. U.S.A. *104*, 19518–19523.

Nakane, D., and Miyata, M. (2009). Cytoskeletal asymmetrical-dumbbell structure of a gliding mycoplasma, *Mycoplasma gallisepticum*, revealed by negative-staining electron microscopy. J. Bacteriol. *191*, 3256–3264.

Nakane, D., and Miyata, M. (2012). *Mycoplasma mobile* cells elongated by detergent and their pivoting movements in gliding. J. Bacteriol. *194*, 122–130.

Nakane, D., Adan-Kubo, J., Kenri, T., and Miyata, M. (2011). Isolation and characterization of P1 adhesin, a leg protein of the gliding bacterium *Mycoplasma pneumoniae*. J. Bacteriol. *193*, 715–722.

Nonaka, T., Adan-Kubo, J., and Miyata, M. (2010). Triskelion structure of Gli521 protein involved in gliding mechanism of *Mycoplasma mobile*. J. Bacteriol. *191*, 636–642.

Ohtani, N., and Miyata, M. (2007). Identification of a novel nucleoside triphosphatase from *Mycoplasma mobile*: a prime candidate for motor of gliding motility. Biochem. J. *403*, 71–77.

Papazisi, L., Gorton, T.S., Kutish, G., Markham, P.F., Browning, G.F., Nguyen, D.K., Swartzell, S., Madan, A., Mahairas, G., and Geary, S.J. (2003). The complete genome sequence of the avian pathogen *Mycoplasma gallisepticum* strain R(low). Microbiology *149*, 2307–2316.

Peterson, S.N., Bailey, C.C., Jensen, J.S., Borre, M.B., King, E.S., Bott, K.F., and Hutchison, C.A., 3rd (1995). Characterization of repetitive DNA in the *Mycoplasma genitalium* genome: possible role in the generation of antigenic variation. Proc. Natl. Acad. Sci. U.S.A. *92*, 11829–11833.

Pich, O.Q., Burgos, R., Ferrer-Navarro, M., Querol, E., and Pinol, J. (2006). *Mycoplasma genitalium* mg200 and mg386 genes are involved in gliding motility but not in cytadherence. Mol. Microbiol. *60*, 1509–1519.

Pich, O.Q., Burgos, R., Ferrer-Navarro, M., Querol, E., and Pinol, J. (2008). Role of *Mycoplasma genitalium* MG218 and MG317 cytoskeletal proteins in terminal organelle organization, gliding motility and cytadherence. Microbiology *154*, 3188–3198.

Pich, O.Q., Burgos, R., Querol, E., and Pinol, J. (2009). P110 and P140 cytadherence-related proteins are negative effectors of terminal organelle duplication in *Mycoplasma genitalium*. PLoS One *4*, e7452.

Popham, P.L., Hahn, T.W., Krebes, K.A., and Krause, D.C. (1997). Loss of HMW1 and HMW3 in noncytadhering mutants of *Mycoplasma pneumoniae* occurs post-translationally. Proc. Natl. Acad. Sci. U.S.A. *94*, 13979–13984.

Razin, S., Yogev, D., and Naot, Y. (1998). Molecular biology and pathogenicity of mycoplasmas. Microbiol. Mol. Biol. Rev. *62*, 1094–1156.

Relich, R.F., Friedberg, A.J., and Balish, M.F. (2009). Novel cellular organization in a gliding mycoplasma, *Mycoplasma insons*. J. Bacteriol. *191*, 5312–5314.

Roberts, D.D., Olson, L.D., Barile, M.F., Ginsburg, V., and Krivan, H.C. (1989). Sialic acid-dependent adhesion of *Mycoplasma pneumoniae* to purified glycoproteins. J. Biol. Chem. *264*, 9289–9293.

Rosengarten, R., and Kirchhoff, H. (1987). Gliding motility of *Mycoplasma* sp. nov. Strain 163K. J. Bacteriol. *169*, 1891–1898.

Rosengarten, R., Klein-Struckmeier, A., and Kirchhoff, H. (1988). Rheotactic behavior of a gliding mycoplasma. J. Bacteriol. *170*, 989–990.

Sato, C., Manaka, S., Nakane, D., Nishiyama, H., Suga, M., Nishizaka, T., Miyata, M., and Maruyama, Y. (2012). Rapid imaging of mycoplasma in solution using Atmospheric Scanning Electron Microscopy (ASEM). Biochem. Biophys. Res. Commun. *417*, 1213–1218.

Seto, S., and Miyata, M. (2003). The attachment organelle formation represented by localization of cytadherence protein and formation of electron-dense core in the wild-type and mutant strains of *Mycoplasma pneumoniae*. J. Bacteriol. *185*, 1082–1091.

Seto, S., Layh-Schmitt, G., Kenri, T., and Miyata, M. (2001). Visualization of the attachment organelle and cytadherence proteins of *Mycoplasma pneumoniae* by immunofluorescence microscopy. J. Bacteriol. *183*, 1621–1630.

Seto, S., Kenri, T., Tomiyama, T., and Miyata, M. (2005a). Involvement of P1 adhesin in gliding motility of *Mycoplasma pneumoniae* as revealed by the inhibitory effects of antibody under optimized gliding conditions. J. Bacteriol. *187*, 1875–1877.

Seto, S., Uenoyama, A., and Miyata, M. (2005b). Identification of 521-kilodalton protein (Gli521) involved in force generation or force transmission for *Mycoplasma mobile* gliding. J. Bacteriol. *187*, 3502–3510.

Seybert, A., Herrmann, R., and Frangakis, A.S. (2006). Structural analysis of *Mycoplasma pneumoniae* by cryo-electron tomography. J. Struct. Biol. *156*, 342–354.

Shimizu, T., and Miyata, M. (2002). Electron microscopic studies of three gliding mycoplasmas, *Mycoplasma mobile, M. pneumoniae*, and *M. gallisepticum*, by using the freeze-substitution technique. Curr. Microbiol. *44*, 431–434.

Sokurenko, E.V., Vogel, V., and Thomas, W.E. (2008). Catch-bond mechanism of force-enhanced adhesion: counterintuitive, elusive, but... widespread? Cell Host Microbe *4*, 314–323.

Springer, T.A. (2009). Structural basis for selectin mechanochemistry. Proc. Natl. Acad. Sci. U.S.A. *106*, 91–96.

Spuesens, E.B., Oduber, M., Hoogenboezem, T., Sluijter, M., Hartwig, N.G., van Rossum, A.M., and Vink, C. (2009). Sequence variations in RepMP2/3 and RepMP4 elements reveal intragenomic homologous DNA recombination events in *Mycoplasma pneumoniae*. Microbiology *155*, 2182–2196.

Stadtlander, C.T., and Kirchhoff, H. (1995). Attachment of *Mycoplasma mobile* 163 K to piscine gill arches and rakers-light, scanning and transmission electron microscopic findings. Br. Vet. J. *151*, 89–100.

Stadtlander, C.T., Lotz, W., Korting, W., and Kirchhoff, H. (1995). Piscine gill epithelial cell necrosis due to *Mycoplasma mobile* strain 163 K: comparison of *in-vivo* and *in-vitro* infection. J. Comp. Pathol. *112*, 351–359.

Uenoyama, A., and Miyata, M. (2005a). Gliding ghosts of *Mycoplasma mobile*. Proc. Natl. Acad. Sci. U.S.A. *102*, 12754–12758.

Uenoyama, A., and Miyata, M. (2005b). Identification of a 123-kilodalton protein (Gli123) involved in machinery for gliding motility of *Mycoplasma mobile*. J. Bacteriol. *187*, 5578–5584.

Uenoyama, A., Kusumoto, A., and Miyata, M. (2004). Identification of a 349-kilodalton protein (Gli349) responsible for cytadherence and glass binding during gliding of *Mycoplasma mobile*. J. Bacteriol. *186*, 1537–1545.

Uenoyama, A., Seto, S., Nakane, D., and Miyata, M. (2009). Regions on Gli349 and Gli521 protein molecules directly involved in movements of *Mycoplasma mobile* gliding machinery suggested by inhibitory antibodies and mutants. J. Bacteriol. *191*, 1982–1985.

Weisburg, W.G., Tully, J.G., Rose, D.L., Petzel, J.P., Oyaizu, H., Yang, D., Mandelco, L., Sechrest, J., Lawrence, T.G., Van Etten, J., *et al.* (1989). A phylogenetic analysis of the mycoplasmas: basis for their classification. J. Bacteriol. *171*, 6455–6467.

Wilson, M.H., and Collier, A.M. (1976). Ultrastructural study of *Mycoplasma pneumoniae* in organ culture. J. Bacteriol. *125*, 332–339.

Wu, H.N., Kawaguchi, C., Nakane, D., and Miyata, M. (2012). 'Mycoplasmal antigen modulation,' a novel surface variation suggested for a lipoprotein specifically localized on *Mycoplasma mobile*. Curr. Microbiol. *64*, 433–440.

Wu, H.N., and Miyata, M. (2012). Whole surface image of *Mycoplasma mobile*, suggested by protein identification and immunofluorescence microscopy. J. Bacteriol. *194(21)*, 5848–5855.

# Biofilm Formation by Mycoplasmas

13

Laura McAuliffe

### Abstract

Previously it has been difficult to explain how mycoplasmas manage to cause such severe and chronic infections given their paucity of virulence factors. In many other bacterial species adherence to a solid surface and biofilm formation are important steps in the initiation of disease. The possibility of environmental persistence and virulence in the host could also be explained by biofilm formation in mycoplasmas. Biofilms are sessile bacterial communities that live attached to each other and/or surfaces enclosed in a sugary exopolysaccharide matrix. Biofilm structure is highly variable and dependent on a number of factors, including the organism, the surface, the surrounding nutrient environment and the rate of flow of any aqueous interface. Biofilms are formed by the vast majority of mycoplasma species studied to date and the capacity to form biofilms is found in diverse species from all phylogenetic groups of mycoplasmas. Intriguingly, mycoplasmas lack all of the known regulatory systems that are involved in biofilm formation in other bacterial species, but recent research is beginning to unravel the genetic basis of biofilm formation. The growth of biofilms *in vitro* and the use of biofilm model systems are also discussed.

## Introduction

Biofilms form when bacteria adhere to a surface and secrete a polysaccharide matrix. They were first described by Anthony Van Leeuwenhoek in 1684 when he remarked on the vast accumulation of microorganisms in dental plaque in a report to the Royal Society of London: 'the number of these animalcules in the scurf of a man's teeth are so many that I believe they exceed the number of men in a kingdom' (Van Leeuwenhoek, 1684). Biofilms, which are virtually ubiquitous in nature and have been found in almost every bacterial species investigated to date, have fuelled extensive research in recent years. They are beneficial in many contexts, including in the correct function of the mammalian digestive system, in water pre-treatment systems and in the bioremediation of contaminated soils. However, biofilms are more often extremely detrimental and much research has focused on their role in chronic infections, particularly in association with indwelling medical devices, in the failure of wound healing and also in industrial contexts, where they contribute a huge economic burden due to micro-organism-induced corrosion, bio-fouling and blockage of industrial pipes.

A number of definitions of a biofilm have evolved over the years and most engender controversy within the field. A definition should be simple, broad and all encompassing to reflect the influence of different microbial communities, nutrient limitations and environmental factors on the resulting biofilm structure and morphology. A well-used definition, based on biofilm morphology, is that of Costerton (1995) who described them as: 'complex communities of microorganisms attached to a surface or interface enclosed in an exopolysaccharide matrix of microbial and host origin to produce a spatially organized three-dimensional structure'. This definition reflects the complexity of biofilm structure and morphology in many bacterial infections and is particularly well suited to biofilms found on indwelling medical devices such as catheters and stents. This definition may be too prescriptive for many purposes as it excludes *in vitro* biofilms, which lack the host component of the polysaccharide, and more simplistic monolayer biofilms found in some infections. A more general definition of 'bacterial cells attached to a surface, or each other, surrounded by a polysaccharide matrix' is probably preferable. It is crucial to remember that, regardless of the biofilm definition used, the most important aspect of a biofilm is that it is phenotypically different from planktonic cells, particularly in terms of resistance and persistence, and the physical structure and morphology of the biofilm is of secondary importance to this.

## Biofilms may help explain Mycoplasma persistence and chronicity

Despite their small size and tiny genome, mycoplasmas cause a wide range of disease in humans and animals. In animals particularly, mycoplasmas are typically causes of arthritis, mastitis, reproductive and respiratory disease of a chronic and persistent nature. Mycoplasmas are intriguing as they have very few known virulence factors, such as the toxins, cytolysins and invasins seen in other bacteria, yet they can cause severe disease. Despite extensive research, virulence factors have only been found in a few mycoplasma species and these include the production of hydrogen peroxide (Miles, *et al.*, 1991), which causes oxidative damage, the carbohydrate capsule (Almeida *et al.*, 1992), which can be inflammatory, the ability to scavenge and thus deprive host cells of arginine (Sasaki *et al.*, 1984), T-cell mitogens (Tu *et al.*, 2005), which disrupt the immune system and can cause toxic shock, a secreted sialidase (May and Brown, 2009) and a cytotoxic nuclease, which may induce apoptosis (Somaranjan *et al.*, 2010). Despite this, many highly pathogenic mycoplasma species still have no known virulence factors. Previously, it has been difficult to explain how mycoplasmas manage to cause such severe and chronic infections given their paucity of virulence factors. In many other bacterial species adherence to a solid surface and biofilm formation are important steps in the initiation of disease. The possibility of environmental persistence and virulence in the host could also be explained by biofilm formation in mycoplasmas.

## Mycoplasma biofilms – a view from history

Even today very little is known about how mycoplasmas cause such severe disease and how they persist in the host or the environment. Biofilm formation by mycoplasmas was indirectly alluded to hundreds of years ago, even before mycoplasmas had been discovered by Nocard and Roux (1898). In 1896 when the cause of the disease was unknown, there were indications that *Mycoplasma mycoides* subsp. *mycoides* SC (MmmSC), the cause of contagious bovine pleuropneumonia, may persist in the environment. Salmon (1896) wrote, in a US government report on cattle disease, that 'many stables have been found in which the disease would appear and reappear after the slaughter of affected herds, and in spite of any precautions which were adopted. In every one of these cases the destruction of the stable, the burning of the lumber, the removal of the accumulations beneath the floors, and thorough disinfection prevented the recurrence of the plague in new stables built on the same premises. This experience conclusively shows that under certain conditions, at least, stables may retain the infection for a considerable time, and that when restocked the disease may break out again from such infection.' Similarly, contagious agalactia, a disease of sheep and goats characterized by mastitis, arthritis and keratoconjunctivitis and caused by *Mycoplasma agalactiae,* was known as 'mal di sito' ('disease of the place') in Italy because of its ability to contaminate the environment and infect successive flocks on a farm (Nicholas *et al.*, 2010). It has never been explained how mycoplasmas, which are seemingly so fragile and lack a rigid cell wall, could survive in the environment, but biofilm formation may provide an explanation.

## Biofilms represent a natural phenotype

Traditionally microbiologists have employed the use of liquid media for the culture and maintenance of bacteria in the laboratory. However, this method of cell growth is likely to be completely unnatural and offers an unrealistic model of the conditions the vast majority of bacteria experience in their natural context, either in the host, or in the environment. In the host and in the natural environment, bacteria do not grow in a nutrient-rich, stable and unchanging liquid environment, but more likely are exposed to stress of various forms and ever changing environmental conditions with fluctuations in nutrient concentrations. They would also usually be attached to some surface. Although it is not feasible to accurately model *in vitro* the changes in environmental conditions and fluctuations in nutrient availability that bacteria experience under natural circumstances, it is possible to offer a more realistic environment by providing a solid substratum to enable biofilm growth.

## Biofilm formation by Mycoplasma species

Studies of biofilm formation have shown that biofilms are formed by the vast majority of *Mycoplasma* species (McAuliffe *et al.*, 2006). At least 20 mycoplasma species from all phylogenetic groups of mycoplasmas have the ability to form biofilms. Mycoplasma species that form prolific biofilms include *M. capricolum* subspecies *capricolum*, *M. putrefaciens*, *M. bovis*, *M. yeatsii*, *M. cottewii*, *M. gallisepticum* and *M. pullorum* (summarized in Table 13.1). Biofilm formation has also been demonstrated in clinical isolates of *Ureaplasma urealyticum* and *Ureaplasma parvum* (Garcia-Castillo *et al.*, 2008) and in *M. fermentans*, which has been found to form a biofilm on intrauterine devices *in vitro* (Rivera *et al.*, 2009). Biofilm formation is not simply a laboratory phenomenon, but is likely to be an important step in disease initiation in the host, as it has recently been shown to occur *in vivo* in the mouse pathogen *M. pulmonis* (Simmons and Dybvig, 2009). Similarly, more recently *M. salivarium*, preferentially an inhabitant of the human oral cavity, has been found, together with *Candida glabrata*, in a multi-species biofilm in an occluded biliary stent of an icteric, cholestatic patient (Henrich *et al.*, 2010).

No confirmed link has yet been found between the ability to form a biofilm and virulence, with some non-pathogenic species, such as *M. cottewii* and *M. yeatsii*, forming prolific biofilms. Intriguingly some highly virulent species, which are known to have a polysaccharide capsule, did not form prolific biofilms in some biofilm systems, leading to the suggestion that the capsule may be hydrophobic and prevent adherence. The highly pathogenic species MmmSC does not form a prolific biofilm when cultured using a biofilm model system with an air–liquid interface. However, it does form a biofilm when grown on a membrane placed on an agar plate and exhibits all the characteristics of a 'true' biofilm, including a different phenotype with altered morphology, differential gene expression and increased resistance to stress (McAuliffe *et al.*, 2008).

Intriguingly, Minga (1981) described two colony types for MmmSC: a smooth colony type that was thought to be more virulent, causing more widespread disease, with arthritis and weight loss in addition to respiratory signs; and a rough colony type, which caused more localized disease signs, with only respiratory involvement. Rough colony types formed 'filaments and comets' in liquid media, whereas smooth colony types resulted in uniform turbidity. It may be that the rough colony type may have been capable of forming a biofilm and that the filaments and comets described by Minga were possibly the first account of biofilm formation by a mycoplasma species.

## Biofilm life cycle and morphology

A typical life cycle of a biofilm is illustrated in Fig. 13.1. The life cycle begins with the adherence of planktonic cells to a solid surface, or, in some instances, to each other. If conditions are favourable, cell multiplication occurs and cells begin to produce a polysaccharide matrix to surround the newly forming biofilm. In many bacterial species morphological differentiation then occurs, with cells forming a three-dimensional structure of stacks and channels, although in other bacterial species this differentiation may be more subtle or even absent, resulting in a simple biofilm monolayer. Cells may persist in a stable biofilm for a long period, possibly for many months or years;

**Table 13.1** Biofilm characteristics of various *Mycoplasma* species

| Mycoplasma species | Model system used | Biofilm formation ability/comments | Reference |
|---|---|---|---|
| *M. agalactiae* | GC, MWP, MP | Prolific growth at air–liquid interface, up to 6.2 × $10^6$ cfu/ml in GC model. Ability to form biofilm often lost upon passage. Large degree of intraspecific variation in biofilm ability seen | McAuliffe *et al.* (2006) |
| *M. bovis* | GC, MWP | Prolific growth at air–liquid interface, up to 1.3 ×$10^7$ cfu/ml in GC model. Ability to form biofilm often lost upon passage. Large degree of intraspecific variation in biofilm ability seen | McAuliffe *et al.* (2006) |
| *M. leachii* | GC | Moderate growth and staining at air–liquid interface with up to 2.8 × $10^3$ cfu/ml in GC model | McAuliffe *et al.* (2006) |
| *M. cotewii* | GC | Very prolific growth and strong staining at air–liquid interface with up to 4.6 × $10^7$ cfu/ml in GC model | McAuliffe *et al.* (2006) |
| *M. c. capricolum* | GC, MWP, MP | Moderate staining at air–liquid interface in GC model with up to 2.6 × $10^3$ cfu/ml. Prolific biofilm seen in some strains using MWP with collagen coating and CV staining | McAuliffe *et al.* (2006) |
| *M. ovipneumoniae* | GC | Very little staining at air–liquid interface seen using CV but up to 1.8 × $10^4$ cfu/ml in GC model | McAuliffe *et al.* (2006) |
| *M. m. capri* | GC, MWP, MP | Moderate staining at air–liquid interface with up to 5.5 × $10^5$ cfu/ml in GC model | McAuliffe *et al.* (2006) |
| *Mmm*SC | GC | Very little staining at air–liquid interface with only 2.1 × $10^1$ cfu/ml in GC model. Only grows well in model lacking an air–liquid interface. Biofilm morphology varies considerably dependent on strain | McAuliffe *et al.* (2006, 2008) |
| *M. putrefaciens* | GC, MWP, MP | Very prolific growth in GC model with dense staining using CV. Up to 8.7 × $10^6$ cfu/ml in GC model | McAuliffe *et al.* (2006) |
| *M. yeatsii* | GC | Very prolific growth in GC model with dense staining using CV. Up to 3.8 × $10^7$ cfu/ml in GC model | McAuliffe *et al.* (2006) |
| *M. gallisepticum* | GC, MWP, MP | Moderate staining using CV in GC model and staining below air–liquid interface. Up to 1 × $10^5$ cfu/ml | McAuliffe (unpublished) |
| *M. iowae* | GC | Moderate staining using CV in GC model with up to 1 × $10^4$ cfu/ml | McAuliffe (unpublished) |
| 355A05 – new species from penguins | GC | Very prolific growth in GC model with dense staining using CV. Up to 1 × $10^7$ cfu/ml | McAuliffe (unpublished) |
| *M. gallinarum* | GC | Relatively poor staining using CV at air–liquid interface with up to 1 × $10^4$ cfu/ml | McAuliffe (unpublished) |
| *M. gallopavonis* | GC | Good staining at air–liquid interface with 1x$10^6$ cfu/ml | McAuliffe (unpublished) |
| *M. pullorum* | GC | Good staining at air–liquid interface using CV and up to 1 × $10^7$ cfu/ml | McAuliffe (unpublished) |
| *M. lipofaciens* | GC | Moderate biofilm formation | McAuliffe (unpublished) |
| *M. cloacale* | GC | Poor staining using CV at air–liquid interface but moderate growth using cell counts with up to 1 × $10^4$ cfu/ml | McAuliffe (unpublished) |
| *M. columborale* | GC | Poor staining using CV at air–liquid interface but moderate growth using cell counts with up to 1 × $10^4$ cfu/ml | McAuliffe (unpublished) |
| *M. collumbinum* | GC | Poor staining using CV at air–liquid interface but moderate growth using cell counts with up to 1 × $10^4$ cfu/ml | McAuliffe (unpublished) |
| *M. hyopneumoniae* | GC | Poor staining using CV and only moderate growth, up to 1 × $10^5$ cfu/ml. Few field strains studied and may be affected by passage | McAuliffe (unpublished) |
| *M. hyorhinis* | GC | Poor staining using CV and only moderate growth, up to 1 × $10^4$ cfu/ml | McAuliffe (unpublished) |

| Mycoplasma species | Model system used | Biofilm formation ability/comments | Reference |
|---|---|---|---|
| *M. salivarium* | Polyethylene stents | Prolific growth seen on stents using SEM | Henrich *et al.* (2010) |
| *M. pulmonis* | GC, MWP, Tracheal Organ culture | Prolific growth in model systems, ability to form biofilm affected by length of Vsa protein | Simmons *et al.* (2004, 2007, 2009) |
| *M. fermentans* | GC, Intrauterine device (*in vitro*) | Growth seen on glass coverslips using CV staining | Riviera *et al.* (2009) |
| *Ureaplasmas* | MWP | | Garcia-Castillo *et al.* (2008) |

GC, glass coverslip; MWP, multi-well plate; MP, membrane on plate.

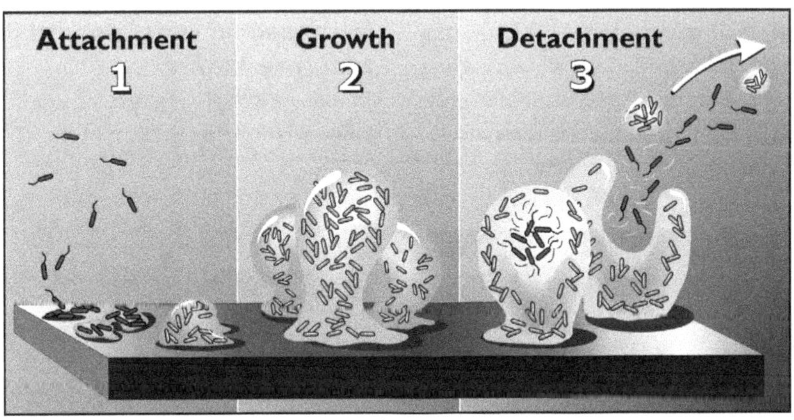


**Figure 13.1** The biofilm life cycle illustrated in three steps: initial attachment events, the growth of complex biofilms, and detachment events by clumps of bacteria or by a 'swarming' phenomenon within the interior of bacterial clusters, resulting in 'seeding dispersal' (reproduced with permission from Peg Dirckx, CBE, Montana State University, USA). A colour version of this figure is available in the plate section at the back of the book.

in a host this may be linked to chronic disease, as cells can lie dormant, protected from host cell defences and chemotherapy. The final stage in the maturing biofilm is dispersal, when cells revert to a planktonic form and leave the mature biofilm. In the host, this stage of dispersal may result in reversion from a chronic to an acute disease status, as newly released planktonic cells may cause infections in new sites or systemic infection. In many bacterial species cell signalling is thought to result in the transition between the different biofilm stages, but quorum sensing signals have not been discovered in mycoplasmas, so the role of bacterial communication in mycoplasma biofilms is yet to be established.

Biofilm structure is highly variable and dependent on a number of factors, including the organism, the surface, the surrounding nutrient environment and the rate of flow of any aqueous interface (Costerton *et al.*, 1999). Biofilms may vary in configuration, from sparse amorphous masses to highly complex, organized structures with mushroom-like cell stacks interspersed with fluid-filled channels. These channels have been compared with circulatory systems, leading to the view that biofilms can be considered analogous to primitive multicellular organisms (Costerton *et al.*, 1999).

Studies of the morphology of mycoplasma biofilms using confocal scanning laser microscopy have shown that they have a highly differentiated structure, like that seen in biofilms formed by 'higher' bacteria (McAuliffe *et al.*, 2006, 2008) (Fig. 13.2). In mycoplasma species which formed prolific biofilms in an air–liquid interface model system, such as *M. putrefaciens*, *M. bovis* and *M. c. capricolum*, stacks several hundred cells high, with channels in between, were seen in mature biofilms (McAuliffe *et al.*, 2006, 2008; unpublished). In many mycoplasma species cells initially adhered to the solid surface at and below the air–liquid interface and multiplied over a period of 48–72 hours to form small microcolonies visible using light microscopy and crystal violet staining. Over a period of 4–12 days these microcolonies grew to form towers, and in some species the space in between microcolonies became filled with a monolayer of cells.

Similarly, in *M. pulmonis*, the biofilm is described as having honeycombed regions consisting of thin layers of mycoplasmas pocketed with cavities. Interspersed throughout and arising out of the honeycombed regions are towers (Simmons *et al.*, 2007). Intriguingly, in *M. pulmonis* it has been found that, when biofilms are formed on glass, cells initially adhere as individual cells and then form small towers, followed by larger tower structures, and then finally a stage of continuous film is achieved (Simmons *et al.*, 2007). A final stage of continuous film following on from tower structures has not been seen in any other mycoplasma species studied to date.

## Intraspecific variation in ability to form a biofilm

In many mycoplasma species significant intraspecific variation can be seen in the capacity to form

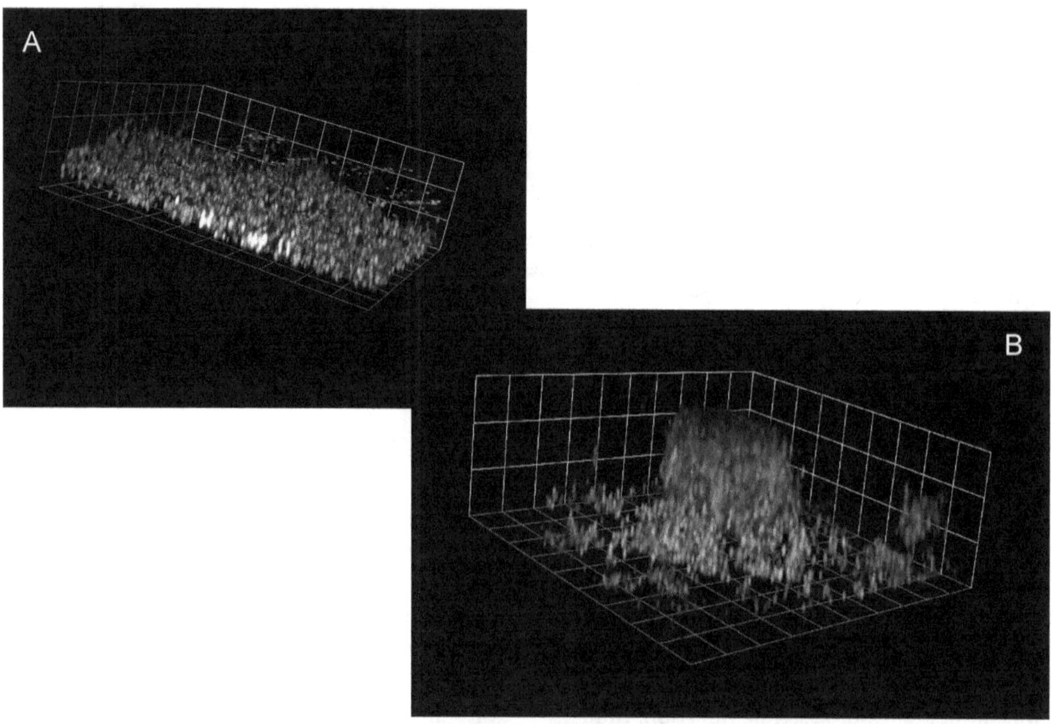


**Figure 13.2** Confocal scanning light micrographic image of Live/Dead stained *M. putrefaciens* biofilm showing (A) early stages of a biofilm with a monolayer of cells and (B) a more mature biofilm with stacks. A colour version of this figure is available in the plate section at the back of the book.

biofilms. MmmSC strains varied considerably in their ability to grow as an adherent layer on a membrane on an Eaton's agar plate. There were also differences in biofilm morphology between the strains examined, with strain Afadé and several others, including N6, growing very profusely and demonstrating an unusual morphology, with large micro-colonies or stack-like structures, many cells deep, interspersed amongst a flatter layer of cells covering the membrane. Strain V5 produced a densely packed, profuse biofilm, with smaller stacks than strain Afadé and very little space between stacks, which was, unusually, covered in a thick polysaccharide material containing DNA that was detected with a nucleic acid stain. Some European strains, including 2091, 6526 and 6512, and the African strain IS31, produced a flatter biofilm, which was still several cells deep with only a few microcolonies around the edge. Other European strains, including 138/5, B103, 197, Madrid and Segovia, grew poorly on membranes, with strains 138/5 and B103 exhibiting the poorest growth of all, with only a few adherent cells and no stacks or microcolonies, while strains 197, Madrid and Segovia produced sparse biofilms with a very thin monolayer (as shown in Fig. 13.3).

Similarly, studies with *M. bovis* and *M. agalactiae* have found intraspecific variation in capacity to form biofilms, with some field strains able to form a prolific biofilm, whereas others cannot. Interestingly, the type strains of both of these species only form moderate biofilms, and the ability to form a biofilm is often lost in field strains if they are sub-cultured two or more times (McAuliffe, unpublished). As there is such striking intraspecific variation in the capacity to form biofilms in

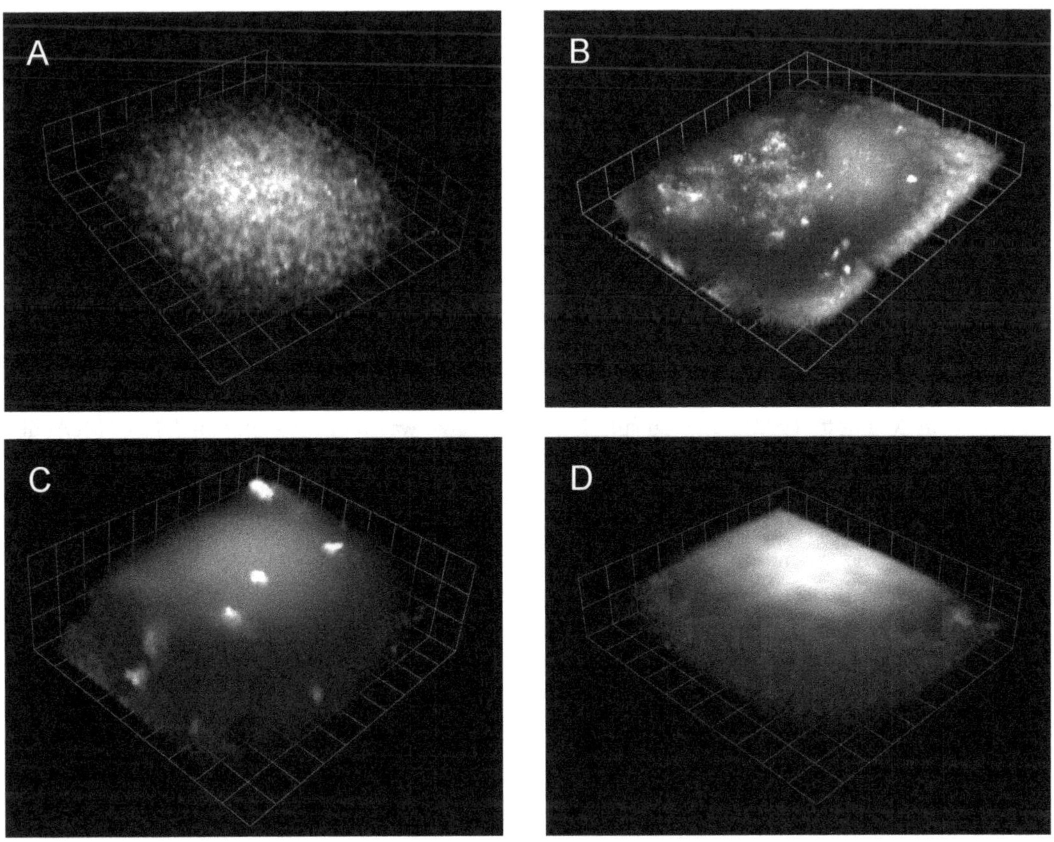


**Figure 13.3** Confocal scanning light micrographic images that illustrate intraspecific variation seen in MmmSC biofilms grown using a membrane on plate method without an air–liquid interface. (A) strain IS31, (B) strain Madrid, (C) strain N6, and (D) strain V5. A colour version of this figure is available in the plate section at the back of the book.

*M. bovis*, studies have been undertaken to ascertain whether there is a link between genotype, as determined using molecular epidemiological analysis, and the capacity to form a biofilm in *M. bovis*. It appears that strains with a good capacity to form a biofilm are often closely grouped in pulsed field gel electrophoretic analyses, indicating that there may be a genetic basis for the capacity to form a biofilm in *M. bovis* (Gosney, 2008).

## Role of polysaccharide

The exopolysaccharide (EPS) or glycocalyx layer is an essential component of the biofilm. It helps protect cells from desiccation, delays penetration of antibiotics and host cell defences and it has other structural and physiological roles. The EPS is often described as a 'slime' layer in other bacterial species and is not attached to the cell surface. The role and structure of Mollicute EPS has been subject to very little study over the years and has not been characterized in detail. It is thought that there is a high degree of variation in the chemical composition of EPS produced by different mycoplasma species.

In addition to EPS, some mycoplasmas produce a capsular polysaccharide that is more intimately associated with the cell. The capsule ranges in depth from 24 to 40 nm and contains high molecular weight polysaccharide that is attached to the cell surface, and in some cases extracellular fibrillar material may be seen (reviewed in Maniloff, 1992). The capsule is a polyanionic carbohydrate or lipid structure that enables binding to negatively charged surfaces and may help cells resist phagocytosis. It may also have a role in pathogenesis, inducing a toxic effect on host cells. Capsular structures have been described in MmmSC, *M. dispar*, *M. hyopneumoniae*, *M. gallisepticum*, *M. hominis*, *M. meleagridis*, *M. pneumoniae* and *U. urealyticum* (reviewed in Maniloff, 1992). Interestingly, studies using ruthenium red staining found that *M. leachii* (Boatman, 1979) and *M. bovoculi* (Salih and Rosenbusch, 1988) produced a thin capsule-like extracellular structure. In addition, *M. bovis* has been described as having a 73-kDa thermo-stable polysaccharide complex, consisting of glucose, glucosamine or

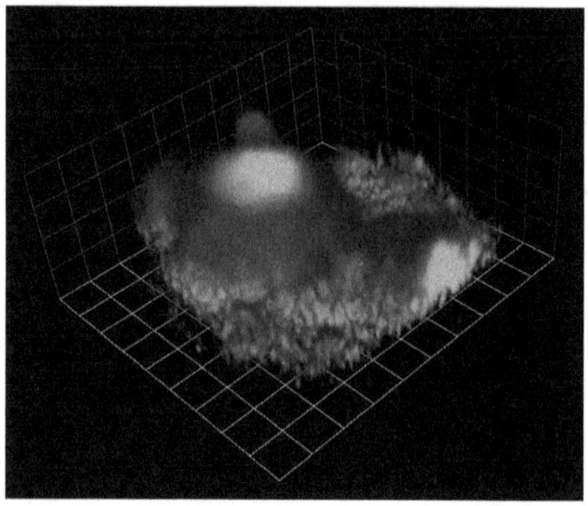

**Figure 13.4** Confocal scanning light micrographic with calcofluor white EPS staining of *M. putrefaciens* biofilm (L. McAuliffe). A colour version of this figure is available in the plate section at the back of the book.

galactosamine, and a heptose, localized on the cytoplasm membrane and tightly associated with proteins (Geary *et al.*, 1981).

All mycoplasma biofilms studied to date have been covered with a polysaccharide layer. The stain calcofluor white can be used to visualize EPS surrounding biofilms using confocal scanning laser microscopy (Fig. 13.4). Staining with calcofluor white indicates that the EPS specifically contains 1–4-β-D-glucan-based polysaccharides (Wood, 1980). In some mycoplasma species, including certain strains of MmmSC, the glycocalyx was found to contain DNA (McAuliffe *et al.*, 2008). Future studies to accurately define the precise nature of the biofilm glycocalyx are crucial. Studies using fluorescently labelled lectins have shown that in MmmSC the glycocalyx binds wheat germ agglutinin-conjugated (WGA) lectins (McAuliffe, unpublished results). In some strains of *M. bovis* an unusual striated structure was seen in lectin-stained biofilms, suggesting waves of cell growth over time (Fig. 13.5), almost akin to the swarming phenomenon seen in other bacterial species.

The role of EPS in the murine pathogen *M. pulmonis* has been studied in greater detail. *M. pulmonis* was found to bind *Griffonia simplicifolia* lectin I (GS-I), which is specific for terminal

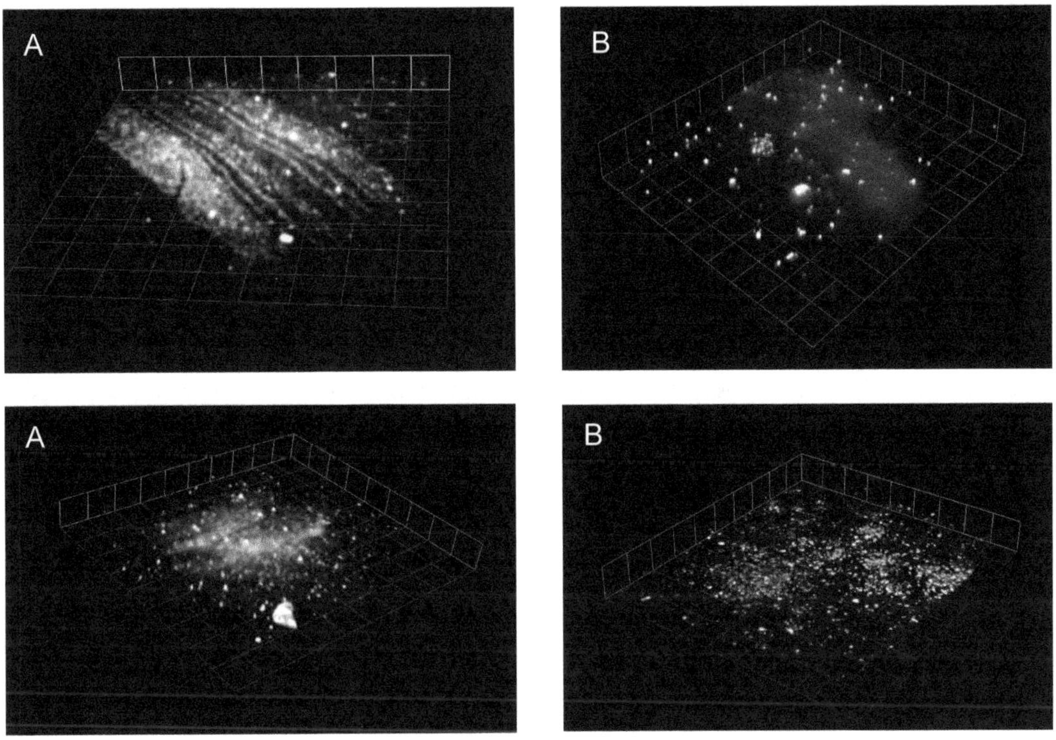


**Figure 13.5** Confocal scanning light micrographic image showing *M. bovis* biofilm with (A) conjugated wheat germ agglutinin lectin staining, and (B) Live/Dead staining (L. McAuliffe). A colour version of this figure is available in the plate section at the back of the book.

β-linked galactose residues. Mutants that failed to produce the EPS bound by GS-I were isolated from a transposon library. These mutants lacked overlapping genes that are predicted to code for a heterodimeric pair of ABC transporter permeases and may code for part of a new pathway for synthesis of EPS. It was demonstrated that the wild-type mycoplasma produced an EPS (EPS-I) composed of equimolar amounts of glucose and galactose that was lacking in the mutants. Phenotypic analysis revealed that the mutants had an increased propensity to form a biofilm on glass surfaces and colonized mouse lung and trachea efficiently, but had decreased capacity to bind to the A549 lung cell line. Confounding the interpretation of these results is the observation that the mutants lacking EPS-I had an eightfold overproduction of a second EPS (EPS-II) containing N-acetylglucosamine (Daubenspeck *et al.*, 2009).

It has also been proposed that in some instances the capsular polysaccharide may have an inhibitory role in biofilm formation. Some species, including MmmSC and *M. ovipneumoniae,* which produce a capsular polysaccharide, were unable to form a prolific biofilm in a model system using an air–liquid interface. It is feasible that the galactan-based capsule of MmmSC inhibits adherence *in vitro*. *M. ovipneumoniae* is also thought to produce a galactan-based capsule similar to MmmSC (Niang *et al.*, 1995). However, it has previously been suggested that the capsule may actually facilitate the adhesion of *M. ovipneumoniae* to the epithelium (Niang *et al.*, 1995). The idea that the polysaccharide capsule may actually inhibit adhesion has been proposed previously. In *Neisseria meningitidis* an inverse correlation is seen between capsule and biofilm formation (Yi *et al.*, 2004), and in *Vibrio vulnificans* capsule production has also been shown to prevent biofilm formation (Joseph and Wright, 2004). Further studies to ascertain the chemical composition and role of polysaccharide in biofilm formation are essential.

## Genetic basis of biofilm formation

It is often thought that in more complex bacteria programmed changes in gene expression result in a specific biofilm phenotype that is more resistant to stress and exhibits differential gene expression compared with planktonic cells. Most bacteria have enormous adaptive capacity and can modulate and reprogramme gene expression in response to a changing environment. In other bacteria master regulators that can switch on a different subset of genes, such as *rpo*S in *Escherichia coli* (Prigent-Combaret *et al.*, 2001), *spoA* of *Bacillus subtilis* (Hamon and Lazazzera, 2001), *sar*A of *Staphylococcus epidermidis* (Valle *et al.*, 2003) and *sin*R of *Bacillus subtilis* (Kearns *et al.*, 2005), are crucial in biofilm formation. Bacterial communication via quorum sensing is essential for gene regulation during biofilm formation in many bacterial species. Intriguingly, mycoplasmas lack all of the known regulatory systems that are involved in biofilm formation in other bacterial species; they only have one sigma factor for RNA polymerase and they lack quorum sensing systems and any of the specific genes linked to biofilm growth in other bacterial species. This raises the question of the genetic basis of biofilm formation in an organism with a minimal genome, as they clearly employ complex processes to regulate biofilm formation and are capable of regulating gene expression in order to modulate different modes of growth.

It has previously been shown, using proteomic analysis, that several proteins are up-regulated when MmmSC forms an adherent biofilm, including elongation factor Tu, the PTS system glucose-specific transporter IIB component, phosphoenolpyruvate protein phosphotransferase, fructose-bisphosphate aldolase class II and pyruvate dehydrogenase (McAuliffe *et al.*, 2008). Interestingly, pyruvate dehydrogenase and elongation factor Tu are thought to play a role in the binding of *M. pneumoniae* to the extracellular matrix component fibronectin (Dallo *et al.*, 2002) and are also thought to be important parts of the cytoskeleton of *M. pneumoniae*, and, as such, are linked to cell adhesion (Layh-Schmitt *et al.*, 2000). PTS system proteins and EF-Tu have also been linked to the stress response in *M. pulmonis* (Fehri *et al.*, 2005). The other proteins found were involved in carbohydrate catabolism. Previously, it has been shown that some glycolytic enzymes are not simply limited to substrate turnover and other functions have been ascribed to them. A recent study of protein expression in *Streptococcus mutans* biofilms showed that glycolytic enzymes analogous to those up-regulated in MmmSC were highly expressed during early biofilm formation (Welin *et al.*, 2004).

Recently it has been shown in *M. pulmonis* that biofilm formation occurs stochastically as a result of rearrangements in variable surface antigens and is not due to complex changes in programmed gene expression (Simmons *et al.*, 2007). In *M. pulmonis* the length of the tandem repeat region of the Vsa protein modulates biofilm formation and the susceptibility of individual cells to complement (Simmons and Dybvig, 2003, Simmons *et al.*, 2004, 2007). *M. pulmonis* producing a short Vsa protein with few tandem repeats form prolific biofilms, while those with long tandem repeats and thus a longer Vsa do not form biofilms (Simmons *et al.*, 2007). It has also been reported previously that in *M. bovis* there is some correlation between variable surface protein expression and ability to form a biofilm (McAuliffe *et al.*, 2006). It has been well documented that other mycoplasma species, including MmmSC, possess variable surface proteins (Westburg *et al.*, 2004) so it will interesting to discover whether these influence biofilm formation in a similar manner.

Proteomics studies in *M. bovis* found several proteins that differ between 'good' and 'poor' biofilm forming strains. Proteins of interest identified included the p48 lipoprotein, dihydrolipoamide dehydrogenase, thioredoxin dehydrogenase, the p80 lipoprotein, EF-Tu and several unnamed lipoproteins (McAuliffe, unpublished).

Studies using a panel of transposon mutants in *M. c. capricolum* suggest that the genes that are implicated in biofilm formation in *M. c. capricolum* are mainly membrane proteins or hypothetical proteins, although the pyruvate dehydrogenase complex E1 beta subunit appears to be crucial for *M. c. capricolum* biofilm formation, in parallel with the implied role of the alpha subunit in MmmSC (MmmSC) biofilm formation (McAuliffe *et al.*, 2008). In addition, mutants lacking the PTS system IIBC, which is also involved in MmmSC biofilm formation (McAuliffe *et al.*, 2008), are unable to form a prolific biofilm. Mutants with disrupted ABC transporters, various lipoprotein genes, including *lppA*, *lppQ*, MSC_0500 and numerous putative lipoprotein genes, membrane protein genes that showed some similarity to genes encoding proteins of the *Mycoplasma arthritidis* virulence signal family and numerous other genes linked to the PTS system, including the mannitol-specific IIBC component were also unable to form a prolific biofilm. Mutants that formed more prolific biofilms than the wild type strain included mutants with disrupted genes for lipoproteins, hypothetical membrane proteins, ABC transporters and the methyltransferase *gidB*, as well as numerous genes within the integrative conjugative elements (ICE) and the putative rRNA methyltransferase RsmB gene (McAuliffe *et al.*, 2010).

In summary, studies in both MmmSC and *M. c. capricolum* have indicated that the PTS system may be important in biofilm formation. The PTS system is a multicomponent sugar transport system that phosphorylates sugars as they enter the cell. In *Escherichia coli*, components of the PTS fulfil many regulatory roles, including regulation of nutrient scavenging and catabolism, chemotaxis, glycogen utilization, catabolite repression and inducer exclusion (Houot *et al.*, 2008). It has also been shown in *Vibrio cholerae* that components of the PTS are coregulated with the *vps* genes, which are required for synthesis of the biofilm matrix exopolysaccharide, and that the phosphorylated form of enzyme I has a novel role in specific regulation of biofilm-associated growth (Houot *et al.*, 2008). The mannitol IIAB component of the PTS system was found to be crucial for biofilm formation in *Streptococcus mutans* (Abranches *et al.*, 2008). Similarly, mutants lacking the mannitol components of the PTS system were found to develop very poor biofilms compared with their wild-type counterpart.

Although the role of the PTS system in mycoplasma biofilm formation has not previously been explored, a crucial role for the PTS system has been described in *M. pneumoniae*, in regulation of cytadherence proteins, including HMW1, HMW3, the major adhesin P1 and the surface protein MPN474, by phosphorylation (Schimdl *et al.*, 2010). Therefore, it is feasible that mycoplasmas use protein phosphorylation as a means of regulating gene expression, with post translational modification playing the role that alternate sigma factors play in other bacteria, allowing a subset of genes to be switched on and enabling rapid adaptation to change.

## The role of biofilms in resistance and persistence

Perhaps the most intriguing and significant property of biofilms is that they often display a completely different phenotype to planktonic cells, not just in terms of their structure and morphology, but also in terms of their virulence, persistence and susceptibility to stress.

Perhaps some of the most important recent findings relate to the persistence of mycoplasmas when grown as a biofilm. Certain strains of MmmSC were found to survive for over 20 weeks when grown as an adherent biofilm on a bare surface exposed to the atmosphere (McAuliffe *et al.*, 2008). Obviously, this may have implications for disease control, as it has previously been assumed that MmmSC, and other mycoplasma species, would not survive in the environment; in fact, they may be capable of surviving outside of their host for considerable amounts of time, particularly, perhaps, in the temperate climate of Europe.

It has also been shown that other *Mycoplasma* spp., including *M. bovis*, can persist in the environment, as *M. bovis* was found in recycled bedding sand originating from a dairy experiencing an outbreak of clinical mycoplasma mastitis. Intriguingly, it was found that mycoplasmas survived in the sand pile for 8 months and that the concentration of mycoplasmas within the sand pile was directly related to temperature and precipitation (Justice-Allen *et al.*, 2010).

Confocal imaging has also provided further evidence of the contribution of biofilm growth to the persistence of some mycoplasma species. When mature biofilms that were starved of nutrients were observed using confocal microscopy and Live/Dead staining, it was found that live cells persisted in the centre of biofilm microcolonies, even after all other cells around the edge

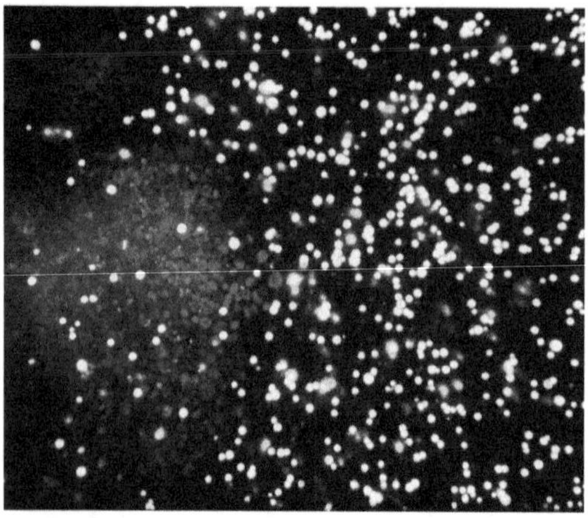

**Figure 13.6** Live/dead stained confocal scanning light micrographic image of a mature, starved *M. putrefaciens* biofilms, demonstrating persistence of cells within the interior of the biofilm (L. McAuliffe). A colour version of this figure is available in the plate section at the back of the book.

of the biofilm had died (McAuliffe *et al.*, 2006). It seems likely that mycoplasmas in the centre of the biofilm are more protected from environmental stress than free-living counterparts or those at the edge of the biofilm (Fig. 13.6).

## Resistance to antimicrobial agents and host cell defences

The most important and widely studied property of biofilms is their vastly increased resistance to antimicrobials and host defences. Compared with planktonic cells, biofilms are commonly 10–1000 times more resistant (Mah and O' Toole, 2001). Unattached bacteria can be cleared by antibodies and phagocytes and are susceptible to antibiotics. However, adherent biofilm cells are resistant to antibiotics, antibodies and phagocytes. In addition, biofilms can cause host damage, as phagocytes are attracted, but phagocytosis is frustrated, leading to the release of phagocytic enzymes, which damage surrounding tissue and exacerbate the effects of infection. As well as enabling chronic infection of hosts, biofilms may cause bouts of acute infection when planktonic cells are periodically released from the biofilm.

Studies with ureaplasmas have shown that biofilm cells were more resistant (as determined by global resistance percentages) to erythromycin, telithromycin, ciprofloxacin, levofloxacin and tetracycline, but that all strains tested were still fully susceptible to clarithromycin in both planktonic and biofilm types of growth (Garcia-Castillo *et al.*, 2008).

Studies on *M. bovis* found that there was no significant difference in the minimum inhibitory concentrations of any of the antibiotics tested for planktonic and biofilm grown cells. However, biofilm growth did have some phenotypic effects on *M. bovis* cells. Planktonic-grown *M. bovis* was inhibited from producing polysaccharide film at much lower concentrations of oxytetracycline than biofilm-grown cells (McAuliffe *et al.*, 2006).

The method of antibiotic susceptibility testing is of crucial significance in studies of the effect of biofilm growth on susceptibility. There is a real need for further studies to develop methods to test the minimum biofilm eradication concentration for antibiotics, rather than simply testing the minimum inhibitory concentration using resuspended biofilm cells, as resuspended cells obviously lose many of the characteristic properties that make them different from planktonic cells, as they are no longer attached to a surface or protected within a polysaccharide matrix. Other problems associated with using minimum inhibitory concentrations are that they simply test if the organisms are inhibited (that is, not actively growing) but it does not test if the culture is viable. Testing minimum mycoplasmacidal concentrations (MMC) are more meaningful, as it can then be determined whether the organisms are capable of multiplication once the antimicrobial agent is removed, a situation that much better reflects the scenario *in vivo*. It seems likely that, even if minimum inhibitory concentrations do not differ when cells form a biofilm, the kinetics of killing may well be different, as antimicrobials will take longer to penetrate to the centre of the biofilm and cells there may take longer to kill even if the minimum inhibitory concentration does not differ.

Biofilm formation may also be significant in helping mycoplasmas to evade host defences and also resist attack by the host. This has been demonstrated recently in *M. pulmonis*, where biofilms formed *in vitro* protected mycoplasmas from the lytic effects of complement and the small antimicrobial peptide gramicidin (Simmons and Dybvig, 2007). When encased within a biofilm, cells of *M. pulmonis* were more resistant to complement and gramicidin than mycoplasmas that were dispersed. The ability to resist killing by complement was only found in cells within tower regions and cells were sensitive to gramicidin when dispersed from the biofilm, thus indicating that the resistance imparted by the biofilm resides in its structure. As cell density and polysaccharide production are higher in towers than in the honeycombed regions, it is plausible that, akin to findings in other bacterial species, either (or both) of these factors may play an important role in biofilm resistance. Regardless of the precise mechanism of resistance, these studies have shown that biofilm formation may be a mechanism that protects mycoplasmas from host immune responses (Simmons and Dybvig, 2007).

## Resistance to other stress

Biofilms are also generally much more resistant to other stresses that they would encounter in their natural environment, including heat, oxidative and desiccation stress (McAuliffe *et al.*, 2006, 2008). In studies of oxidative stress it was found that when MmmSC was exposed to 100 mM peroxide, biofilm grown cells were much more resistant than planktonic cells, with all planktonic cells killed within 20 minutes, but over 2% of biofilm cells still remaining viable at this time. Similarly, when MmmSC and *M. bovis* were exposed to 53°C, biofilms were significantly more resistant (by at least one log cycle of killing) than with planktonic cells.

In studies of desiccation at 20°C, where cells were left exposed to the air to completely dry out, biofilm-grown MmmSC were more resistant than planktonic cells, but there was intraspecific variation in the response to drying. Biofilm-grown *M. bovis* were also significantly more resistant to drying, with no detectable planktonic survivors after 24 hours, but between 0.001 and 0.01% of biofilm-grown cells were still viable, depending on the strain.

Studies of the response of mycoplasma biofilms to biocides or detergents are limited, but it has been shown that, when MmmSC was exposed to 0.05% sodium dodecyl sulphate, biofilms were much more resistant, with less than one log cycle of killing seen after 40 minutes, compared with 4 log cycles of killing for planktonic cells (McAuliffe *et al.*, 2008).

## Genetic basis for resistance – why are biofilms different?

Despite years of research, it is still not fully understood why biofilms are so much more resistant to stress than their planktonic counterparts, but a number of mechanisms are thought to act synergistically (reviewed by Mah and O' Toole, 2001). It has been proposed that multiple mechanisms of resistance exist that vary in importance, depending on the species of bacteria. These resistance mechanisms include the failure of antimicrobials/host cell defences to penetrate the biofilm, a decreased growth rate, induction of stress responses, the presence of persister cells, high cell density and induction of a specific 'biofilm phenotype' (Stewart, 2002). Some of these mechanisms are illustrated in Fig. 13.7.

### Failure of antimicrobials/host cell defences to penetrate the biofilm

Biofilms are, by definition, enclosed in an exopolysaccharide matrix that can both reduce the diffusion of and bind to antimicrobial agents (Stewart, 1996). This mechanism is thought to be most effective in protecting biofilm cells against the action of large, host-produced molecules such as lysozyme and complement (Lewis, 2001). Delayed penetration may provide some protection against degradable antimicrobials, as the antimicrobial can be rendered harmless before it can damage the biofilm cells. The diffusion barrier provided by the polysaccharide matrix may also provide protection against exposure to hydrogen peroxide (Elkins *et al.*, 1999). In summary, although the polysaccharide matrix may contribute to biofilm resistance, this mechanism alone cannot explain the resistance of biofilms to antimicrobial agents (Mah and O'Toole, 2001).

### Decreased growth rate

Biofilms typically have a much slower growth rate than planktonic cells and it is well documented

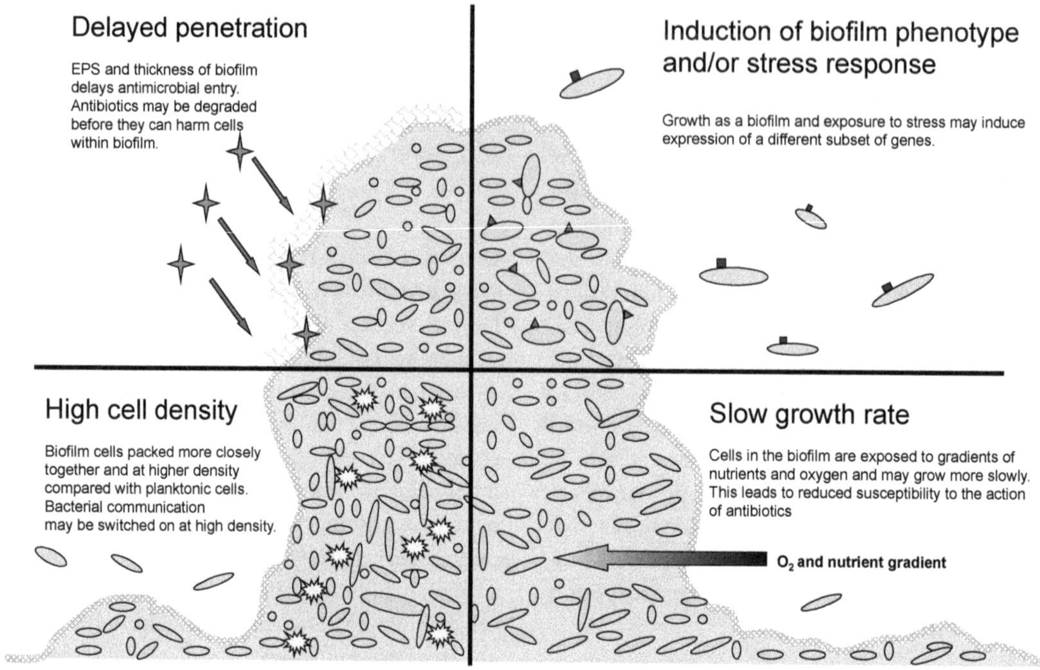


**Figure 13.7** Illustration of the different biofilm characteristics that contribute to increased resistance to antimicrobial agents and host cell defences (copyright L. McAuliffe). A colour version of this figure is available in the plate section at the back of the book.

that almost all antimicrobial agents are more effective in killing rapidly growing cells (Mah and O'Toole, 2001). Several studies have shown that sensitivity to antimicrobials increases simultaneously with growth rate in biofilm cells (Evans *et al.*, 1991), but, as the same phenomenon is seen in planktonic cells, this alone cannot explain biofilm resistance. Although slow growth undoubtedly contributes to biofilm resistance, other properties also influence the resistance of biofilms to stress.

## High cell density

By definition cell density is high in a closely packed, compact, adherent biofilm, compared with free floating planktonic cells in liquid medium. High cell density in bacterial populations is explicably linked to quorum sensing. Quorum sensing describes the phenomenon of bacterial communication using diffusible chemical signals (termed autoinducers). Bacteria produce autoinducer chemicals and, when a threshold level of autoinducer is reached within the bacterial population, it signals that the population is 'quorate', and therefore, capable of making co-ordinated changes in gene expression.

Quorum sensing pathways have been discovered in numerous bacterial species, with Gram-negative bacteria generally using systems based on acylated homoserine lactones and autoinducer 2 (AI2) (reviewed in Van Bodman *et al.*, 2008). Gram-positive bacteria employ different systems, based on small diffusible peptides (Thoendel and Horswill, 2010). No quorum sensing systems have been described in mycoplasmas and searches of currently published mycoplasma genomes have yielded no genes with sequence similarity to known quorum sensing systems. However, the recent findings that several putative ABC transporters are linked to biofilm formation in *M. c. capricolum* are intriguing, as transporters of this type often play a role in bacterial communication systems (McAuliffe *et al.*, 2010).

## Development of 'persister' cells

Bacterial populations produce persisters, cells that neither grow nor die in the presence of bactericidal agents, and thus exhibit multidrug tolerance.

Persisters are 'shut down' cells that have entered a state of dormancy, almost analogous to sporulation in spore-forming bacteria. The mechanism of resistance and the nature of persisters, which were discovered in 1944, have remained elusive. It has been proposed that persisters are largely responsible for the chronicity of infections caused by bacterial biofilms (reviewed in Lewis, 2005). Gene expression profiles of persister *E. coli* cells have indicated elevated expression of toxin-antitoxin modules and other genes that can block important cellular functions, such as translation. Inhibition of translation leads to a shutdown of other cellular functions as well, preventing antibiotics from affecting their targets, giving rise to tolerant persister cells. Overproduction of chromosomally encoded 'toxins', such as RelE, an inhibitor of translation, or HipA, causes a sharp increase in persisters. It has been suggested that the function of 'toxins' is the exact opposite of the term, namely, to protect the cell from lethal damage (Lewis, 2005). It appears that stochastic fluctuations in the levels of multiple drug tolerance proteins lead to formation of rare persister cells. Persisters are essentially altruistic cells that forfeit propagation in order to ensure survival of kin cells in the presence of lethal factors.

## Induction of a specific 'biofilm phenotype'

The final possibility exists that biofilms have a distinct phenotype with different regulation of resistance-related genes compared with planktonic culture. The emerging idea is that a biofilm-specific phenotype is induced in a subpopulation of the biofilm community and that it results in the expression of genes that initiate active mechanisms to combat the effects of specific stresses and/or antimicrobial agents (Costerton, 1999; Mah and O'Toole, 2001). Biofilm-specific global changes in gene expression have been seen in a number of bacterial species, including *E. coli* and *S. aureus*. Microarray analysis has shown that biofilm-grown cells show altered expression of 38% of genes compared with planktonic cells (Prigent-Combaret *et al.*, 1999). In *S. aureus* at least five genes were differentially regulated in biofilm grown cells, including important regulators linked to stress resistance (Becker *et al.*, 2001). Although studies have not yet been undertaken to ascertain how mycoplasma biofilm gene expression alters in response to stress, studies have indicated that a significant proportion of mycoplasma genes are differentially expressed in biofilms compared with planktonic cells (McAuliffe *et al.*, 2006, 2008, 2010).

## Biofilm methods and model systems

A number of *in vitro* biofilm model systems have been developed and applied to mycoplasmas. The method of choice will depend on the species to be studied, as different species are better suited to different model systems, and on the purpose of the study, as some methods are more suited to microscopic analysis and some are ideal for generating large cell volumes. Often it is necessary to analyse strains using several methods in parallel.

Commonly used methods include the use of glass or plastic coverslips placed in 6- or 12-well polystyrene plates to provide a solid substratum for biofilm attachment (Simmons *et al.*, 2007) or the use of glass coverslips placed vertically in 50 ml conical tubes to provide an air–liquid interface half way up the glass coverslip for biofilm attachment and growth (McAuliffe *et al.*, 2006). The use of glass coverslips is well suited to resistance studies as coverslips can be removed and exposed to stress and/or antimicrobial agents and then the bacteria enumerated. Coverslip models are also well suited tp studies of biofilm structure and morphology and can visualized directly using various microscopic techniques.

For larger scale culture purposes biofilms have also been grown in tissue culture flasks (25 cm$^2$ polystyrene) and can then be scraped off using a sterile cell scraper (Simmons *et al.*, 2004). For large scale culture for proteomic analysis and for mycoplasma species that do not form a biofilm in a model with an air–liquid interface (such as MmmSC) a simple membrane on an agar plate can be used (McAuliffe *et al.*, 2008). Sterile hydrophobic-edged membranes with a 0.2 μm pore size (Sartorius) are preferable. The membrane on plate method has advantages as the membrane (and the attached biofilm) can be removed and placed onto a new agar plate to ensure that nutrients do not become limiting and it enables the growth as biofilms of species that cannot be studied using other methods. It is also easy to remove adherent cells by vortexing, so that they can be analysed using proteomic methods or by enumeration. The main disadvantage of this method is that it may not represent the situation *in vivo*, as cells are not exposed to a liquid interface. Care must also be taken using this method to ensure that cells are not subjected to desiccation, so a moist incubation environment is imperative. Biofilms also tend to form a flatter, less differentiated structure using this method than would be formed using an air–liquid interface model.

For high throughput screening of strains, for example for screening mutant libraries, models using 96-well plates are best. Various types of 96-well plates have been used for biofilm studies and their suitability varies widely, depending on the manufacturer, plate composition and the plate coating. Many mycoplasma species that do not adhere well to uncoated plate surfaces can be studied if coated plates are used. Collagen, gelatine and poly D-lysine plates have proved to be useful for mycoplasma biofilm studies (McAuliffe, unpublished). Many species can be grown in uncoated plates, but the yield is often not high and it is not suited to culturing biofilms for downstream proteomics or imaging purposes. Multi-well plates often need to be washed several times in sterile distilled water prior to use, as unwashed plates can inhibit mycoplasma attachment (McAuliffe, unpublished).

Perhaps the most exciting development in terms of biofilm model systems has recently been developed for *M. pulmonis*, with the application of a mouse tracheal organ culture, which has, for the first time, enabled *ex vivo* culture of mycoplasma biofilms (Simmons *et al.*, 2009). The use of organ culture will enable us to study mycoplasma biofilms under conditions that are highly analogous to the *in vivo* situation without the need to utilize whole animal models.

Regardless of the biofilm model system used, care must be taken to ensure that cells do not become nutrient limited and that the medium is sufficiently buffered. Glucose fermenting mycoplasmas, in particular, require frequent changes of medium to ensure that the pH of the growth medium remains relatively neutral and that cells are not exposed to an acidic environment. It is also imperative that planktonic and biofilm cells are in the same growth phase if comparisons are to be made between the two, so often it is more practical to study stationary phase cells, rather than exponential phase cells.

## Future of biofilm research

Biofilm research is a growing area of interest within the mycoplasma community and there is a pressing need to explore the role of biofilm formation in the pathogenesis and persistence of mycoplasma infections. There is also a need for focus on the infectious potential of mycoplasma biofilms in the environment and whether they can act as a reservoirs for disease. Finally, research should focus on trying to model biofilms under conditions that accurately reflect the situation *in vivo*, and for this there is a need to explore the role of multi-species biofilms in infections, both in terms of multiple mycoplasma species and the relationships between mycoplasmas, other bacteria, viruses and fungi in infection.

## References

Abranches, J., Candella, M.M., Wen, Z.T., Baker, H.V., and Burne, R.A. (2006). Different roles of EIIABMan and EIIGlc in regulation of energy metabolism, biofilm development, and competence in *Streptococcus mutans*. J. Bacteriol. *188*, 3748–3756.

Almeida, R.A., and Rosenbusch, R.F. (1991). Capsule like surface material of *Mycoplasma dispar* induced by in vitro growth in culture with bovine cells is antigenically related to similar structures expressed in vivo. Infect. Immun. *59*, 3119–3125.

Becker, P., Hufnagle, W., Peters, G., and Herrmann, M. (2001). Detection of differential gene expression in biofilm-forming versus planktonic populations of *Staphylococcus aureus* using micro-representational-difference analysis. Appl. Environ. Microbiol. *67*, 2958–2965.

Boatman, E.S. (1979). Morphology and ultrastructure of the mycoplasmatales, In The Mycoplsmas, Vol 1, Cell biology, Barile, M.F., and Razin, S., eds. (Academic press Inc, New York, USA), pp. 63–102.

Costerton, J., Stewart, W.P.S., and Greenberg, E.P. (1999). Bacterial biofilms: a common cause of persistent infections. Science *284*, 1318–1322.

Costerton, J.W., Lewandowski, Z., Caldwell, D.E., Korber, D.R., and Lappin-Scott, H.M. (1995). Microbial Biofilms. Ann. Rev. Microbiol. *49*, 711–745.

Daubenspeck, J.M., Bolland, J.R., Luo, W., Simmons, W.L., and Dybvig, K. (2009). Identification of exopolysaccharide-deficient mutants of *Mycoplasma pulmonis*. Mol. Microbiol. *72*, 1235–1245.

Elkins, J.G., Hassett, D.J., Stewart, P.S., Schweizer, H.P., and McDermott, T.R., (1999). Protective role of catalase in *Pseudomonas aeruginosa* biofilm resistance to hydrogen peroxide. Appl. Environ. Microbiol. *65*, 4594–4600.

Evans, D.J., Allison, D.G., Brown, M.R., and Gilbert, P. (1991). Susceptibility of *Pseudomonas aeruginosa* and *Escherichia coli* biofilms towards ciprofloxacin: effect of specific growth rate. J. Antimicrob. Chemother. *27*, 177–184.

Fehri, L.F., Sirand-Pugnet, P., Gourgues, G., Jan, G., Wróblewski, H., and Blanchard, A. (2005). Resistance to antimicrobial peptides and stress response in *Mycoplasma pulmonis*. Antimicrob. Agents Chemother. *49*, 4154–4165.

García-Castillo, M., Morosini, M.I., Gálvez, M., Baquero, F., del Campo, R., and Meseguer, M.A. (2008). Differences in biofilm development and antibiotic susceptibility among clinical *Ureaplasma urealyticum* and *Ureaplasma parvum* isolates. J. Antimicrob. Chemother. *62*, 1027–1030.

Geary, S.J., Tourtellotte, M.E., and Camerons, J.A. (1981). Inflammatory toxin from *Mycoplasma bovis*: Isolation and characterization. Science *212*, 1032.

Gosney, F. (2008). The role of biofilm formation in economically important Mycoplasma species. MSc thesis. The Royal Veterinary College, University of London, UK.

Hamon, M.A., and Lazazzera, B.A. (2001). The sporulation transcription factor Spo0A is required for biofilm development in *Bacillus subtilis*. Mol. Microbiol. *42*, 1199–1209.

Henrich, B., Schmitt, M., Bergmann, N., Zanger, K., Kubitz, R., Häussinger, D., and Pfeffer, K. (2010). *Mycoplasma salivarium* detected in a microbial community with *Candida glabrata* in the biofilm of an occluded biliary stent. J. Med. Microbiol. *59*, 239–241.

Houot, L., and Watnick, P.I. (2008). A novel role for Enzyme I of the *Vibrio cholerae* Phosphoenolpyruvate Phosphotransferase System in regulation of growth in a biofilm. J. Bacteriol. *190*, 311–320.

Joseph, L.A., and Wright, A.C. (2004). Expression of *Vibrio vulnificus* capsular polysaccharide inhibits biofilm formation. J. Bacteriol. *186*, 889–893.

Justice-Allen, A., Trujillo, J., Corbett, R., Harding, R., Goodell, G., and Wilson, D. (2010). Survival and replication of Mycoplasma species in recycled bedding sand and association with mastitis on dairy farms in Utah. J. Dairy Sci. *93*, 192–202.

Kearns, D.B., Chu, F., Branda, S.S., Kolter, R., and Losick, R. (2005). A master regulator for biofilm formation by *Bacillus subtilis*. Mol. Microbiol. *55*, 739–749.

Layh-Schmitt, G., Podtelejnikov, A., and Mann, M. (2000). Proteins complexed to the P1 adhesin of *Mycoplasma pneumoniae*. Microbiology *146*, 741–747.

Lewis, K. (2001). Riddle of biofilm resistance. Antimicrob. Agents Chemother. *45*, 999–1007.

Lewis, K. (2005). Persister cells and the riddle of biofilm survival. Biochemistry *70*, 267–274.

Mah, T.F., and O'Toole, G.A. (2001). Mechanisms of biofilm resistance to antimicrobial agents. Trends Microbiol. *9*, 34–39.

May, M., and Brown, D.R. (2009). Secreted sialidase activity of canine mycoplasmas. Vet. Microbiol. *137*, 380–383.

McAuliffe, L., Ellis, R.J., Miles, K., Ayling, R.D., and Nicholas, R.A. (2006). Biofilm formation by mycoplasma species and its role in environmental persistence and survival. Microbiology *152*, 913–922.

McAuliffe, L., Ayling, R.D., Ellis, R.J., and Nicholas, R.A. (2008). Biofilm-grown *Mycoplasma mycoides* subsp. *mycoides* SC exhibit both phenotypic and genotypic variation compared with planktonic cells. Vet. Microbiol. *129*, 315–324.

Miles, R.J., Taylor, R.R., and Varsani, H. (1991). Oxygen uptake and $H_2O_2$ production by fermentative Mycoplasma spp. Med. Microbiol. *34*, 219–223.

Minga, U.M. (1981). Smooth and rough colony types of *Mycoplasma mycoides* subsp. *mycoides*: In vitro characterisation and pathogenicity in cattle. Tanz. Vet. Bull. *3*, 53–62.

Niang, M., Rosenbusch, R.F., Andrews, J.J., and Kaeberle, M.L. (1998). Demonstration of a capsule on *Mycoplasma ovipneumoniae*. Am. J. Vet. Res. *59*, 557–562.

Nocard, E.I.E., and Roux, E. (1990). Le microbe de la péripneumonie. Ann. Inst. Pasteur *12*, 240–262. (Translated as 'The microbe of pleuropneumonia'). Rev. Infect. Dis. *12*, 354–358.

Prigent-Combaret, C., Brombacher, E., Vidal, O., Ambert, A., Lejeune, P., Landini, P., and Dorel, C. (2001). Complex regulatory network controls initial adhesion and biofilm formation in *Escherichia coli* via regulation of the *csgD* gene. J. Bacteriol. *183*, 7213–7223.

Razin, S., Yogev, D., and Naot, Y. (1998). Molecular biology and pathogenicity of mycoplasmas. Microbiol. Mol. Biol. Rev. *62*, 1094–1156.

Rivera, A., Cedillo, L., Giono, S., and Rodriguez, N. (2009). Biofilm formation by *Mycoplasma fermentans* on intrauterine devices. Int. J. Med. Sci. *1*, 267–271.

Rosenbusch, R.F., and Minion, F.C. (1992). Cell Envelope: morphology and biochemistry. In Mycoplasmas: Molecular Biology and Pathogenesis, Maniloff, J., McElhaney, R.N., Finch, L.R., and Baseman, J.B., eds. (American Society of Microbiology, Washington, D.C.), pp. 549–559.

Salih, B.A., and Rosenbusch, R.F. (1998). Identification and localization of a 94 kDa membrane protein found in *Mycoplasma bovoculi* strains. Comp. Immunol. Microbiol. Infect. Dis. *21*, 281–290.

Salmon, D.E. (1896). Special Report on Diseases of Cattle and Cattle Feeding (Washington Government Printing Office, Washington D.C.).

Sasaki, T., Shintani, M., and Kihara, K. (1984). Inhibition of growth of mammalian cell cultures by extracts of arginine-utilizing mycoplasmas. In Vitro *20*, 369–375.

Schmidl, S.R., Gronau, K., Pietack, N., Hecker, M., Becher, D., and Stülke, J. (2010). The Phosphoproteome of the

Minimal Bacterium *Mycoplasma pneumoniae*. Mol. Cell. Prot. *9*, 1228–1242.

Simmons, W.L., and Dybvig, K. (2003). The Vsa proteins modulate susceptibility of *Mycoplasma pulmonis* to complement killing, hemadsorption, and adherence to polystyrene. Infect. Immun. *71*, 5733–5738.

Simmons, W.L., and Dybvig, K. (2009). Mycoplasma biofilms *ex vivo* and *in vivo*. FEMS Microbiol. Lett. *295*, 77–81.

Simmons, W.L., Denison, A.M., and Dybvig, K. (2004). Resistance of *Mycoplasma pulmonis* to complement lysis is dependent on the number of Vsa tandem repeats: shield hypothesis. Infect. Immun. *72*, 6846–6851.

Simmons, W.L., Bolland, J.R., Daubenspeck, J.M., and Dybvig, K. (2007). A Stochastic Mechanism for Biofilm Formation by *Mycoplasma pulmonis*. J. Bacteriol. *189*, 1905–1913.

Somarajan, S.R., Kannan, T.R., and Baseman, J.B. (2010). *Mycoplasma pneumoniae* Mpn133 is a cytotoxic nuclease with a glutamic acid-, lysine- and serine-rich region essential for binding and internalization but not enzymatic activity. Cell. Microbiol. *12*, 1821–1831.

Stewart, P.S. (1996). Theoretical aspects of antibiotic diffusion into microbial biofilms. Antimicrob. Agents Chemother. *40*, 2517–2522.

Stewart, P.S. (2002). Mechanisms of antibiotic resistance in bacterial biofilms. Int. J. Med. Microbiol. *292*, 107–113.

Thoendel, M., and Horswill, A.R. (2010). Biosynthesis of peptide signals in Gram-positive bacteria. Adv. Appl. Microbiol. *71*, 91–112.

Tu, A.H., Clapper, B., Schoeb, T.R., Elgavish, A., Zhang, J., Liu, L., Yu, H., and Dybvig, K. (2005). Association of a major protein antigen of *Mycoplasma arthritidis* with virulence. Infect. Immun. 73, 245–249.

Valle, J., Toledo-Arana, A., Berasain, C., Ghigo, J.M., Amorena, B., Penades, J.R., and Lasa, I. (2003). SarA and not sigmaB is essential for biofilm development by *Staphylococcus aureus*. Mol. Microbiol. *48*, 1075–1087.

Van Leeuwenhoek, A. (1684). An abstract of a letter from Mr. Anthony Leevvenhoek at Delft, dated Sep. 17, 1683, Containing some microscopical observations, about animals in the scurf of the teeth, the substance call'd worms in the nose, the cuticula consisting of scales'. Phil. Trans. *14*, 568–574.

von Bodman, S.B., Willey, J.M., and Diggle, S.P. (2008). Cell–cell communication in bacteria: united we stand. J. Bacteriol. *190*, 4377–4391.

Welin, J., Wilkins, J.C., Beighton, D., and Svensater, G. (2004). Protein expression by *Streptococcus mutans* during initial stage of biofilm formation. Appl. Environ. Microbiol. *70*, 3736–3741.

Westberg, J., Persson, A., Holmberg, A., Goesmann, A., Lundeberg, J., Johansson, K.E., Pettersson, B., and Uhlen, M. (2004). The genome sequence of *Mycoplasma mycoides* subsp. *mycoides* SC type strain PG1T, the causative agent of contagious bovine pleuropneumonia (CBPP). Genome Res. *14*, 221–227.

Wood, P.J. (1980). Specificity in the interaction of direct dyes with polysaccharides. Carbohyd. Res. *85*, 271–287.

Yi, K., Rasmussen, A.W., Gudlavalleti, S.K., Stephens, D.S., and Stojiljkovic, I. (2004). Biofilm formation by *Neisseria meningitidis*. Infect. Immun. 72, 6132–6138.

# Host Immune Responses to Mycoplasmas

# 14

Steven M. Szczepanek and Lawrence K. Silbart

## Abstract

The atypical characteristics of mycoplasmas are often associated with dysregulated host immune responses, which allow these pathogens to occupy an ecological niche associated with mucosal surfaces (and occasionally beyond). The lack of a cell wall and the presence of variable surface lipoproteins are paramount amongst these uncommon features, and host defence includes an intricate innate immune signalling system that includes TLRs, cytokines, and chemotactic molecules to attract, alert and activate leucocytes during infection. Mycoplasmas are adept at manipulating many of these signals to their own advantage, resulting in commensal relationships in some cases, or insidious chronic infections in others. Successful resolution of disease usually depends on robust humoral and cell-mediated immune responses, but such responses can take weeks to develop, thereby allowing the bacteria an opportunity to adapt to their environment and gain a foothold in colonized tissues. For this reason, successful prophylactic vaccines block initial colonization, thereby preventing the subsequent over-exuberant inflammatory and dysregulated adaptive immune responses. This chapter examines the intricate interplay between highly evolved host immune responses versus the highly adaptable mycoplasmas, with an eye towards identifying gaps in our knowledge that must be addressed in future research.

## Introduction

Identifying guiding principles and relationships that govern interactions between mycoplasmas and their host organism(s) is challenging due to the high degree of molecular tailoring that has occurred during their co-evolution. Most mycoplasmologists would agree that mycoplasmas are primarily surface pathogens or commensals, which in some cases have evolved mechanisms to invade host tissues and thereby establish additional ecological niches. When these mechanisms involve colonization of host mucosal surfaces or invasion into host tissues and systemic distribution, entirely different immune mechanisms are required to limit bacterial replication and ameliorate disease. One fairly common trait of pathogenic mycoplasmas is immune activation, which begins with stimulation of the innate immune system, but frequently involves subsequent activation of cells of the adaptive immune system. Typical inflammatory diseases associated with mycoplasmal infection include atypical pneumonia, mastitis, non-gonococcal urethritis, salpingitis and arthritis (Baseman and Tully, 1997). These inflammatory conditions are mediated at least in part by cytokines/chemokines released by epithelial cells and leucocytes in response to mycoplasmas, and result in a misdirected (dysregulated) or maladaptive immune response, thereby allowing the organism to establish a stable ecological niche that is often associated with chronic infection. Following initial recognition, inflammatory cytokines and chemokines are released, often in association with NF-κB activation, resulting in the influx of inflammatory leucocytes (including granulocytes, macrophages and lymphocytes. Traditional lymphocyte activation can occur, however polyclonal activation (viz. mitogenic or superantigen activation) can also lead to inappropriate or over-exuberant immune activation, resulting in immunopathology. Immune evasion is another mechanism employed by various mycoplasmas, and often occurs in association with antigenic variation (including size variation) and/or phase variation of surface-exposed variable lipoproteins. This chapter will examine what is known about these mechanisms. However, the reader is cautioned that our current knowledge of mycoplasmal host–pathogen interactions is quite patchy, and that generalization from one host–pathogen interaction to another is a dubious exercise.

## Innate immune responses to *Mycoplasma* lipoproteins

The initial site of mycoplasma–host interaction typically occurs at the mucosal surfaces of the respiratory, reproductive or conjunctival epithelium. As these surfaces are coated with mucus, the mycoplasma must have some level of motility in order to bind to cilia/adherent mucus, or risk being swept away from their target cells via the mucociliary escalator (Jordan *et al.*, 2007). This enables the organism to interact with cilia or other components of epithelial cells, including sialo-oligosaccharide receptors (Loveless *et al.*, 1992), and facilitates colonization. Many mycoplasmas have specific lipoproteins/lipopeptides (e.g. VlhAs) and/or a highly evolved and fairly conserved tip structure that facilitate a tight attachment to epithelial cells, but which also may engage TLR receptors and initiate bacterial recognition by the host. For many mycoplasmas the specific cell receptors for the attachment organelle are not known, so understanding subsequent host signalling is not yet possible. However, when isogenic mutants are created that are incapable of forming tight epithelial attachments, both colonization and the host response are dramatically reduced, as seen with *M. gallisepticum* (Papazisi *et al.*, 2000; Javed *et al.*, 2005).

### TLR ligation

Mycoplasmas express a multitude of lipoproteins (LP) on their surface, many of which can interact with the ciliated epithelial cells and leucocytes of the host. Innate immune cells have evolved to recognize molecular patterns found on microbes (called pathogen associated molecular patterns (PAMPs)), through their interaction with specialized receptors. Such receptors are called pattern recognition receptors (PRRs) and include the Toll-like receptors (TLRs), NOD-like receptors, as well as others [reviewed in (Medzhitov, 2007)]. TLRs are often the first molecule to interact with PAMPs and initiate specific signalling pathways within a host cell to respond to a foreign invader (Kumar and Yerneni, 2009). Since many common PAMPs are absent in mycoplasmas (including lipoteichoic acid, flagellin, and lipopolysaccharide), most research has focused on TLRs 1, 2, and 6, as they are known to bind to bacterial LPs.

The first mycoplasmal LP that was demonstrated to bind to TLRs is macrophage activating lipopeptide-2 (MALP-2) of *M. fermentans*. TLR 2 and MyD88 (but not TLR 4) knockout (KO) mice were not stimulated by MALP-2 (as evidenced by a lack of cytokine and nitric oxide (NO) production), in contrast to vigorous responses of wild-type strains. Furthermore, over a 100-fold increase in cytokine production was observed when the MALP-2 R stereoisomer was used rather than the S stereoisomer, indicating that the conformation of the LP is crucial for interaction with the TLRs (Takeuchi *et al.*, 2000). This was the first report of TLRs (and the downstream cell signalling pathways) playing a role in host defences against mycoplasmas. The interaction of TLR2 with either TLR1 or TLR6 was later discovered, and the TLR1/2 or TLR2/6 heterodimers were found to bind to triacylated or diacylated LPs, respectively (Okusawa *et al.*, 2004; Shimizu *et al.*, 2007). Interestingly, the affinity of TLRs for mycoplasmal LPs can be altered by substitution of amino acids or fatty acids, thereby leading to significantly reduced downstream host cell signalling (Okusawa *et al.*, 2004). This again reflects the importance of LP structure in TLR recognition and how subtle changes can alter the subsequent response to a particular LP. In some cases additional TLRs may be involved in pathogen sensing, including TLR4 (Peltier *et al.*, 2007).

### Leucocyte chemotaxis

Most mycoplasma-induced lesions are characterized by the presence of granulocytic (e.g.

neutrophils in mammals, heterophils in birds) and myeloid cells, especially macrophages, early post-infection. Over time, these cells diminish in number and are replaced by lymphocytes, and the lesions are referred to as 'lymphoproliferative'. What is less clear and not well-studied is the contribution of lymphocyte migration from blood into the inflamed tissues vs. proliferation of lymphocytes *in situ*. Several studies have examined chemotactic factors released from mycoplasma-infected tissues or cell cultures. CC and CXC chemokines such as IL-8, GRO-α, MCP-1, MIP-1α, MIP-1β (Kaufmann *et al.*, 1999) lymphotactin (Lam and DaMassa, 2000) and, CXCL13, CXCL14, RANTES (Mohammed *et al.*, 2007), as well as CCL19 (unpublished observations) and MIP-2 (Deiters and Mühlradt, 1999) are involved in this process. However, in some cases, mycoplasmal lipoproteins such as MALP-2 actually inhibit neutrophil migration upon TLR 2/6 ligation (Chin, 2009), and this inhibition is further exacerbated by treatment with an anti-CD47 antibody. This mechanism may be used to signal the arrival of neutrophils at the site of microbial infection, thereby facilitating their antimicrobial function (Chin, 2009).

Monocyte chemotactic protein 2 (MCP-2) is a potent chemokine expressed by pneumocytes during infection by mucosal pathogens and is in part responsible for attracting monocytic cells (including T-cells) to the site of inflammation. Another important cytokine expressed during airway infections is macrophage inflammatory protein 1β (MIP-1β), which is secreted by macrophages and activates many leucocytes, including T-cells. MIP-1β is often detected in airways exhibiting a T helper 2 ($T_h2$) skewed response and is associated with an IgA and IgE humoral immune response. CCR5 is a common receptor for both MCP-2 and MIP-1β and is often expressed by CD4+ T-cells. Indeed, during *M. pulmonis* infection of BALB/c mice both cytokines are found to be up-regulated and the CD4+ cells accumulated in the lungs express CCR5. These $T_h$ cells appear to contribute to the immunopathology associated with mycoplasmal disease in the lungs of infected mice (Sun *et al.*, 2006). Additionally, newborns exhibiting signs of bronchopulmonary dysplasia, and infected with *Ureaplasma urealyticum*, appear to have increased levels of MIP-1α and MIP-1β in tracheal aspirates (Baier *et al.*, 2004). These data indicate that the contribution of the MIPs (and possibly CD4+ T-cells) to disease can be observed in clinical cases. The role of MIP-1β in pathogenesis is not only seen in mammals, as chicken macrophages in culture that have been stimulated with *M. gallisepticum* secrete high levels of this cytokine (Lam and DaMassa, 2000). The release of MIP-1β has chemotactic effects on chicken heterophils and lymphocytes (Lam and DaMassa, 2000), and appears to contribute to the immunopathogenesis associated with influx of lymphocytes (including $CD4^+$ cells) found during experimental avian mycoplasmosis (Mohammed *et al.*, 2007). The exact role of the up-regulation of MCP-2, MIP-1β, and CCR5 by $T_h$ cell subsets during mycoplasmal respiratory infection has yet to be elucidated; however, accumulating evidence seems to indicate that each plays a role in the dysregulated immune response driven by these pathogens and may be targets for future intervention strategies.

In some studies, purified mycoplasmal products such as di- and triacylated lipoproteins/ lipopeptides have been shown to induce leucocyte infiltration into respiratory tissues on their own. For example, synthetic lipopeptides derived from the *M. genitalium* b subunit of the $F_0F_1$-type ATPase, as well as NALP-1 and 2, not only activate NF-κB activity but can also induce the infiltration of leucocytes into tissues following intranasal administration to C57B/6 mice (Shimizu *et al.*, 2008). The recently identified CARDS toxin of *M. pneumoniae* has also been shown to be sufficient for the lymphocytic and eosinophilic infiltration of the lungs that is typically observed during infection ((Medina *et al.*, 2012). Such findings suggest that much of the inflammation observed during mycoplasma infection is driven by the presence of specific components of these bacteria and may influence the development of future therapies to treat disease.

## Cytokine signalling

The differential recognition of di- and triacylated LPs by host cells is determined by their binding to preformed heterodimers of TLR1/2/CD14 (Kataoka *et al.*, 2006) or TLR2/6/CD36 (Hoebe *et al.*, 2005). However, LP binding to either heterodimer results in the activation of the same downstream signalling pathways (Farhat *et al.*, 2008), although the ligand specificity is broadened by these multiple interactions. The interaction of TLRs is not dependent on ligand binding, whereas the interaction with CD36 does require receptor ligation. Both TLR1/2 and TLR2/6 traffic to lipid rafts that are subsequently shuttled to the Golgi apparatus, ultimately leading to NF-κB activation in the cell's nucleus (Triantafilou *et al.*, 2006). Curiously, the interaction with the Golgi is not required for activation. In any case, activation of this transcription factor induces the expression of proinflammatory cytokines and chemokines. Indeed, MALP-2 is capable of inducing human monocytes to secrete TNF-α, IL-6, MIP-1β, IL-8, GROα, MCP-1, and MIP-1α (Kaufmann *et al.*, 1999). Furthermore, human trophoblast cells appear to express cyclo-oxygenase 2 (COX-2) and prostaglandin E2 (PGD2) when stimulated with MALP-2 in a NF-κB-dependent manner (Mitsunari *et al.*, 2006), indicating that mycoplasma LP may affect this cell type as well. In addition to stimulation of TNF-α, IL-6, and MIP-1β by MALP-2 in human cells, these cytokines are also up-regulated when chickens are infected with the $R_{low}$ strain of *M. gallisepticum* (Mohammed *et al.*, 2007). Curiously, the *M. gallisepticum* homologue of MALP-2, P47, does not appear to have an effect on attachment, growth, or virulence as there is no difference in the lesions caused by deletion mutants and the wild-type strain in tracheal organ cultures (Markham *et al.*, 2003). However, cytokine production caused by inoculation of chickens with P47 has never been tested so this LP may still play a role in the immunopathogenesis of the disease caused by *M. gallisepticum*. Another chicken pathogen, *M. synoviae*, expresses a variable LP called VlhA which is critical for attachment *in vivo*. VlhA is post-translationally cleaved into two proteins, and MSPB (the N-terminal LP fragment of VlhA) is often expressed in a truncated form *in vivo* (tMSPB). When inoculated with tMSPB, chicken monocytes (MDM cells) secrete NO and the cytokines IL-6 and IL-1β (Lavric *et al.*, 2007). The purified subunit b of the $F_0F_1$-ATPase (AtpD) of *M. pneumoniae* was found to be a LP (Pyrowolakis *et al.*, 1998), and lipopeptides derived from AtpD cause inflammation in wild-type C57Bl/6 mice, but not TLR2 KO mice. Upon stimulation of TLR2 by AtpD lipopeptides *in vivo*, several cytokines are up-regulated, including TNF-α and IL-6 (Shimizu *et al.*, 2008). This was the first experimental evidence indicating that cytokine expression in infected animals is associated with a specific mycoplasmal virulence factor.

## Apoptosis

Apoptosis is a well-established host mechanism for killing intracellular pathogens (Kovacs-Simon *et al.*, 2011), but it is not always simple to assess whether this is to the advantage of the pathogen or the host. In some instances mycoplasmas can induce host cells to undergo apoptosis during infection. Immune signalling via TLRs usually leads to NF-κB activation, resulting in an anti-apoptotic state. However, this NF-κB-induced anti-apoptotic state can be circumvented by some pathogens, and apoptotic cells are frequently observed in tissues infected with these bacteria. In some cells, mycoplasma infection can be either pro- or anti-apoptotic depending on the mycoplasma species (Zhang and Lo, 2007). It is also possible for MyD88 to directly activate caspase-8 through interactions with FADD (Aliprantis *et al.*, 2000), so it may be possible that expression of the pro-apoptotic proteins Bcl-$X_s$, Bax, and Bad is induced by NF-κB through MyD88 signalling, thereby resulting in apoptosis of the eukaryotic cell (Shou *et al.*, 2002).

LPs from *M. fermentans* (LPfer) and *M. salivarium* (LPsal) have been shown to interact with TLR2 and drive lymphocytes and monocytes to undergo necrotic or apoptotic death (Into *et al.*, 2002a,b). HL-60 cells (a monocyte cell line) appear to die by apoptosis, whereas lymphocytes and TPH-1 cells (a different monocyte cell line) die by necrosis. In HL-60 cells exposed to LPfer and LPsal, activation of caspase-3 can be demonstrated and inhibition of caspase 3 or 8 activity abrogates apoptosis. In contrast, TPH-1 cells exposed to *M. genitalium* LPs NF-κB appear to be activated and the cells die by apoptosis (Wu *et al.*, 2008). The authors attributed the NF-κB induction of apoptosis to the stimulation of the pro-apoptotic proteins Bax and Bad (although this was not determined experimentally). Inhibition of

NF-κB activity prevented apoptosis in this study, thereby supporting their conclusion. It has been observed that apoptosis signal-regulating kinase 1 (ASK-1) phosphorylates p38 MAPK via TLR2 activation by mycoplasma LPs, and the transcription factors NF-κB and AP-1 become active under these conditions. Cells stimulated in this manner appear to die by apoptosis, and ASK-1 mutations significantly diminish this phenomenon (Into and Shibata, 2005). A contrasting report indicates that *M. fermentans* LPs are, in-fact, anti-apoptotic in nature and act through suppression of caspase 8 (likely due to the activation of NF-κB) (Gerlic *et al.*, 2007). Interestingly, this discrepancy may be resolved by the finding that NF-κB activation by simultaneous stimulation of TLR2 and TLR6 by mycoplasmal LPs depends on both IRAK-4 (pro-inflammatory) and FADD (pro-apoptotic) signalling pathways. IRAK-4 signalling appears to dominate early in infection, while FADD signalling is observed later in infection (Into *et al.*, 2004). The complex dynamics between pro- and anti-apoptotic signals elicited by TLR recognition of mycoplasmal LPs early and late in infection may be a mechanism to allow infected hosts an opportunity to clear an active infection while protecting them from overreacting to transient colonization.

Lastly, mycoplasmal LPs have been shown to trigger apoptosis in host cells through mechanisms that require more than just TLR binding. HeLa cells lose ATP when inoculated with the ecto-ATPase LP OppA of *M. hominis* and undergo apoptosis, but do not readily die from necrosis (Hopfe and Henrich, 2008). Into *et al.* observed that extracellular ATP produced by inoculation of LPfer and LPsal with host cells was associated with apoptotic death (Into *et al.*, 2002c). It was postulated that this process is driven by interactions between extracellular ATP and P2Z/P2$X_7$ receptors on host cells and stimulates the endogenous apoptotic pathway.

## Cell-mediated immune responses to mycoplasmas

Cell-mediated immunity (CMI) is an essential component of the adaptive immune system that requires the activation of CD8$^+$ cytotoxic T lymphocytes (CTL) along with the appropriate CD4$^+$ helper T lymphocytes ($T_h$). This branch of the immune system is often utilized to fight intracellular pathogens such as viruses, and may be particularly important for those *Mycoplasma* spp. which invade host cells. However, the cell-mediated immune response to mycoplasma infections is not well understood and many current studies seek to determine the roles these cells play in host defence. In some cases, biasing the host immune response towards a type I response may contribute to aberrant inflammatory responses that are associated with pathogenesis, but not mycoplasmal clearance, in keeping with the immune dysregulation characteristic of this genus.

### IFN-γ production and protection from disease by T-cells

CTLs are the primary cell type involved in CMI, often identifying infected host cells in an antigen-specific manner via interactions with MHC-I in concert with pathogen-specific processed peptides. While the role of antigen-specific CTL responses has yet to be determined in many mycoplasma diseases, T-cells have been found to make a significant contribution to disease pathogenesis, and this involvement appears to be dependent on IFN-γ. During *M. pulmonis* infection, there is an increase in the total number of T-cells infiltrating the lungs and lymph nodes of mice, and a dramatic increase in the number of both CD4+ and CD8+ T-cells in the lung, with a slightly higher proportion of CD4$^+$ than CD8+ cells (Jones *et al.*, 2002). It also appears that, in an antigen-specific manner, the lower respiratory tract lymph nodes exhibit a $T_h$2 response (IL-4), while the splenic response is $T_h$1 dominant (IFN-γ), and the lung response is mixed but skewed towards $T_h$1 (Jones *et al.*, 2002). An increased influx of T-cells and a mixed $T_h$1/$T_h$2 response (with a $T_h$1 skew) was also observed in the lungs of gnotobiotic mice infected multiple times with *M. pneumoniae* (Hayakawa *et al.*, 2002). It appears that the cytokines induced by T-cells in response to infection by mycoplasmas are varied and complex, indicating that multiple cell types in different tissues are essential to control disease.

Determining the role of T-cell subsets and their associated cytokines is crucial for understanding the pathogenesis of and recovery from mycoplasma infection, as well as for vaccine development. *In vivo* depletion of $CD8^+$ cells dramatically increases the severity of lesions in the lower respiratory tract and results in reduced body weight of infected animals, while depletion of $CD4^+$ cells results in decreased pathology but no change in *M. pulmonis* colonization of the lungs. Loss of either $CD4^+$ or $CD8^+$ cells dramatically reduces IFN-γ levels and depletion of both cell types drastically (but not totally) reduces levels of this cytokine, indicating that T-cells (and to a lesser degree other cell types, such as NK cells) are responsible for its production (Jones *et al.*, 2002). A follow-up study was conducted in which IL-4 and IFN-γ KO BALB/c mice were utilized to determine the role these cytokines play in T-cell responses and disease. Loss of either cytokine had little effect on disease or bacterial burden in the upper respiratory tract (URT), but IL-4 KO mice had greater infiltration of macrophages and $CD8^+$ cells into the URT. However, in the lungs the absence of IFN-γ (but not IL-4) resulted in both increased disease and mycoplasma colonization, indicating that a $T_h1$ response is essential to control infection (Woolard *et al.*, 2004). Interestingly, it appears that, while there was more immunopathology in IFN-γ KO mice, the increased population of lymphocytes was not sufficient to combat the pathogen. The same was observed with $CD8^+$ depleted mice in the earlier study. The authors inferred that the presence of $T_h2$ cells, in the absence of $T_h1$ cells, was potentially responsible for the observed immunopathology, in part due to the absence of early innate IFN-γ responses from NK, NKT and γδ T-cells, resulting in defective macrophage activation. Taken together, these results indicate that CTL and IFN-γ responses are essential to control inflammation in the lungs, but may not contribute to the decreasing bacterial load in mice. This conclusion may seem counterintuitive, as IFN-γ and CTL responses are typically thought of as 'inflammatory' and would likely increase immunopathology. However, as is well known from the asthma literature, a $T_h2$-biased inflammatory response can be very potent (Woodruff *et al.*, 2009) due to the influx of leucocytes into the respiratory tissues. This inflammation is exacerbated in the absence of IFN-γ (at least in part produced by CTLs) which sometimes counteracts the effects of $T_h2$ cytokines.

## Activation of CTLs by MAM superantigen

Currently, there is no evidence that $CD8^+$ cells directly respond to (or recognize) host cells infected with mycoplasmas, even in light of the potential of an intracellular niche exploited by certain species within the genera. However, mitogenic activation of lymphocytes (including CTLs) can be induced by the superantigen MAM of the murine pathogen *M. arthritidis* via binding to MHC-II molecules on target cells. It was determined that human $TCR\alpha/\beta^+/CD8^+$ CTLs are among the cells that proliferate and exhibit cytotoxic effects when co-incubated with MAM and MHC-II expressing accessory/target cells. This interaction is independent of antigen presented by MHC and was blocked upon addition of anti-CD2, CD3, or TCRα/β antibodies (Matthes *et al.*, 1988). These findings were further verified when multiple bacterial superantigens were compared and MAM was found to be capable of activating $CD8^+$ cells in the presence of MHC-II bearing accessory/target cells (Fleischer, 1991).

Only a small subset of human $CD8^+$ T-cells are affected by MAM, but a large proportion of murine CTLs become activated when in contact with this superantigen (Fleischer, 1991). Most of the human MAM reactive T-cells express the TCR Vβ17 gene (although other genes have been shown to be involved) and greater than 60 per cent of these cells are CTLs (Friedman *et al.*, 1991). Alvarez-Ossorio *et al.* further elucidated the molecular interactions between MHC, MAM, and the Vβ restricted TCR by collecting sera from arthritis patients and determining which individuals produced high levels of cytokines. HLA-DR4, DR7, and DR12 restricted patients showed the highest levels of cytokine secretion in cell-free supernatants collected from peripheral blood stimulated with MAM. Similarly, when murine T-cells bearing a Vβ8.1 TCR were co-stimulated with MAM, only HLA-DR4 and DR7 expressing B-cells stimulated proliferation of the murine CTLs, indicating that $CD8^+$ activation is HLA-dependent and that these HLA subtypes (or similar subtypes of mice) may play a role in rheumatic inflammation in MAM-infected animals (Alvarez-Ossorio *et al.*, 1997).

## Role of CMI in *Mycoplasma gallisepticum* infection

Avian mycoplasma infections are found most commonly in the respiratory and reproductive tracts, the conjunctiva, and may also involve systemic tissues such as the brain and blood. Tracheitis is an important component of the inflammatory response of chickens to *M. gallisepticum* infection. Disease progresses slowly and lesions/mucosal thickness peaks at approximately 8–12 days post infection, which coincides with a decline in the mycoplasmal burden in this tissue. As noted above, early lesions are typically composed of heterophils and macrophages, and then later by large numbers of lymphocytes. A high proportion of the infiltrating lymphocytes are T-cells, but their subtypes have not been well-characterized. Blood leucocyte migration inhibition (LMI) has been shown when tested *in vitro* following antigen exposure (Chhabra and Goel, 1981). This response was well correlated with haemagglutination inhibition (HI) titres. LMI in tracheal washes also paralleled HI titres and was inversely correlated with lesion score and organism recovery. In addition, all seven *M. gallisepticum* infected chickens demonstrated a delayed-type hypersensitivity response. The role of CMI in mycoplasma clearance has also been inferred from studies in which thymectomized birds were treated with anti-thymocyte antisera. Although some residual T-cell activity was noted, the birds demonstrated an impaired capacity to clear *M. gallisepticum* infection and developed more severe tracheal lesions (Tiwary, 1986). It is not clear if this failure relates to aberrant regulation of humoral immune responses mediated by T helper cells or due to defects in cytokine/chemokine/inflammatory regulation via $TCR^{-/-}$, NK like cells (Gaunson *et al.*, 2006). Defects in T helper responses are likely to have contributed to the disease exacerbation, as HI titres were dramatically reduced in the T-cell deficient birds. However, no analysis of individual cell populations was performed (Tiwary, 1986). The distribution of $CD4^+$, $CD8^+$, and TCRα/β 1/2+ T-cells has been examined at varying times post-infection. These cells are found in the tracheas of chickens throughout the course of infection, with $CD4^+$ cells dispersed throughout the mucosa, while $CD8^+$ cells form follicles (Gaunson *et al.*, 2000). A subsequent study found that $CD8^+$ cells infiltrating the trachea during the first week post-infection do not express a TCR and are the predominant cells observed in these follicles. In the following weeks, $CD8^+$ cells become $TCR^+$ and the centres of the follicles predominantly contained central B-cells, while the $CD8^+$ cells spread to the periphery (Gaunson *et al.*, 2006). Interestingly, *M. gallisepticum* infection studies in chickens conducted by our laboratory have shown that $CD8^+$ cells tend to be dispersed throughout the mucosa, and follicles only form when birds are vaccinated and then challenged. In these studies, vaccinated and challenged chickens produce follicles comprised predominantly of B-cells and $CD4^+$ T-cells (even during the first week of infection) (Javed *et al.*, 2005). Differences in the observations by the two groups may be attributed to the different challenge strains of *M. gallisepticum* (Ap3AS versus $R_{low}$), local differences in the genetic background of the birds, and/or the route of inoculation (aerosol versus intratracheal). Taken together, it appears that cell-mediated immunity is involved with clearance and host-resistance at some level, but the cells involved and their roles remain undefined. In unpublished work in our laboratory, we were unable to detect mycoplasma antigen by immunohistochemical analysis or mycoplasma ribosomal RNA by *in situ* hybridization in tracheal sub-epithelial tissues, even in the presence of intense staining at the mucosal surface. Thus, the role of CTLs and other T-cell subsets may be restricted to immunoregulation rather than cell-mediated killing. Conversely, soluble factors or antigenic fragments may have been present in sub-epithelial compartments, but may not have been detected in our assay system.

## $CD4^+$ memory responses to MMMSC

The role of $CD4^+$ T-cells in memory responses to antigens has been characterized after infection with *M. mycoides* subsp. *mycoides small colony* (MMMSC). $T_h$ cells appear to be the primary PBMC type found in blood after infection and recovering animals have $T_h$ cells that produce large amounts of IFN-γ following re-stimulation in culture and proliferate in response to heat-killed MMMSC (Dedieu *et al.*, 2005). Animals that succumbed to the disease did not show as robust a response in terms of IFN-γ production from $CD4^+/CD25^+$ lymphocytes. Indeed, $CD4^+/CD25^+$ cells collected from the lymph nodes of recovered animals retained the ability to proliferate and secrete IFN-γ in response to the MMMSC lipoprotein LppA one year after infection (Dedieu *et al.*, 2010). The long duration of the response to specific antigens after animals became ill indicates that the cattle maintained memory $T_h$ cells. Two subsets of $CD4^+$ memory cells were identified in recovered animals; CD45RO(+)CD45R(-)CD62L(-) short-lived effector memory cells that secrete IFN-γ (accounting for about 2/3 of IFN-γ secreting $T_h$ cells), as well as longer-lived CD45RO(+)CD45R(-)CD62L(+) central memory cells that were the major proliferating cell type in response to specific antigen. Effector memory $T_h$ cells up-regulated the CCR5 receptor and reduced expression of CCR7/CD62L in response to MMMSC, a phenotype associated with migration into infected tissues and the expression of effector cytokines such as IFN-γ. Conversely, central memory cells proliferated in response to MMMSC antigen, but did not produce IFN-γ, consistent with recirculation within the lymphatic system (Totte *et al.*, 2010a). In contrast, a recent study showed that $CD4^+$ T-cell depletion in cattle prior to infection with MMMSC resulted in a mild increase in death, indicating that other immune mechanisms are likely to be important in mediating protection (Sacchini *et al.*, 2011).

Identification of antigenic proteins that elicit $T_h$ central memory responses are of crucial importance in vaccine design, and thus far three reactive MMMSC proteins have been found. Lipoprotein LppA has already been mentioned, but potent $CD4^+$ antigens have also been found in the glucose transporter PtsG, and to a lesser extent in the ABC transporter MSC_0804 (Totte *et al.*, 2010b). All three proteins are intercalated into the membrane of MMMSC (they were selected for this study because they had previously been shown to contain B-cell epitopes) and each is capable of stimulating proliferation in effector and central memory cells. Taken together, these data indicate that surface proteins of MMMSC can elicit B-cell responses and are also recognized by $T_h$ cells.

## $T_h$17 responses to Mycoplasma infection

In addition to traditional $CD4^+$ T-cell cytokine responses ($T_h$1 and $T_h$2), several additional T-cell subsets are now recognized, including $T_h$17 cells, which are characterized by the secretion of IL-17A upon stimulation by IL-23. $T_h$17 cytokines are pro-inflammatory and are often associated with bacterial infections and autoimmune disease states. $T_h$17 cells (or their associated cytokines) have been found in mice infected with several different species of mycoplasmas. Indeed, at both 4 and 24 hours post-infection with *M. pneumoniae,* IL-23 is up-regulated in the lungs of mice, while IL-17 is expressed within 24 hours and persists for over a week (Wu *et al.*, 2007). From this study, alveolar macrophages appear to be the major source of IL-23 while $CD4^+$ T-cells appear to be the primary producers of IL-17, indicating that alveolar macrophages may initially respond the mycoplasma and drive $CD4^+$ T-cells to commit to a $T_h$17 lineage. Conversely, our own work has shown that significant IL-17 levels are not observed in BAL fluid unless mice are re-exposed to *M. pneumoniae,* and is also associated with increased disease severity and a switch from neutrophils to eosinophils as the predominant BAL leucocyte population (Szczepanek *et al.*, 2012). Thus, it is clear that IL-17 is important during *M. pneumoniae* infection but its precise role remains undefined.

While the mechanism is not entirely clear, it appears that NK cells may influence lineage decisions of $T_h$ cells, including $T_h$17. Lymphocytes isolated from the lungs of mice inoculated with *M. pulmonis* antigens and depleted of NK cells have reduced IFN-γ, IL-4, and IL-17 secretion upon re-stimulation by antigen, while lymphocytes isolated from the lymph nodes produce significant amounts of IFN-γ, but lower concentrations of IL-4, IL-13, and IL-17 (Bodhankar *et al.*, 2009). NK cell depleted mice were significantly more resistant to *M. pulmonis* infection, with a 100-fold reduction in organisms found in lungs at 7 and 14 days post infection compared to controls. Thus, NK cells may suppress $T_h17$ responses in the lymph nodes by increasing expression of IFN-γ. Another point pertaining to helper T-cells made by Bodhankar *et al.* is that depletion of $CD4^+$ cells eliminates protection afforded by immunization with *M. pulmonis* antigens, indicating that these cells are necessary for protective immunity (Bodhankar *et al.*, 2009).

In addition to various cells of the immune system, virulence factors of mycoplasmas may play a regulatory role in the differentiation of $CD4^+$ cells into the $T_h17$ lineage. Recently, the *M. arthritidis* MAM superantigen was shown to delay the expression of IL-17 in TLR4 deficient mice (C3H/HeJ) compared to controls. The higher expression of IL-6 in wild type mice and higher expression of IL-1β in TLR4 deficient mice appears to be responsible for the downstream differential expression of IL-17. Not surprisingly, blocking the expression of TLR4 in MAM-inoculated wild type mice results in reduced detection of IL-6 and IL-17 (but not IL-23). However, TLR2 KO mice produce higher levels of IL-6 and IL-17 than wild type mice in response to MAM (Mu *et al.*, 2011). The effect that ligation of *M. arthritidis* LPs to TLR2 in the presence of MAM has on cytokine secretion by $T_h$ cells has not been studied, but it is possible that the differential use of TLRs by these ligands contributes to immune dysregulation during infection.

While undoubtedly common in the field, until recently, polymicrobial diseases have rarely been studied in the laboratory. This is unfortunate because the complex dynamics that occur in a host during multiple infections can teach us much about host–pathogen interactions. Interestingly, infection of mice with *M. pulmonis* appears to enhance clearance of *Listeria monocytogenes* 72 hours post-challenge. This cross-protection was attributed to the production of IL-17 by $CD4^+$ T-cells, along with the corresponding infiltration of neutrophils associated with secretion of this cytokine. This is supported by the reduced clearance of both *M. pulmonis* and *L. monocytogenes* in either neutrophil depleted or IL-17R KO mice (Sieve *et al.*, 2009). Hence, identification of the molecular factors that drive a $T_h17$ response following *M. pulmonis* infection could lead to the development of therapeutic agents to prevent infection with this and other bacterial pathogens.

## Antibody responses to Mycoplasmas

The generation of effective antibody responses directed against many mycoplasma species appears to correlate with protection against infection and disease. However, many mycoplasmas have developed mechanisms to counteract host defences, and may even render them maladaptive. Expanding our knowledge of the development of protective versus pathogenic (or ineffectual) antibody responses will be crucial for the development of rationally designed next-generation vaccines.

### Mucosal antibody responses to *Mycoplasma gallisepticum*

Antibody responses to *Mycoplasma gallisepticum* are crucial for controlling disease (Tiwary, 1986) and for the development of effective vaccines (Javed *et al.*, 2005). Unfortunately, this pathogen has evolved ways to dysregulate the immune system in such a manner that humoral responses are often ineffective. *M. gallisepticum* infection in the tracheas of infected chickens induces a robust antibody response. Early studies suggested that antibody responses are critically important for immune protection, as bursectomized chickens were more susceptible to disease than thymectomized birds (Adler *et al.*, 1973; Lam and Lin, 1984), and hyperimmune serum transferred to naïve animals conferred protection upon challenge (Lin, 1984). IgG titres peak at about two weeks post-inoculation (PI) in tracheal washes of infected birds, and at 4 weeks in serum, and are still detectable in serum 25 weeks after infection and at least 16 weeks after infection in tracheal washes (Elfaki *et al.*, 1992). Tracheal wash HI titres are detectable at around three weeks after infection and peak at six weeks after infection (Chhabra and Goel, 1981). Serum HI and leucocyte migration inhibition scores both peak at seven weeks after infection and are inversely correlated with lesion scores. Our own work (and that of others) on the correlates of immune protection against *M. gallisepticum* has shown that infected chickens produce a thick and diffuse lymphoproliferative response in the trachea that includes many IgG and IgA secreting B-cells, a large proportion of which are not *M. gallisepticum* specific (Javed *et al.*, 2005). However, birds that are vaccinated have far fewer lymphocytic infiltrates and these cells often appear to be arranged in well-ordered follicles. Vaccination protects chickens from the immunopathology observed in sham-vaccinated birds as these animals produce a higher proportion of *M. gallisepticum* specific B-cells (Javed *et al.*, 2005). These results indicate that natural infection produces a dysregulated and exaggerated immune response that is not effective for controlling disease. Furthermore, vaccination alone appears to produce little transudated IgG and no secretory IgA in tracheal washes, although subsequent challenge with a virulent strain results in high titres of *M. gallisepticum* specific antibodies (Papazisi *et al.*, 2002b). Interestingly, in this study the sham-vaccinated birds produced high titres of secretory IgA upon challenge, which did little to clear the infection. Later studies conducted by our group showed that both IgG and IgA tracheal wash titres are much higher in sham-vaccinated and challenged chickens than in those that are vaccinated before challenge (likely due to the high number of infiltrating B-cells in the sham vaccinated birds, as was observed by Javed *et al.*, 2005). As noted above, vaccinated birds produce much higher titres of IgG upon challenge and this appears to protect the animals from disease. These data collectively indicate that *M. gallisepticum*-specific transudated IgG, ostensibly produced locally by resident plasma cells, and not secretory IgA, is well correlated with vaccine-induced protection, while secretory IgA (and some elements of cell mediated immunity) (Tiwary, 1986) may be more important for recovery from primary infection. In this regard, it is important to note that a recent study indicates that the CysP protein expressed by *M. gallisepticum* is capable of papain-like cleavage of chicken IgG (Cizelj *et al.*, 2011). It is not known if this factor is secreted or surface expressed, so its precise role in immune evasion/dysregulation/pathogenesis is unknown.

B-cells elicit antibody responses against numerous *M. gallisepticum* proteins during infection. Young chicks appear to preferentially produce IgA antibodies, as two to three times as many *M. gallisepticum* proteins are recognized by IgA than IgG in tracheal washes. However, as the birds age this trend reverses and chickens tend to target more *M. gallisepticum* proteins using transudated IgG antibodies than secretory IgA. Over time, fewer anti-*M gallisepticum* antibodies of either class can be found in tracheal washes, but there appears to be an increase in the number of *M. gallisepticum* proteins recognized in serum (Ellakany *et al.*, 1998).

The functional role of mucosal antibodies are often different from those directed against microbes found in circulation, as they are not very efficient at opsonizing the pathogen, but may block binding to host cell receptors. Indeed, washings from either the URT or LRT of chickens previously infected with *M. gallisepticum* block *ex vivo* attachment of co-incubated *M. gallisepticum* to tracheal rings. A correlation was noted in this experiment that blocking is correlated with the *M. gallisepticum*-specific IgG concentrations, but not the IgA concentrations (Avakian and Ley, 1993). Other studies show that anti-lipoprotein Lp64 (VlhA) sera co-incubated with *M. gallisepticum* is sufficient to block attachment to tracheal rings (Avakian and Ley, 1993; Forsyth *et al.*, 1992). These studies diverge on findings of the ability of anti-Lp64 sera to also block haemagglutination, but such a phenotype is logical as VlhA is the major haemagglutinin of *M. gallisepticum*. Additionally, antiserum directed against the cytadhesin GapA has been shown to reduce growth of *M. gallisepticum* in tracheal organ cultures by 64% (Goh *et al.*, 1998). Insertion of the *gapA* gene into the high passage attenuated strain, $R_{high}$, resulted in the generation of the efficacious vaccine strain GT5, which was shown to be capable of inducing anti-GapA antibodies (Papazisi *et al.*, 2002a). Given these results, it has been postulated that anti-GapA antibodies induced by vaccination of chickens with GT5 block attachment of virulent strains upon challenge, aiding in protection from disease. Taken together, it appears that transudated IgG is protective during *M. gallisepticum* infection as it targets multiple proteins, including those involved in attachment to host cells.

## *Mycoplasma pneumoniae* infection, serum IgE and asthma

In recent years it has come to light that atopic asthma can be caused (or at least exacerbated) by some respiratory pathogens. *M. pneumoniae* appears to be among these pathogens, and there is a correlation between anti-*M. pneumoniae* serum IgE antibodies and the exacerbation of asthma. Indeed, anti-*M. pneumoniae* IgE, IgG1, and IgG3 antibodies have been detected in large proportions of asthmatic patients previously infected with this pathogen (Seggev *et al.*, 1996). Some studies have further correlated the association of *M. pneumoniae* infection, IgE serum antibodies, and asthma in children that develop wheeze during infection (Nagayama and Sakurai, 1991; Chung *et al.*, 2006), although this has been contradicted in at least one other study (Choi *et al.*, 2009). However, one clinical report suggests that not only can *M. pneumoniae* exacerbate atopic asthma, it may also be an etiologic agent. This 37 year old patient developed asthma one month after infection, which persisted for at least two years based on *M. pneumoniae* specific serum IgE and airway hyper-responsiveness to *M. pneumoniae* protein (Yano *et al.*, 1994). Anti-*M. pneumoniae* serum IgE persists in some asthmatic patients for 2 to 16 months (Tipirneni *et al.*, 1980; Stelmach *et al.*, 2005). Interestingly, total and allergen-specific IgE antibodies tend to wane as children progress from the acute to convalescent phase of *M. pneumoniae* disease (Nagayama *et al.*, 1987), indicating that infection modulates allergy. IgE-mediated antibody responses to *M. pneumoniae* infection tend to be driven by a $T_h2$ cytokine milieu. For example, bronchoalveolar lavage fluids (BALF) from *M. pneumoniae* infected patients have higher levels of IL-4 ($T_h2$) and equivalent levels of IFN-gamma ($T_h1$) compared to *Streptococcus pneumoniae* infected patients or controls (Koh *et al.*, 2001). Furthermore, patients with *M. pneumoniae* lower respiratory tract infection have higher levels of $T_h2$ cytokines and pathogen-specific serum IgE than those who have infections in the upper respiratory tract (Hassan *et al.*, 2008), with IL-5 elevated in both sites, indicating that *M. pneumoniae* infection is associated with the onset of atopic disease. This may be explained by the finding that *M. pneumoniae* specific IgE is capable of binding to the FcεRI found on mast cells, resulting in the release of IL-4 and other Th2 cytokines from these cells (Anand *et al.*, 2012), (Luo *et al.*, 2008). Moreover, children with *M. pneumoniae* infection and wheeze were found to have higher serum IgE titres and increased expression of endothelin-1 (Chung *et al.*, 2006). This vasoconstrictor has been implicated in airway remodelling associated with airway hyper-reactivity, providing another possible mechanism for exacerbation of asthma by *M. pneumoniae*. Chung *et al.* also found increased levels of serum IL-5 in children with *M. pneumoniae* infection and wheeze [and this correlation has been substantiated by others (Esposito *et al.*, 2002)], indicating that high levels of serum IL-5 are associated with *M. pneumoniae* exacerbated asthma.

## Anti-mycoplasma antibody mediated arthritis

The sequelae associated with mycoplasmoses are vast and varied. Infection with several species is associated with the development of arthritic disease; some through direct colonization of synovial tissues and others through antibody-mediated mechanisms. One mycoplasma that is notorious for the induction of antibody-mediated arthritic disease is *M. arthritidis*. Virulent strains of this pathogen are acutely arthritogenic in infected rats, while attenuated strains take much longer to induce arthritis (Matsumoto *et al.*, 1998; Washburn and Ramsay, 1989). This implies that even attenuated strains of *M. arthritidis* are capable of persisting *in vivo* for prolonged periods of time. It has been found that sera from patients with rheumatoid arthritis (RA) tend to have increased titres of anti-MAM IgG antibodies when compared to controls (including patients with other immune based diseases). This increase in serum anti-MAM IgG was also observed (and shown to be specific) when antibodies from the same RA patients were tested for reactivity with the staphylococcal enterotoxin, indicating that the reactions are not an experimental artefact (Sawitzke *et al.*, 2000). However, a recent study reported that RA patients do not have higher IgG or IgM antibody titres directed against MAM than controls. Serum from RA patients (but not patients with systemic lupus erythematosus) has been shown to react with LPs from *M. fermentans* and *M. hominis* (da Rocha Sobrinho *et al.*, 2011). Indeed, another study found that blood from 23% of RA patients harboured *M. fermentans* detectable by PCR and/or culture. Not surprisingly, these RA patients had increased levels of anti-*M. fermentans* IgG and IgM antibodies when compared to controls (Gil *et al.*, 2009).

There appears to be an even stronger association between antibody-mediated arthritic disease and infection with *M. pneumoniae*. One patient was reported as presenting with polyarthritis, with synovial fluid both PCR and IgG positive to *M. pneumoniae*. No other pathogens could be directly or indirectly detected in these samples (Chaudhry *et al.*, 2003). Furthermore, one case–control study found a correlation between anti-*M. pneumoniae* serum IgG and RA (Ramirez *et al.*, 2005). The association between *M. pneumoniae* and arthropathy is often diagnosed in children, and can be confirmed by observation of clinical disease in these patients and the presence of anti-*M. pneumoniae* ELISA titres (Cimolai *et al.*, 1989). Some evidence indicates that infection with *M. pneumoniae* is a direct cause of reactive arthritis in children, as infection precedes the autoimmune response. This is evidenced by high levels of anti-*M. pneumoniae* IgM in the serum of a large proportion of these patients (Harjacek *et al.*, 2006). Altogether, these results suggest that human mycoplasmas have a strong association with RA and other arthritic ailments and are possible etiologic agents of these diseases.

The mechanism by which mycoplasmas cause antibody-mediated arthritic disease is not well understood, but a few examples suggest that it may be a result of the development of circulating immune complexes (ICs). A rabbit model of mycoplasma-induced chronic arthritis provides evidence that arthritic disease is caused by ICs. Synovial fluid from infected animals harbours B cells that produce antibodies involved in the development of ICs (Washburn *et al.*, 1980).

In rats, *M. arthritidis* infection results in the production of anti-*M. arthritidis* antibodies and polyarthritis that can last up to nine months. ICs associated with these antibodies, however, can be detected as soon as four days post-infection and appear to be present in the absence of rheumatoid factors (Kirchhoff *et al.*, 1983). The presence of these *M. arthritidis*-associated ICs has also been attributed to ocular inflammation. In fact, such ICs have been found in the circulation and ocular vasculature of infected rats. This inflammation does not appear to have an effect on the vision of affected animals (Thirkill *et al.*, 1992). Thus, it is not surprising that ICs can be purified from the synovial fluid from RA patients. Antibodies coupled with these ICs react with proteins from several species of mycoplasmas (Clark *et al.*, 1988), suggesting a mechanism by which these pathogens indirectly contribute to RA. Such studies point to one possible means by which mycoplasmas can cause arthritic disease, but much more research is needed to fully understand the pathogenesis of these complex diseases.

## Conclusions

Many mycoplasmas have evolved elaborate and diverse strategies to manipulate and evade host immune responses, and in so doing have carved out unique ecological niches. Absent a cell wall, their parasitic lifestyle is often not much more than a nuisance for the host, but in other cases can lead to important diseases due to invasive and pathogenic characteristics. Understanding the details of these fascinating host–pathogen interactions is the key to the development of prophylactic vaccines and therapeutic agents, and research to date has already shed considerable light on many evolutionary strategies. The correlates of immune protection remain undefined for all but a few mycoplasmas, and much work lies ahead to define the role of the innate and adaptive immune responses, including the role of humoral antibodies and cell-mediated immunity. The development of genetic tools, including synthetic biological approaches for manipulating mycoplasma genomes, will hasten the rate of discovery of these mechanisms and is anticipated to lead to a deeper understanding of this important genus.

## Acknowledgements

We thank Debra Rood for proofreading the manuscript and managing the references. We apologize to any authors whose papers were not referenced due to space considerations.

## References

Adler, H.E., Bryant, B.J., Cordy, D.R., Shifrine, M., and DaMassa A.J. (1973). Immunity and mortality in chickens infected with *Mycoplasma gallisepticum*: influence of the bursa of fabricius. J. Infect. Dis. *127*, S61-S66.

Aliprantis, A.O., Yang, R.B., Weiss, D.S., Godowski, P., and Zychlinsky, A. (2000). The apoptotic signaling pathway activated by Toll-like receptor-2. EMBO J. *19*, 3325–3336.

Alvarez-Ossorio, L., Johannsen, M., Russlies, M., Kirchner, H., and Rink, L. (1997). HLA-dependent heterogeneous T-cell response to *Mycoplasma arthritidis*-derived superantigen (MAS). Med. Microbiol. Immunol. *185*, 245–251.

Anand, P., Singh, B., Jaggi, A.S., and Singh, N. (2012). Mast cells: an expanding pathophysiological role from allergy to other disorders. Naunyn-Schmiedeberg's Arch. Pharmacol. *385*, 657–670.

Avakian, A.P., and Ley, D.H. (1993). Protective immune response to *Mycoplasma gallisepticum* demonstrated in respiratory-tract washings from *M. gallisepticum*-infected chickens. Avian Dis. *37*, 697–705.

Baier, R.J., Majid, A., Parupia, H., Loggins, J., and Kruger, T.E. (2004). CC chemokine concentrations increase in respiratory distress syndrome and correlate with development of bronchopulmonary dysplasia. Pediatr. Pulmonol. *37*, 137–148.

Baseman, J.B., and Tully, J.G. (1997). Mycoplasmas: sophisticated, reemerging, and burdened by their notoriety. Emerg. Infect. Dis. *3*, 21–32.

Bodhankar, S., Woolard, M.D., Sun, X., and Simecka, J.W. (2009). NK cells interfere with the generation of resistance against mycoplasma respiratory infection following nasal-pulmonary immunization. J. Immunol. *183*, 2622–2631.

Chaudhry, R., Nisar, N., Malhotra, P., Kumar, A., and Chauhan, V.S. (2003). Polymerase chain reaction confirmed *Mycoplasma pneumoniae* arthritis: a case report. Indian J. Pathol. Microbiol. *46*, 433–436.

Chhabra, P.C., and Goel, M.C. (1981). Immunological response of chickens to *Mycoplasma gallisepticum* infection. Avian Dis. *25*, 279–293.

Chin, A.C., Fournier, B., Peatman, E.J., Reaves, T.A., Lee, W.Y., and Parkos, C.A. (2009). CD47 and TLR-2 cross-talk regulates neutrophil transmigration. J. Immunol. *183*, 5957–5963.

Choi, I.S., Byeon, J.H., Yoo, Y., Lee, K.C., and Choung, J.T. (2009). Increased serum interleukin-5 and vascular endothelial growth factor in children with acute mycoplasma pneumonia and wheeze. Pediatr. Pulmonol. *44*, 423–428.

Chung, H.L., Kim, S.G., and Shin, I.H. (2006). The relationship between serum endothelin (ET)–1 and wheezing status in the children with *Mycoplasma pneumoniae* pneumonia. Pediatr. Allergy Immunol. *17*, 285–290.

Cimolai, N., Malleson, P., Thomas, E., and Middleton, P.J. (1989). *Mycoplasma pneumoniae* associated arthropathy: confirmation of the association by determination of the antipolypeptide IgM response. J. Rheumatol. *16*, 1150–1152.

Cizelj, I., Bercic, R.L., Dusanic, D., Narat, M., Kos, J., Dovc, P., and Bencina, D. (2011). *Mycoplasma gallisepticum* and *Mycoplasma synoviae* express a cysteine protease CysP, which can cleave chicken IgG into Fab and Fc. Microbiology *157*, 362–372.

Clark, H.W., Coker-Vann, M.R., Bailey, J.S., and Brown, T.M. (1988). Detection of mycoplasmal antigens in immune complexes from rheumatoid arthritis synovial fluids. Ann. Allergy *60*, 394–398.

da Rocha Sobrinho, H.M., Jarach, R., da Silva, N.A., Shio, M.T., Jancar, S., Timenetsky, J., Oliveira, M.A., Dorta, M.L., and Ribeiro-Dias, F. (2011). Mycoplasmal lipid-associated membrane proteins and *Mycoplasma arthritidis* mitogen recognition by serum antibodies from patients with rheumatoid arthritis. Rheumatol. Int. *31*, 951–957.

Dedieu, L., Balcer-Rodrigues, V., Yaya, A., Hamadou, B., Cisse, O., Diallo, M., and Niang, M. (2005). Gamma interferon-producing CD4 T-cells correlate with resistance to *Mycoplasma mycoides* subsp. *mycoides* S.C. infection in cattle. Vet. Immunol. Immunopathol. *107*, 217–233.

Dedieu, L., Totte, P., Rodrigues, V., Vilei, E.M., and Frey, J. (2010). Comparative analysis of four lipoproteins from *Mycoplasma mycoides* subsp. *mycoides* Small Colony identifies LppA as a major T-cell antigen. Comp. Immunol. Microbiol. Infect. Dis. *33*, 279–290.

Deiters, U., and Mühlradt, P.F. (1999). Mycoplasmal lipopeptide MALP-2 induces the chemoattractant proteins macrophage inflammatory protein 1à (MIP-1à), monocyte chemoattractant protein 1, and MIP – 2 and promotes leukocyte infiltration in mice. Infect. Immun. *67*, 3390–3398.

Elfaki, M.G., Ware, G.O., Kleven, S.H., and Ragland, W.L. (1992). An enzyme-linked immunosorbent assay for the detection of specific IgG antibody to *Mycoplasma gallisepticum* in sera and tracheobronchial washes. J. Immunoassay *13*, 97–126.

Ellakany, H., Fabian, K., Nemeth, I., and Stipkovits, L. (1998). Antibody response detected by immunoblot in respiratory tract washings of chickens after infection with *Mycoplasma gallisepticum*. Avian Pathol. *27*, 547–554.

Esposito, S., Droghetti, R., Bosis, S., Claut, L., Marchisio, P., and Principi, N. (2002). Cytokine secretion in children with acute *Mycoplasma pneumoniae* infection and wheeze. Pediatr. Pulmonol. *34*, 122–127.

Farhat, K., Riekenberg, S., Heine, H., Debarry, J., Lang, R., Mages, J., Buwitt-Beckmann, U., Roschmann, K., Jung, G., Wiesmuller, K.H., *et al.* (2008). Heterodimerization of TLR2 with TLR1 or TLR6 expands the ligand spectrum but does not lead to differential signaling. J. Leukoc. Biol. *83*, 692–701.

Fleischer, B. (1991). Stimulation of the immune system by microbial 'superantigens'. Immun. Infekt. *19*, 8–11.

Forsyth, M.H., Tourtellotte, M.E., and Geary, S.J. (1992). Localization of an immunodominant 64 kDa lipoprotein (LP 64) in the membrane of *Mycoplasma gallisepticum* and its role in cytadherence. Mol. Microbiol. *6*, 2099–2106.

Friedman, S.M., Crow, M.K., Tumang, J.R., Tumang, M., Xu, Y.Q., Hodtsev, A.S., Cole, B.C., and Posnett, D.N. (1991). Characterization of human T-cells reactive with the *Mycoplasma arthritidis*-derived superantigen (MAM): generation of a monoclonal antibody against V beta 17, the T-cell receptor gene product expressed by a large fraction of MAM-reactive human T-cells. J. Exp. Med. *174*, 891–900.

Gaunson, J.E., Philip, C.J., Whithear, K.G., and Browning, G.F. (2000). Lymphocytic infiltration in the chicken trachea in response to *Mycoplasma gallisepticum* infection. Microbiology *146*, 1223–1229.

Gaunson, J.E., Philip, C.J., Whithear, K.G., and Browning, G.F. (2006). The cellular immune response in the tracheal mucosa to *Mycoplasma gallisepticum* in vaccinated and unvaccinated chickens in the acute and chronic stages of disease. Vaccine *24*, 2627–2633.

Gerlic, M., Horowitz, J., Farkash, S., and Horowitz, S. (2007). The inhibitory effect of *Mycoplasma fermentans* on tumour necrosis factor (TNF)-alpha-induced apoptosis resides in the membrane lipoproteins. Cell. Microbiol. *9*, 142–153.

Gil, C., Rivera, A., Banuelos, D., Salinas, S., Garcia-Latorre, E., and Cedillo, L. (2009). Presence of *Mycoplasma fermentans* in the bloodstream of Mexican patients with rheumatoid arthritis and IgM and IgG antibodies against whole microorganism. BMC Musculoskelet. Disord. *10*, 97.

Goh, M.S., Gorton, T.S., Forsyth, M.H., Troy, K.E., and Geary, S.J. (1998). Molecular and biochemical analysis of a 105 kDa *Mycoplasma gallisepticum* cytadhesin (GapA). Microbiology *144*, 2971–2978.

Harjacek, M., Ostojic, J., and Djakovic Rode, O. (2006). Juvenile spondyloarthropathies associated with *Mycoplasma pneumoniae* infection. Clin. Rheumatol. *25*, 470–475.

Hassan, J., Irwin, F., Dooley, S., and Connell, J. (2008). *Mycoplasma pneumoniae* infection in a pediatric population: analysis of soluble immune markers as risk factors for asthma. Hum. Immunol. *69*, 851–855.

Hayakawa, M., Taguchi, H., Kamiya, S., Fujioka, Y., Watanabe, H., Kawai, S., and Kobayashi, H. (2002). Animal model of *Mycoplasma pneumoniae* infection using germfree mice. Clin. Diagn. Lab. Immunol. *9*, 669–676.

Hoebe, K., Georgel, P., Rutschmann, S., Du, X., Mudd, S., Crozat, K., Sovath, S., Shamel, L., Hartung, T., Zahringer, U., *et al.* (2005). CD36 is a sensor of diacylglycerides. Nature *433*, 523–527.

Hopfe, M., and Henrich, B. (2008). OppA, the ecto-ATPase of *Mycoplasma hominis* induces ATP release and cell death in HeLa cells. BMC Microbiol. *8*, 55.

Into, T., and Shibata, K. (2005). Apoptosis signal-regulating kinase 1-mediated sustained p38 mitogen-activated protein kinase activation regulates mycoplasmal lipoprotein- and staphylococcal peptidoglycan-triggered Toll-like receptor 2 signalling pathways. Cell. Microbiol. *7*, 1305–1317.

Into, T., Fujita, M., Okusawa, T., Hasebe, A., Morita, M., and Shibata, K. (2002a). Synergic effects of mycoplasmal lipopeptides and extracellular ATP on activation of macrophages. Infect. Immun. *70*, 3586–3591.

Into, T., Nodasaka, Y., Hasebe, A., Okuzawa, T., Nakamura, J., Ohata, N., and Shibata, K. (2002b). Mycoplasmal lipoproteins induce Toll-like receptor 2- and caspases-mediated cell death in lymphocytes and monocytes. Microbiol. Immunol. *46*, 265–276.

Into, T., Okada, K., Inoue, N., Yasuda, M., and Shibata, K. (2002c). Extracellular ATP regulates cell death of lymphocytes and monocytes induced by membrane-bound lipoproteins of *Mycoplasma fermentans* and *Mycoplasma salivarium*. Microbiol. Immunol. *46*, 667–675.

Into, T., Kiura, K., Yasuda, M., Kataoka, H., Inoue, N., Hasebe, A., Takeda, K., Akira, S., and Shibata, K. (2004). Stimulation of human Toll-like receptor (TLR) 2 and TLR6 with membrane lipoproteins of *Mycoplasma fermentans* induces apoptotic cell death after NF-kappa B activation. Cell. Microbiol. *6*, 187–199.

Javed, M.A., Frasca, S., Jr., Rood, D., Cecchini, K., Gladd, M., Geary, S.J., and Silbart, L.K. (2005). Correlates of immune protection in chickens vaccinated with *Mycoplasma gallisepticum* strain GT5 following challenge with pathogenic *M. gallisepticum* strain R(low). Infect. Immun. *73*, 5410–5419.

Jones, H.P., Tabor, L., Sun, X., Woolard, M.D., and Simecka, J.W. (2002). Depletion of $CD8^+$ T-cells exacerbates $CD4^+$ Th cell-associated inflammatory lesions during murine mycoplasma respiratory disease. J. Immunol. *168*, 3493–3501.

Jordan, J.L., Chang, H.Y., Balish, M.F., Holt, L.S., Bose, S.R., Hasselbring, B.M., Waldo, R.H., 3rd, Krunkosky, T.M., and Krause, D.C. (2007). Protein P200 is dispensable for *Mycoplasma pneumoniae* hemadsorption but not gliding motility or colonization of differentiated bronchial epithelium. Infect. Immun. *75*, 518–522.

Kataoka, H., Yasuda, M., Iyori, M., Kiura, K., Narita, M., Nakata, T., and Shibata, K. (2006). Roles of N-linked glycans in the recognition of microbial lipopeptides and lipoproteins by TLR2. Cell. Microbiol. *8*, 1199–1209.

Kaufmann, A., Muhlradt, P.F., Gemsa, D., and Sprenger, H. (1999). Induction of cytokines and chemokines in human monocytes by *Mycoplasma fermentans*-derived lipoprotein MALP-2. Infect. Immun. *67*, 6303–6308.

Kirchhoff, H., Heitmann, J., Mielke, H., Dubenkropp, H., and Schmidt, R. (1983). Studies of polyarthritis caused by *Mycoplasma arthritidis* in rats. II. Serological investigation of rats experimentally infected with M. arthritidis ISR 1. Zentralbl. Bakteriol. Mikrobiol. Hyg. A *254*, 275–280.

Koh, Y.Y., Park, Y., Lee, H.J., and Kim, C.K. (2001). Levels of interleukin-2, interferon-gamma, and interleukin-4 in bronchoalveolar lavage fluid from patients with Mycoplasma pneumonia: implication of tendency toward increased immunoglobulin E production. Pediatrics *107*, E39.

Kovacs-Simon, A., Titball, R.W., and Michell, S.L. (2011). Lipoproteins of bacterial pathogens. Infect. Immun. *79*, 548–561.

Kumar, A., and Yerneni, L.K. (2009). Semi-automated relative quantification of cell culture contamination with mycoplasma by Photoshop-based image analysis on immunofluorescence preparations. Biologicals *37*, 55–60.

Lam, K.M., and DaMassa, A.J. (2000). *Mycoplasma gallisepticum* -induced release of macrophage inflammatory protein-1 beta from chicken monocytes-macrophages. J. Comp. Pathol. *122*, 35–42.

Lam, K.M., and Lin, W. (1984). Resistance of chickens immunized against *Mycoplasma gallisepticum* is mediated by bursal dependent lymphoid cells. Vet. Microbiol. *9*, 509–514.

Lavric, M., Bencina, D., Kothlow, S., Kaspers, B., and Narat, M. (2007). *Mycoplasma synoviae* lipoprotein MSPB, the N-terminal part of VlhA haemagglutinin, induces secretion of nitric oxide, IL-6 and IL-1beta in chicken macrophages. Vet. Microbiol. *121*, 278–287.

Lin, M.Y., and Kleven,S.H. (1984). Transferred humoral immunity in chickens to *Mycoplasma gallisepticum*. Avian Dis. *28*, 79–87.

Loveless, R.W., Griffiths, S., Fryer, P.R., Blauth, C., and Feizi, T. (1992). Immunoelectron microscopic studies reveal differences in distribution of sialo-oligosaccharide receptors for *Mycoplasma pneumoniae* on the epithelium of human and hamster bronchi. Infect. Immun. *60*, 4015–4023.

Luo, D., Dai, Y., Duffy, L.B., and Atkinson, T.P. (2008). Inhibition of message for FcepsilonRI alpha chain blocks mast cell IL-4 production induced by co-culture with *Mycoplasma pneumoniae*. Microb. Pathog. *44*, 286–292.

Markham, P.F., Kanci, A., Czifra, G., Sundquist, B., Hains, P., and Browning, G.F. (2003). Homologue of macrophage-activating lipoprotein in *Mycoplasma gallisepticum* is not essential for growth and pathogenicity in tracheal organ cultures. J. Bacteriol. *185*, 2538–2547.

Matsumoto, M., Nishiguchi, M., Kikkawa, S., Nishimura, H., Nagasawa, S., and Seya, T. (1998). Structural and functional properties of complement-activating protein M161Ag, a *Mycoplasma fermentans* gene product that induces cytokine production by human monocytes. J. Biol. Chem. *273*, 12407–12414.

Matthes, M., Schrezenmeier, H., Homfeld, J., Fleischer, S., Malissen, B., Kirchner, H., and Fleischer, B. (1988). Clonal analysis of human T-cell activation by the *Mycoplasma arthritidis* mitogen (MAS). Eur. J. Immunol. *18*, 1733–1737.

Medina, J.L., Coalson, J.J., Brooks, E.G., Winter, V.T., Chaparro, A., Principe, M.F., Kannan, T.R., Baseman, J.B., and Dube, P.H. (2012). *Mycoplasma pneumoniae* CARDS toxin induces pulmonary eosinophilic and lymphocytic inflammation. Am. J. Respir. Cell Mol. Biol. *46*, 815–822.

Medzhitov, R. (2007). Recognition of microorganisms and activation of the immune response. Nature *449*, 819–826.

Mitsunari, M., Yoshida, S., Shoji, T., Tsukihara, S., Iwabe, T., Harada, T., and Terakawa, N. (2006). Macrophage-activating lipopeptide-2 induces cyclooxygenase-2 and prostaglandin E(2) via Toll-like receptor 2 in human placental trophoblast cells. J. Reprod. Immunol. *72*, 46–59.

Mohammed, J., Frasca, S., Jr., Cecchini, K., Rood, D., Nyaoke, A.C., Geary, S.J., and Silbart, L.K. (2007). Chemokine and cytokine gene expression profiles in chickens inoculated with *Mycoplasma gallisepticum* strains Rlow or GT5. Vaccine *25*, 8611–8621.

Mu, H.H., Hasebe, A., Van Schelt, A., and Cole, B.C. (2011). Novel interactions of a microbial superantigen with TLR2 and TLR4 differentially regulate IL-17 and Th17-associated cytokines. Cell. Microbiol. *13*, 374–387.

Nagayama, Y., Sakurai, N., Kojima, S., and Funabashi, S. (1987). Total and specific IgE responses in the acute and recovery phases of respiratory infections in children. J. Asthma *24*, 159–166.

Nagayama, Y., and Sakurai, N. (1991). Clinical observations on lower respiratory tract infections with special reference to serum IgE levels. Pediatr. Pulmonol. *11*, 44–48.

Okusawa, T., Fujita, M., Nakamura, J., Into, T., Yasuda, M., Yoshimura, A., Hara, Y., Hasebe, A., Golenbock, D.T., Morita, M., *et al.* (2004). Relationship between structures and biological activities of mycoplasmal diacylated lipopeptides and their recognition by Toll-like receptors 2 and 6. Infect. Immun. *72*, 1657–1665.

Papazisi, L., Troy, K.E., Gorton, T.S., Liao, X., and Geary, S.J. (2000). Analysis of cytadherence-deficient, GapA-negative *Mycoplasma gallisepticum* strain R. Infect. Immun. *68*, 6643–6649.

Papazisi, L., Frasca, S., Jr., Gladd, M., Liao, X., Yogev, D., and Geary, S.J. (2002a). GapA and CrmA coexpression is essential for *Mycoplasma gallisepticum* cytadherence and virulence. Infect. Immun. *70*, 6839–6845.

Papazisi, L., Silbart, L.K., Frasca, S., Rood, D., Liao, X., Gladd, M., Javed, M.A., and Geary, S.J. (2002b). A modified live *Mycoplasma gallisepticum* vaccine to protect chickens from respiratory disease. Vaccine *20*, 3709–3719.

Peltier, M.R., Freeman, A.J., Mu, H.H., and Cole, B.C. (2007). Characterization of the macrophage-stimulating activity from *Ureaplasma urealyticum*. Am. J. Reprod. Immunol. *57*, 186–192.

Pyrowolakis, G., Hofmann, D., and Herrmann, R. (1998). The subunit b of the F0F1-type ATPase of the bacterium *Mycoplasma pneumoniae* is a lipoprotein. J. Biol. Chem. *273*, 24792–24796.

Ramirez, A.S., Rosas, A., Hernandez-Beriain, J.A., Orengo, J.C., Saavedra, P., de la Fe, C., Fernandez, A., and Poveda, J.B. (2005). Relationship between rheumatoid arthritis and *Mycoplasma pneumoniae*: a case–control study. Rheumatology *44*, 912–914.

Sacchini, F., Naessens, J., Awino, E., Heller, M., Hlinak, A., Haider, W., Sterner-Kock, A., and Jores, J. (2011). A minor role of $CD4^+$ T lymphocytes in the control of a primary infection of cattle with *Mycoplasma mycoides* subsp. *mycoides*. Vet. Res. *42*, 77.

Sawitzke, A., Joyner, D., Knudtson, K., Mu, H.H., and Cole, B. (2000). Anti-MAM antibodies in rheumatic disease: evidence for a MAM-like superantigen in rheumatoid arthritis? J. Rheumatol. *27*, 358–364.

Seggev, J.S., Sedmak, G.V., and Kurup, V.P. (1996). Isotype-specific antibody responses to acute *Mycoplasma pneumoniae* infection. Ann. Allergy Asthma Immunol. *77*, 67–73.

Shimizu, T., Kida, Y., and Kuwano, K. (2007). Triacylated lipoproteins derived from *Mycoplasma pneumoniae* activate nuclear factor-kappaB through Toll-like receptors 1 and 2. Immunology *121*, 473–483.

Shimizu, T., Kida, Y., and Kuwano, K. (2008). A triacylated lipoprotein from *Mycoplasma genitalium* activates NF-kappaB through Toll-like receptor 1 (TLR1) and TLR2. Infect. Immun. *76*, 3672–3678.

Shou, Y., Li, N., Li, L., Borowitz, J.L., and Isom, G.E. (2002). NF-kappaB-mediated up-regulation of Bcl-X(S) and Bax contributes to cytochrome c release in cyanide-induced apoptosis. J. Neurochem. *81*, 842–852.

Sieve, A.N., Meeks, K.D., Bodhankar, S., Lee, S., Kolls, J.K., Simecka, J.W., and Berg, R.E. (2009). A novel IL-17-dependent mechanism of cross protection: respiratory infection with mycoplasma protects against a secondary listeria infection. Eur. J. Immunol. *39*, 426–438.

Stelmach, I., Podsiadlowicz-Borzecka, M., Grzelewski, T., Majak, P., Stelmach, W., Jerzynska, J., Poplawska, M., and Dziadek, J. (2005). Humoral and cellular immunity in children with *Mycoplasma pneumoniae* infection: a 1-year prospective study. Clin. Diagn. Lab. Immunol. *12*, 1246–1250.

Sun, X., Jones, H.P., Hodge, L.M., and Simecka, J.W. (2006). Cytokine and chemokine transcription profile during *Mycoplasma pulmonis* infection in susceptible and resistant strains of mice: macrophage inflammatory protein 1beta (CCL4) and monocyte chemoattractant protein 2 (CCL8) and accumulation of $CCR5^+$ Th cells. Infect. Immun. *74*, 5943–5954.

Szczepanek, S.M., Majumder, S., Sheppard, E.S., Liao, X., Rood, D., Tulman, E.R., Wyand, S., Krause, D.C., Silbart, L.K., and Geary, S.J. (2012). Vaccination of BALB/c mice with an avirulent *Mycoplasma pneumoniae* P30 mutant results in disease exacerbation upon challenge with a virulent strain. Infect. Immun. *80*, 1007–1014.

Takeuchi, O., Kaufmann, A., Grote, K., Kawai, T., Hoshino, K., Morr, M., Muhlradt, P.F., and Akira, S. (2000). Cutting edge: preferentially the R-stereoisomer of the mycoplasmal lipopeptide macrophage-activating lipopeptide-2 activates immune cells through a Toll-like receptor 2- and MyD88-dependent signaling pathway. J. Immunol. *164*, 554–557.

Thirkill, C.E., Tyler, N.K., and Roth, A.M. (1992). Circulating and localized immune complexes in experimental mycoplasma-induced arthritis-associated ocular inflammation. Infect. Immun. *60*, 401–405.

Tipirneni, P., Moore, B.S., Hyde, J.S., and Schauf, V. (1980). IgE antibodies to *Mycoplasma pneumoniae* in asthma and other atopic diseases. Ann. Allergy *45*, 1–7.

Tiwary, B.K., and Goel, M.C. (1986). Experimental *Mycoplasma gallisepticum* infection in normal,IgA-and CMI-deficient chickens. Indian J. Anim. Sci. *56*, 719–725.

Totte, P., Duperray, C., and Dedieu, L. (2010a). CD62L defines a subset of pathogen-specific bovine CD4 with central memory cell characteristics. Dev. Comp. Immunol. *34*, 177–182.

Totte, P., Mather, A., Reslan, L., Boublik, Y., Niang, M., Du Plessis, D., and Dedieu, L. (2010b). Identification of *Mycoplasma mycoides* subsp. *mycoides* small colony genes coding for T-cell antigens. Clin. Vaccine Immunol. *17*, 1211–1216.

Triantafilou, M., Gamper, F.G., Haston, R.M., Mouratis, M.A., Morath, S., Hartung, T., and Triantafilou, K. (2006). Membrane sorting of Toll-like receptor (TLR)-2/6 and TLR2/1 heterodimers at the cell surface determines heterotypic associations with CD36 and intracellular targeting. J. Biol. Chem. *281*, 31002–31011.

Washburn, L.R., and Ramsay, J.R. (1989). Experimental induction of arthritis in LEW rats and antibody response to four *Mycoplasma arthritidis* strains. Vet. Microbiol. *21*, 41–55.

Washburn, L.R., Cole, B.C., and Ward, J.R. (1980). Chronic arthritis of rabbits induced by mycoplasmas. II. Antibody response and the deposition of immune complexes. Arthritis Rheum. *23*, 837–845.

Woodruff, P.G., Modrek, B., Choy, D.F., Jia, G., Abbas, A.R., Ellwanger, A., Koth, L.L., Arron, J.R., and Fahy, J.V. (2009). T-helper type 2-driven inflammation defines major subphenotypes of asthma. Am. J. Respir. Crit. Care Med. *180*, 388–395.

Woolard, M.D., Hodge, L.M., Jones, H.P., Schoeb, T.R., and Simecka, J.W. (2004). The upper and lower respiratory tracts differ in their requirement of IFN-gamma and IL-4 in controlling respiratory mycoplasma infection and disease. J. Immunol. *172*, 6875–6883.

Wu, Q., Martin, R.J., Rino, J.G., Breed, R., Torres, R.M., and Chu, H.W. (2007). IL-23-dependent IL-17 production is essential in neutrophil recruitment and activity in mouse lung defense against respiratory *Mycoplasma pneumoniae* infection. Microbes Infect. *9*, 78–86.

Wu, Y., Qiu, H., Zeng, Y., You, X., Deng, Z., Yu, M., and Zhu, C. (2008). *Mycoplasma genitalium* lipoproteins induce human monocytic cell expression of proinflammatory cytokines and apoptosis by activating nuclear factor kappaB. Mediators Inflamm. *2008*, 195427.

Yano, T., Ichikawa, Y., Komatu, S., Arai, S., and Oizumi, K. (1994). Association of *Mycoplasma pneumoniae* antigen with initial onset of bronchial asthma. Am. J. Respir. Crit. Care Med. *149*, 1348–1353.

Zhang, S., and Lo, S.C. (2007). Effect of mycoplasmas on apoptosis of 32D cells is species-dependent. Curr. Microbiol. *54*, 388–395.

# Emerging Antimicrobial Resistance in Mycoplasmas of Humans and Animals 15

Ken B. Waites, Inna Lysnyansky and Cécile M. Bébéar

## Abstract

Antimicrobial resistance has emerged in many types of bacteria and has spread worldwide, often as a result of selective pressure caused by overuse and misuse of antimicrobial agents in humans and animals. Clinically significant resistance to drugs such as tetracyclines, fluoroquinolones, and macrolides, has also developed in mycoplasmas and ureaplasmas of humans and animals, and appears to be increasing. These changes in susceptibility patterns have led to a renewed interest in development of standardized and reproducible methods for antimicrobial susceptibility testing to guide individual case management; surveillance for resistance locally, nationally, and internationally; and for evaluation of new antimicrobial agents. *In vitro* studies have been performed to induce resistance by stepwise selection followed by nucleic acid sequencing and analysis of the resistant microbes genetically to elucidate the molecular mechanisms involved. Clinical isolates proven to be resistant to various drugs phenotypically have also been characterized genetically and compared with mutants selected *in vitro* to clarify further the resistance mechanisms that are operative in a natural setting. In many instances, the same mechanisms have been shown to occur naturally and *in vitro*. In this chapter we have summarized antimicrobial agents useful for treatment of mycoplasma and ureaplasma infections of humans and animals and the current trends in development of antimicrobial resistance in these organisms. Evidence for the molecular basis of antimicrobial resistance is also discussed along with descriptions of methods for determination of antimicrobial susceptibilities.

## Introduction

During the past several years, antimicrobial resistance has emerged in many types of bacteria and has spread worldwide, often as a result of selective pressure caused by overuse and misuse of antimicrobial agents in humans and animals. Clinically significant resistance to several different drugs has also developed in mycoplasmas and ureaplasmas of humans and animals over the past several years and appears to be increasing. These changes in susceptibility patterns have led to a renewed interest in development of standardized and reproducible methods for antimicrobial susceptibility testing to guide individual case management; surveillance for resistance locally, nationally, and internationally; and for evaluation of new antimicrobial agents.

Concurrent with the development of *in vivo* antimicrobial resistance among the Mollicutes of humans and animals, numerous studies have also been performed *in vitro* in an attempt to induce resistance by stepwise selection followed by nucleic acid sequencing and analysis of the resistant microbes genetically to elucidate the molecular mechanisms involved in causing the elevation of minimal inhibitory concentrations (MICs) for specific drugs. Clinical isolates proven to be resistant to various drugs have also been characterized genetically and compared with mutants selected *in vitro* to clarify further the resistance mechanisms that are operative in a natural setting.

In this chapter we have summarized the types of antimicrobial agents useful for treatment of mycoplasma and ureaplasma infections of humans and animals and the current trends in development of clinical antimicrobial resistance in these organisms. Evidence for the molecular basis of antimicrobial resistance is also discussed along with descriptions of methods for determination of antimicrobial susceptibilities *in vitro*. Other sources exist for description of clinical conditions that may be caused by or associated with the most important microbes in this group that affect humans or animals (Jordan and Pattison, 1996; Waites and Talkington, 2004; Waites *et al.*, 2005a; Jensen, 2006; Nicholas *et al.*, 2009), so the epidemiology and pathogenesis of specific diseases will not be discussed here. With the notable exception of *Ureaplasma* spp., the other Mollicutes of humans and animals that are discussed in this chapter belong to the genus *Mycoplasma*, hereafter abbreviated *M.* followed by the species name.

## *In vitro* determination of antimicrobial susceptibilities

### Mycoplasmas of humans

#### The importance of *in vitro* susceptibility testing

Despite recent increased interest in the performance of *in vitro* susceptibility testing of mycoplasmas and ureaplasmas isolated from humans, such testing has not previously achieved widespread usage for several reasons. These include the facts that the choices of drugs have been quite limited; testing methods were not standardized; no interpretative MIC breakpoints or quality control guidelines were designated by any regulatory or advisory organization; no commercially sold test systems were approved for use in the USA, though some have been commercialized in Europe for several years; the time necessary to obtain a result may require several days up to several weeks due to the slow growth *in vitro* of some species; unfamiliarity of most clinical laboratories with how to perform the tests; unavailability of clinical isolates obtained by culture; and the belief that activities of most drugs are predictable since acquired resistance is uncommon.

It is now widely appreciated that acquired resistance to drugs often used to treat human mycoplasmal and ureaplasmal infections may occur commonly in some populations and is rapidly increasing in various countries as discussed in subsequent sections of this chapter. Drug resistance may vary geographically. It may also be clinically significant as well as unpredictable based on clinical presentation. Therefore, it is reasonable to recommend that *in vitro* susceptibility testing for all drugs that are being considered for therapeutic use should be performed whenever systemic infection occurs; especially if the host is immunosuppressed; if there is clinical treatment failure with a drug class that is usually active against these organisms; if a patient has had prolonged antimicrobial exposure; or if there is a known history of a drug-resistant infection. Knowing the susceptibility patterns for pathogenic organisms at the national or local level through targeted surveillance is also valuable in order to make general recommendations regarding empiric treatment since microbiological diagnosis may not be attempted or successful. The development and spread of antimicrobial resistance in many diverse bacterial pathogens of the respiratory and urogenital tracts of humans that may produce clinically similar manifestations to those caused by mycoplasmas and ureaplasmas has led to development of some new drugs in the past several years. For any drug that might be used empirically to treat an infection that could be due to mycoplasmas or ureaplasmas, accurate data regarding *in vitro* susceptibilities is prerequisite for such a drug to be approved for use against these infections.

#### The need for standardization of *in vitro* susceptibility testing

Numerous publications since the 1960s have described methods for antimicrobial susceptibility testing of human and animal mycoplasmas using both agar and broth-based techniques, but until recently no methods had been standardized. Lack of a uniform method of testing and guidelines for quality control led to widely disparate results, inaccurate conclusions, and considerable confusion regarding the activities of various drugs for some species. An attempt was made in the early 1990s to develop a consensus method for MIC determination for *Ureaplasma* spp. by the International Program on Comparative Mycoplasmology (IRPCM) Chemotherapy

Working Team, but several important aspects of this endeavour that were lacking contributed to its failure. These included a lack of infrastructure to organize and monitor the work being performed; no documentation of intra-laboratory reproducibility of test results; no standardized media or test protocol; and no designated reference strains with known MICs for quality control and inter-laboratory comparisons.

In 2002 an international multi-laboratory group working under the direction of the newly formed Mycoplasma Antimicrobial Susceptibility Testing Subcommittee of the Clinical and Laboratory Standards Institute (CLSI) began to develop standardized and reproducible methods for agar and broth-based susceptibility test methods for *M. pneumoniae, M. hominis,* and *Ureaplasma urealyticum.* The techniques developed by this working group specify how organism inocula and antimicrobial solutions are prepared; the media to be used for the respective agar and broth microdilution assays; incubation conditions; how MIC endpoints are determined; mandated controls for each assay; reference strains of each species with defined MIC ranges for several drugs; and tentative interpretive MIC breakpoints for some drugs. The above information is included in the CLSI document Approved Guideline for Antimicrobial Susceptibility Testing of Human Mycoplasmas (CLSI, 2011). It is beyond the scope of this chapter to provide details of antimicrobial susceptibility testing protocols and quality control parameters that are now available in the CLSI document. However, brief mention will be made concerning the essential components of *in vitro* susceptibility test techniques for *M. pneumoniae, M. hominis,* and *Ureaplasma* spp. Very few data are available concerning susceptibility testing and activities of antimicrobial agents against other human mycoplasmas such as *M. fermentans* or *M. genitalium* and the methodology published in the CLSI document is not intended to be applied to test any organisms other than those specifically named.

## Animal Mycoplasmas

In contrast to Mollicutes of humans, with a limited number of clinically important pathogenic species, there are many species of veterinary mycoplasmas with a wide variation in nutritional requirements, growth conditions and time of incubation. The diverse nature of these organisms hampers the development of a universal veterinary mycoplasmal MIC assay. In 2000, Hannan (Hannan, 2000) published comprehensive guidelines for *in vitro* susceptibility testing for animal mycoplasmas based on a consensus of methods employed by members of the IRPCM. This document still provides a very useful reference for veterinary laboratories doing *in vitro* testing. Careful adherence to recommended test procedures can help avoid pitfalls in the tests which may lead to erroneous results. Differences in methodology make it hard to compare results from different laboratories.

As a result of the successful efforts of the CLSI Mycoplasma Antimicrobial Susceptibility Testing Subcommittee, a Veterinary Mycoplasma Working Group was subsequently established with the eventual goals of performing multi-laboratory analyses to establish optimum growth media; test methods and reproducibility; antimicrobial concentrations to be tested; MIC endpoints; and identification of quality control strains with designated MIC ranges. These objectives are comparable to what have been developed for the human mycoplasmas. The scope of the Veterinary Working group is initially being limited to testing *M. bovis* with florfenicol. Preliminary studies of 50 isolates using various media with phenol red or alamar blue indicate that choice of media and growth indicator will be a critical component in the development of standardized antimicrobial susceptibility testing procedures for mycoplasmas from animals. Further studies to include other *Mycoplasma* species and antimicrobial agents will depend on the ability to obtain sponsorship for the extensive studies that will be required to meet CLSI requirements.

## Conventional methods for *in vitro* susceptibility testing

### Broth microdilution

The broth microdilution technique to determine MICs is the most practical and widely used method employed in diagnostic laboratories testing mycoplasmas and ureaplasmas of humans and animals. The test is based on the principle that a constant number of microorganisms is added to serial doubling concentrations of antimicrobial agents diluted in broth in a 96-well microtitre plate. It allows several antimicrobials to be tested in the same microtitre plate and minimal bactericidal or mycoplasmacidal concentrations (MBCs) can be measured by obtaining a subculture from the same system. In broth microdilution, the MIC is defined as the lowest concentration of antimicrobial in which no evidence of growth indicated by colour change in broth is apparent when examined at the same time colour change is first evident in a drug-free growth control well with a defined organism inoculum. For mycoplasmas and ureaplasmas of humans, the recommended inoculum is $10^4$–$10^5$ CFU/ml. The colour change is based on an increase or decrease in pH in the presence of an indicator such as phenol red which results from metabolic activity of the organism being tested, i.e. fermentation of glucose by glycolytic organisms such as *M. pneumoniae*, hydrolysis of arginine by organisms such as *M. hominis* or urea by *Ureaplasma* spp. Despite its popularity as a method of susceptibility testing, broth microdilution also has some disadvantages. Specifically, the preparation of antimicrobial dilutions is labour intense and the endpoint tends to shift over time. Furthermore, the acidic pH (6.0–6.5) necessary for growth of ureaplasmas can affect the activity of some drugs such as macrolides when tested *in vitro*. The latter problem also affects agar-based testing methods. Depending on the species being evaluated, requirements for growth media, pH, and length of incubation will vary for all test methods. The length of time to obtain an MIC by broth microdilution or other methods is dependent on the intrinsic growth rates of the organism being tested. These times range from 1 to 2 days for *Ureaplasma* spp., 2 to 4 days for *M. hominis*, and 5 days or longer for *M. pneumoniae* (Waites *et al.*, 2001; CLSI, 2011). Whether testing is performed by broth microdilution or agar dilution, it is essential that the appropriate quality control reference strains for each relevant species with defined MIC ranges for several drugs must be included in each assay as outlined in the CLSI document (CLSI, 2011). The media used in the assays must also be standardized. For broth microdilution testing of ureaplasmas, Shepard's 10B broth prepared in-house or purchased commercially is recommended. For testing *M. hominis*, a modified version of Hayflick's broth with arginine is used. SP4 glucose broth prepared in-house or purchased commercially, or modified Hayflick's broth with glucose, is recommended for *M. pneumoniae*. Detailed formulations of these media are available in the CLSI document (CLSI, 2011).

For broth microdilution MIC testing of mycoplasmas from animals, there are several examples of media which have been used to carry out MIC assays as described by Hannan (Hannan, 2000). Selection of the growth medium depends on the ability of animal mycoplasmas to break down metabolizable substrates present in the medium with the resulting pH change visible as a colour change of a pH indicator similar to what is done with those species isolated from humans. Glucose is the metabolic substrate used for testing *M. gallisepticum*, *M. synoviae* or mycoplasmas of the mycoides cluster, whereas pyruvate phosphate is utilized in the case of *M. bovis* or *M. agalactiae*. Other *Mycoplasma* spp. may utilize arginine and *Ureaplasma* spp. metabolize urea. Rosenbusch *et al.* (Rosenbusch *et al.*, 2005), utilized 5% alamar blue (Biosource, Camarillo, CA) as redox colour indicator (blue to red shift) to measure growth of *M. bovis* in a microbroth test for MIC determination. The 5% alamar blue may be used to determine MICs for mycoplasmas for which growth in liquid media does not result in a visible colour change. In addition, the microtitre plates can be centrifuged for three minutes at 800g to concentrate the cells at the bottom of the wells. The plates are examined with an inverted mirror, and the growth or absence of growth is recorded for each well as was described previously by Ayling *et al.* (Ayling *et al.*, 2000).

There are several pitfalls with the microdilution method when used for testing animal mycoplasmas, just as there are for using this method for human mycoplasmas. Clinical samples may contain mixed mycoplasma flora; colour change is not consistently sharp making reading the results somewhat subjective; and variation in titre of the inoculum can have a major impact on the results. Furthermore, differences in the time of reading the results such as comparison to a fixed pH control; reading at a fixed time; or at the endpoint of colour change make it impossible to compare results between laboratories since these methods are not employed uniformly. Duplicates within an assay and repetition of the test including a laboratory reference strain are highly recommended to increase the reliability of the results. Thus far, there are no universally defined reference strains recommended for quality control of animal mycoplasma MIC assays making the choice an individual decision by each laboratory performing the tests. Such a reference strain should be chosen based on consistent reliable growth in the test system and stable MICs for the drugs of interest.

### Agar dilution

Agar dilution has been used extensively as a reference method for susceptibility testing for many types of bacteria. It has the advantages of a relatively stable endpoint over time; it allows detection of mixed cultures readily; and is most suitable for testing larger numbers of clinical isolates to a single agent or a few drugs at the same time. However, this technique is not practical for testing small numbers of strains or occasional isolates which may be encountered in diagnostic laboratories and it is not possible to perform MBC testing in the same system. The preferred inoculum for testing mycoplasmas and ureaplasmas from humans is $10^4$–$10^5$ CFU/ml. If too many cells are present in the inoculum colonies may not form on agar and the test will be uninterpretable. While commercial media is acceptable for broth microdilution testing, the agar dilution method requires preparation of fresh media on site since the antimicrobial concentrations are incorporated directly into the appropriate agar plates. The recommended media are A8 agar for *Ureaplasma* spp., modified Hayflick's agar with arginine for *M. hominis* and SP4 glucose agar or modified Hayflick's agar with glucose for *M. pneumoniae* (CLSI, 2011). Agar dilution testing is performed by inoculating a defined organism inoculum on a single agar plate using a multipoint replicator, micropipettor or calibrated loop to deliver a specified volume, usually 10 µl. The goal is to obtain 30 to 300 colonies per spot of inoculum on the agar plate after incubation. The MIC is defined as the lowest concentration of antimicrobial that prevents colony formation when examined under a stereomicroscope at the same time the drug-free control plate demonstrates growth as described in various other reference texts and the current CLSI document (Waites *et al.*, 2001; CLSI, 2011).

For animal mycoplasmas, a major advantage of the agar dilution method, especially when dealing with clinical isolates, is that it is possible to detect the presence of more than one species of mycoplasma in the test culture by colony morphology. It is also the best method to detect 'tailing' which may result from a culture which is homogeneous with respect to species but contains strains varying in MIC. The techniques used for agar dilution testing of animal mycoplasmas are generally similar to what is done for mycoplasmas from humans with specific modifications as needed to fit the nutritional requirements and growth characteristics of the individual species.

### Other susceptibility testing methods and commercial kits

Agar disk diffusion is not useful for testing mycoplasmas since there has been no correlation between inhibitory zones and MICs. Furthermore, the relatively slow growth of some of these organisms further limits this technology since the antimicrobial agent would be likely to diffuse throughout the agar before visible microbial growth could be detected. However, studies using the Etest (bioMérieux, Durham, NC) agar gradient diffusion technique for detection of tetracycline resistance in *M. hominis* yielded results comparable to microbroth dilution and susceptibilities of ureaplasmas to various antimicrobials (Waites *et al.*, 2001). The Etest has the advantages of simplicity of agar-based testing, has an endpoint which does not shift over time, does not have a large inoculum effect, and can easily be adapted for testing single isolates. This method is sometimes useful when results are needed quickly in laboratories that do not perform susceptibility testing frequently since Etest strips are available commercially for most commonly used antimicrobial agents and have a long shelf life. The Etest is best suited for *M. hominis* and some of the more rapidly growing large colony animal mycoplasmas and is not suitable for slow-growing more fastidious species such as *M. pneumoniae*. It is technically demanding when used with ureaplasmas because the colonies are so small and it is difficult to get a sufficient inoculum to demonstrate the ellipse on agar. To perform an Etest a 0.5 ml suspension containing approximately $10^5$ bacteria per ml is inoculated onto an appropriate agar plate. A8 agar can be used for *M. hominis* or *Ureaplasma* spp. SP4 agar can also be used for *M. hominis* with similar results since the test is not standardized. The agar plate is rotated to spread the microorganisms evenly over the surface and allowed to dry for 15 minutes before the Etest strip is applied. Agar plates are incubated at 37 ° C in air plus 5% $CO_2$ until growth is visible under a stereomicroscope and an ellipse can be detected. The MIC is read in the same manner as for conventional bacteria as the value where the ellipse of growth inhibition intersects the strip. Mycoplasmas will not make a confluent lawn of growth comparable to conventional bacteria, but it is usually possible to see a definite ellipse when viewing the agar plate through a stereomicroscope. Staining the plates with Dienes methylene blue stain helps visualize the colonies and identify the margin of the ellipse. The Etest technique is not recommended for quantitative reference work, but can be useful to detect high level resistance to a drug that is generally used to treat mycoplasmal infections such as a tetracycline or fluoroquinolone for which acquired resistance sometimes occurs.

A variety of commercial products that perform detection, quantitation, identification, and sometimes antimicrobial susceptibility testing of *Ureaplasma* spp. and *M. hominis* from urogenital specimens have been developed. These kits consist of microwells containing dried antimicrobial agents in various concentrations corresponding to breakpoints for conventional bacteria. Some of these products such as the Mycoplasma IST2 (bioMérieux, Marcy l'Etoile, France), Mycoplasma SIR (Bio-Rad, Hercules, CA), and Mycofast Evolution 3 (ELItech, Salon-de-Provence, France) are used to some extent in various European countries. Limited evaluations suggest these kits give results comparable to those obtained by non-commercial MIC determination by broth microdilution (Waites *et al.*, 2002). Although some products have been developed for use with direct specimen inoculation, lack of a defined inoculum which can only be determined by subculture can potentially lead to errors in detection of antimicrobial resistance since MICs can be influenced by the numbers of organisms present (Bébéar and Bébéar, 2002).

The Etest has also been applied for determination of MICs for various animal mycoplasmas. Excellent agreement between results of Etest and the previously described tests for *M. gallisepticum* and *M. synoviae* strains with MICs in the susceptible range has been reported (Gerchman *et al.*, 2008). Notably, this agreement was not found for isolates with diminished susceptibility to enrofloxacin. It was not possible to determine the MIC endpoint for *M. gallisepticum* field strains with MICs >1 μg/ml or for *M. bovis* strains with MICs >2.5 μg/ml to enrofloxacin (Gerchman *et al.*, 2008, 2009). Similarly, it was not possible to determine the MIC endpoint for spectinomycin and streptomycin in *M. bovis* field strains (Francoz *et al.*, 2005; Gerchman *et al.*, 2009). The differences between Etest and other methods have also been reported for fluoroquinolones in *Streptococcus pneumoniae* with reduced susceptibilities to these drugs (Kays and Graff, 2002). It should be noted that the lack of a measurable Etest endpoint should not change the classification of the isolate as susceptible or non-susceptible since it only occurs in the high range of MIC values. One problem with Etest in veterinary medicine that is available only for a limited number of antibiotics such as enrofloxacin, tetracycline and several aminoglycosides, but not for macrolides such as tylosin, tilmicosin, tulathromycin, or for pleuromutilins, tiamulin and valnemulin.

## Mycoplasmacidal testing

Determination of whether an antimicrobial agent actually kills a microorganism or merely inhibits its growth is not standardized and is rarely performed for routine diagnostic purposes. However, it can be beneficial to know about this property of a drug when selecting treatment for invasive infections in immunocompromised hosts or in the initial evaluation of investigational agents. Some drugs are known to be bactericidal against mycoplasmas and ureaplasmas, such as the fluoroquinolones, but most others, including the tetracyclines and macrolides are merely bacteriostatic.

In classical bacteriology, the MBC is determined by measuring the time-kill reaction for individual drug–organism combinations. This is accomplished by adding the drug to an actively growing culture and incubating it along with a drug-free control. At various time points, samples are taken from experimental and control cultures and diluted in broth to stop the killing reaction. Dilutions are plated onto agar and the colony counts are compared at each time point. The usual end point for a time-kill reaction of this type for a bactericidal agent is when a 99.9% (3 $\log_2$) reduction in numbers of colonies from the starting point is detected. However, 99% killing has been recommended in some references because of the limited growth of mycoplasmas compared to conventional bacteria (Waites *et al.*, 2001). Mycoplasmacidal activities can also be measured directly from a broth microdilution MIC system by subculturing each of the dilutions above the MIC, diluting in broth to stop the killing, and examining for colour change. The MBC is defined as the lowest concentration of antimicrobial at which there is no evidence of a colour change and no colonies are formed on subculture to agar. This latter means to measure the MBC is simpler and more convenient than performing the time-kill reaction, but it does not provide information regarding how long an exposure to an antimicrobial is required to kill the bacteria.

## Molecular methods for determination of antimicrobial susceptibilities

*M. hominis* and *Ureaplasma* spp. can be grown quickly, within 1 to 2 days, and *M. pneumoniae* can be grown within a matter of several days, even though culture is not particularly sensitive for direct detection of the latter organism in clinical specimens. In contrast, *M. genitalium* presents the greatest challenge among mycoplasmas that are pathogenic for humans, for direct detection and in determination of antimicrobial susceptibilities, because it is so difficult to cultivate *in vitro*. Growth, when it occurs, may require up to several weeks so performance of *in vitro* susceptibility tests using a culture-based approach would require even more additional time beyond that. Considering these limitations of culture-based susceptibility testing, alternative molecular-based approaches have been investigated for measurement of antimicrobial susceptibilities for *M. genitalium* as well as *M. pneumoniae*. Such efforts are increasingly important in view of the proven occurrence and clinical significance of acquired antimicrobial resistance in these two mycoplasmas as discussed elsewhere in this chapter.

### Real-time PCR-based methods to measure MICs

A TaqMan 5′ nuclease real-time PCR assay has been described to measure antimicrobial susceptibilities in *M. genitalium* (Hamasuna *et al.*, 2005). This assay still requires culture of the mycoplasma *in vitro* using a Vero cell line in which the DNA load is determined by amplification of a 78-bp fragment of the *mgB* adhesin gene. Trypsinized Vero cells are resuspended in Eagle's medium and dispensed in 24-well tissue culture plates. Aliquots of Eagle's medium with or without antibiotic dilutions are added to the plates followed by Vero cells adjusted to contain a defined inoculum of *M. genitalium* and incubated at 37 ° C in air plus 5% $CO_2$. Real-time PCR assays to measure DNA load are then performed after 1, 2, and 3 weeks. The inhibition rate (%) of individual antimicrobials are calculated by the formula [(average DNA loads in control wells – DNA load in test well)/ average DNA loads in control wells)] × 100. The MIC is defined as the lowest concentration of antibiotic causing 99% inhibition and the MBC is defined as the lowest concentration causing 99.9% inhibition. These investigators (Hamasuna *et al.*, 2005) compared the MICs obtained by the real-time PCR method with those obtained on the same *M. genitalium* strains by broth microdilution and found similar results.

### Real-time PCR-based methods to detect resistance genes and mutations

Thus far, all naturally occurring macrolide-resistant *M. pneumoniae* evaluated have contained mutations in a few different loci in 23S rRNA. Investigators in Europe and the USA have developed real-time PCR assays to detect these mutations in clinical isolates or directly in clinical specimens (Wolff *et al.*, 2008; Li *et al.*, 2009; Peuchant *et al.*, 2009). This method of direct detection of resistance genes is based on the fact that nucleic acid will melt at a precise temperature which is related to the nucleotide base composition. The presence of one or more point mutations in 23S rRNA that impair antimicrobial agent attachment to the bacterial ribosome will be detected by this extremely sensitive method which can be completed in just a few hours. Chinese investigators have reported a rapid and inexpensive method that combines nested PCR, single-strand conformation polymorphisms (SSCPs), and capillary electrophoresis (CE) to detect macrolide-resistant mutants directly from throat swabs. (Lin *et al.*, 2010). Pyrosequencing technology has also been applied for detection of macrolide resistance in *M. pneumoniae* as well as for molecular strain typing (Spuesens *et al.*, 2010).

Detection of other resistance markers such as the gene for the *tet*(M) transposon in *M. hominis* or *Ureaplasma* spp. (Blanchard *et al.*, 1992) or mutations in DNA gyrase and/or topoisomerase IV mediating fluoroquinolone resistance can also be used to detect acquired resistance in these organisms (Bébéar *et al.*, 1999; Duffy *et al.*, 2006) as well as *M. genitalium* (Shimada *et al.*, 2010a,b). Detection of resistance genes directly for *M. hominis* and *Ureaplasma* spp. is not as critical since these organisms can be cultured fairly easily and quickly so that MICs can be determined in the traditional culture-based manner as described above.

Recently, two molecular assays *parC*-PCR-RFLP and single-nucleotide-polymorphism (SNP) real-time PCR have been published to discriminate between enrofloxacin-susceptible and resistant *M. bovis* strains (Lysnyansky *et al.*, 2009; Ben Shabat *et al.*, 2010). Both methods are designed to detect nucleotide changes leading to amino acid substitutions in the quinolone resistance determining regions (QRDRs) of the *parC* gene in enrofloxacin-resistant *M. bovis* strains. The TaqMan SNP real-time PCR assay is highly specific for *M. bovis*, with a detection limit of 5 fg/μl (about 5 *M. bovis* genomes) and allows detection of decreased susceptibility of *M. bovis* to fluoroquinolones directly from field samples (Ben Shabat *et al.*, 2010).

## Defining antimicrobial resistance

Most publications from around the world over the past several years have merely described quantitative MICs, but some have gone further and arbitrarily assigned interpretations to MIC results and actually gave percentages of susceptibility for various antimicrobial agents. Some authors simply extrapolated interpretive breakpoints based on what had been assigned by organizations such as the CLSI, US Food and Drug Administration (FDA) or European Committee on Antimicrobial Susceptibility Testing (EUCAST) to conventional bacteria. This practice is not recommended because it can be erroneous owing to the fact that the media, growth conditions, pH, incubation, replication time, and other variables for mycoplasmas may be different. Among different species of conventional bacteria, the MIC breakpoints for various drugs can even be different, depending on the type of bacteria being considered and how the drug is metabolized.

The process of establishing MIC interpretive criteria or breakpoints for a drug for conventional bacteria is complex and when such criteria are recognized and approved by organizations such as the FDA and/or CLSI, considerable data must be presented and analysed. This includes evaluation of *in vitro* MIC determinations that include organisms with and without well-characterized resistance mechanisms that affect the activities of the drug, pharmacokinetic and pharmacodynamic parameters, and clinical and bacteriological outcomes of patients enrolled in large clinical trials. Unfortunately, it is unlikely that a sufficient body of data will ever be available for mycoplasmas to meet these rigorous standards for designating MIC breakpoints. In the CLSI Approved Guideline for Antimicrobial Susceptibility Testing of Human Mycoplasmas (CLSI, 2011), for the very first time MIC interpretations specific for *M. pneumoniae*, *M. hominis* and *U. urealyticum* are proposed for some drugs. To the extent possible, the interpretive criteria proposed in that document and summarized in Table 15.1 have been derived by relating MICs to the presence or absence of resistance determinants such as *tet*(M) (tetracyclines); 23S rRNA mutations (macrolides); mutations in QRDRs of DNA (fluoroquinolones); or by relating MICs for mycoplasmas and ureaplasmas for specific drugs to those of other bacteria for which interpretive criteria for the same drugs have been established. These recommendations are considered tentative

**Table 15.1** Proposed interpretive guidelines for *in vitro* susceptibility testing of *M. pneumoniae*, *M. hominis*, and *U. urealyticum*[a]

| | *M. pneumoniae* MIC (µg/ml) | | *M. hominis* MIC (µg/ml) | | *U. urealyticum*[c] MIC (µg/ml) | |
|---|---|---|---|---|---|---|
| Antimicrobial Agent | Susceptible | Resistant | Susceptible | Resistant | Susceptible | Resistant |
| Levofloxacin | ≤1 | –[b] | ≤1 | ≥2 | ≤2 | ≥4 |
| Moxifloxacin | ≤0.5 | –[b] | ≤0.25 | ≥0.5 | ≤2 | ≥4 |
| Erythromycin | ≤0.5 | ≥1 | NA | NA | ≤8 | ≥16 |
| Azithromycin | ≤0.5 | ≥1 | NA | NA | NA | NA |
| Telithromycin | NA | NA | NA | NA | ≤4 | –[b] |
| Tetracycline | ≤2 | –[b] | ≤4 | ≥8 | ≤1 | ≥2 |
| Clindamycin | NA | NA | ≤0.25 | ≥0.5 | NA | NA |

[a]Interpretive guidelines have been proposed based on the presence or absence of resistance genes or mutations, ranges of MIC values obtained in natural populations of clinical isolates, and/or designations from other bacterial species which are related to drug metabolism. In order for interpretations to be valid, testing has to conform with CLSI guidelines (CLSI, 2011), using appropriate inocula, media, incubation conditions, and quality control reference strains that yield MICs within the appropriate ranges. [b]Absence of well-documented naturally occurring clinical resistance precludes defining any results other than susceptible. [c]Results are also valid for *U. parvum*. NA: not available indicating that there are insufficient data to designate a breakpoint for this organism/drug combination and/or the drug is not appropriate for testing against the species.

and may be subject to modification if additional pertinent data are collected. The presence or absence of resistance determinants such as *tet*(M) in *M. hominis* and *Ureaplasma* spp. correlates closely with non-overlapping MICs, so it is possible to designate a breakpoint based on its presence or absence (Fig. 15.1).

## Intrinsic resistance to antimicrobial agents in Mycoplasmas and Ureaplasmas

Due to their lack of cell wall, major antimicrobial classes such as the beta-lactams, glycopeptides and fosfomycin are of no value in the treatment of infections caused by mycoplasmas or ureaplasmas from humans or animals. Additionally, sulfonamides, trimethoprim, rifampin, polymixins, and nalidixic acid and linezolid are also inactive (Bébéar and Bébéar, 2002). Some species will have intrinsic resistance to certain drugs within a class while maintaining susceptibility to others as in the case of *M. hominis* which is usually resistant to 14 and 15-membered macrolides such as erythromycin, clarithromycin, and azithromycin, while it is usually susceptible to 16-membered macrolides such as josamycin and to lincosamides such as clindamycin (Table 15.2). The intrinsic resistance of *M. hominis* to erythromycin was first investigated by Furneri *et al.* (Furneri *et al.*, 2000). They determined that a $G_{2057}A$ transition mutation (*Escherichia coli* numbering) in the central loop of domain V of 23S rRNA, similar to the location of mutations conferring macrolide resistance in other organisms, was likely responsible for the reduced activity of this agent against *M. hominis*. Further studies by Pereyre *et al.* (Pereyre *et al.*, 2002) confirmed this mutation was associated with intrinsic macrolide resistance in *M. hominis* as well as in *M. fermentans*. An additional $C_{2610}U$ change was also found in the sequence of *M. hominis* when compared to *M. pneumoniae*. *In vitro* selection studies of *M. hominis* with several macrolides and related antibiotics confirmed the involvement of the 2057–2611 pair in the intrinsic resistance of *M. hominis* to macrolides (Pereyre *et al.*, 2006). Indeed, the $C_{2611}U$ transition selected in the presence of clindamycin and the quinupristin–dalfopristin combination was associated with decreased MICs of erythromycin, azithromycin and telithromycin, leading to a loss of the intrinsic resistance of *M. hominis* to erythromycin and azithromycin.

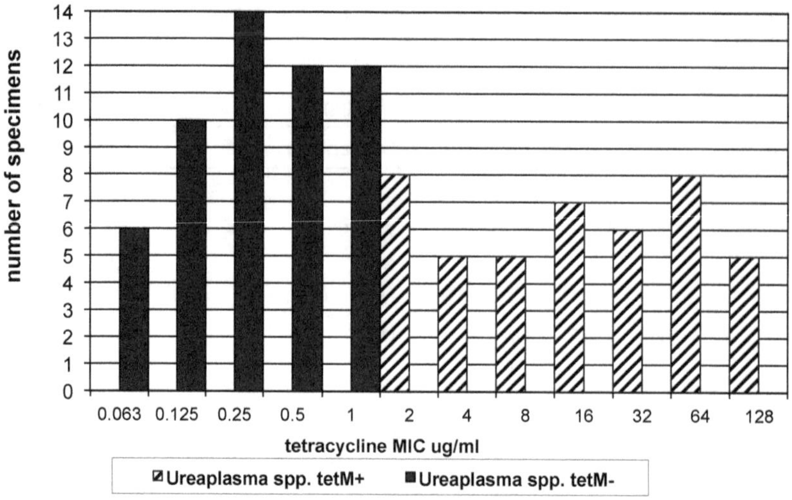


**Figure 15.1** MIC distribution for tetracycline in 100 non-duplicate clinical isolates of *Ureaplasma* spp. collected over a five year period between 2000 and 2004 from various geographic regions of the USA. Isolates were also tested for the *tet*(M) transposon gene by PCR. In this collection, 45 isolates the contained the *tet*(M) gene as shown in the graph, all of which had tetracycline MICs of ≥2µg/ml, suggesting this is the appropriate interpretive breakpoint for tetracycline resistance since this is the only naturally occurring mechanism that has been described in ureaplasmas *in vivo*.

Several *Mycoplasma* spp. present in food animals are also intrinsically resistant to erythromycin but susceptible to 16-membered macrolides such as tylosin and tilmicosin and to lincosamides as shown in Table 15.3. It was previously published that *M. pulmonis*, *M. flocculare* and *M. hyopneumoniae* contain the same nucleotide $G_{2057}A$ substitution in domain V of 23S rRNA that is putatively responsible for intrinsic resistance to erythromycin in *M. hominis* (Furneri, *et al.*, 2000; Pereyre *et al.*, 2002). Screening of nucleotide sequences of domain V of 23S rRNAs in the published genomes revealed the presence of the $G_{2057}A$ substitution in *M. bovis*, *M. agalactiae*, *M. conjunctivae*, *M. arthritidis*, *M. synoviae*, *M. crocodyli* and in *Acholeplasma laidlawii*, pointing to intrinsic lack of susceptibility to erythromycin in these species (Vasconcelos *et al.*, 2005; Sirand-Pugnet *et al.*, 2007; Dybvig *et al.*, 2008; Calderon-Copete *et al.*, 2009; Brown *et al.*, 2011; Wise *et al.*, 2011). In contrast, *M. gallisepticum*, *M. leachii*, *M. capricolum* subsp. *capricolum*, *M. hyorhinis* and *M. mycoides* subsp. *mycoides* SC have a guanine at position 2057, indicating susceptibility of those mycoplasmas to erythromycin (Papazisi *et al.*, 2003; Westberg *et al.*, 2004; Liu *et al.*, 2010a; Wise *et al.*, 2012; genBank accession no. NC_007633) Like the human ureaplasmal species, the bovine *U. diversum* is susceptible to macrolides but resistant to lincosamides (ter Laak *et al.*, 1993).

In addition, the molecular mechanism for intrinsic resistance to rifampin has been investigated for the plant pathogen *Spiroplasma citri* (Gaurivaud *et al.*, 1996). It was shown that resistance to rifampin in *S. citri* has been associated with the presence of the amino acid asparagine (N) at position 526 of the β subunit of RNA polymerase. Indeed, construction of a chimeric *rpoB* gene, in which *E. coli* Rif region (encompassing RpoB codons 500–575) was replaced by the equivalent region from *S. citri*, revealed that a single amino acid substitution H526N was able to provide high-level (≤ 500 mg/ml) rifampin resistance in *E. coli* (Gaurivaud *et al.*, 1996). The N526 was identified also in RpoB of *M. gallisepticum* and *M. genitalium* (Gaurivaud *et al.*, 1996). In addition, alignment of amino acid sequences of the RpoB proteins from mycoplasmas species, available in GenBank, showed the presence of N526 in *M. hyorhinis*, *M. bovis*, *M. agalactiae*, *M. conjunctivae*, *M. arthritidis*, *M. synoviae*, *M. gallisepticum*, *M. crocodyli*, *M. capricolum* subsp. *capricolum*, *M. mycoides* subsp.

**Table 15.2** MICs for various antimicrobial agents active against human mycoplasmas and ureaplasmas[a]

| Antimicrobial | *M. pneumoniae* | *M. hominis* | *M. genitalium* | *M. fermentans* | *Ureaplasma* spp. |
|---|---|---|---|---|---|
| *Tetracyclines* | | | | | |
| Tetracycline | 0.12–1 | 0.2–2 | ≤0.01–0.05 | 0.1–1 | 0.05–2 |
| Doxycycline | 0.03–0.5 | 0.1–2 | ≤0.01–0.3 | 0.05–1 | 0.02–1 |
| Minocycline | 0.06–0.5 | 0.03–1 | ≤0.01–0.2 | ND | 0.06–1 |
| Tigecycline | 0.06–0.25 | 0.125–0.5 | ND | ND | 1–16 |
| *MLSK*[b] | | | | | |
| Erythromycin | ≤0.004–0.06 | 32–>1,000 | ≤0.01 | 0.5–64 | 0.02–16 |
| Roxithromycin | ≤0.01 | >16 | 0.01 | 32–64 | 0.1–2 |
| Dirithromycin | ≤0.015–0.5 | >64 | ≤0.015–0.125 | ≥64 | 0.25–4 |
| Clarithromycin | ≤0.004–0.125 | 16–>256 | ≤0.01–0.06 | 1–64 | ≤0.004–2 |
| Azithromycin | ≤0.004–0.01 | 4–>64 | ≤0.01–0.03 | ≤0.003–0.05 | 0.06–4 |
| Josamycin | ≤0.01–0.03 | 0.05–2 | 0.01–0.02 | 0.12–0.5 | 0.03–4 |
| Pristinamycin | 0.02–0.05 | 0.1–0.5 | ≤0.01–0.02 | ND | 0.1–1 |
| Quinupristin/Dalfopristin | 0.008–0.12 | 0.03–0.3 | 0.05 | 0.12–0.5 | 0.05–0.5 |
| Clindamycin | ≤0.008–2 | ≤0.008–2 | 0.2–1 | 0.01–0.25 | 0.2–64 |
| Telithromycin | ≤0.008–0.06 | 2–32 | ≤0.015 | 0.06–0.25 | ≤0.015–0.25 |
| Cethromycin | ≤0.001–0.016 | ≤0.008–0.031 | ND | ≤0.008 | ≤0.008–0.031 |
| Solithromycin | ≤0.000000063–0.000125 | 0.002–0.008 | ≤0.000032 | ≤0.008 | 0.002–0.063 |
| *Fluoroquinolones* | | | | | |
| Ofloxacin | 0.05–2 | 0.1–4 | 1–2 | 0.02–0.25 | 0.2–4 |
| Ciprofloxacin | 0.5–2 | 0.1–4 | 2 | 0.02–>64 | 0.1–16 |
| Levofloxacin | 0.5–1 | 0.1–2 | 0.5–1 | 0.05–1 | 0.2–2 |
| Moxifloxacin | 0.06–0.125 | 0.06–0.125 | 0.03–0.06 | ≤0.015–0.06 | 0.125–1 |
| Gemifloxacin | 0.05–0.125 | 0.0025–0.01 | 0.05–0.125 | 0.001–0.01 | 0.03–0.5 |
| Phenicols | | | | | |
| Chloramphenicol | 2–10 | 2–25 | 0.5–25 | 0.5–10 | 0.4–8 |
| Aminoglycosides | | | | | |
| Gentamicin | 4 | 2–16 | ND | 0.25–>500 | 0.1–13 |

[a]Data were compiled from multiple published studies or reference texts in which different methodologies, and often different antimicrobial concentrations were used and represents an update of Table 15.1 from Bébéar and Kempf, 2005. MICs listed are indicative of drug activities for microorganisms that have not been shown to have resistance to the various drug classes. [b]MLSK: macrolide-lincosamide, streptogramin-ketolide group of antimicrobials. ND: no data available.

*mycoides* SC (Papazisi *et al.*, 2003; Westberg *et al.*, 2004a; Vasconcelos *et al.*, 2005; Sirand-Pugnet *et al.*, 2007; Dybvig *et al.*, 2008; Calderon-Copete *et al.*, 2009; Liu *et al.*, 2010b; Wise *et al.*, 2011; Brown *et al.*, 2011; genBank accession no. NC_007633). The RpoB of *M. pulmonis* and *M. hyopneumoniae* contains T526 and S526, respectively (Vasconcelos *et al.*, 2005; Chambaud *et al.*, 2001). Since T, S and N are related to the same group of polar amino acids with uncharged side chains, it is likely that *M. pulmonis* and *M. hyopneumoniae* are also resistant to rifampin.

**Table 15.3** MICs for various antimicrobial agents active against *Mycoplasma* species of food animals[a]

| Antimicrobial | *MH* | *MHP* | *MHS* | *MB* | *MALK* | *MBOV* | *MBOVIR* | *MA* | *MPUT* | *MmycLC* | *MG* | *MS* |
|---|---|---|---|---|---|---|---|---|---|---|---|---|
| Tetracyclines | | | | | | | | | | | | |
| Tetracycline | ≤0.03–2 | ≤0.03–8.1 | 0.015–10 | <0.03–>256 | ND | ND | ND | 0.125–4 | ND | ND | 0.08–0.16 | 0.08–2 |
| Oxytetracycline | 0.05–10 | ≤0.025–12.5 | 0.1–10 | 0.1–128 | 4–128 | 0.125–8 | 0.39–128 | 0.06–4 | 0.25–8 | 0.125–4 | 0.05–>16 | 0.025–>100 |
| Chlortetracycline | 0.2–12.5 | 0.12–≥100 | 0.05–50 | 0.25–256 | 3.12–128 | ≤0.125–8 | ≤0.125–100 | 0.125–8 | 2–16 | 0.5–1 | 0.2–6.25 | 0.39–>12.5 |
| Doxycycline | ≤0.03–0.5 | ≤0.03–2 | ND | 0.023–64 | ND | ND | ND | 0.008–1 | 0.125–2 | 0.03–0.06 | 0.1–0.79 | 0.015–0.125 |
| Minocycline | ≤0.03–2 | ≤0.03–0.06 | ND | 0.5–8 | ND | ND | ND | ND | ND | ND | ND | ND |
| *MLSK group*[b] | | | | | | | | | | | | |
| Erythromycin | >16–>100 | 1.2–>64 | 25–>100 | 4–>512 | 25–>512 | 16–>512 | 12.5–512 | 6–64 | 0.008–2 | 0.015–0.25 | ≤0.03–>80 | 4–>40 |
| Josamycin | 0.2–100 | 0.04–0.32 | 0.2–25 | 0.78–12.5 | 0.1–1.56 | ND | 0.2–3.12 | ND | ND | ND | ≤0.03–>50 | 0.125–1.5 |
| Spiramycin | 0.06–200 | 0.06–0.6 | 3–100 | 0.12–64 | ND | ND | ND | 0.125–4 | ND | 0.125–0.25 | 0.02–>20 | 0.04–1.25 |
| Kitasamycin | 0.4–>100 | 0.12–1.25 | 0.2–25 | 0.5–8 | ND | ND | ND | ND | ND | ND | 0.03–11 | 0.4–5.6 |
| Clindamycin | 0.12–1 | 0.063–>64 | ND | 0.094–>256 | ND | ND | ND | ND | ND | ND | ND | ND |
| Lincomycin | ≤0.25–200 | ≤0.0125–>64 | 0.03–25 | 0.12–>64 | 0.05–16 | 0.25–16 | 0.1–16 | 0.125–1 | 0.25–16 | 0.125–0.5 | 0.1–>64 | 0.1–2 |
| Tylosin | 0.06–100 | ≤0.0125–16 | 0.025–>10 | 0.025–>512 | 0.1–128 | ≤0.125–2 | <0.05–128 | 0.03–2 | 0.015–2 | 0.03–1 | 0.0025–12.5 | ≤0.0025–50 |
| Tilmicosin | ≤0.25–8 | ≤0.0125–>16 | ND | ≤0.06–>512 | 16–>512 | 0.25–2 | 0.25–>512 | ND | ND | ND | ≤0.0125–>25 | 0.006–0.125 |
| *Pleuromutilins* | | | | | | | | | | | | |
| Tiamulin | ≤0.03–1.56 | ≤0.0125–1 | 0.0025–0.1 | ≤0.015–2 | <0.05–0.78 | ND | >0.05–1.56 | 0.05–0.25 | ND | ND | 0.0005–1 | ≤0.03–1 |
| Valnemulin | ND | 0.00025–0.001 | 0.0001–0.00025 | ND | ND | ND | ND | ND | ND | ND | ≤0.008–0.125 | ND |

| | | | | | | | | | | | | |
|---|---|---|---|---|---|---|---|---|---|---|---|---|
| *Fluoroquinolones* | | | | | | | | | | | | |
| Flumequine | 2.5–25 | 0.25–>16 | 5–>10 | >10–256 | 32–64 | ≤0.125–16 | 0.5–128 | 5–100 | ND | ND | 2.5–>10 | 5–64 |
| Ofloxacin | 2–4 | ≤0.025–0.1 | 0.5–1 | ND | ND | ND | 0.2–50 | ND | ND | ND | ND | ND |
| Norfloxacin | ND | 0.025–0.1 | 2.5–5.0 | ND | ND | ND | 0.1–50 | 0.5–2 | 0.5–4 | 0.5–1 | ND | ND |
| Enrofloxacin | 0.06–4 | ≤0.015–>1 | ≤0.03–2 | 0.032–8 | ≤0.125–1 | ≤0.125–2 | ≤0.125–32 | 0.05–1 | 0.06–0.25 | 0.06–0.5 | 0.01–10 | 0.024–10 |
| Ciprofloxacin | 0.25–0.5 | 0.01–0.1 | 0.05–1 | ND | ND | ND | ND | 0.06–0.5 | 0.06–0.5 | 0.06–0.125 | ND | ND |
| Difloxacin | ND | ND | ND | ND | ND | ND | 0.05–12.5 | ND | ND | ND | 0.025–5 | ≤0.015–10 |
| Danofloxacin | 0.25–1 | 0.01–0.062 | 0.1–1 | 0.08–8 | ≤0.125–0.25 | ≤0.125–2 | ≤0.125–16 | 0.05–2.5 | ND | ND | 0.01–0.5 | 0.064–1 |
| Marbofloxacin | ND | ≤0.03–0.5 | ND | 0.25–8 | ND | ND | ND | ND | ND | ND | ND | ND |
| Aminoglycosides | | | | | | | | | | | | |
| Gentamycin | 1–4 | 0.12–>50 | 0.25–6.25 | 2–>64 | ND | ND | ND | 0.–16 | 16–64 | 16–64 | >10–>50 | 1 |
| Kanamycin | 0.4–25 | 0.16–2.5 | 0.78–25 | 3–64 | 6.25–100 | 0.125–100 | 2–512 | ND | ND | ND | ND | ND |
| Spectinomycin | ≤1–8 | 0.12–1 | 1.6–6.25 | 0.03–>1024 | ≤0.125–32 | 0.25–4 | 0.25–8 | 1–8 | 8–32 | 8–16 | 0.5–10 | 0.5–2 |
| Neomycin | ND | 0.16–16.3 | ND | 16–64 | ND | ND | ND | ND | ND | ND | ND | 2–>8 |
| Streptomycin | ND | 6.3–25 | 50–400 | 4–>1024 | ND | ND | ND | 1–64 | 4–64 | 8–64 | ND | ND |
| Apramycin | ND | ND | ND | ND | 0.2–100 | ND | 0.1–25 | ND | ND | ND | >16 | 2–4 |
| Spiramycin | ND | ND | ND | 0.25–>100 | ND | ND | ND | ND | 0.06–2 | 4–8 | 0.25–10 | ND |
| *Phenicols* | | | | | | | | | | | | |
| Chloramphenicol | 0.4–4 | 0.5–4.11 | 0.78–3 | 4–64 | <0.05–8 | 0.5–8 | 1–16 | 1–4 | 1–8 | 1–8 | ND | ND |
| Thiamphenicol | 0.2–12.5 | 0.1–1.56 | 1.6–6.25 | 0.78–512 | 0.39–>512 | 16–512 | 1.56–>512 | ND | ND | ND | ND | ND |
| Florfenicol | ND | 0.12–0.5 | ND | 0.06–64 | ND | ND | ND | ND | 0.5–4 | 1–4 | ND | ND |

[a]Data were compiled from multiple published studies or reference texts in which different methodologies, and often different antimicrobial concentrations were used and represents an update of Table 15.2 from Bébéar and Kempf, 2005. MICs listed are indicative of drug activities for microorganisms that have not been shown to have resistance to the various drug classes. [b]MLSK: macrolide-lincosamide, streptogramin-ketolide group of antimicrobials. MH: *M. hyorhinis*; MHP: *M. hyopneumoniae*; MHS: *M. hyosynoviae*; MB: *M. bovis*; MALK: *M. alkalescens*; MBOV: *M. bovoculi*; MBOVIR: *M. bovirhinis*; MA: *M. agalactiae*; MPUT: *M. putrefaciens*; MmycLC: *M. mycoides* subsp. *mycoides* LC; MG: *M. gallisepticum*; MS: *M. synoviae*; ND: no data available.

## Antimicrobial agents active against Mycoplasmas and Ureaplasmas

The most common antimicrobials active against these organisms are included in three major drug classes: tetracyclines, fluoroquinolones, and those agents included in the MLSK group (macrolides, lincosamides, streptogramins and ketolides). The inhibitory activities of the various classes of antimicrobials act in the same manner as they do for conventional bacteria. With regards to each of these drug classes, some representatives may be more active than others within the same class. Moreover, among the MLSK group, there are differences among the various species in their inherent susceptibilities and resistance as discussed in previous sections. Although it can be generally stated that tetracyclines, fluoroquinolones and the MLSK drugs are active against mycoplasmas and ureaplasmas, the occurrence and spread of acquired resistance in each of these classes has occurred among humans and animals and is becoming more prevalent. Tables 15.2 and 15.3 show MIC ranges for a large number of antimicrobial agents that may be used to treat mycoplasmal infections in humans or animals. Specific comments concerning each of the major antimicrobial groups useful for treatment of infections caused by these organisms are provided below. Recommended treatment regimens for human infections are summarized in other reference texts (Waites, 2008).

Unlike in human medicine, antimicrobial agents in food-producing animals are used not only for prevention and control of bacterial infections, but also to promote growth and improve weight gain (growth promoters). As result, relative to human medicine, very large amounts of antibiotics are used in food animals. In addition to tetracyclines, fluoroquinolones and MLSK drugs, there are other classes of antibiotics such as pleuromutilins and aminoglycosides which are used against mycoplasmas in veterinary medicine. In many cases there are different members of the same class approved for use in veterinary in comparison to human medicine. For example, certain fluoroquinolones, like enrofloxacin, danofloxacin, difloxacin, sarafloxacin, or marbofloxacin, or macrolides, such as tilmicosin or tylosin, are used exclusively in veterinary medicine. In addition, the pleuromutilins tiamulin and valnemulin are uniquely used in veterinary medicine. Moreover, they are used in some countries, but not in others due to concerns about increased toxicity when administered in combination with ionophores that are widely used as coccidiostats or growth promoters in food animals (Nogueira *et al.*, 2009).

Most animal mycoplasmoses are chronic diseases with high morbidity and relatively low mortality. The presence of infected animals that do not show any clinical signs of diseases but play an important role as carriers is important in perpetuation of infection and is often the main obstacle in the control and eradication of mycoplasmoses. Usually, the treatment against mycoplasmas in food animals is performed on a herd or flock basis and not individually. Use of antibiotics may be very helpful in preventing clinical signs and lesions, as well as reducing economic losses, but not necessarily for complete elimination of infection from a herd or a flock and therefore cannot be a satisfactory long-term solution. Since mycoplasma infections in food animals are often multifactorial, the antibiotic of choice is often a broad-spectrum drug, such as a tetracycline or fluoroquinolone, which may also be effective against other bacterial pathogens in the disease complex.

### Tetracyclines

Tetracyclines exert bacteriostatic activity by binding reversibly with the 30S ribosomal subunit involving 16S rRNA and several ribosomal proteins to prevent association of aminoacyl-tRNA at the acceptor site. This action prevents bacterial protein synthesis. There are several representatives in this class which are commonly used to treat human infections. In general, doxycycline has lower MICs than tetracycline. Newer representatives of this class include the glycylcyclines, represented by tigecycline. The glycylcyclines have added substituents that interfere with efflux pumps and ribosomal protection proteins such as *tet*(M) which may be present in some isolates of *M. hominis* and *Ureaplasma* spp.

Historically, the tetracyclines have been among the most widely used drugs to treat mycoplasmal and ureaplasmal infections of the urogenital tracts and *M. pneumoniae* respiratory infections in adults. Despite potential for bone and tooth toxicity, these agents have been used successfully to treat infections of the central nervous system in neonates when there were no other feasible alternatives (Waites *et al.*, 2005a).

Tetracyclines are also frequently used for treatment of mycoplasmal infections in food animals due to their broad-spectrum activity, capacity to attain effective concentrations in most tissue, relative safety, ability to be administrated by many routes, and relatively low cost. Conventional or long-acting formulations are indicated for treatment of bovine or swine pneumonia, mastitis, arthritis, or genital tract infections (Giguere, 2007c). Oxytetracycline maintains a higher concentration in the diseased lung than in healthy lungs and milk drug concentrations are higher than plasma concentrations when there is inflammation of the mammary gland (Gunn, 2002). Despite the fact that mastitis caused by mycoplasmas is usually unresponsive to the treatment, it was shown that repeated intramammary administration of tylosin (500 mg) cured experimentally induced *M. colifornicum* mastitis in cows (Ball and Campbell, 1989).

## Macrolides-lincosamides-streptogramins and ketolides

Members of the MLSK group of drugs are similar overall in that they share the same mode of action. However, individual agents in this diverse group may show variations *in vitro* activity according to species as mentioned in previous sections. Unlike most *Mycoplasma* species, ureaplasmas are not consistently inhibited by lincosamides administered at the usual dosages but they are susceptible to most macrolides (Bébéar and Bébéar, 2002). With respect to other human mycoplasmas of clinical importance, activities of MLSK drugs against *M. genitalium* are generally similar to those of *M. pneumoniae* while the susceptibilities of *M. fermentans* are similar to *M. hominis* (Bébéar and Bébéar, 2002).

Like the tetracyclines, most of the MLSK drugs are primarily bacteriostatic agents, with the exception of streptogramin combinations such as quinupristin/dalfopristin and certain ketolides (Bébéar and Bébéar, 2002). MLSK drugs bind to specific nucleotides in domains II and/or V of 23S ribosomal RNA in the 50S ribosomal subunit blocking protein synthesis by causing premature dissociation of peptidyl-tRNA from the ribosome.

Much information is now forthcoming concerning resistance mechanisms which affect the activity of the MLSK drugs. For those *Mycoplasma* spp. that are susceptible to MLSK agents, the MICs are typically the lowest of any antimicrobial class. A new investigational ketolide, solithromycin (CEM-101) is the most potent antimicrobial ever tested against *M. pneumoniae* and has MICs as low as 0.5 µg/ml for isolates with high level resistance to azithromycin (Waites *et al.*, 2009).

Macrolides have historically been the drugs of choice for treatment of respiratory infections due to *M. pneumoniae* in children and they are also widely used to treat neonatal infections due to *Ureaplasma* spp. Macrolides may also be useful in situations where tetracyclines or fluoroquinolones cannot be used. Clindamycin is useful in neonates or young children with *M. hominis* infection and adults with urogenital infections, but is not as active against *M. pneumoniae* as the macrolides and is not generally considered a first-line therapeutic agent for this mycoplasma.

Macrolides approved for veterinary use include erythromycin, tylosin, tilmicosin, spiramycin, and tulathromycin. Many mycoplasmas found in food animals are intrinsically resistant to erythromycin as described above. However, the 16-membered macrolide tylosin and its semisynthetic derivate tilmicosin are widely used against avian, swine or cattle mycoplasmosis although the latter is not approved for use in poultry in the USA. In pigs, where tylosin is also used as a growth promoter, its use in the prevention and treatment of swine mycoplasmal infection is being replaced by tiamulin (Giguere, 2007b). In Europe, the label claim for tulathromycin also includes treatment and prevention of *M. bovis* and *M. hyopneumoniae* infections (Giguere, 2007b). It was shown that a single dose of tulathromycin was as effective as three daily administrations of enrofloxacin for treatment of pigs inoculated experimentally with *M. hyopneumoniae* (McKelvie *et al.*, 2005). Interestingly, tulathromycin was also effective in the treatment of calves experimentally infected with

*M. bovis* irrespective of the MIC of the challenge strain (1 or >64 mg/ml) (Godinho *et al.*, 2005).

### Fluoroquinolones

Fluoroquinolones are bactericidal agents that target DNA gyrase and topoisomerase IV which are required for DNA replication. DNA gyrase is a tetrameric enzyme that catalyses negative supercoiling of DNA. It is comprised of subunits A and B that are encoded by *gyrA* and *gyrB* genes. Topoisomerase IV is necessary for decatenation of DNA following its replication and is encoded by *parC* and *parE* genes. Various members of the fluoroquinolone class may target one or the other of these enzymes preferentially and they also have different degrees of activity against mycoplasmas and ureaplasmas. Nalidixic acid, a non-fluorinated member of the quinolone class, is inactive against Gram-positive bacteria and mycoplasmas (Bébéar and Bébéar, 2002). One of the earliest fluoroquinolones, ciprofloxacin, is primarily active against Gram-negative bacteria and possesses only modest activity against the mycoplasmas and ureaplasmas, with MICs often > 1 µg/ml, which is the lower limit of susceptibility for conventional bacteria. This contrasts with other agents such as moxifloxacin for which MICs are usually ≤ 0.125 µg/ml for all of the pathogenic human mycoplasmas and ureaplasmas (Table 15.2).

Fluoroquinolones have become very popular for treatment of a wide array of bacterial infections, besides those caused by mycoplasmas and ureaplasmas. Since they are bactericidal, fluoroquinolones may be important therapeutic alternatives for invasive infections in immunocompromised hosts where bacteriostatic agents often prove to be ineffective. These drugs have been used very sparingly in infants and children because they are not approved for use in these patient populations and concerns regarding potential toxicity.

Veterinary fluoroquinolones (Table 15.3) such as enrofloxacin, danofloxacin, difloxacin, sarafloxacin, or marbofloxacin are indicated for treatment of respiratory infections in cattle, swine or poultry. For example, it was shown that medication of naturally infected birds with enrofloxacin was highly effective in reducing or eliminating upper respiratory infection with *M. gallisepticum*, but had little effect on populations of *M. synoviae* (Stanley *et al.*, 2001). However, in the USA, because of concern about resistance in zoonotic foodborne pathogens, enrofloxacin as well as danofloxacin are approved only for the treatment of acute pneumonia in beef cattle and sarafloxacin has been withdrawn from use in poultry (Walker and Dowling, 2007).

### Other drug classes

Drugs such as the aminoglycosides and chloramphenicol sometimes demonstrate activity *in vitro*. However, their toxicities and availability of better agents means they are not normally considered as suitable antimycoplasmal agents in humans with the exception of occasional use of chloramphenicol for treatment of systemic infections in neonates caused by *M. hominis* or *Ureaplasma* spp. in the setting of tetracycline resistance or clinical failure with other agents (Waites *et al.*, 2005a; Bébéar, 2010). A new class of antimicrobial agents with potential use for treatment of mycoplasmal and ureaplasmal infections is the peptide deformylase inhibitor group which block the actions of this enzyme that is involved in early stages of polypeptide synthesis in a bacteriostatic manner. The investigational agent LBM415 was shown to be highly active against *M. pneumoniae*, but showed no activity against *M. hominis* and *M. fermentans* and only modest activity against *Ureaplasma* spp. (Waites *et al.*, 2005b). No drugs in this class have been approved for clinical use thus far.

Pleuromutilins, which are used exclusively in veterinary medicine, inhibit protein synthesis by binding to the 50S subunit of the bacterial ribosome. Chemical footprinting revealed that tiamulin and valnemulin bind at the peptidyl transferase centre in the 23S rRNA, inhibiting the peptide bond formation (Poulsen *et al.*, 2001). Pleuromutilins are used mostly in the swine industry for treatment of respiratory diseases and dysentery in pigs (Giguere, 2007a). There are also a few reports of successful use of tiamulin or valnemulin in cattle in field or experimental conditions (Stipkovits *et al.*, 2001a,b, 2005). In addition, Ball and McCaughey found that a single subcutaneous dose of aqueous tiamulin, eliminated *Ureaplasma* spp. from the genital tract of 18/22 infected sheep (Ball and McCaughey, 1987).

## Mechanisms of antimicrobial resistance and clinical implications

Among the various mechanisms of acquired resistance, the only ones described *in vivo* for mycoplasmas are antimicrobial target modification or protection. An active efflux mechanism has also been demonstrated *in vitro* for fluoroquinolones. Resistance is mediated either by chromosomal mutations or acquisition of a transposon. Mycoplasmas are characterized by high mutation rates. Sequencing studies of several mycoplasma genomes including *M. pneumoniae* have revealed that only a small amount of genetic information is dedicated to DNA repair (Rocha and Blanchard, 2002). It has been shown in other bacteria that the lack of some DNA repair systems like the *mut* gene is associated to a mutator phenotype. Thus, a link could be hypothesized between high mutation rates and antimicrobial resistance in mycoplasmas, as it has been found for *Pseudomonas aeruginosa* (Oliver *et al.*, 2000). Resistance through mutation concerns all classes of antimicrobial agents used to treat mycoplasmal infections (Bébéar, 2010). Regarding the acquisition of new resistance genes from other bacteria, no extrachromosomal element has been described in human mycoplasmas or ureaplasmas. However, transposons carrying antibiotic resistance genes have been found, the main example being the *tet*(M) determinant conferring tetracycline resistance in *M. hominis* and *Ureaplasma* spp.

The genetic basis of acquired antimicrobial resistance in mycoplasmas and ureaplasmas has been investigated in several published studies by *in vitro* selection of antimicrobial resistant mutants through cultivation of the organisms in successively high concentrations of the drugs of interest followed by gene sequencing to determine the presence of gene mutations in antimicrobial targets. Summaries of the current knowledge of these resistance mechanisms based on *in vitro* studies have been published elsewhere (Bébéar, 2010; Bébéar and Bébéar, 2002; Bébéar and Kempf, 2005). In many instances these mechanisms of resistance determined *in vitro* have been confirmed in clinical isolates that exhibit naturally occurring resistance of the same phenotype.

Studies on resistance mechanisms concern not only *M. hominis* and *Ureaplasma* spp., two species long known to have tetracycline-resistant strains, but also *M. pneumoniae* and *M. genitalium*, two species with high antibiotic susceptibility in which macrolide resistance is now becoming a problem (Bébéar and Pereyre, 2005; Bébéar, 2010; Bébéar *et al.*, 2011). However, as *M. pneumoniae* and especially *M. genitalium* are rarely isolated from clinical specimens, and *in vitro* susceptibility testing is even less often used for patient management purposes, the number of clinical strains tested is limited and the prevalence of acquired resistance is poorly documented worldwide. Newer data about acquired resistance in these two species have come from investigations utilizing molecular-based methods to assess antimicrobial activities or detect resistance genes.

A discussion of pertinent mechanisms of acquired resistance as they affect the various antimycoplasmal drug classes follows. Tables 15.4 and 15.5 summarize the known mechanisms of antimicrobial resistance that occur naturally and those that have been induced by *in vitro* selection in various species isolated from humans or animals. Table 15.6 summarizes the prevalence of macrolide resistance in *M. pneumoniae* in various countries.

## Tetracyclines

### Human urogenital Mycoplasmas and Ureaplasmas

Acquired resistance to tetracyclines has long been known in *Ureaplasma* spp. and *M. hominis* and is due to the presence of the *tet*(M) determinant (Bébéar and Kempf, 2005). In mycoplasmas this determinant is located on the conjugative transposon Tn916. It codes for the Tet(M) protein which protects the ribosome from the action of tetracyclines. The Tet(M) protein is homologous to elongation factors eF-Tu and eF-G and upon binding to the ribosome induces a conformational change which is thought to prevent tetracycline binding without altering protein synthesis.

High-level tetracycline resistance (MIC ≥8 µg/ml) is associated with the presence of the *tet*(M) determinant which confers cross-resistance to all tetracyclines. Such strains are easily detected in susceptibility tests with tetracycline, doxycycline,

**Table 15.4** Acquired antimicrobial resistance in pathogenic mycoplasmas and ureaplasmas of humans

| Antimicrobial class | Resistance *In vitro* | *In vivo* | Mechanism | MIC Range for resistant isolates (µg/ml) |
|---|---|---|---|---|
| *M. pneumoniae* | | | | |
| MLSK | Yes | Yes | 23S rRNA mutations at positions 2611, 2058, 2059, and 2062[a]. Mutations, insertions or deletions in L4 and L22 ribosomal proteins (*in vitro* only) | 64–>256 (erythromycin) |
| Tetracycline | Yes | No | 16S rRNA mutations at position 968 and 1193 (*in vitro* only) | 2 (tetracycline) |
| Fluoroquinolones | Yes | No | Mutations in *gyrA*, *gyrB*, *parC* or *parE* genes of QRDRs | 2–16 (levofloxacin), 8–128 (ciprofloxacin) |
| *M. hominis* | | | | |
| MLSK | Yes | Yes | 23S rRNA mutations at positions 2610, 2611, 2057, 2059, and 2062 | 16–64 (clindamycin) |
| Tetracycline | Yes | Yes | *tet*(M) ribosomal protection. 16S rRNA mutations at positions 346, 965, 966, 967, and 1054 (*in vitro* only) | 8–>64 (tetracycline), 2–8 (tetracycline) |
| Fluoroquinolones | Yes | Yes | Mutations in *gyrA*, *gyrB*, *parC* or *parE* genes of QRDRs. Drug efflux (*in vitro* only increasing ciprofloxacin and norfloxacin MICs) | 2–32 (levofloxacin), 4–8 (ciprofloxacin) |
| *M. genitalium* | | | | |
| MLSK | No | Yes | 23S rRNA mutations at positions 2058 or 2059. Mutations in L4 ribosomal protein | 16–>64 (erythromycin) |
| Tetracycline | No | No | Treatment failures have been reported but no resistant genes have been identified | ND |
| Fluoroquinolones | No | Yes | Mutations in *gyrA*, *gyrB*, *parC* or *parE* genes of QRDRs | ND |
| *Ureaplasma* spp. | | | | |
| MLSK | Yes | Yes | Deletions or insertions in L4 ribosomal proteins and/or 23S rRNA mutations at position 2056, 2057, and 2058. Ribosomal methylation mediated by *ermB*[b]. Drug efflux mediated by *msrA*, *msrB* or *msrD*[b] | 64–>128 (erythromycin) |
| Tetracycline | Yes | Yes | *tet*(M) ribosomal protection | 2–>32 |
| Fluoroquinolones | Yes | Yes | Mutations in *gyrA*, *gyrB*, *parC* or *parE* genes of QRDRs | 4–32 (levofloxacin) |

[a]*E. coli* numbering system. [b]Macrolide *erm* and efflux genes have been detected in only one published study (Lu *et al.*, 2010) and has not been confirmed by other investigators. MLSK: macrolide, lincosamide, streptogramin, ketolide group of antimicrobials; QRDRs: quinolone resistance determining regions of DNA; ND: not determined.

and minocycline. Genotypic detection can be performed by PCR amplification of the *tet*(M) gene (de Barbeyrac *et al.*, 1996). Glycylcyclines such as tigecycline retain activity against *M. hominis* containing *tet*(M), but not *Ureaplasma* spp. *M. pneumoniae* strains with slightly reduced tetracycline susceptibility (MIC = 2 µg/ml) have been obtained *in vitro* in the presence of increasing tetracycline concentrations (Dégrange *et al.*, 2008) and were due to mutations of 16S rRNA in the tetracycline binding pocket of the bacterial ribosome. Even though tetracycline resistance has been described in many mycoplasmas of veterinary importance, the molecular mechanisms of resistant strains isolated *in vivo* or selected *in vitro* has not been elucidated as they have in humans.

The prevalence of acquired tetracycline resistance among *M. hominis* and *Ureaplasma* spp. varies according to the country and the antimicrobial exposure of the population. Resistance was estimated to be present in about 10% of patients consulting for a sexually transmitted infection in the United Kingdom and 3% in France in the early 1990s (Bébéar, 2010). Later studies from France

**Table 15.5** Acquired antimicrobial resistance in pathogenic mycoplasmas of animals

| Antimicrobial class | Resistance: *In vitro* | Resistance: Natural *in vivo* | Mechanism | MIC range for resistant isolates (μg/ml) |
|---|---|---|---|---|
| *M. hyorhinis* | | | | |
| MLSK | Yes | Yes | Mutation at the position 2059[a] of 23S rRNA (*in vivo*) | 10–100 (tylosin), 25–>100 (lincomycin) |
| | | | Mutation at the position 2059 of 23S rRNA (*in vitro*) | >100 (tylosin) |
| | | | Mutations at the positions 2597, 2611 and the insertion of an adenine at the pentameric adenine sequence of 23S rRNA (*in vitro*) | 50 (lincomycin) |
| | | | Mutations at the positions 2062, 2597, 2611 and the insertion of an adenine at the pentameric adenine sequence of 23S rRNA (*in vitro*) | 100 (tylosin)[b], 50 (lincomycin) |
| Tetracyclines | No | Yes | ND | 12.5 (chlortetracycline) |
| Fluoroquinolones | No | Yes | ND | 1–4 (enrofloxacin) |
| *M. hyopneumoniae* | | | | |
| MLSK | No | Yes | Mutation at the position 2058 of 23S rRNA (only one *in vivo* mutant was studied) | >64 (lincomycin) |
| Tetracyclines | Yes | Yes | ND | 12.5–≥100 (chlortetracycline) |
| Fluoroquinolones | Yes | Yes | Mutations in the QRDR of *parC* (*in vivo*) | 0.25–>1 (enrofloxacin) |
| *M. bovis* | | | | |
| MLSK | Yes | Yes | Mutations at the positions of 2058 of *rrl_3* and 748 and 2058 at *rrl_4* (tylosin) or only mutation 2058 in *rrl_3* (tilmicosin) genes of 23S rRNA (only one *in vitro* mutant per each antimicrobial was studied) | >1024 (tylosin), >256 (tilmicosin) |
| | | | Mutations at the positions 748, 752, 2058, or 2059 of one or both 23S rRNA genes (*in vivo*). Mutations in L4 and L22 ribosomal proteins | 8–1024 (tylosin), 32–>256 (tilmicosin) |
| Tetracyclines | Yes | Yes | ND | >32 (oxytetracycline) |
| Fluoroquinolones | Yes | Yes | Mutations in the QRDRs of *gyrA* and *parC in vivo* | 2.5–32 (enrofloxacin) |
| *M. gallisepticum* | | | | |
| MLSK | Yes | Yes | Mutations at the positions 2058 or 2059 of *rrnA* (MGA_r01) gene (*in vivo*) | 0.63–5 (tylosin), 1.25–>10 (tilmicosin) |
| | | | Mutations at the positions 2058, 2503 of one of 23S rRNA genes (*in vitro*) | 256–512 (tilmicosin), 256–≥512 (erythromycin) |
| Tetracyclines | Yes | Yes | ND | 5–>16 (oxytetracycline) |
| Fluoroquinolones | Yes | Yes | Mutations in the QRDRs of *gyrA*, *gyrB*, *parC* and *parE* (*in vitro*) | 1–32 (enrofloxacin) |
| | | | Mutations in the QRDRs of *gyrA*, *gyrB* and *parC* (*in vivo*) | 1–10 (enrofloxacin) |

[a]*E. coli* numbering system. [b]Lincomycin-resistant strain had undergone 11 additional serial passages with tylosin. MLSK: macrolide, lincosamide, streptogramin, ketolide group of antimicrobials; QRDRs: quinolone resistance determining regions; ND: not determined.

conducted between 1999 and 2002 reported that 19% of *M. hominis* isolates had reduced tetracycline susceptibility (Bébéar, 2010). On the other hand, no such increase was found in *Ureaplasma* spp. for which the percentage of tetracycline-resistant isolates was 2% over the same period (Bébéar, 2010). These results have been confirmed recently in Germany where significantly more *M. hominis*

**Table 15.6** Prevalence of naturally occurring acquired resistance in *M. pneumoniae* in various countries

| Country | Year | No. resistant strains (or samples PCR-positive for a resistant strain)/no. susceptible strains (or samples PCR-positive for a susceptible strain) (%) | MIC (µg/ml) | 23S rRNA mutations[a] | Reference |
|---|---|---|---|---|---|
| Japan | 1988–1997 | 1/141 strains (0.07) | >400 (ERY) | $A_{2058}G$ | Okazaki *et al.* (2001) |
| | 1986–1999 | 0/296 strains (0) | ND | None | Okazaki *et al.* (2007) |
| | 2000–2006 | 15/85 strains (17.6) | ND | $A_{2058}G$, $A_{2059}G$ | Okazaki *et al.* (2007) |
| | 2000–2003 | 13/76 strains (17.1) | 0.03–64 (ERY) | $A_{2058}G/C$, $A_{2059}G$, $C_{2611}G$ | Matsuoka *et al.* (2004) |
| | | 23/94 samples (24) | NA | $A_{2058}G$ | Matsuoka *et al.* (2004) |
| | 2002–2004 | 12/195 strains (6.1) | 16–>64 (AZM) | $A_{2058}G$, $A_{2059}G$ | Morozumi *et al.* (2005) |
| | 2002–2006 | 55/380 strains (14.5) | 16–>64 (AZM) | $A_{2058}G$, $A_{2059}G$ | Morozumi *et al.* (2008) |
| | 2002–2006 | 30/94 strains (31.9) | 16–>64 (AZM) | $A_{2058}G$, $A_{2059}G$ | Matsubara *et al.* (2009) |
| USA, Europe | 1995–1999 | 2/41 strains (4.9) | 4–>8 (AZM) | ND | Critchley *et al.* (2002) |
| USA | 1991–2008 | 5/100 strains (5) | 128 (ERY) | $A_{2058}G$, $A_{2059}G$ | Wolff *et al.* (2008) |
| | 2006–2007 | 5/30 samples (16.7) | NA | $A_{2059}G$ | Wolff *et al.* (2008) |
| Germany | 1991–2009 | 3/99 strains (3) | 100–>200 (ERY) | $A_{2058}G$, $A_{2059}G$ | Dumke *et al.* (2010) |
| | 2003–2008 | 2/167 samples (1.2) | NA | $A_{2058}C$, $A_{2059}G$ | Dumke *et al.* (2010) |
| France | 1994–2006 | 2/155 strains (1.3) | 64–256 (ERY) | $A_{2058}G$, $A_{2059}G$ | Pereyre *et al.* (2007a) |
| | 1998–2004 | 0/86 samples (0) | NA | None | Peuchant *et al.* (2009) |
| | 2005–2007 | 5/51 samples (9.8) | NA | $A_{2058}G$, $A_{2059}G$, $C_{2611}G$ | Peuchant *et al.* (2009) |
| China | 2003–2006 | 46/50 strains (92) | 128–256 (AZM) | $A_{2058}G/C$, $A_{2059}G$ | Xin *et al.* (2009) |
| | 2005–2008 | 11/53 strains (83) | 64–>128 (ERY) | $A_{2058}G$ | Liu *et al.* (2009) |
| | 2008–2009 | 46/67 strains (69) | 0.06–32 (AZM) | $A_{2058}G/T$, $A_{2059}G$ | Cao *et al.* (2010) |
| | 2008–2009 | 90/100 strains (90) | 64–>128 (ERY) | $A_{2058}G/T$, $A_{2059}G$ | Liu *et al.* (2010) |
| | 2009 | 58/64 samples (91) | NA | $A_{2058}G/T$, $T_{2611}C$ | Lin *et al.* (2010) |
| Italy | 2010 | 11/43 samples (26) | NA | $A_{2058}G$, $A_{2059}G$ | Chironna *et al.* (2011) |
| Israel | 2010 | 9/30 samples (30) | NA | $A_{2058}G$ | Averbuch *et al.* (2012) |

[a]*E.* coli numbering. ERY: erythromycin; AZM: azithromycin; ND: not determined; NA: not applicable because macrolide resistance was determined by PCR directly from *M. pneumoniae*-positive samples; no resistant strain was isolated to be studied for MIC determination.

isolates (10–13%) than ureaplasmas (1–3%) were resistant to tetracyclines (Krausse and Schubert, 2010). The low tetracycline resistance rate for *Ureaplasma* spp. was also reported in the UK and in China with less than 5% of *Ureaplasma* isolates being classified as resistant to tetracycline in different *in vitro* susceptibility surveys (Xie and Zhang, 2006; Beeton *et al.*, 2009b). In contrast, surveillance from different regions in the USA between 2000 and 2004 showed that 45% of *Ureaplasma* spp. isolates contained *tet*(M) and were resistant to tetracycline with MICs 2–64 µg/ml (Waites *et al.*, 2005a).

There has been a significant number of treatment failures with tetracyclines in *M. genitalium* urethritis reaching 55% in the randomized clinical trial evaluating the treatment of NGU conducted by Mena (Mena *et al.*, 2009). Another recent study from the USA reported a clearance rate of 30.8% for *M. genitalium* with doxycycline versus 66.7% for azithromycin (Schwebke *et al.*, 2011). No precise mechanism of resistance such as *tet*(M) has been identified thus far and these clinical observation may reflect the fact that this class is simply less active against this mycoplasma. Natural tetracycline resistance is not known to occur in *M. pneumoniae.*

### Animal Mycoplasmas

The presence of acquired resistance to tetracyclines in field strains of several animal mycoplasmas has been demonstrated by *in vitro* susceptibility surveys. Organisms evaluated include *M. alkalescens* (Hirose *et al.*, 2003; Uemura *et al.*, 2010), *M. bovirhinis* (Hirose *et al.*, 2003; Uemura *et al.*, 2010), *M. bovis* (Cooper *et al.*, 1993; ter Laak *et al.*, 1993; Ayling *et al.*, 2000; Hirose *et al.*, 2003; Thomas *et al.*, 2003; Francoz *et al.*, 2005; Rosenbusch *et al.*, 2005; Uemura *et al.*, 2010), *M. gallisepticum* (Cooper *et al.*, 1993; Bradbury *et al.*, 1994; Hannan *et al.*, 1997b), *M. hyopneumoniae* (Etheridge *et al.*, 1979; Inamoto *et al.*, 1994), *M. hyorhinis* (Kobayashi *et al.*, 1996; Hannan *et al.*, 1997b), *M. hyosynoviae* (Hannan *et al.*, 1989, 1997a,b) and *M. synoviae* (Cooper *et al.*, 1993; Bradbury *et al.*, 1994; Hannan *et al.*, 1997b) Although tetracycline resistance has been reported sporadically in avian and swine mycoplasmas, overall these species remain susceptible, especially to the more recently introduced derivatives of this drug class. However, the susceptibility of bovine pathogen *M. bovis* to this group of antimicrobials is significantly decreased, with reported $MIC_{50}$ and $MIC_{90}$ for oxytetracycline of 2 mg/ml and 32 mg/ml (Belgium), 16 mg/ml and 32 mg/ml (Netherlands), 32 mg/ml and 64 mg/ml (UK), 25 mg/ml and 50 mg/ml and 128 mg/ml and 128 mg/ml (Japan) (ter Laak *et al.*, 1993; Ayling *et al.*, 2000; Hirose *et al.*, 2003; Thomas *et al.*, 2003; Uemura *et al.*, 2010). Interestingly, heterogeneity in susceptibility to tetracyclines was observed among cohorts of *M. bovis* strains, isolated in different European countries, with a high percentage of resistant strains in the UK, Belgium and the Netherlands (ter Laak *et al.*, 1993; Ayling *et al.*, 2000; Thomas *et al.*, 2003), but a lower percentage of resistance strains in Italy (Mazzolini *et al.*, 1997).

Several authors have studied *in vitro* selection of tetracycline-resistant mycoplasmas isolated from food animals. *Mycoplasma* isolates with decreased susceptibility to tetracyclines were selected *in vitro* for *M. bovis* (Pilaszek and Truszeczynski, 1980; Ayling *et al.*, 2008), *M. mycoides* subsp. *mycoides SC* (Lee *et al.*, 1987), *M. hyopneumoniae* (Hannan *et al.*, 1997a) and for *M. gallisepticum* (Zanella *et al.*, 1998). Gautier-Bouchardon *et al.* failed to obtain tetracycline-resistant mutants of *M. gallisepticum* or *M. synoviae* after 10 passages with oxytetracycline, while resistant-mutants of *M. iowae* were selected (Gautier-Bouchardon *et al.*, 2002).

## Macrolides-lincosamides-streptogramins and ketolides

Macrolide resistance in mycoplasmas, which have a small number of ribosomal operons, is conferred by mutations in the ribosomal target (23S rRNA and ribosomal proteins L4 and L22). Among mycoplasmas and ureaplasmas of humans, acquired macrolide resistance has been described mainly in *M. pneumoniae,* but resistant strains of *Ureaplasma* spp. and *M. genitalium* have been reported. While *M. hominis* is naturally resistant to 14 and 15-membered macrolides, and usually susceptible to 16-membered drugs, acquired resistance to other MLSK drugs has been described (Pereyre *et al.*, 2002).

### *Mycoplasma pneumoniae*

Most macrolide-resistant strains of *M. pneumoniae* contain a $A_{2058}G$ mutation (*E. coli* numbering) in the peptidyltransferase loop of 23S rRNA (Bébéar *et al.*, 2010), (Fig. 15.2). The other mutations found in clinical strains at positions 2059 and 2611 (Okazaki *et al.*, 2001; Matsuoka *et al.*, 2004) have been identified as macrolide resistance hot spots in other bacteria. No mutations in domain

II or in the ribosomal proteins L4 or L22 genes have been described *in vivo*. The resistant *M. pneumoniae* clinical isolates are typically resistant to macrolides, as well as lincosamides, streptogramin B (phenotype $MLS_B$) and ketolides, with different increases in the MICs according to the mutation, but cross resistance to other drug classes does not occur (Bébéar and Pereyre, 2005). For 16-membered macrolides there was a larger increase in the MICs for the mutation at position 2059 than at position 2058, whereas the reverse was true for ketolides. The mutation at position 2611 was associated with the lowest levels of resistance, while mutations 2058 and 2059 led to a high level resistance to macrolides. Quinupristin-dalfopristin retained activity against the mutants.

Resistant mutants of *M. pneumoniae* have been obtained by selection *in vitro* on erythromycin. Resistance is due to mutations at positions 2058 and 2059 which were previously described *in vivo*. An exhaustive *in vitro* study described the selection of mutants resistant to different macrolides, streptogramins and a ketolide, testing a total of eight antimicrobial agents. The selected mutations were in domain V of 23S rRNA at positions 2611 and 2062 and in the genes encoding ribosomal proteins L4 and L22; these were point mutations, insertions or deletions. No mutations were detected in domain II of 23S rRNA (Pereyre *et al.*, 2004).

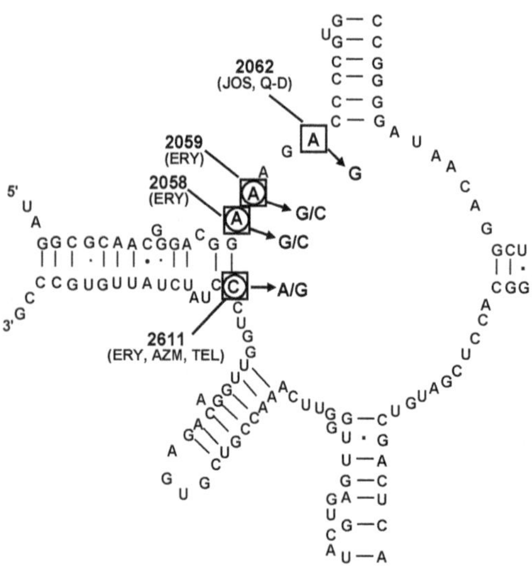


**Figure 15.2** Secondary structure of the peptidyltransferase loop of domain V of 23S rRNA of *M. pneumoniae* (*E. coli* numbering). Squared nucleotides indicate positions mutated *in vitro*. Antimicrobials in parentheses (AZM: azithromycin, ERY: erythromycin, JOS: josamycin, Q-D: quinupristin-dalfopristin and TEL: telithromycin) indicate the selective agent. Circled nucleotides indicate positions mutated *in vivo*. This figure is adapted from an earlier publication (Bébéar, 2010).

Prior to the year 2000, very few clinical isolates of *M. pneumoniae* were resistant to macrolides. Rare strains resistant to erythromycin were reported in the literature between 1968 and 1999 (Niitu *et al.*, 1970; Stopler *et al.*, 1980; Critchley *et al.*, 2002; Pereyre *et al.*, 2007a). No erythromycin resistance was detected among 150 strains isolated from France and Denmark between 1962 and 1996 (Bébéar and Kempf, 2005; Pereyre *et al.*, 2007a). Two macrolide-resistant isolates have been detected in a collection of 41 *M. pneumoniae* strains from North America and Europe between 1995 and 1999 (Critchley *et al.*, 2002). In contrast, since 2000 several Japanese studies (Matsuoka *et al.*, 2004; Morozumi *et al.*, 2005; Okazaki *et al.*, 2007; Morozumi *et al.*, 2008; Matsubara *et al.*, 2009) have reported a significant increase in macrolide resistance rates in *M. pneumoniae*, affecting more than 40% of strains in 2007 according to Morozumi (Morozumi *et al.*, 2010), (Table 15.5). Some Chinese studies described an even higher percentage of macrolide-resistant isolates of *M. pneumoniae*, ranging from 69% to 92%, obtained from both children and adults between 2003 and 2009 (Liu *et al.*, 2009; Xin *et al.*, 2009; Cao *et al.*, 2010; Lin *et al.*, 2010). In a *M. pneumoniae* epidemic in the USA, three of eleven isolates were resistant to macrolides (Wolff *et al.*, 2008). In Germany, Dumke described three out of 99 strains, isolated between 1991 and 2009, that were resistant to macrolides (Dumke *et al.*, 2010). In France, only two macrolide-resistant clinical isolates were described within a series of 155 strains isolated between 1994 and 2006 (Pereyre *et al.*, 2007a). More recently, however, macrolide resistance in *M. pneumoniae* has been on the rise in France, with a 10% rate (5/51) of resistant genotypes reported between 2005 and 2007 (Peuchant *et al.*, 2009). In Italy, a recent study reported macrolide resistant genotypes in 26% of the 43 *M. pneumoniae*-positive specimens from children (Chironna *et al.*,

2011), a resistance rate of 30% (9 of 30 samples) was reported in Israel during 2010 (Averbuch *et al.*, 2011) (Table 15.5). This increase in resistance has paralleled a similar rise in macrolide resistance in other respiratory pathogens apparently as a result of antibiotic selective pressure in children during a period of extensive macrolide use in many parts of the world, especially in Asia (Liu *et al.*, 2009; Cao *et al.*, 2010; Morozumi *et al.*, 2010).

The isolation of naturally occurring macrolide-resistant *M. pneumoniae* may lead to treatment failure which, for the patients concerned, translates into more febrile days and longer duration of persistent cough than patients with macrolide-susceptible isolates (Suzuki *et al.*, 2006; Matsubara *et al.*, 2009). Furthermore, children with macrolide-resistant *M. pneumoniae* infections required therapeutic changes with substitution of minocycline or levofloxacin because of either persistent symptoms or unresolved or worsening chest radiographic abnormalities (Morozumi *et al.*, 2008). Finally, another study reported that the clinical efficacy rate of macrolide therapy was 91.5% and 22.7% in macrolide-susceptible versus macrolide-resistant *M. pneumoniae* infections, respectively (Matsubara *et al.*, 2009). Thus far, most infections with macrolide resistant *M. pneumoniae* described in the past ten years have occurred in children, but many fewer adults have been evaluated (Peuchant *et al.*, 2009; Morozumi *et al.*, 2010). To date, no difference was found between resistant strains isolated in children or in adults.

Several French (Dégrange *et al.*, 2009), German (Dumke *et al.*, 2010), Japanese (Matsuoka *et al.*, 2004; Morozumi *et al.*, 2005), and Chinese (Liu *et al.*, 2009, 2010; Cao *et al.*, 2010) isolates were molecularly typed either by PCR- restriction fragment length polymorphisms (RFLP) of the adhesin P1 gene or by multi-locus variable-number-tandem-repeat analysis (MLVA). No clear association was observed between the macrolide-resistant isolates and the P1 subtypes and the more discriminating MLVA analysis on French and Japanese isolates did not reveal any link between a particular MLVA type and macrolide resistance (Dégrange *et al.*, 2009). These data confirmed the absence of a particular emerging macrolide-resistant clone.

### Human urogenital Mycoplasmas and Ureaplasmas

Twelve macrolide-resistant clinical isolates of *M. genitalium* were described in Australian and Scandinavian patients with urethritis treated with azithromycin (Bradshaw *et al.*, 2006; Jensen *et al.*, 2008). All isolates contained a $A_{2058}G$ or $A_{2059}G$ mutation. In the majority of cases these macrolide-resistant mutants were selected during azithromycin therapy. Two randomized trials have compared azithromycin to doxycycline (Mena *et al.*, 2009; Schwebke *et al.*, 2011). One included 78 men with *M. genitalium* urethritis and found that azithromycin was superior to doxycycline (microbiological cure 87% vs. 45%, $P = 0.002$) (Mena *et al.*, 2009).The other included 305 urethritis patients and found no significant differences in clinical response rates among the doxycycline (49%) and the azithromycin arm (43.6%). However, the *M. genitalium* clearance rate was 30.8% for the doxycycline arm and 66.7% for the azithromycin arm ($P = 0.002$) (Schwebke *et al.*, 2011).

Among the other two urogenital species, a very few case reports of acquired *M. hominis* resistance to macrolides have been described in the literature. Two *M. hominis* clinical isolates resistant to 16-member macrolides and lincosamides were obtained from respiratory specimens in a patient with chronic obstructive pulmonary disease with multiple antibiotic exposure (Pereyre *et al.*, 2002). These strains were also resistant to fluoroquinolones through target modification and to tetracyclines through acquisition of the *tet*(M) gene. Macrolide resistance was related to the presence of two transitions, $A_{2059}G$ and $C_{2611}U$, in domain V of one of the two ribosomal operons for one strain, while the other strain contained a single mutation at position 2059 of the same operon. These clinical descriptions have been confirmed by *in vitro* selection of resistant mutants of *M. hominis* and *Ureaplasma* spp. in which mutations were found in the 23S rRNA and ribosomal proteins L4 and L22 (Pereyre *et al.*, 2006, 2007b).

High-level macrolide resistance in human ureaplasmas has been known to occur since the late 1980s, possibly due to reduction in macrolide influx and accumulation, as well as binding affinity to the ribosomes. The genetic basis for those initial observations was not investigated given the limited technology available at that time. Recently, one isolate of *U. parvum* was found to be highly resistant to erythromycin (MIC = 64 mg/ml) and contained a two amino acid deletion (R66Q67) from the L4 ribosomal protein (Beeton *et al.*, 2009b). High-level macrolide resistance in *Ureaplasma* spp. (erythromycin MICs 64->256 µg/ml) has recently been described in 4 clinical isolates from the USA which appeared to be due to deletions or insertions in L4 ribosomal proteins or mutations in 23S rRNA at position 2058 (Xiao *et al.*, 2011). Two of these isolates also contained *tet*(M) and had tetracycline MICs ≥4 µg/ml. Several clinical strains with 23S rRNA mutations were also described in China (Dongya *et al.*, 2008). Ribosomal methylation mediated by *ermB* and drug efflux mediated by *msrA*, *msrB* or *msrD* have been reported by a single study from China (Lu *et al.*, 2010), but these mechanisms for macrolide resistance in ureaplasmas have not been confirmed in any other investigations.

The prevalence of acquired macrolide resistance in clinical isolates of *M. hominis* and *Ureaplasma* spp. is unknown but probably very low, except perhaps in China where recent reports suggest that macrolide resistance in ureaplasmas may be common in that part of the world, presumably as a result of selective pressure because of widespread macrolide use (Xie and Zhang, 2006; Dongya *et al.*, 2008). In contrast, macrolide resistance rates appear to be increasing in *M. genitalium*. This drug resistance has resulted in failure of azithromycin treatment (Bradshaw *et al.*, 2006; Jensen *et al.*, 2008) and justifies development of epidemiological surveillance studies using molecular technologies that can detect mutations in the 23S rRNA directly from the specimen.

### Animal Mycoplasmas

In contrast to human mycoplasmas, in which acquired resistance to macrolides is still a rare phenomenon occurring mainly in certain geographic regions, such resistance to macrolides in food animals has been reported for several decades (Levisohn, 1981; Cooper *et al.*, 1993; Levisohn *et al.*, 1993; Bradbury *et al.*, 1994; Jordan and Horrocks, 1996; Kobayashi *et al.*, 1996) and still remains a problem. Indeed, comparison of *in vitro* activity of different antimicrobials against 62 British field strains of *M. bovis* revealed that tilmicosin was ineffective, with 92% of the isolates having MIC values of 128 mg/ml (Ayling *et al.*, 2000). In another study, the $MIC_{50}$ and $MIC_{90}$ of tilmicosin for 223 *M. bovis* strains, isolated from different clinical conditions in the USA, were defined as 64 and 128 mg/ml, respectively (Rosenbusch *et al.*, 2005). Moreover, marked difference in susceptibility profiles to tylosin and tilmicosin of *M. bovis* strains isolated from different geographical regions was recently reported (Gerchman *et al.*, 2009). For example, Israeli strains were significantly more resistant to macrolides than most isolates from animals imported from Hungary, with $MIC_{50}$ of 128 mg/ml vs. 2 mg/ml for tilmicosin and 8 mg/ml vs. 1 mg/ml for tylosin, respectively (Gerchman *et al.*, 2009). Such data emphasize the necessity for antimicrobial susceptibility testing periodically and on a regional basis.

Macrolide-resistant *M. bovis* PG45 mutants selected *in vitro* harboured $A_{2058}G$ in the 23S rRNA gene *rP_4–3* of a tilmicosin-resistant mutant and G748A in domain II of *rrl_4* gene as well as $A_{2058}G$ in both *rrl_3* and *rrl_4* genes of a tylosin-resistant mutant (Ayling *et al.*, 2008). Characterisation of 54 *M. bovis* field isolates showing different susceptibility of tylosin and tilmicosin revealed clear correlation between point mutations G748A, c752T, A2058G, A2058C and A2059C in one or both alleles and MICs of 8–1024 µg/ml to tylosin in 27/27 strains. Although a single mutation in domain II or V could confer resistance, a combination is necessary to achieve high level resistance (> 128 µg/ml) (Lysnyansky *et al.*, unpublished). Multiple amino acid substitutions have been found in the L4 protein of *M. bovis* macrolide-resistant strains as well as at position L22 Glu 90 His. More precise research is needed to clarify the mechanisms responsible for acquired macrolide resistance in *M. bovis*.

Acquired resistance in a field isolate of *M. hyopneumoniae* with MICs of 8–16 mg/ml for tylosin and >64 mg/ml for erythromycin, azithromycin, clindamycin, clarithromycin and lincomycin revealed a point mutation $A_{2058}G$ in domain V of the 23S rRNA (Stakenborg *et al.*, 2005). In another swine pathogen, *M. hyorhinis*, the acquired resistance to macrolides was attributed to the mutation $A_{2059}G$ (Kobayashi *et al.*, 2005). This mutation was found in tylosin-resistant *M. hyorhinis* strains obtained *in vitro* as well as *in vivo*. However, an *in vitro*-selected lincomycin-resistant mutant of the type strain BTS7 (after 15 passages, MIC 50 mg/ml) showed two point mutations at positions $G_{2597}U$ and $C_{2611}U$ in domain V of the 23S rRNA gene and the insertion of an adenine at the pentameric adenine sequence in domain II. Interestingly, when a lincomycin-resistant mutant with MIC 6.25 mg/ml to tylosin was transferred for 11 passages in sub-inhibitory concentrations of tylosin (final MIC of > 100 mg/ml for tylosin), an $A_{2062}G$ mutation in addition to $G_{2597}U$ and $C_{2611}U$ was identified (Kobayashi *et al.*, 2005). The authors suggested that mutations at positions $A_{2059}G$ and $A_{2062}G$ in 23S rRNA may play important role in conferring resistance to 16- member macrolides and lincomycin.

In poultry, *M. gallisepticum* strains isolated during 2004–2005 in Thailand, were reported to be susceptible to tylosin (Pakpinyo and Sasipreeyajan, 2007). However, some of those tylosin-susceptible isolates possessed MICs in the resistant range for tilmicosin and erythromycin. In Israel, in contrast, *in vitro* susceptibility testing of 50 strains of *M. gallisepticum* isolated during the period 1997–2010 revealed that 50% had elevated MICs to tylosin (Gerchman *et al.*, 2011).

Macrolide-resistant (erythromycin, tylosin, tilmicosin) *M. gallisepticum* mutants have been selected *in vitro* (Gautier-Bouchardon *et al.*, 2002; Wu *et al.*, 2005). Molecular characterization identified an $A_{2058}G$ mutation in one of the two 23S rRNA genes in all tylosin-resistant mutants selected *in vitro* by erythromycin (Wu *et al.*, 2005). However, since authors did not use the designation *rrnA* and *rrnB*, it was not possible to clarify in which of two 23S rRNA genes the $A_{2058}G$ mutation was identified (Wu *et al.*, 2005). In addition, among the tylosin-resistant mutants selected by erythromycin, a $G_{2057}A$ mutation and an $A_{2059}G$ mutation were found in one of 23S rRNA genes. In both tilmicosin-selected tylosin-resistant mutants, two mutations, $A_{2058}G$ and $A_{2503}U$ occurred in one of the two 23S rRNA genes and those mutants were characterized by markedly high resistance.

In another study, *in vitro* selection of *M. gallisepticum* mutants resistant to tiamulin, a member of the pleuromutilin family of antibiotics that also bind at the peptidyl transfer site in the 23S rRNA, resulted in nucleotide substitutions within domain V of the 23S rRNA genes (Li *et al.*, 2010). Mutants with $A_{2058}G$ or the $A_{2059}G$ showed cross-resistance to erythromycin, tilmicosin and tylosin. Interestingly, 1/3 of these mutants harboured $A_{2058}G$ substitution in the rrnB gene, while the other two have $A_{2058}G$ or $A_{2059}G$ in the rrnA gene (Li *et al.*, 2010).

The mechanism of acquired resistance to tylosin was recently determined for field strains of *M. gallisepticum* (Gerchman *et al.*, 2011). (The data revealed a clear-cut correlation between single point mutations $A_{2058}G$ or $A_{2059}G$ in domain V of the *rrnA* gene (MGA_r01) and macrolide resistance in this species. Indeed, all isolates with MIC ≥0.63 μg/ml to tylosin and with MIC > 1.25 μg/ml to tilmicosin possess this mutation, suggesting an essential role in decreased susceptibility of *M. gallisepticum* to 16-membered macrolides.

Several studies reported that *M. synoviae* field strains are susceptible to tylosin (Wang *et al.*, 2001; Cerda *et al.*, 2002; Landman *et al.*, 2008). However, the number of strains checked for susceptibility profile is often very small which may impact the final conclusions. *M. synoviae* as well as *M. iowae* mutants have been selected *in vitro* with tylosin and erythromycin (Gautier-Bouchardon *et al.*, 2002). However, the molecular mechanism of macrolide resistance in these species was not elucidated.

## Fluoroquinolones

### Human urogenital Mycoplasmas and Ureaplasmas

Mutations in the target genes *gyrA* and *gyrB* of DNA gyrase and *parC* and *parE* of topoisomerase IV are the main mechanisms conferring fluoroquinolone resistance in mycoplasmas (Bébéar and Kempf, 2005). Naturally occurring resistance has only been described in genital mycoplasmas in humans (Table 15.4). Mutations in the QRDRs of the two targets (*gyrA*, *gyrB*, *parC* and *parE* genes) have been detected in French clinical isolates of *M. hominis* (Bébéar *et al.*, 1999, 2003) and French, Chinese, North American and English clinical isolates of *Ureaplasma* spp. (Bébéar *et al.*, 2000, 2003; Zhang *et al.*, 2002; Duffy *et al.*, 2006; Xie and Zhang, 2006; Beeton *et al.*, 2009a; Shimada *et al.*, 2010a,b; Xiao *et al.*, 2012).

Treatment failure has been described in patients with non-gonococcal urethritis in whom *M. genitalium* was isolated following treatment with levofloxacin. Mutations in the *gyrA* and *parC* genes of *M. genitalium*, and more recently in *gyrB* and *parE*, have been detected by molecular methods (Deguchi *et al.*, 2001; Shimada *et al.*, 2010b).

Resistant clinical isolates of *M. hominis* and *Ureaplasma* spp. show cross-resistance to all fluoroquinolones. The level of resistance depends on the number and positions of the mutations. High-level resistance is associated with target modification on DNA gyrase and topoisomerase IV. It is the newest molecules, like moxifloxacin, which remain most effective against these mutants, although they lose their bactericidal activity *in vitro* (Bébéar *et al.*, 2008). A recent comparison of full *gyrA*, *gyrB*, *parC* and *parE* sequences between all *U. parvum* and *U. urealyticum* serovars enabled separation of true fluoroquinolone resistance mutations from non-resistant polymorphisms (Beeton *et al.*, 2009a). Thus, among all previously reported fluoroquinolone-resistant strains of *Ureaplasma* spp., one third of identified residue substitutions could be attributed to normal species polymorphism.

Similar results have been obtained by *in vitro* selection of resistant mutants of *M. hominis* and *M. pneumoniae* using different fluoroquinolones, including the most recently developed agents (levofloxacin, trovafloxacin, gatifloxacin, moxifloxacin, and gemifloxacin), (Bébéar *et al.*, 1998; Kenny *et al.*, 1999; Gruson *et al.*, 2005). Regardless of which drug was used for selection, fluoroquinolone resistance hot spots in the QRDRs of other bacteria were found to be mutated in the resistant strains. Mutation rates were lower for the newer fluoroquinolones than for the older ones like ofloxacin or ciprofloxacin. An active efflux system of the ATP binding cassette (ABC) transporter type has been described in *M. hominis* strains selected on ethidium bromide displaying a multiresistant phenotype with increased MICs of ciprofloxacin and ethidium bromide (Raherison *et al.*, 2002). Two genes, *md1* and *md2*, encoding putative ABC transporters are constitutively expressed in the reference strain *M. hominis* PG21 and over expressed in the resistant strains (Raherison *et al.*, 2005).

Here again, the prevalence of fluoroquinolone resistance in genital mycoplasmas is unknown but still believed to be relatively low overall, estimated at less than 1% for *Ureaplasma* spp. in France (Bébéar *et al.*, 2003). However, as these drugs have been used much more extensively over the past several years, it is not surprising that cases of clinically significant infections caused by fluoroquinolone-resistant *M. hominis* or *Ureaplasma* spp. are being reported more often, especially in persons who have received the drugs and those who are immunosuppressed (Bébéar, 2010) and in countries such as China where unregulated antimicrobial use and drug resistance is particularly widespread (Xie and Zhang, 2006). Recently, an emergence of clinical strains of *M. genitalium* containing alterations associated with fluoroquinolone resistance in the *parC* gene has been reported and the prevalence may be about 10% in Japan (Shimada *et al.*, 2010a) It should also be noted that fluoroquinolone-resistant strains of *M. hominis* and *M. fermentans* have been isolated from cell cultures infected with these mycoplasmas after treatment with these antimicrobials (Bébéar and Kempf, 2005).

### Animal Mycoplasmas

In contrast to acquired resistance to macrolides in food animals, which has been reported for many decades, the resistance to veterinary fluoroquinolones is a relatively new phenomenon. The emergence of fluoroquinolone-resistance mutants in mycoplasmal species causing disease in food animals has been relatively rapid, especially in light of the fact that fluoroquinolones were introduced into veterinary practice only about two decades ago (Wu *et al.*, 2000; Cerda *et al.*, 2002; Thomas *et al.*, 2003; Vicca *et al.*, 2004; Pakpinyo and Sasipreeyajan, 2007; Gerchman *et al.*, 2008, 2009; Landman *et al.*, 2008; Uemura *et al.*, 2010). In part, relatively rapid development of fluoroquinolone-resistance may be explained by the fact that this group of antibacterial has been used for treatment of a variety of animal as well as poultry diseases in addition to mycoplasmoses due to the fairly wide spectrum of efficacy. For example, in Israel, *M. gallisepticum* enrofloxacin-resistant strains were not detected prior to 2005. However, since then, 23/29 (79%) of the *M. gallisepticum* field strains tested were found to be resistant to enrofloxacin. Moreover, 61% (11/18) of the strains isolated from clinical samples since 2006 are resistant to both enrofloxacin and tylosin. Molecular typing of 50 *M. gallisepticum* field isolates performed by gene-target sequencing (GTS), detected 13 molecular types, designated I-XIII. The predominant type prior to 2006 was type II whereas type IX, first detected in 2008, is currently prevalent. All ten type X strains were resistant to both fluoroquinolones and macrolides, suggesting selective pressure leading to clonal dissemination of resistance. However, this was not a unique event since resistant strains with other GTS molecular were also found (Gerchman *et al.*, 2011). In Thailand, *in vitro* antimicrobial susceptibility testing of 20 *M. gallisepticum* strains isolated from 20 different farms and typed by random amplification of polymorphic DNA analysis (RAPD) into 5 groups, revealed that more than 50% of the strains with different genotypes were resistant to enrofloxacin (Pakpinyo and Sasipreeyajan, 2007). In another study, the $MIC_{50}$ and $MIC_{90}$ values of nine Argentinean *M. synoviae* isolates for enrofloxacin were 1.56 in both cases (Cerda *et al.*, 2002). A small but significant increase in MIC values for enrofloxacin was detected in *M. synoviae* isolated after repeated treatment of experimentally infected chickens (Le Carrou *et al.*, 2006). Moreover, resistant mutants of *M. gallisepticum* were readily selected by *in vitro* passage with increasing concentrations of enrofloxacin (Gautier-Bouchardon *et al.*, 2002; Reinhardt *et al.*, 2002a). Characterization of *in vitro* -selected *M. gallisepticum* mutants revealed the presence of different amino acid substitutions in the QRDRs of GyrA, GyrB, ParC and ParE (Reinhardt *et al.*, 2002a,b). However, Reinhardt suggested that the substitutions of 83- Ser/Argo in GyrA and 80-Ser/Leu or Ser/Trp in ParC QRDRs had the greatest impact on resistance to fluoroquinolones (Reinhardt *et al.*, 2002b). Characterization of the QRDRs of GyrA, GyrB, ParC and ParE in 25 *M. gallisepticum* strains isolated from commercial poultry flocks during 1997–2007 and exhibiting different levels of susceptibility to fluoroquinolones supported the findings with *in vitro*-selected mutants. Clear correlation was found between the presence of mutations in GyrA (83-Ser/Ile or 87-Glu/Lys) and in ParC (80-Ser/Leu) and decreased susceptibility to enrofloxacin (Lysnyansky *et al.*, 2008).

In animal mycoplasmas, the naturally acquired resistance to fluoroquinolones has been characterized in clinical isolates of *M. hyopneumoniae* (Le Carrou *et al.*, 2006; Vicca *et al.*, 2007), *M. bovirhinis* (Hirose *et al.*, 2004) and in *M. bovis* (Lysnyansky *et al.*, 2009). It seems that the most common hot spots for fluoroquinolone resistance in those mycoplasmas are positions 80 [Ser/Tyr (Vicca *et al.*, 2007), Ser/Phe (Le Carrou *et al.*, 2006), Ser/Leu (Hirose *et al.*, 2004)] or 84 [Asn/Asp (Lysnyansky *et al.*, 2009) or Asp/Asn (Le Carrou *et al.*, 2006)] in the ParC as well as positions 83 [Ser/Phe (Lysnyansky *et al.*, 2009)] in the GyrA. Indeed, two fluoroquinolone-resistant *M. bovirhinis* field isolates (as well as four *in vitro* selected mutants) and five enrofloxacin-resistant *M. hyopneumoniae* isolates have 80-Ser/Leu and 80-Ser/Tyr substitutions in ParC, respectively (Hirose *et al.*, 2004; Vicca *et al.*, 2007). One of five *M. hyopneumoniae* field isolates also harboured 83-Ala/Val substitution in GyrA. In addition, mutation at this position (83-Ser/Phe) together with mutation at position 84-Asn/Asp of ParC, were also found in 10/11 enrofloxacin-resistant *M. bovis* clinical isolates. It was suggested that a change in GyrA is sufficient to achieve an intermediate level of susceptibility of *M. bovis* to fluoroquinolone, but a concurrent modification in the ParC protein is required for resistance (Lysnyansky *et al.*, 2009).

## Summary

The information provided in this chapter provides a comprehensive and up to date summary of *in vitro* susceptibility testing, mechanisms of antimicrobial resistance mediated by *in vitro* selection as well as natural occurrence, epidemiology and clinical significance for mycoplasmas and ureaplasmas of humans, as well as many animal species. A least three significant changes have developed in this field of study since the publication of earlier chapters and reviews. One of these changes has been the development and publication of standardized methods and quality control guidelines for *in vitro* susceptibility testing of mycoplasmas and ureaplasmas of humans which may help reduce confusion and reduce potentially misleading information on the actions of new antimicrobials and development of drug resistance, providing the recommended methods are adopted on a wide scale. Indeed, now that such guidelines for *in vitro* testing and tentative interpretive breakpoints have been endorsed by the CLSI, it is hoped that future publications on this topic will adhere to the recommended methods in order to provide the most accurate results which can be compared legitimately to data obtained in various laboratories around the world. A second change that has been observed over the past few years is the emergence and spread of clinically significant antimicrobial resistance in multiple drug classes in Mollicute species of humans as well as many animals. Endeavours to identify the molecular mechanisms of antimicrobial resistance, either through examination of clinical isolates themselves or by performing *in vitro* selection studies have greatly improved understanding of the mechanisms of resistance. The final major change that has occurred in recent years is the development of molecular-based methods for detection of resistance genes mediating specific types of drug resistance in various species. This development is extremely important since it enables direct examination of individual clinical specimens to guide antimicrobial treatment as well as for epidemiologic surveillance for the occurrence of antimicrobial resistance without having to depend on cultivation of fastidious, slow-growing organisms *in vitro*. Development of similar procedures tailored specifically for mycoplasmas of veterinary importance may provide similar advantages in the future. While antimicrobial susceptibilities and documented of acquired resistance has now been described in many mycoplasmas of veterinary importance, there are many more species about which very little is known and there are probably many other species that have never been identified and characterized. Thus, despite many major advances over the past few years in the field of antimicrobial chemotherapy, much is left to be done.

## References


Averbuch, D., Hidalgo-Gross, C., Moses, A.E., Englehard, D., and Nir-Paz, R. (2011). Macrolide resistance in *Mycoplasma pneumoniae* in Israel, 2010. Emerg. Infect. Dis. *17*, 1079–1082.

Ayling, R.D., Baker, S.E., Peek, M.L., Simon, A.J., and Nicholas, R.A. (2000). Comparison of in vitro activity of danofloxacin, florfenicol, oxytetracycline, spectinomycin and tilmicosin against recent field isolates of *Mycoplasma bovis*. Vet. Rec. *146*, 745–747.

Ayling, R.D., Patel, M., Nicholas, R.A.J., and McAuliffe, L. (2008). Induced antimicrobial resistance in *Mycoplasma mycoides* supsp. *mycoides* small colony type. Abstract: 17th Congress of the International Organization for Mycoplasmology (Tianjin China).

Ball, H.J., and Campbell, J.N. (1989). Antibiotic treatment of experimental *Mycoplasma californicum* mastitis. Vet. Rec. *125*, 377–378.

Ball, H.J., and McCaughey, W.J. (1987). Experimental intramuscular inoculation of tylosin in the elimination of ureaplasmas from ewes. Vet. Rec. *120*, 557–558.

Bébéar, C. (2010). *Mycoplasma, Ureaplasma*. In Antibiogram, Courvalin, P., Leclercq, R., and Rice, L.B., eds. (Portland, ESKA publishing, ASM Press), pp. 519–529.

Bébéar, C.M., and Bébéar, C. (2002). Antimycoplasmal agents. In Molecular Biology and Pathogenicity of Mycoplasmas, Razin, S., and Herrmann, R., eds. (London, Kluwer Academic/Plenum Publishers), pp. 545–566.

Bébéar, C.M., and Kempf, I. (2005). Antimicrobial therapy and antimicrobial resistance. In Mycoplasmas: Pathogenesis, Molecular Biology, and Emerging Strategies for Control, Blanchard, A., and Browning, G.F., eds. (Horizon Bioscience, Wymondham), pp. 535–568.

Bébéar, C.M., and Pereyre, S. (2005). Mechanisms of drug resistance in *Mycoplasma pneumoniae*. Curr. Drug Targets Infect. Disord. *5*, 263–271.

Bébéar, C.M., Renaudin, H., Charron, A., Bové, J.M., Bébéar, C., and Renaudin, J. (1998). Alterations in topoisomerase IV and DNA gyrase in quinolone-resistant mutants of *Mycoplasma hominis* obtained in vitro. Antimicrob. Agents Chemother. *42*, 2304–2311.

Bébéar, C.M., Renaudin, J., Charron, A., Renaudin, H., de Barbeyrac, B., Schaeverbeke, T., and Bébéar, C. (1999). Mutations in the *gyrA*, *parC*, and *parE* genes associated with fluoroquinolone resistance in clinical isolates of *Mycoplasma hominis*. Antimicrob. Agents Chemother. *43*, 954–956.

Bébéar, C.M., Renaudin, H., Charron, A., Gruson, D., Lefrancois, M., and Bébéar, C. (2000). In vitro activity of trovafloxacin compared to those of five antimicrobials against mycoplasmas including *Mycoplasma hominis* and *Ureaplasma urealyticum* fluoroquinolone-resistant isolates that have been genetically characterized. Antimicrob. Agents Chemother. *44*, 2557–2560.

Bébéar, C.M., Renaudin, H., Charron, A., Clerc, M., Pereyre, S., and Bébéar, C. (2003). DNA gyrase and topoisomerase IV mutations in clinical isolates of *Ureaplasma* spp. and *Mycoplasma hominis* resistant to fluoroquinolones. Antimicrob. Agents Chemother. *47*, 3323–3325.

Bébéar, C.M., de Barbeyrac, B., Pereyre, S., Renaudin, H., Clerc, M., and Bébéar, C. (2008). Activity of moxifloxacin against the urogenital mycoplasmas *Ureaplasma* spp., *Mycoplasma hominis* and *Mycoplasma genitalium* and *Chlamydia trachomatis*. Clin Microbiol Infect *14*, 801–805.

Bébéar, C.M., Pereyre, S., and Peuchant, O. (2011). *Mycoplasma pneumoniae*: susceptibility and resistance to antibiotics. Future Microbiol. *6*, 423–431.

Beeton, M.L., Chalker, V.J., Kotecha, S., and Spiller, O.B. (2009a). Comparison of full *gyrA*, *gyrB*, *parC* and *parE* gene sequences between all *Ureaplasma parvum* and *Ureaplasma urealyticum* serovars to separate true fluoroquinolone antibiotic resistance mutations from non-resistance polymorphism. J. Antimicrob. Chemother. *64*, 529–538.

Beeton, M.L., Chalker, V.J., Maxwell, N.C., Kotecha, S., and Spiller, O.B. (2009b). Concurrent titration and determination of antibiotic resistance in *Ureaplasma* species with identification of novel point mutations in genes associated with resistance. Antimicrob. Agents Chemother. *53*, 2020–2027.

Ben Shabat, M., Mikula, I., Gerchman, I., and Lysnyansky, I. (2010). Development and evaluation of a novel single-nucleotide polymorphism real-time PCR assay for rapid detection of fluoroquinolone-resistant *Mycoplasma bovis*. J. Clin. Microbiol. *48*, 2909–2915.

Blanchard, A., Crabb, D.M., Dybvig, K., Duffy, L.B., and Cassell, G.H. (1992). Rapid detection of *tet*(M) in *Mycoplasma hominis* and *Ureaplasma urealyticum* by PCR: *tet*(M) confers resistance to tetracycline but not necessarily to doxycycline. FEMS Microbiol. Lett. *71*, 277–281.

Bradbury, J.M., Yavari, C.A., and Giles, C.J. (1994). In vitro evaluation of various antimicrobials against *Mycoplasma gallisepticum* and *Mycoplasma synoviae* by the micro-broth method, and comparison with a commercially-prepared test system. Avian Pathol. *23*, 105–115.

Bradshaw, C.S., Jensen, J.S., Tabrizi, S.N., Read, T.R., Garland, S.M., Hopkins, C.A., Moss, L.M., and Fairley, C.K. (2006). Azithromycin failure in *Mycoplasma genitalium* urethritis. Emerg. Infect. Dis. *12*, 1149–1152.

Brown, D.R., Farmerie, W.G., May, M., Benders, G.A., Durkin, A.S., Hlavinka, K., Hostetler, J., Johnson, J., Miller, R.H., *et al.* (2011). Genome sequences of *Mycoplasma alligatoris* A21JP2T and *Mycoplasma crocodyli* MP145T. J. Bacteriol. *193*, 2892–2893.

Calderon-Copete, S., Wigger, G., Wunderlin, C., Schmidheini, T., Frey, J., Quail, M., and Falquet, L. (2009). The *Mycoplasma conjunctivae* genome sequencing, annotation and analysis. BMC Bioinformatics *10*, S7.

Cao, B., Zhao, C.J., Yin, Y.D., Zhao, F., Song, S.F., Bai, L., Zhang, J.Z., Liu, Y.M., Zhang, Y.Y., Wang, H., *et al.* (2010b). High prevalence of macrolide resistance in *Mycoplasma pneumoniae* isolates from adult and adolescent patients with respiratory tract infection in China. Clin. Infect. Dis. *51*, 189–194.

Cerda, R.O., Giacoboni, G.I., Xavier, J.A., Sansalone, P.L., and Landoni, M.F. (2002). In vitro antibiotic susceptibility of field isolates of *Mycoplasma synoviae* in Argentina. Avian Dis. *46*, 215–218.

Chambaud, I., Heilig, R., Ferris, S., Barbe, V., Samson, D., Galisson, F., Moszer, I., Dybvig, K., Wróblewski, H., Viari, A., *et al.* (2001). The complete genome sequence of the murine respiratory pathogen *Mycoplasma pulmonis*. Nucleic Acids Res. *29*, 2145–2153.

Chironna, M., Sallustio, A., Esposito, S., Perulli, M., Chinellato, I., Di Bari, C., Quarto, M., and Cardinale, F. (2011). Emergence of macrolide-resistant strains during an outbreak of *Mycoplasma pneumoniae* infections in children. J. Antimicrob. Chemother. *66*, 734–737.

CLSI (2011). Antimicrobial susceptibility testing of human mycoplasmas: approved guideline (Clinical and Laboratory Standards Institute, Wayne, PA).

Cooper, A.C., Fuller, J.R., Fuller, M.K., Whittlestone, P., and Wise, D.R. (1993). In vitro activity of danofloxacin, tylosin, and oxytetracycline against mycoplasmas of veterinary importance. Res. Vet. Sci. *54*, 329–334.

Critchley, I.A., Jones, M.E., Heinze, P.D., Hubbard, D., Engler, H.D., Evangelista, A.T., Thornsberry, C., Karlowsky, J.A., and Sahm, D.F. (2002). In vitro activity of levofloxacin against contemporary clinical isolates of *Legionella pneumophila*, *Mycoplasma pneumoniae* and *Chlamydia pneumoniae* from North America and Europe. Clin Microbiol Infect *8*, 214–221.

de Barbeyrac, B., Dupon, M., Rodriguez, P., Renaudin, H., and Bébéar, C. (1996). A Tn1545-like transposon carries the tet(M) gene in tetracycline resistant strains of *Bacteroides ureolyticus* as well as *Ureaplasma urealyticum* but not *Neisseria gonorrhoeae*. J. Antimicrob. Chemother. *37*, 223–232.

Dégrange, S., Renaudin, H., Charron, A., Pereyre, S., Bébéar, C., and Bébéar, C.M. (2008). Reduced susceptibility to tetracyclines is associated in vitro with the presence of 16S rRNA mutations in *Mycoplasma hominis* and *Mycoplasma pneumoniae*. J. Antimicrob. Chemother. *61*, 1390–1392.

Dégrange, S., Cazanave, C., Charron, A., Renaudin, H., Bébéar, C., and Bébéar, C.M. (2009). Development of Multiple-Locus Variable-Number Tandem-Repeat Analysis for the Molecular Typing of *Mycoplasma pneumoniae*. J. Clin. Microbiol. *47*, 914–923.

Deguchi, T., Maeda, S., Tamaki, M., Yoshida, T., Ishiko, H., Ito, M., Yokoi, S., Takahashi, Y., and Ishihara, S. (2001). Analysis of the *gyrA* and *parC* genes of *Mycoplasma genitalium* detected in first-pass urine of men with non-gonococcal urethritis before and after fluoroquinolone treatment. J. Antimicrob. Chemother. *48*, 742–744.

Dongya, M., Wencheng, X., Xiaobo, M., and Lu, W. (2008). Transition mutations in 23S rRNA account for acquired resistance to macrolides in *Ureaplasma urealyticum*. Microb. Drug Resist. *14*, 183–186.

Duffy, L., Glass, J., Hall, G., Avery, R., Rackley, R., Peterson, S., and Waites, K. (2006). Fluoroquinolone resistance in *Ureaplasma parvum* in the United States. J. Clin. Microbiol. *44*, 1590–1591.

Dumke, R., von Baum, H., Luck, P.C., and Jacobs, E. (2010). Occurrence of macrolide-resistant *Mycoplasma pneumoniae* strains in Germany. Clin Microbiol Infect *16*, 613–616.

Dybvig, K., Zuhua, C., Lao, P., Jordan, D.S., French, C.T., Tu, A.-H.T., and Loraine, A.E. (2008). The genome of *Mycoplasma arthritidis*. Infect. Immun. *76*, 4000–4008.

Etheridge, J.R., Lloyd, L.C., and Cottew, G.S. (1979). Resistance of *Mycoplasma hyopneumoniae* to chlortetracycline. Aust. Vet. J. *55*, 40.

Francoz, D., Fortin, M., Fecteau, G., and Messier, S. (2005). Determination of *Mycoplasma bovis* susceptibilities against six antimicrobial agents using the E test method. Vet. Microbiol. *105*, 57–64.

Furneri, P.M., Rappazzo, G., Musumarra, M.P., Tempera, G., and Roccasalva, L.S. (2000). Genetic basis of natural resistance to erythromycin in *Mycoplasma hominis*. J. Antimicrob. Chemother. *45*, 547–548.

Gaurivaud, P., Laigret, F., and Bove, J.M. (1996). Insusceptibility of members of the class Mollicutes to rifampin: studies of the *Spiroplasma citri* RNA polymerase beta-subunit gene. Antimicrob. Agents Chemother. *40*, 858–862.

Gautier-Bouchardon, A.V., Reinhardt, A.K., Kobisch, M., and Kempf, I. (2002). In vitro development of resistance to enrofloxacin, erythromycin, tylosin, tiamulin and oxytetracycline in *Mycoplasma gallisepticum*, *Mycoplasma iowae* and *Mycoplasma synoviae*. Vet. Microbiol. *88*, 47–58.

Gerchman, I., Lysnyansky, I., Perk, S., and Levisohn, S. (2008). In vitro susceptibilities to fluoroquinolones in current and archived *Mycoplasma gallisepticum* and *Mycoplasma synoviae* isolates from meat-type turkeys. Vet. Microbiol. *131*, 266–276.

Gerchman, I., Levisohn, S., Mikula, I., and Lysnyansky, I. (2009). In vitro antimicrobial susceptibility *Mycoplasma bovis* isolated in Israel from local and imported cattle. Vet. Microbiol. *137*, 268–275.

Gerchman, I., Levisohn, S., Mikula, I., Manso-Silván, L., and Lysnyansky, I. (2011). Characterization of in vitro-acquired resistance to macrolides of *Mycoplasma gallisepticum* strains isolated from poultry. Vet. Res. *42*, 90.

Giguere, S. (2007a). Lincosamides, pleuromutilins, and streptogramins. In Antimicrobial Therapy in Veterinary Medicine, Giguere, S., Prescott, J.F., Baggot, J.D., Walker, R.D., and Dowling, P.M., eds. (Wiley Blackwell, Oxford, UK), pp. 179–190.

Giguere, S. (2007b). Macrolides, azalides, and ketolides. In Antimicrobial Therapy in Veterinary Medicine, Giguere, S., Prescott, J.F., Baggot, J.D., Walker, R.D., and Dowling, P.M., eds. (Wiley Blackwell, Oxford, UK), pp. 191–206.

Giguere, S. (2007c). Tetracyclines and glycycyclines. In Antimicrobial Therapy in Veterinary Medicine, Giguere, S., Prescott, J.F., Baggot, J.D., Walker, R.D., and Dowling, P.M., eds. (Wiley Blackwell, Oxford, UK), pp. 231–240.

Godinho, K.S., Rae, A., Windsor, G.D., Tilt, N., Rowan, T.G., and Sunderland, S.J. (2005). Efficacy of tulathromycin in the treatment of bovine respiratory disease associated with induced *Mycoplasma bovis* infections in young dairy calves. Vet. Ther. *6*, 96–112.

Gruson, D., Pereyre, S., Charron, A., Renaudin, H., Bébéar, C., and Bébéar, C.M. (2005). In vitro development of resistance to six and four fluoroquinolones in *Mycoplasma pneumoniae* and in *Mycoplasma hominis*. Antimicrob. Agents Chemother. *49*, 1190–1193.

Gunn, A. (2002). Rational use of antimicrobial therapy in cattle. Abstract: World Small Animal Veterinary Association (Grenada, Spain).

Hamasuna, R., Osada, Y., and Jensen, J.S. (2005). Antibiotic susceptibility testing of *Mycoplasma genitalium* by TaqMan 5′ nuclease real-time PCR. Antimicrob. Agents Chemother. *49*, 4993–4998.

Hannan, P.C. (2000). Guidelines and recommendations for antimicrobial minimum inhibitory concentration (MIC) testing against veterinary mycoplasma species. International Research Programme on Comparative Mycoplasmology. Vet. Res. *31*, 373–395.

Hannan, P.C., Windsor, H.M., and Ripley, P.H. (1997a). In vitro susceptibilities of recent field isolates of *Mycoplasma hyopneumoniae* and *Mycoplasma hyosynoviae* to valnemulin (Econor), tiamulin and enrofloxacin and the in vitro development of resistance to certain antimicrobial agents in *Mycoplasma hyopneumoniae*. Res. Vet. Sci. *63*, 157–160.

Hannan, P.C.T., O'Hanolon, P.J., and Rogers, N.H. (1989). In vitro evaluation of various quinolone antibacterial agents against veterinary mycoplasmas and porcine respiratory bacterial pathogens. Res. Vet. Sci. *46*, 202–211.

Hannan, P.C.T., Windsor, G.D., de Jong, A., Schmeer, N., and Stegemann, H. (1997b). Comparative susceptibilities of various animal-pathogenic mycoplasmas to fluoroquinolones. Antimicrob. Agents Chemother. *41*, 2037–2040.

Hirose, K., Kobayashi, H., Ito, N., Kawasaki, Y., Zako, M., Kotani, K., Ogawa, H., and Sato, H. (2003). Isolation of mycoplasmas from nasal swabs of calves affected with respiratory diseases and antimicrobial susceptibility of their isolates. J. Vet. Med. B *50*, 347–351.

Hirose, K., Kawasaki, Y., Kotani, K., Abiko, K., and Sato, H. (2004). Characterization of a point mutation in the *par*C gene of *Mycoplasma bovirhinis* associated with fluoroquinolone resistance. J. Vet. Med. B *51*, 169–175.

Inamoto, T., Takahashi, H., Yamamoto, K., Nakai, Y., and Ogimoto, K. (1994). Antibiotic susceptibility of *Mycoplasma hyopneumoniae* isolated from swine. J. Vet. Med. Sci. *56*, 393–394.

Jensen, J.S. (2006). *Mycoplasma genitalium* infections. Diagnosis, clinical aspects, and pathogenesis. Dan. Med. Bull. *53*, 1–27.

Jensen, J.S., Bradshaw, C.S., Tabrizi, S.N., Fairley, C.K., and Hamasuna, R. (2008). Azithromycin treatment failure in *Mycoplasma genitalium*-positive patients with nongonococcal urethritis is associated with induced macrolide resistance. Clin. Infect. Dis. *47*, 1546–1553.

Jordan, F.T.W., and Horrocks, B.K. (1996). The minimum inhibitory concentration of Tilmicosin and Tylosin for *Mycoplasma gallisepticum* and *Mycoplasma synoviae* and a comparison of their efficacy in the control of *Mycoplasma gallisepticum* infection in broiler chicks. Avian Dis. *40*, 326–334.

Jordan, F.T.W. (1996). Avian mycoplasmosis. In Poultry Diseases, F.T.W. Jordan and M. Pattison, eds. (W.B. Saunders, London, UK), pp. 81–93.

Kays, M.B., and Graff, M.A. (2002). Broth microdilution and E-test for determining fluoroquinolone activity against *Streptococcus pneumoniae*. Ann. Pharmacother. *36*, 416–422.

Kenny, G.E., Young, P.A., Cartwright, F.D., Sjostrom, K.E., and Huang, W.M. (1999). Sparfloxacin selects gyrase mutations in first-step *Mycoplasma hominis* mutants, whereas ofloxacin selects topoisomerase IV mutations. Antimicrob. Agents Chemother. *43*, 2493–2496.

Kobayashi, H., Morozumi, T., Munthall, G., Mitani, K., Ito, N., and Yamamoto, K. (1996). Macrolide susceptibility of *Mycoplasma hyorhinis* isolated from piglets. Antimicrob. Agents Chemother. *40*, 1030–1032.

Kobayashi, H., Nakajima, H., Shimizu, Y., Eguchi, M., Hata, E., and Yamamoto, K. (2005). Macrolides and lincomycin susceptibility of *Mycoplasma hyorhinis* and variable mutation of domain II and V in 23S ribosomal RNA. J. Vet. Med. Sci. *67*, 795–800.

Krausse, R., and Schubert, S. (2010). In-vitro activities of tetracyclines, macrolides, fluoroquinolones and clindamycin against *Mycoplasma hominis* and *Ureaplasma* ssp. isolated in Germany over 20 years. Clin Microbiol Infect *16*, 1649–1655.

Landman, W.J.M., Mevius, D.J., Veldman, K.T., and Feberwee, A. (2008). In vitro antibiotic susceptibility of Dutch *Mycoplasma synoviae* field isolates originating from joint lesions and the respiratory tract of commercial poultry. Avian Pathol. 37, 415–420.

Le Carrou, J., Reinhardt, A.K., Kempf, I., and Gautier-Bouchardon, A.V. (2006). Persistence of *Mycoplasma synoviae* in hens after two enrofloxacin treatments and detection of mutations in the *parC* gene. Vet. Res. *37*, 145–154.

Lee, D.H., Miles, R.J., and Inal, J.R. (1987). Antibiotic sensitivity and mutation rates to antibiotic resistance in *Mycoplasma mycoides* ssp. *mycoides*. Epidemiol. Infect. *98*, 361–368.

Levisohn, S. (1981). Antibiotic sensitivity patterns in field isolates of *Mycoplasma gallisepticum* as a guide to chemotherapy. Isr. J. Med. Sci. *17*, 661–666.

Levisohn, S., Berman, E., and Weisman, Y. (1993). Effect of treatment of turkeys under field conditions with tylosin or enrofloxacin on frequency of isolation and sensitivity patterns of *Mycoplasma meleagridis*. Public Health Rev. *20*, 183.

Li, B.B., Shen, J.Z., Cao, X.Y., Wang, Y., Dai, L., Huang, S.Y., and Wu, C.M. (2010). Mutations in 23S rRNA gene associated with decreased susceptibility to tiamulin and valnemulin in *Mycoplasma gallisepticum*. FEMS Microbiol. Lett. *308*, 144–149.

Li, X., Atkinson, T.P., Hagood, J., Makris, C., Duffy, L.B., and Waites, K.B. (2009). Emerging macrolide resistance in *Mycoplasma pneumoniae* in children: detection and characterization of resistant isolates. Pediatr. Infect. Dis. J. *28*, 693–696.

Lin, C., Li, S., Sun, H., Zhao, H., Feng, Y., Cao, L., Yuan, Y., and Zhang, T. (2010). Nested PCR-linked capillary electrophoresis and single-strand conformation polymorphisms for detection of macrolide-resistant *Mycoplasma pneumoniae* in Beijing, China. J. Clin. Microbiol. *48*, 4567–4572.

Liu, W., Fang, L., Li, S., Li, Q., Zhou, Z., Feng, Z., Luo, R., Shao, G., Wang, L., Chen, H., and Xiao, S. (2010a). Complete genome sequence of *Mycoplasma hyorhinis* strain HUB-1. J. Bacteriol. *192*, 5844–5845.

Liu, W., Fang, L., Li, S., Li, Q., Zhou, Z., Feng, Z., Luo, R., Shao, G., Wang, L., Chen, H., *et al.* (2010b). Complete genome sequence of *Mycoplasma hyorhinis* strain HUB-1. J. Bacteriol. *192*, 5844–5845.

Liu, Y., Ye, X., Zhang, H., Xu, X., Li, W., Zhu, D., and Wang, M. (2009). Antimicrobial susceptibility of *Mycoplasma pneumoniae* isolates and molecular analysis of macrolide-resistant strains from Shanghai, China. Antimicrob. Agents Chemother. *53*, 2160–2162.

Liu, Y., Ye, X., Zhang, H., Xu, X., Li, W., Zhu, D., and Wang, M. (2010). Characterization of macrolide resistance in *Mycoplasma pneumoniae* isolated from children in Shanghai, China. Diagn. Microbiol. Infect. Dis. *67*, 355–358.

Lu, C., Ye, T., Zhu, G., Feng, P., Ma, H., Lu, R., and Lai, W. (2010). Phenotypic and genetic characteristics of macrolide and lincosamide resistant *Ureaplasma urealyticum* isolated in Guangzhou, China. Curr. Microbiol. *61*, 44–49.

Lysnyansky, I., Gerchman, I., Perk, S., and Levisohn, S. (2008). Molecular characterization and typing of enrofloxacin-resistant clinical isolates of *Mycoplasma gallisepticum*. Avian Dis. *52*, 685–689.

Lysnyansky, I., Mikula, I., Gerchman, I., and Levisohn, S. (2009). Rapid detection of a point mutation in the *parC* gene associated with decreased susceptibility to fluoroquinolones in *Mycoplasma bovis*. Antimicrob. Agents Chemother. *53*, 4911–4914.

Lysnyansky, I., Gerchman, I., Mikula, I., and Levisohn, S. (2010). Emergence and molecular analysis of macrolide and fluoroquinolone resistance in *Mycoplasma gallisepticum* strains isolated in Israel (1997–2009). Abstract: 18th Congress of the International Organization for Mycoplasmology (Chianciano Terme, Italy).

Matsubara, K., Morozumi, M., Okada, T., Matsushima, T., Komiyama, O., Shoji, M., Ebihara, T., Ubukata, K., Sato, Y., Akita, H., *et al.* (2009). A comparative clinical study of macrolide-sensitive and macrolide-resistant *Mycoplasma pneumoniae* infections in pediatric patients. J. Infect. Chemother. *15*, 380–383.

Matsuoka, M., Narita, M., Okazaki, N., Ohya, H., Yamazaki, T., Ouchi, K., Suzuki, I., Andoh, T., Kenri, T., Sasaki, Y., *et al.* (2004a). Characterization and molecular analysis of macrolide-resistant *Mycoplasma pneumoniae* clinical isolates obtained in Japan. Antimicrob. Agents Chemother. *48*, 4624–4630.

Mazzolini, E., Agnoletti, F., and Frisco, S. (1997). Farmaco-sensibilita dei micoplasmi respiratori del bovino. Obiettivi e Documenti Veterinari *18*, 61–66.

McKelvie, J., Morgan, J.H., Nanjiani, I.A., Sherington, J., Rowan, T.G., and Sunderland, S.J. (2005). Evaluation of tulathromycin for the treatment of pneumonia following experimental infection of swine with *Mycoplasma hyopneumoniae*. Vet. Ther. *6*, 197–202.

Mena, L.A., Mroczkowski, T.F., Nsuami, M., and Martin, D.H. (2009). A randomized comparison of azithromycin and doxycycline for the treatment of *Mycoplasma genitalium*-positive urethritis in men. Clin. Infect. Dis. *48*, 1649–1654.

Morozumi, M., Iwata, S., Hasegawa, K., Chiba, N., Takayanagi, R., Matsubara, K., Nakayama, E., Sunakawa, K., and Ubukata, K. (2008). Increased macrolide resistance of *Mycoplasma pneumoniae* in pediatric patients with community-acquired pneumonia. Antimicrob. Agents Chemother. *52*, 348–350.

Morozumi, M., Takahashi, T., and Ubukata, K. (2010). Macrolide-resistant *Mycoplasma pneumoniae*: characteristics of isolates and clinical aspects of community-acquired pneumonia. J. Infect. Chemother. *16*, 78–86.

Morozumi, T., Hasegawa, K., Kobayashi, R., Inoue, N., Iwata, S., Kuroki, H., Kawamura, N., Nakayama, E., Tajima, T., Shimizu, K., *et al.* (2005). Emergence of macrolide-resistant *Mycoplasma pneumoniae* with a 23S rRNA gene mutation. Antimicrob. Agents Chemother. *49*, 2302–2306.

Nicholas, R.A.J., Ayling, R., and McAuliffe, L. (2009). Mycoplasma Diseases of Ruminants (CABI, Wallingford, UK).

Niitu, Y., Hasegawa, S., Suetake, T., Kubota, H., Komatsu, S., and Horikawa, M. (1970). Resistance of *Mycoplasma pneumoniae* to erythromycin and other antibiotics. J. Pediatr. *76*, 438–443.

Nogueira, V.A., França, T.N., and Peixoto, P.V. (2009). Ionophore poisoning in animals. Pesquisa Veterinária Brasileira *29*, 191–197.

Okazaki, N., Narita, M., Yamada, S., Izumikawa, K., Umetsu, M., Kenri, T., Sasaki, Y., Arakawa, Y., and Sasaki, T. (2001). Characteristics of macrolide-resistant *Mycoplasma pneumoniae* strains isolated from patients and induced with erythromycin in vitro. Microbiol. Immunol. *45*, 617–620.

Okazaki, N., Ohya, H., and Sasaki, T. (2007). *Mycoplasma pneumoniae* isolated from patients with respiratory infection in Kanagawa Prefecture in 1976–2006: emergence of macrolide-resistant strains. Jpn J. Infect. Dis. *60*, 325–326.

Oliver, A., Canton, R., Campo, P., Baquero, F., and Blazquez, J. (2000). High frequency of hypermutable *Pseudomonas aeruginosa* in cystic fibrosis lung infection. Science *288*, 1251–1254.

Pakpinyo, S., and Sasipreeyajan, J. (2007). Molecular characterization and determination of antimicrobial resistance of *Mycoplasma gallisepticum* isolated from chickens. Vet. Microbiol. *125*, 59–65.

Papazisi, L., Gorton, T.S., Kutish, G., Markham, P.F., Browning, G.F., Nguyen, D.K., Swartzell, S., Madan, A., Mahairas, G., and Geary, S.J. (2003). The complete genome sequence of the avian pathogen *Mycoplasma gallisepticum* strain R(low). Microbiology *149*, 2307–2316.

Pereyre, S., Gonzalez, P., de Barbeyrac, B., Darnige, A., Renaudin, H., Charron, A., Raherison, S., Bébéar, C., and Bébéar, C.M. (2002). Mutations in 23S rRNA account for intrinsic resistance to macrolides in *Mycoplasma hominis* and *Mycoplasma fermentans* and for acquired resistance to macrolides in *M. hominis*. Antimicrob. Agents Chemother. *46*, 3142–3150.

Pereyre, S., Guyot, C., Renaudin, H., Charron, A., Bébéar, C., and Bébéar, C.M. (2004). In vitro selection and characterization of resistance to macrolides and related antibiotics in *Mycoplasma pneumoniae*. Antimicrob. Agents Chemother. *48*, 460–465.

Pereyre, S., Renaudin, H., Charron, A., Bébéar, C., and Bébéar, C.M. (2006). Emergence of a 23S rRNA mutation in *Mycoplasma hominis* associated with a loss of the intrinsic resistance to erythromycin and azithromycin. J. Antimicrob. Chemother. *57*, 753–756.

Pereyre, S., Charron, A., Renaudin, H., Bébéar, C., and Bébéar, C.M. (2007a). First report of macrolide-resistant strains and description of a novel nucleotide sequence variation in the P1 adhesin gene in *Mycoplasma pneumoniae* clinical strains isolated in France over 12 years. J. Clin. Microbiol. *45*, 3534–3539.

Pereyre, S., Métifiot, M., Cazanave, C., Renaudin, H., Charron, A., Bébéar, C., and Bébéar, C.M. (2007b). Characterisation of in vitro-selected mutants of *Ureaplasma parvum* resistant to macrolides and related antibiotics. Int. J. Antimicrob. Agents *29*, 207–211.

Peuchant, O., Menard, A., Renaudin, H., Morozumi, M., Ubukata, K., Bébéar, C.M., and Pereyre, S. (2009). Increased macrolide resistance of *Mycoplasma pneumoniae* in France directly detected in clinical specimens by real-time PCR and melting curve analysis. J. Antimicrob. Chemother. *64*, 52–58.

Pilaszek, J., and Truszczynski, M. (1980). Possibility of acquiring and loosing resistance to antibiotics in bovine *Mycoplasma* strains. II. Acquiring and loosing resistance to chlorophenicol, oxytetracycline, neomycin and streptomycin. Bull. Vet. Inst. Pulawy. *24*, 31–41.

Poulsen, S.M., Karlsson, M., Johansson, L.B., and Vester, B. (2001). The pleuromutilin drugs tiamulin and valnemulin bind to the RNA at the peptidyl transferase centre on the ribosome. Mol. Microbiol. *41*, 1091–1099.

Raherison, S., Gonzalez, P., Renaudin, H., Charron, A., Bébéar, C., and Bébéar, C.M. (2002). Evidence of an active efflux in resistance to ciprofloxacin and ethidium bromide by *Mycoplasma hominis*. Antimicrob. Agents Chemother. *46*, 672–679.

Raherison, S., Gonzalez, P., Renaudin, H., Charron, A., Bébéar, C., and Bébéar, C.M. (2005). Increased expression of two multidrug transporter-like genes is associated with ethidium bromide and ciprofloxacin resistance in *Mycoplasma hominis*. Antimicrob. Agents Chemother. *49*, 421–424.

Reinhardt, A.K., Bébéar, C.M., Kobisch, M., Kempf, I., and Gautier-Bouchardon, A.V. (2002a). Characterization of mutations in DNA gyrase and topoisomerase IV involved in quinolone resistance of *Mycoplasma gallisepticum* mutants obtained in vitro. Antimicrob. Agents Chemother. *46*, 590–593.

Reinhardt, A.K., Kempf, I., Kobisch, M., and Gautier-Bouchardon, A.V. (2002b). Fluoroquinolone resistance in *Mycoplasma gallisepticum*: DNA gyrase as primary target of enrofloxacin and impact of mutations in topoisomerases on resistance level. J. Antimicrob. Chemother. *50*, 589–592.

Rocha, E.P.C., and Blanchard, A. (2002). Genomic repeats, genome plasticity and the dynamics of *Mycoplasma* evolution. Nucleic Acids Res. *30*, 2031–2042.

Rosenbusch, R.F., Kinyon, J.M., Apley, M., Funk, N.D., Smith, S.C., and Hoffman, L.J. (2005). In vitro antimicrobial inhibition profiles of *Mycoplasma bovis* isolates recovered from various regions of the United States from 2002 to 2003. J. Vet. Diagn. Invest. *17*, 436–441.

Schwebke, J.R., Rompalo, A., Taylor, S., Sena, A.C., Martin, D.H., Lopez, L.M., Lensing, S., and Lee, J.Y. (2011). Re-evaluating the treatment of nongonococcal urethritis: emphasizing emerging pathogens--a randomized clinical trial. Clin. Infect. Dis. *52*, 163–170.

Shimada, Y., Deguchi, T., Nakane, K., Masue, T., Yasuda, M., Yokoi, S., Ito, S., Nakano, M., and Ishiko, H. (2010a). Emergence of clinical strains of *Mycoplasma genitalium* harbouring alterations in ParC associated with fluoroquinolone resistance. Int. J. Antimicrob. Agents *36*, 255–258.

Shimada, Y., Deguchi, T., Yamaguchi, Y., Yasuda, M., Nakane, K., Yokoi, S., Ito, S., Nakano, M., and Ishiko, H. (2010b). *gyrB* and *parE* mutations in urinary *Mycoplasma genitalium* DNA from men with non-gonococcal urethritis. Int. J. Antimicrob. Agents *36*, 477–478.

Sirand-Pugnet, P., Lartigue, C., Marenda, M., Jacob, D., Barre, A., Barbe, V., Schenowitz, C., Mangenot, S., Couloux, A., Segurens, B., *et al.* (2007). Being path ogenic, plastic, and sexual while living with a nearly minimal bacterial genome. PLoS Genet. 3, e75.

Spuesens, E.B., Hoogenboezem, T., Sluijter, M., Hartwig, N.G., van Rossum, A.M., and Vink, C. (2010). Macrolide resistance determination and molecular typing of *Mycoplasma pneumoniae* by pyrosequencing. J. Microbiol. Methods 82, 214–222.

Stakenborg, T., Vicca, J., Butaye, P., Maes, D., Minion, F.C., Peeters, J., De Kruif, A., and Haesebrouck, F. (2005). Characterization of in vivo acquired resistance of *Mycoplasma hyopneumoniae* to macrolides and lincosamides. Microb. Drug Resist. *11*, 290–294.

Stanley, W.A., Hofacre, C.L., Speksnijder, G., Kleven, S.H., and Aggrey, S.E. (2001). Monitoring *Mycoplasma gallisepticum* and *Mycoplasma synoviae* infection in breeder chickens after treatment with enrofloxacin. Avian Dis. *45*, 534–539.

Stipkovits, L., Miller, D., Glavits, R., Fodor, L., and Burch, D. (2001a). Treatment of pigs experimentally infected with *Mycoplasma hyopneumoniae, Pasteurella multocida,* and *Actinobacillus pleuropneumoniae* with various antibiotics. Can. J. Vet. Res. *65*, 213–222.

Stipkovits, L., Ripley, P.H., Varga, J., and Palfi, V. (2001b). Use of valnemulin in the control of *Mycoplasma bovis* infection under field conditions. Vet. Rec. *148*, 399–402.

Stipkovits, L., Ripley, P.H., Tenk, M., Glavits, R., Molnar, T., and Fodor, L. (2005). The efficacy of valnemulin (EconorR) in the control of disease caused by experimental infection of calves with *Mycoplasma bovis*. Res. Vet. Sci. *78*, 207–215.

Stopler, T., Gerichter, C.B., and Branski, D. (1980). Antibiotic-resistant mutants of *Mycoplasma pneumoniae*. Isr. J. Med. Sci. *16*, 169–173.

Suzuki, S., Yamazaki, T., Narita, M., Okazaki, N., Suzuki, I., Andoh, T., Matsuoka, M., Kenri, T., Arakawa, Y., and Sasaki, T. (2006). Clinical evaluation of macrolide-resistant *Mycoplasma pneumoniae*. Antimicrob. Agents Chemother. *50*, 709–712.

ter Laak, E.A., Noordergraaf, J.H., and Verschure, M.H. (1993). Susceptibilities of *Mycoplasma bovis, Mycoplasma dispar,* and *Ureaplasma diversum* strains to antimicrobial agents in vitro. Antimicrob. Agents Chemother. *37*, 317–321.

Thomas, A., Nicolas, C., Dizier, I., Mainil, J., and Linden, A. (2003). Antibiotic susceptibilities of recent isolates of *Mycoplasma bovis* in Belgium. Vet. Rec. *153*, 428–431.

Uemura, R., Sueyoshi, M., and Nagatomo, H. (2010). Antimicrobial susceptibilities of four species of mycoplasma isolated in 2008 and 2009 from cattle in Japan. J. Vet. Med. Sci. 72, 1661–1663.

Vasconcelos, A.T.R., Ferreira, H.B., Bizarro, C.V., Bonatto, S.L., Carvalho, M.O., Pinto, P.M., Almeida, D.F., Almeida, L.G.P., Almeida, R., Alves-Filho, L., *et al.* (2005). Swine and poultry pathogens: the complete genome sequences of two strains of *Mycoplasma hyopneumoniae* and a strain of *Mycoplasma synoviae*. J. Bacteriol. *187*, 5568–5577.

Vicca, J., Stakenborg, T., Maes, D., Butaye, P., Peeters, J., de Kruif, A., and Haesebrouck, F. (2004). In vitro susceptibilities of *Mycoplasma hyopneumoniae* field isolates. Antimicrob. Agents Chemother. *48*, 4470–4472.

Vicca, J., Maes, D., Stakenborg, T., Butaye, P., Minion, F., Peeters, J., de Kruif, A., Decostere, A., and Haesebrouck, F. (2007). Resistance mechanism against fluoroquinolones in *Mycoplasma hyopneumoniae* field isolates. Microb. Drug Resist. *13*, 166–170.

Waites, K. (2008). Mycoplasma. In Clinical Infectious Disease, Schlossberg, D., ed. (Cambridge University Press, Cambridge, UK), pp. 1145–1159.

Waites, K.B., Bébéar, C.M., Robertson, J.A., Talkington, D.F., and Kenny, G.E., eds. (2001). Cumitech 34, Laboratory diagnosis of mycoplasmal infections (American Society for Microbiology, Washington, D.C.).

Waites, K.B., Talkington, D.F., and Bébéar, C.M. (2002). Mycoplasmas. In Manual of Commercial Methods in Clinical Microbiology, A.L. Truant, ed. (ASM Press, Washington, D.C.), pp. 201–224.

Waites, K.B., and Talkington, D.F. (2004). *Mycoplasma pneumoniae* and its role as a human pathogen. Clin. Microbiol. Rev. *17*, 697–728.

Waites, K.B., Katz, B., and Schelonka, R.L. (2005a). Mycoplasmas and ureaplasmas as neonatal pathogens. Clin. Microbiol. Rev. *18*, 757–789.

Waites, K.B., Reddy, N.B., Crabb, D.M., and Duffy, L.B. (2005b). Comparative in vitro activities of investigational peptide deformylase inhibitor NVP LBM-415 and other agents against human mycoplasmas and ureaplasmas. Antimicrob. Agents Chemother. *49*, 2541–2542.

Waites, K.B., Crabb, D.M., and Duffy, L.B. (2009). Comparative in vitro susceptibilities of human mycoplasmas and ureaplasmas to a new investigational ketolide, CEM-101. Antimicrob. Agents Chemother. *53*, 2139–2141.

Walker, R.D., and Dowling, P.M. (2007). Fluoroquinolones. In Antimicrobial Therapy in Veterinary Medicine, Giguere, S., Prescott, J.F., Baggot, J.D., Walker, R.D., and Dowling, P.M., eds. (Wiley Blackwell, Oxford, UK), pp. 263–284.

Wang, C., Ewing, M., and Aarabi, S.Y. (2001). In vitro susceptibility of avian mycoplasmas to enrofloxacin, sarafloxacin, tylosin, and oxytetracycline. Avian Dis. *45*, 456–460.

Westberg, J., Persson, A., Holmberg, A., Goesmann, A., Lundeberg, J., Johansson, K.E., Pettersson, B., and Uhlen, M. (2004). The genome sequence of *Mycoplasma mycoides subsp. mycoides SC* type strain PG1T, the causative agent of contagious bovine pleuropneumonia (CBPP). Genome Res. *14*, 221–227.

Wise, K.S., Calcutt, M.J., Foecking, M.F., Roske, K., Madupu, R., and Methe, B.A. (2011). Complete genome sequence of *Mycoplasma bovis* type strain PG45 (ATCC 25523). Infect. Immun. *79*, 982–983.

Wise, K.S., Calcutt, M.J., Foecking, M.F., Madupu, R., Deboy, R.T., Roske, K., Huinden, M.L., Martin, T.R., Durkin, A.S., Glass, J.I., and Methe, B.A. (2012). Complete genome sequences of *Mycoplasma leachii* strain PG50T and the pathogenic *Mycoplasma mycoides subsp. mycoides Small Colony* biotype strain Gladysdale. J. Bacteriol. *194*, 4448–4449.

Wolff, B.J., Thacker, W.L., Schwartz, S.B., and Winchell, J.M. (2008). Detection of Macrolide Resistance in *Mycoplasma pneumoniae* by Real-Time PCR and High Resolution Melt Analysis. Antimicrob. Agents Chemother. *52*, 3542–3549.

Wu, C.C., Shryock, T.R., Lin, T.L., Faderan, M., and Veenhuizen, M.F. (2000). Antimicrobial susceptibility of *Mycoplasma hyorhinis*. Vet. Microbiol. *76*, 25–30.

Wu, C.M., Wu, H., Ning, Y., Wang, J., Du, X., and Shen, J. (2005). Induction of macrolide resistance in *Mycoplasma gallisepticum* in vitro and its resistance-related mutations within domain V of 23S rRNA. FEMS Microbiol. Lett. *247*, 199–205.

Xiao, L., Crabb, D.M., Duffy, L.B., Paralanov, V., Glass, J.I., Hamilos, D.L., and Waites, K.B. (2011). Mutations in ribosomal proteins and ribosomal RNA confer macrolide resistance in human *Ureaplasma* spp. Int. J. Antimicrob. Agents 37, 377–379.

Xiao, L., Crabb, D.M., Duffy, L.B., Glass, J.I., and Waites, K.B. (2012). Chromosomal mutations responsible for fluoroquinolone resistance in *Ureaplasma* species in the United States. Antimicrob. Agents Chemother. *56*, 2780–2783.

Xie, X., and Zhang, J. (2006). Trends in the rates of resistance of *Ureaplasma urealyticum* to antibiotics and identification of the mutation site in the quinolone resistance-determining region in Chinese patients. FEMS Microbiol. Lett. *259*, 181–186.

Xin, D., Mi, Z., Han, X., Qin, L., Li, J., Wei, T., Chen, X., Ma, S., Hou, A., Li, G., *et al.* (2009). Molecular mechanisms of macrolide resistance in clinical isolates of *Mycoplasma pneumoniae* from China. Antimicrob. Agents Chemother. *53*, 2158–2159.

Zanella, A., Martino, P.A., Pratelli, A., and Stonfer, M. (1998). Development of antibiotic resistance in *Mycoplasma gallisepticum* in vitro. Avian Pathol. *27*, 591–596.

Zhang, W., Wu, Y., Yin, W., and Yu, M. (2002). Study of isolation of fluoroquinolone-resistant *Ureaplasma urealyticum* and identification of mutant sites. Chin. Med. J. *115*, 1573–1575.

# Index

## D



## E



## F



## G

## N



## O

## Q



## R



## S

## T

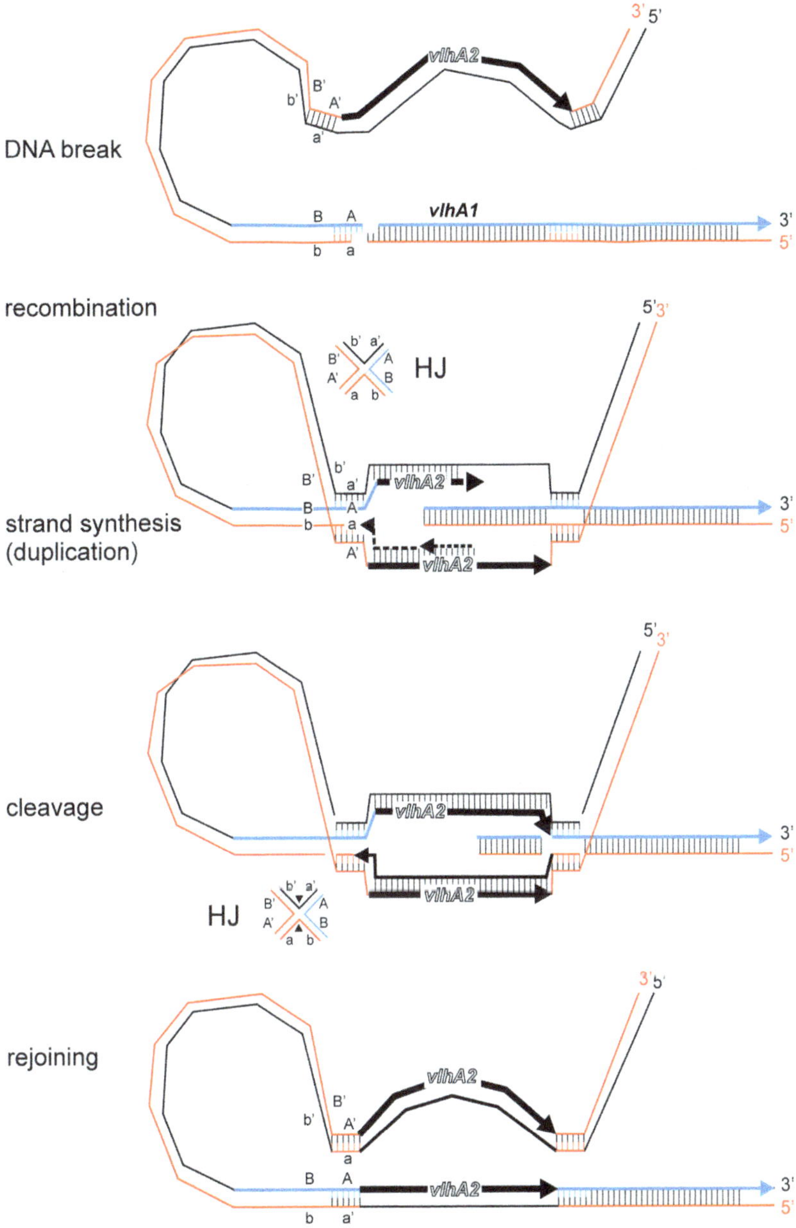


**Figure 9.4** A putative gene conversion event as in the *vlhA* locus of *M. synoviae*. Sequences within the conserved regions of the *vlhA1* gene might be broken, followed by 5′→3′ exonuclease digestion. The free 3′ ssDNA tail that is generated by this process scans the pseudogenes (e.g. *vlhA2*) for homologous sequences. After homologous recombination, the pseudogene region *vlhA2* is duplicated through DNA extension on both strands. During this process, the *vlhA1* sequence to be exchanged might be degraded by nucleases. Holliday junctions (HJ) are cleaved (▲) by an HJ resolvase and nicked ends are relegated.

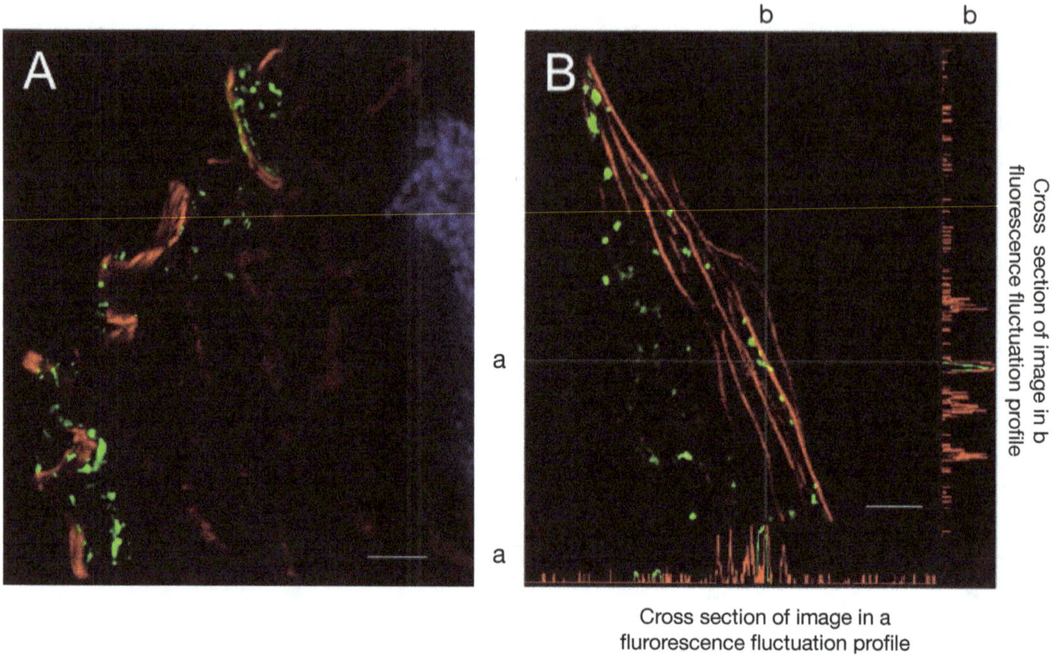


**Figure 10.2** Spiroplasmas (green) and actin filaments (red) in *C. haematoceps* salivary gland (A) and *S. citri*-infected Ciha-1 (B) after infection with *S. citri* GII3, as visualized by immuno-fluorescence confocal laser microscopy. Infected cells nuclei were stained with DAPI (blue). Scale bar, 8μm. Spiroplasmas are localized along the actin cytoskeleton in insect infected cells.

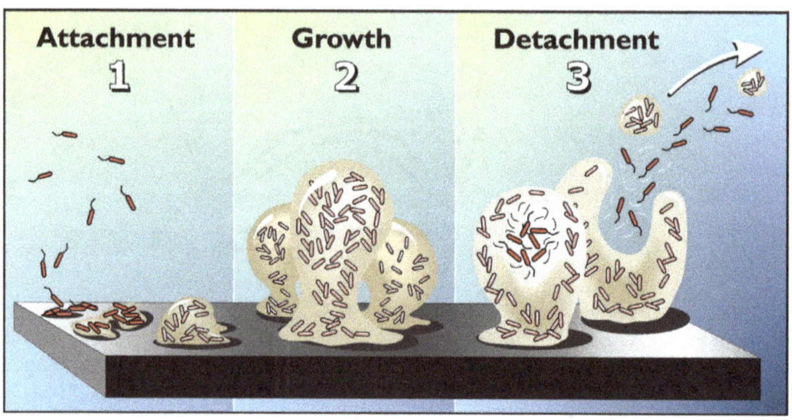


**Figure 13.1** The biofilm life cycle illustrated in three steps: initial attachment events, the growth of complex biofilms, and detachment events by clumps of bacteria or by a 'swarming' phenomenon within the interior of bacterial clusters, resulting in 'seeding dispersal' (reproduced with permission from Peg Dirckx, CBE, Montana State University, USA).

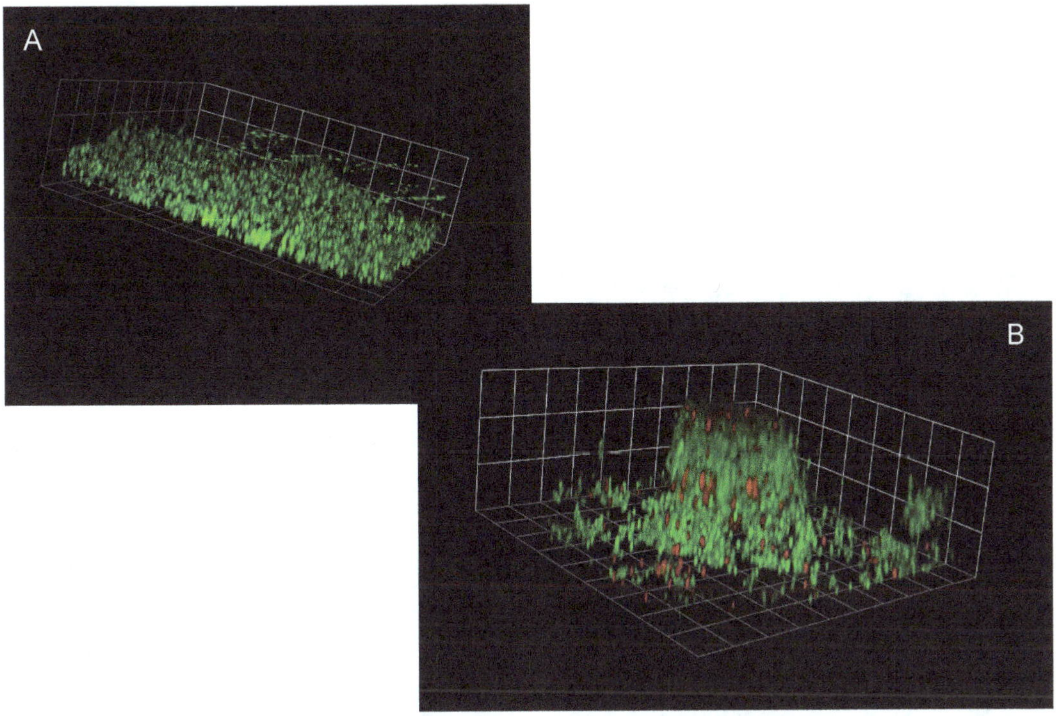


**Figure 13.2** Confocal scanning light micrographic image of Live/Dead stained *M. putrefaciens* biofilm showing (A) early stages of a biofilm with a monolayer of cells and (B) a more mature biofilm with stacks.

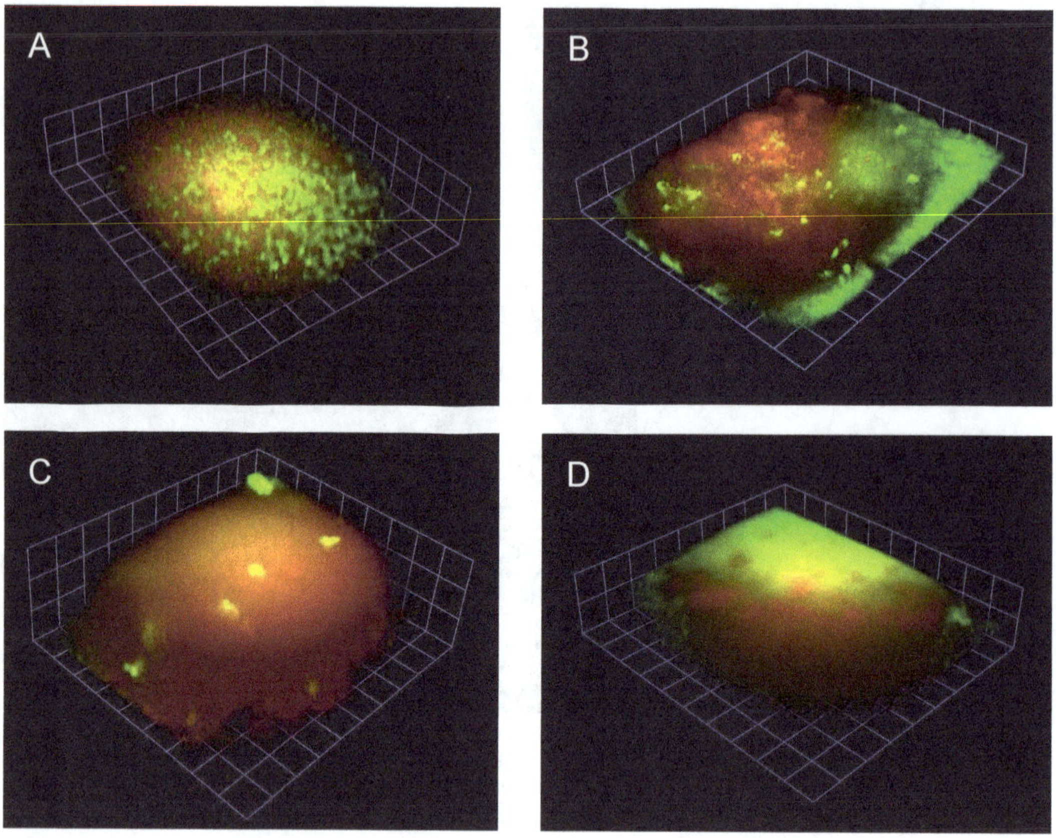


**Figure 13.3** Confocal scanning light micrographic images that illustrate intraspecific variation seen in MmmSC biofilms grown using a membrane on plate method without an air–liquid interface. (A) strain IS31, (B) strain Madrid, (C) strain N6, and (D) strain V5.

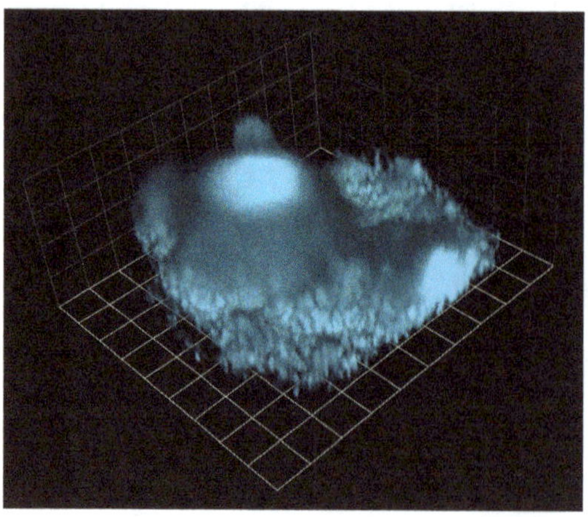

**Figure 13.4** Confocal scanning light micrographic with calcofluor white EPS staining of *M. putrefaciens* biofilm (L. McAuliffe).

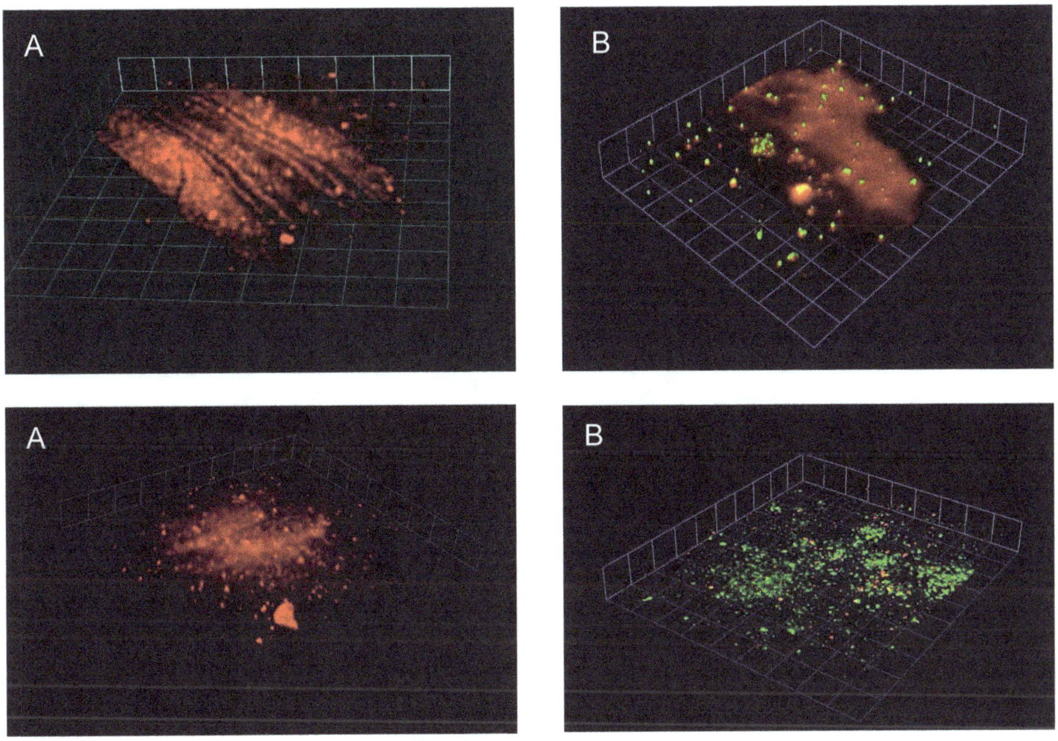


**Figure 13.5** Confocal scanning light micrographic image showing *M. bovis* biofilm with (A) conjugated wheat germ agglutinin lectin staining, and (B) Live/Dead staining (L. McAuliffe).

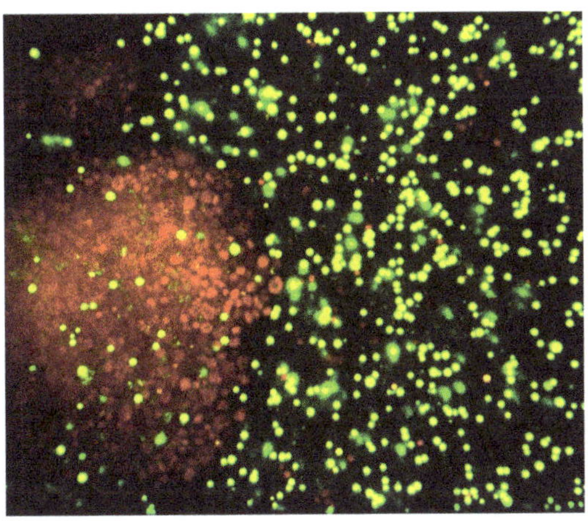

**Figure 13.6** Live/dead stained confocal scanning light micrographic image of a mature, starved *M. putrefaciens* biofilms, demonstrating persistence of cells within the interior of the biofilm (L. McAuliffe).

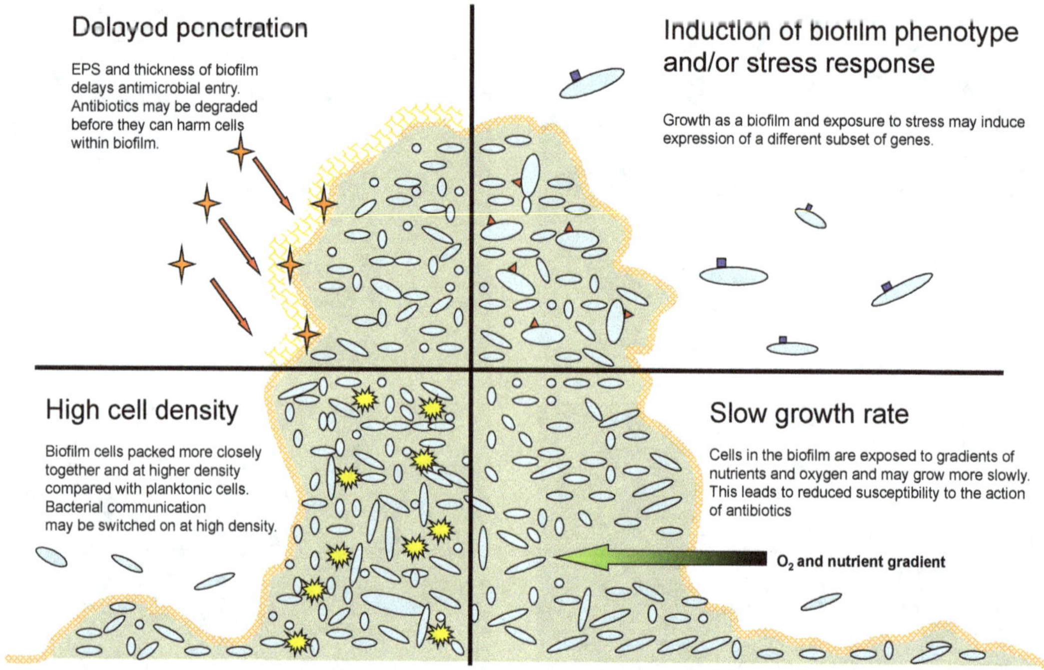


**Figure 13.7** Illustration of the different biofilm characteristics that contribute to increased resistance to antimicrobial agents and host cell defences (copyright L. McAuliffe).

www.ingramcontent.com/pod-product-compliance
Lightning Source LLC
LaVergne TN
LVHW062218200926
848484LV00006B/1427